# Chemical Processes:
# Design, Synthesis and Analysis

# Chemical Processes: Design, Synthesis and Analysis

Editor: Rose Torres

NY RESEARCH PRESS

New York

Published by NY Research Press
118-35 Queens Blvd., Suite 400,
Forest Hills, NY 11375, USA
www.nyresearchpress.com

Chemical Processes: Design, Synthesis and Analysis
Edited by Rose Torres

International Standard Book Number: 978-1-63238-836-0 (Hardback)

**Cataloging-in-Publication Data**

Chemical processes : design, synthesis and analysis / edited by Rose Torres.
    p. cm.
Includes bibliographical references and index.
ISBN 978-1-63238-836-0
1. Chemical processes. 2. Chemical engineering. 3. Chemistry, Analytic. I. Torres, Rose.
TP155.7 .C44 2022
660.28--dc23

# Contents

# Preface

A chemical process is a method used to change the composition of one or more chemicals or materials. In a chemical process, one or several chemical unit operations may be involved. These may include oxidation, reduction, hydrolysis, dehydration, alkylation, esterification, polymerization, nitrification, catalysis, etc. Process design, chemical synthesis and chemical analysis are central to chemical engineering and chemical processes. While chemical synthesis involves the selection of compounds and reactions to synthesize a product, process design determines the sequencing of units for the desired transformation of a material. Chemical analysis is concerned with the identification, separation and quantification of matter. The objective of this book is to give a general view of the different aspects of chemical processes and their significance. It includes some of the vital pieces of work being conducted across the world, on various topics related to process design, chemical synthesis and chemical analysis. The topics covered in this book offer the readers new insights in the field of chemical engineering.

The information shared in this book is based on empirical researches made by veterans in this field of study. The elaborative information provided in this book will help the readers further their scope of knowledge leading to advancements in this field.

Finally, I would like to thank my fellow researchers who gave constructive feedback and my family members who supported me at every step of my research.

Editor

# Al-Waste-Based Zeolite Adsorbent used for the Removal of Ammonium from Aqueous Solutions

Ruth Sánchez-Hernández ⓘ,[1] Isabel Padilla ⓘ,[1] Sol López-Andrés ⓘ,[2] and Aurora López-Delgado ⓘ[1]

[1]National Centre for Metallurgical Research (CSIC), Madrid 28040, Spain
[2]Department of Mineralogy and Petrology, Faculty of Geology, University Complutense of Madrid, Madrid 28040, Spain

Correspondence should be addressed to Aurora López-Delgado; alopezdelgado@cenim.csic.es

Academic Editor: Eric Guibal

This work evaluates the use of a synthetic NaP1 zeolite obtained from a hazardous Al-containing waste for the removal of ammonium ($NH_4^+$) from aqueous solutions by batch experiments. Experimental parameters, such as pH (6–8), contact time (1–360 min), adsorbent dose (1–15 g/L), and initial $NH_4^+$ concentration (10–1500 mg/L), were evaluated. Adsorption kinetic models and equilibrium isotherms were determined by using nonlinear regression. The kinetic was studied by applying both the pseudo-first-order and pseudo-second-order models. The equilibrium isotherms were analyzed according to two-parameter equations (Freundlich, Langmuir, and Temkin) and three-parameter equations (Redlich–Peterson, Sips, and Toth). The results showed that the $NH_4^+$ uptake on NaP1 was fast (15 min) leading to a high experimental sorption capacity (37.9 mg/g). The $NH_4^+$ removal on NaP1 was a favorable process that followed the pseudo-first-order kinetic model. The $NH_4^+$ adsorption was better described by the Sips (54.2 mg/g) and Toth (58.5 mg/g) models. NaP1 zeolite from Al-waste showed good $NH_4^+$ sorption properties, becoming a potential adsorbent to be used in the treatment of contaminated aqueous effluents. Thus, a synergic effect on the environmental protection can be achieved: the end of waste condition of a hazardous waste and the water decontamination.

## 1. Introduction

Global industrial development linked to the population growth is promoting the pollution of natural resources, in particular, aquatic environment. Surface and ground waters present significant amounts of harmful substances for human health and environment. Among such substances, excess nitrogen compounds like ammonium ($NH_4^+$) can cause eutrophication depriving of oxygen to aquatic organisms in lakes and rivers. In addition, the $NH_4^+$ ionized form moves much more slowly than other nitrogen compounds, thus persisting in groundwater for long periods of time after it enters the subsurface [1]. Diverse anthropogenic activities, more specifically, fossil-fuel combustion, septic systems, sewage sludge, landfill leachate, and agricultural practices (including chemical fertilizers or animal manures), lead to an increase in the amount of ammonium in waters [2–4]. Ammonium is also one of the soluble nitrogen species found in airborne particulate matter in urban areas, representing up to approximately 23% [5]. The Spanish legislation (RD 817/2015) on the quality of water establishes concentrations of ammonium ranged between 0.2 and 1 mg/L as acceptable limits. However, higher ammonium concentrations can be found in surface waters, generally, exceeding the limits established by water quality standards [6]. Therefore, the development and improvement of more effective sorbents and water treatment processes to remove contaminants is of vital environmental importance. Biological treatment processes are commonly employed in wastewater treatment facilities. However, several water treatment techniques can be necessary to reduce the content of contaminants in water. Thus, adsorption is a relatively feasible and simple technology that contributes to the elimination of a wide range of substances in aquatic environment. Activated carbons are commonly used as sorbents for the uptake of different compounds from waters. However, such materials must be previously treated by activation methods that usually imply high operating temperatures to enhance their porosity and specific surface area [7]. During the last two decades, other potential sorbents like zeolites, that is, porous crystalline materials, have been also used for removal of diverse contaminants [8]. Structurally, both natural and synthetic zeolites are characterized by frameworks built from tetrahedral $TO_4$ primary units (mainly, T = Si and Al) connected through their oxygen atoms in different ways, resulting in a variety of structures that contain cages, channels, and cavities. The isomorphous substitution of $Si^{4+}$ by $Al^{3+}$ in the tetrahedral units leads to formation of negatively charged zeolitic structures that are balanced by the introduction of extra-framework cations (e.g., $Na^+$, $K^+$, and $Ca^{2+}$). Zeolites, known as molecular sieves and ion exchangers, can be employed as low-cost sorbents compared with activated carbons. Although some zeolites have low specific surface areas as a consequence of their mesoporous character [9, 10], it has been found that the ammonium sorption may be related to the sorbent CEC [11]. So, it is likely that one of the most important properties for the elimination of some compounds from water is the CEC. Unlike activated carbons, some zeolites can be used directly in water treatment applications without any previous activation process. Clinoptilolite, a natural zeolite with a CEC ranged between 0.6 and 2.3 meq/g [12], is widely used due to its good selectivity to ammonium [13–15]. Among synthetic zeolites, NaP1 is also utilized in water treatment applications [16] due to its 3-dimensional channels with pore sizes ranged between $4.5 \times 3.1$ and $4.8 \times 2.8$ Å (according to the International Zeolite Association). Recently, NaP1-type zeolite was synthesized from Al-containing solid wastes leading to a promising sorption capacity (2.4 meq/g) [10], which is quite similar to commercial Na-type zeolite (2.7 meq/g) from chemical reagents by Spanish manufacturers. The recovery processes of such Al-waste, included in the European Waste Catalogue (code 10 03 21), represented around 12–63% in Spain during the period 2010–2015, according to the Spanish Register of Emissions and Pollutant Sources. This Al-waste is usually deposited in secure deposits due to its environmental hazards. The synthesis of NaP1 from a hazardous Al-waste was performed by an eco-friendly bench-scale hydrothermal process [17], contributing to reduce the consumption of raw materials (i.e., water and NaOH) and environmental impact. Although both natural zeolites and zeolites from fly ash have been widely used to remove ammonium from waters [13, 14, 18, 19], studies conduced for zeolites from highly Al-enriched wastes have not been reported, to our knowledge, in the literature.

Thus, this work attempts to evaluate the removal of $NH_4^+$ cation from aqueous solutions onto NaP1 synthesized from a hazardous Al-waste. The effect of experimental parameters, such as pH, contact time, adsorbent dose, and initial $NH_4^+$ concentration, on the removal efficiency and adsorption capacity of Al-waste-NaP1 was studied by batch tests. In order to better study the uptake process of the adsorbate ($NH_4^+$) by the adsorbent (Al-waste-NaP1), adsorption kinetic and equilibrium isotherms were also evaluated by applying several nonlinear models and error functions.

## 2. Materials and Methods

*2.1. Adsorbent and Adsorbate.* The adsorbent used in this work was a Na-type zeolite, NaP1, whose theoretical formula is $Na_6Al_6Si_{10}O_{32} \cdot 12H_2O$. It was synthesized from a hazardous aluminum waste that is generated from slag milling processes within aluminum industries. The preparation of the zeolite from this Al-waste was developed by a bench-scale hydrothermal synthesis process at 120°C for 6 h according to [17]. The evaluation of the main mineralogical and morphological properties of the zeolite was analyzed using X-ray diffraction (XRD) and scanning electron microscope (SEM) as shown in Figure 1. Al-waste-NaP1 was characterized by gismondine-type zeolite showing a tetragonal structure. The zeolite presents a high CEC (2.4 meq/g) but a low $S_{BET}$ (15.93 $m^2$/g) that was quite similar to the $S_{EXT}$ (15.08 $m^2$/g) due to its mesoporous character, its $N_2$ adsorption/desorption isotherm, and pore size distribution (as shown in Figure 2(a)). The total pore volume estimated at the relative pressure of 0.99 was 0.04828 $cm^3$/g. The mesopore and micropore volumes were 0.04785 and 0.00044 $cm^3$/g, respectively, and the average pore diameter of the zeolite, estimated assuming cylindrical pore shape, was 12.12 nm. The zeta potential ($\zeta$-potential) of Al-waste-NaP1 indicated that its surface is negatively charged within a wide pH range (Figure 2(b)); accordingly, the as-obtained zeolite may be considered as a good candidate for sorption of cations like ammonium in waters.

Aqueous solutions containing the adsorbate ($NH_4^+$) were prepared by dissolving a certain amount of a 1000 mg/L stock solution of ammonium chloride (Panreac) in deionized water (resistivity ~18.2 MΩ·cm).

*2.2. Adsorption Procedure.* Adsorption experiments of $NH_4^+$ on NaP1 were performed by the batch process under ambient conditions ($28 \pm 2$°C) using adsorbate aqueous solutions with a fixed volume (100 mL). The adsorption ability of NaP1 was only studied under ambient conditions since the increase of temperature can lead to the decrease of the $NH_4^+$ adsorption capacity [20]. Before adsorption, blank tests were prepared in order to discard possible contamination resulting from the adsorbent, the reagents or the equipment used during sample processing including filtration. All samples were placed in 150 mL glass conical flaks which were kept covered and stirred at constant speed (125 rpm) using an orbital and horizontal shaker (Selecta, Rotabit) with speed and time control. The solution pH was maintained constant during each adsorption experiment by adding small volumes of dilute NaOH or HCl aqueous solutions. Thus, the effect of pH, contact time, and adsorbent dose on the ammonium removal efficiency and adsorption capacity was studied using fixed initial concentrations (50 mg/L). The influence of initial concentration on the $NH_4^+$ uptake by NaP1 was evaluated in a wide initial concentration range (10–1500 mg/L).

*2.2.1. Effect of pH.* The influence of pH on the ammonium uptake was studied as it influences the ionization of surface

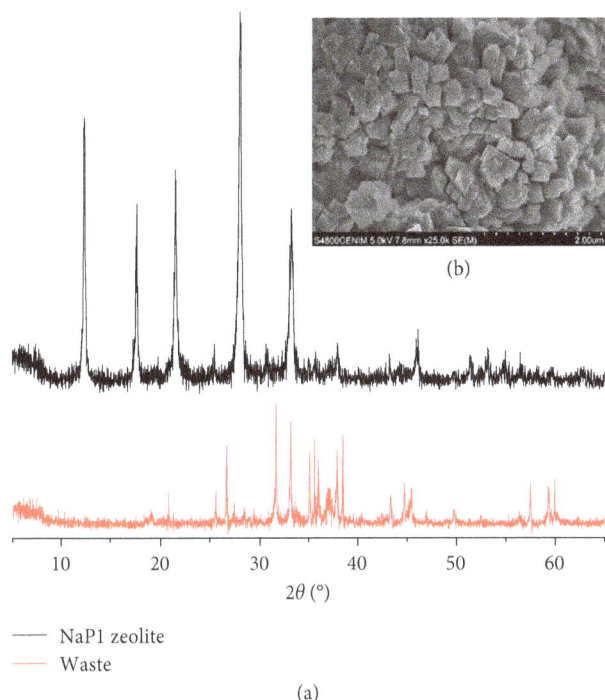

FIGURE 1: (a) X-ray diffraction patterns of Al-waste and Al-waste-NaP1. (b) Morphology of Al-waste-NaP1.

groups of the adsorbent and the speciation of different ions present in the adsorption system. Thus, the effect of the pH on the $NH_4^+$ cation adsorption by NaP1 was evaluated at pH ranged between 6 and 8, simulating pH conditions similar to leachates from landfills, which contain high concentrations of ammonium [3]. For the pH experiments, an adsorbent dose of 5 g/L (i.e., 0.5 g of zeolite per 100 mL of $NH_4^+$ aqueous solution) and an initial adsorbate concentration of 50 mg/L were selected for a fixed contact time (15 min).

### 2.2.2. Effect of Contact Time.
Before developing equilibrium experiments, the required contact time for reaching the adsorption equilibrium was determined. Thus, the effect of the contact time on the $NH_4^+$ adsorption efficiency of NaP1 was studied by varying the time (1, 2, 5, 10, 15, 30, 60, 120, and 360 min) and using 0.5 g of zeolite per 100 mL of adsorbate aqueous solution with an initial $NH_4^+$ concentration of 50 mg/L at pH 7.5. The adsorption kinetic on the cation adsorption capacity was also studied by evaluating the pseudo-first-order and pseudo-second-order adsorption rate models. Subsequent adsorption tests were developed according to the optimized parameters.

### 2.2.3. Effect of Adsorbent Dose.
Once selected the optimal conditions of pH and contact time for the $NH_4^+$ removal on NaP1, the adsorbent dose was evaluated by varying the zeolite mass (0.1, 0.25, 0.5, 0.75, 1, 1.25, and 1.5 g) in contact with 100 mL aqueous solutions of $NH_4^+$, with an initial adsorbate concentration of 50 mg/L, at pH 7.5 for the selected equilibrium time (15 min).

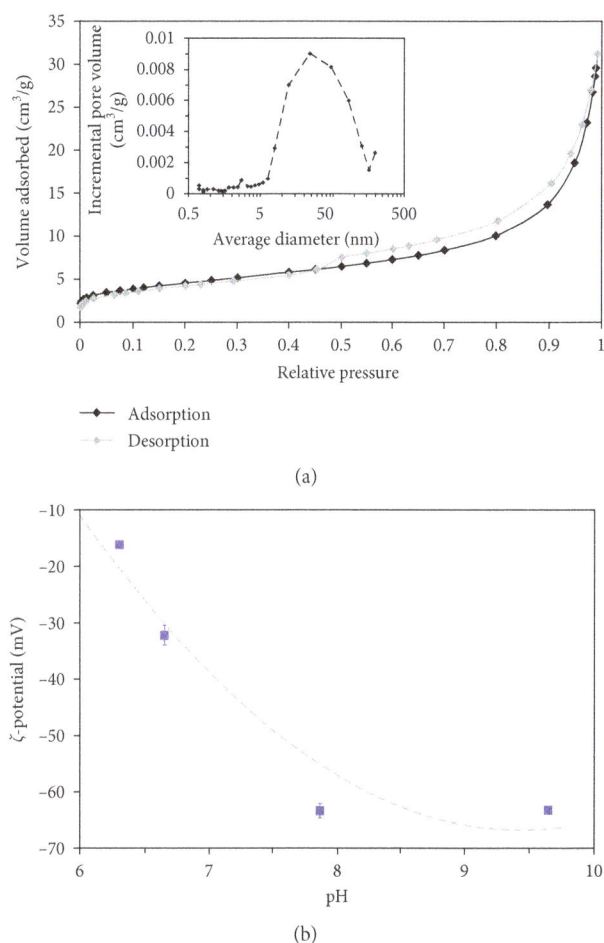

FIGURE 2: (a) Nitrogen adsorption/desorption isotherm and pore size distribution and (b) Zeta potential of Al-waste-NaP1.

### 2.2.4. Effect of Initial Adsorbate Concentration.
The effect of the initial ammonium concentration was studied for the optimal adsorption conditions (i.e., pH 7.5; 15 min; 10 g/L). The adsorption equilibrium experiments were determined by varying the initial adsorbate concentration between 10 and 1500 mg/L and evaluating different isotherm models.

After the adsorption process, the zeolite was separated from the adsorbate solutions by filtration. In order to ensure reliability and reproducibility of the experimental data, all the adsorption tests performed by duplicate and average values are reported in this work. The ammonium removal efficiency (expressed in %) and the amount of ammonium adsorbed on NaP1 at any time ($q_t$, expressed in mg/g) and at equilibrium ($q_e$, in mg/g) were calculated as follows:

$$\text{Removal} (\%) = \left(\frac{C_o - C_t}{C_o}\right) \cdot 100,$$

$$q_t = \frac{(C_o - C_t)}{m} \cdot V, \tag{1}$$

$$q_e = \frac{(C_o - C_e)}{m} \cdot V,$$

where $C_o$ (mg/L) is the initial concentration of ammonium, $C_t$ and $C_e$ (mg/L) are the ammonium concentrations at contact time $t$ and at equilibrium, respectively, $m$ (g) is the mass of zeolite, and $V$ (L) is the volume of adsorbate solution.

### 2.3. Kinetic and Isotherm Modeling.

The parameters and constants of the applied kinetic and isotherm models were determined by the nonlinear regression method, which is more accurate than the linear method, using the GRG nonlinear solving method of Microsoft Excel Solver. Applying nonlinear models can give more reliable results, minimizing the bias between the adsorption capacity values obtained from experimental data and those calculated from model equations.

#### 2.3.1. Adsorption Kinetics.

The study of the adsorption kinetic of ammonium on NaP1 was mathematically analyzed by applying the pseudo-first-order [21] and pseudo-second-order [22] models, according to the following nonlinearized equations:

$$q_t = q_e\left(1 - e^{k_1 t}\right),$$

$$q_t = \frac{k_2 q_e^2 t}{1 + k_2 q_e t}, \tag{2}$$

where $q_e$ (mg/g) and $q_t$ (mg/g) are the amount of ammonium adsorbed per mass of zeolite at equilibrium and at any time $t$ (min), respectively, and $k_1$ (1/min) and $k_2$ (g/mg min) are the rate constants for the pseudo-first-order and pseudo-second order models.

#### 2.3.2. Adsorption Isotherms.

The relationship between the ammonium concentration in solution (liquid phase) and the zeolite (solid phase) at constant pH and temperature was studied by several adsorption isotherm models. To determine the best isotherm model fit, the experimental equilibrium data were analyzed by applying isotherm models with two-parameter equations (Freundlich, Langmuir, and Temkin) and with three-parameter equations (Redlich–Peterson, Sips, and Toth) widely reported in the literature [23, 24].

Concerning the two-parameter models, the Freundlich isotherm is widely used and it describes heterogeneous adsorption systems [25]. The nonlinearized form of the Freundlich isotherm is expressed by the following equation:

$$q_e = k_F C_e^n, \tag{3}$$

where $q_e$ (mg/g) is the amount of ammonium adsorbed on the zeolite at equilibrium, $C_e$ (mg/L) is the ammonium concentration at equilibrium, $k_F$ (mg/g)/(mg/L)$^n$ is the Freundlich constant, and $n$ (dimensionless) is the Freundlich intensity parameter which indicates the magnitude of the adsorption driving force or the surface heterogeneity.

The Langmuir isotherm has been commonly used for the removal of a wide variety of compounds from waters using different adsorbents. It assumes monolayer coverage of adsorbate over a homogeneous adsorbent surface [26]. The nonlinearized model of Langmuir is described as follows:

$$q_e = \frac{q_{max} k_L C_e}{1 + k_L C_e}, \tag{4}$$

where $q_e$ (mg/g) is the amount of ammonium adsorbed on the zeolite at equilibrium, $q_{max}$ (mg/g) is the maximum monolayer adsorption capacity of the zeolite, $C_o$ and $C_e$ (mg/L) are the initial and equilibrium concentrations of ammonium, and $k_L$ (L/mg) is a constant related to the affinity between the adsorbent and adsorbate. The Langmuir isotherm model expressed extensively in terms of the separation factor or equilibrium parameter $(R_L)$ [27] can be given by the following expression:

$$R_L = \frac{1}{1 + k_L C_o}. \tag{5}$$

The nature of the adsorption can be estimated through the isotherm profile according to the values of $R_L$ and the Freundlich exponent $(n)$, being irreversible $(R_L = 0)$, favourable $(0 < R_L < 1)$, linear $(R_L = 1)$, or unfavorable $(R_L > 1)$ [28]. The Langmuir theory can be applied to homogeneous adsorption where each adsorbed species involves the same sorption activation energy.

The Temkin model [29] is related to the effects of indirect interactions between adsorbent and adsorbate and is characterized by a uniform distribution of binding energies according to the following expression:

$$q_e = \frac{RT}{b} \ln\left(A C_e\right), \tag{6}$$

where A is the equilibrium binding constant (L/g), $b$ is related to the heat of adsorption (J/mol), $R$ is the gas constant (8.314 J/K mol), and $T$ is the temperature (K).

Among the isotherm models with three parameters, the Redlich–Peterson isotherm [30] can be applied for both homogeneous and heterogeneous adsorption systems in a wide concentration range according to the following equation:

$$q_e = \frac{k_{RP} C_e}{1 + a_{RP} C_e^\beta}, \tag{7}$$

where $k_{RP}$ (L/g) and $a_{RP}$ (mg/L)$^{-\beta}$ are the Redlich–Peterson constants and $\beta$ (dimensionless) is an exponent whose value must lie between 0 and 1. This model tends to the Langmuir isotherm when the exponent $\beta = 1$, while it is described by the Freundlich isotherm when $k_{RP}$ and $a_{RP}$ are higher than 1 and $\beta$ is 1 [23].

The Sips isotherm [31] combines the Langmuir and Freundlich isotherms, leading to the Freundlich isotherm at low adsorbate concentrations, while it approaches Langmuir isotherm at high concentrations. The nonlinearized expression of the Sips isotherm can be represented by the following equation:

$$q_e = \frac{q_{max}\left(k_S C_e\right)^n}{1 + \left(k_S C_e\right)^n}, \tag{8}$$

where $q_{max}$ (mg/g) is the maximum adsorption capacity of the zeolite, $k_S$ (L/mg) is the Sips constants, $C_e$ (mg/L) is the ammonium concentration at equilibrium, and $n$ (dimensionless) is an exponent.

The Toth isotherm [32] can be considered as the improved form of the Langmuir and Freundlich models, given by the following nonlinear expression:

$$q_e = \frac{q_{max}k_T C_e}{\left[1 + (k_T C_e)^n\right]^{1/n}},\qquad(9)$$

where $q_{max}$ (mg/g) is the maximum adsorption capacity of the zeolite, $C_e$ (mg/L) is the ammonium concentration at equilibrium, $k_T$ (L/mg) is the Toth isotherm constant, and $n$ (dimensionless) is the Toth exponent [33].

*2.3.3. Error Functions.* The goodness of fit of the kinetic and isotherm equations to the experimental data was evaluated using the coefficient of determination ($R^2$), chi-square test ($\chi^2$), root-mean-square error (RMSE), and hybrid error function (HYBRID), according to the following equations:

$$R^2 = 1 - \left[\frac{\sum_{i=1}^{n}\left(q_{i,exp} - q_{i,model}\right)^2}{\sum_{i=1}^{n}\left(q_{i,exp} - \overline{q_{i,exp}}\right)^2}\right],$$

$$\chi^2 = \sum_{i=1}^{n}\frac{\left(q_{i,exp} - q_{i,model}\right)^2}{q_{i,exp}},$$

$$(10)$$

$$RMSE = \sqrt{\left(\frac{1}{n-p}\right)\sum_{i=1}^{n}\left(q_{i,exp} - q_{i,model}\right)^2},$$

$$HYBRID = \left(\frac{100}{n-p}\right)\sum_{i=1}^{n}\frac{\left(q_{i,exp} - q_{i,model}\right)^2}{q_{i,exp}},$$

where $n$ is the number of experimental values in a dataset, $q_{i,exp}$ and $q_{i,model}$ are the experimental and calculated adsorption capacities, respectively, and $p$ is the number of parameters contained within the model.

All the parameters and constants of the applied models were determined by maximizing the error function in the case of using $R^2$ or minimizing the error values for $\chi^2$, RMSE, and HYBRID.

*2.4. Analysis Techniques.* Crystalline and morphological properties of Al-waste-NaP1 were analyzed by XRD (D8 Advance, Bruker) and SEM (S4800, Hitachi). Its textural characterization was evaluated by $N_2$ adsorption/desorption analysis at 77 K (ASAP 2010, Micromeritics), previously outgassing at 350°C for 24 h. The BET specific surface area ($S_{BET}$) and external area ($S_{EXT}$) of the zeolite were obtained by the BET method and t-plot analysis. The pore size distribution was calculated using the Barrett–Joyner–Halenda method. The $\zeta$-potential of the zeolite surface was measured (ZetaSizer Nano, Malvern) using the Smoluchowski approximation by preparing aqueous suspensions that contain 0.05 g of zeolite in 100 mL of aqueous solutions at different pH values. Before the $\zeta$-potential measurements, the zeolite suspensions were stirred in an ultrasonic bath (60 min) and then kept in contact for a long time (>15 h) to achieve the

suspensions homogenization and stabilization. Suspensions with absolute $\zeta$-potential values of 30 mV can be considered electrically stable, resulting in an adequately separation of low charged surfaces from highly charged surfaces [34]. The variation of the concentrations of $NH_4^+$ before and after the adsorption experiments was determined colorimetrically with Nessler reagent by UV-Vis spectroscopy (Varian, Cari 1E) monitoring the absorbance at a wavelength of maximum absorbance (420 nm). The pH values of all experiments were adjusted by adding aqueous solutions of dilute NaOH or HCl, using a pH meter (Crison, MM41). The immobilization of the adsorbate on the adsorbent was studied by Fourier transform infrared (FTIR) spectroscopy (Nicolet Nexus 670–870) on KBr discs.

# 3. Results and Discussion

*3.1. Effect of pH.* The pH plays a very important role in adsorption processes since it influences the chemical equilibrium between the ammonium ion ($NH_4^+$) and ammonia ($NH_3$), according to the reversible reaction [35]:

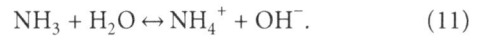

$$NH_3 + H_2O \leftrightarrow NH_4^+ + OH^-.\qquad(11)$$

Likewise, the pH also promotes electrostatic interactions between sorbent materials and ions to be adsorbed, being an essential control parameter in adsorption processes. The influence of pH on the ammonium adsorption capacity and removal efficiency for NaP1 was evaluated in the pH range of 6–8 (Figure 3) in order to select the most adequate conditions for further adsorption experiments. Although the highest ammonium uptake was reached at pH 7.5 leading to the highest adsorption capacity (8.76 mg/g) and removal efficiency (87.6%), similar results were also obtained at pH 7 and 8. Thus, the $NH_4^+$ adsorption capacity for NaP1 was of 8.70 and 8.69 mg/g at pH 7 and 8, involving removal percentages of 87.0 and 86.9%, respectively. Similarly, the $NH_4^+$ adsorption onto NaA zeolite from fly ash using the same adsorbate initial concentration (50 mg/L) also showed the best results at pH ranged between 7 and 8 (removal efficiency around 60%) [36]. In our case, using Al-waste-NaP1 led to higher removal efficiency (87.6%). Thus, the removal of the adsorbate from aqueous medium onto NaP1 could take place effectively at pH 7.5 ± 0.5. The adsorption ability of NaP1 decreased slightly (only 3% less than the highest removal efficiency) as the pH decreased from 7.5 to 6. This can be associated with the potential competition between $H^+$ protons and the $NH_4^+$ cations to be adsorbed onto NaP1. Similar tendencies were shown for the ammonium sorption onto other adsorbent materials, such as volcanic tuff whose main component is clinoptilolite [37] and NaA zeolite from halloysite mineral prepared by a two-step synthesis treatment (alkaline fusion followed by hydrothermal synthesis) [38]. Lower pH conditions were not considered in this work since previous studies have reported mass losses by dissolution as well as dealumination of zeolites at low pH values, especially at pH < 4 [14]. Strong basic conditions (pH > 8) were not evaluated since the $NH_4^+$ concentration could decrease and the chemical equilibrium would be directed to the $NH_3$ (g) formation. Therefore, the fixed pH for further

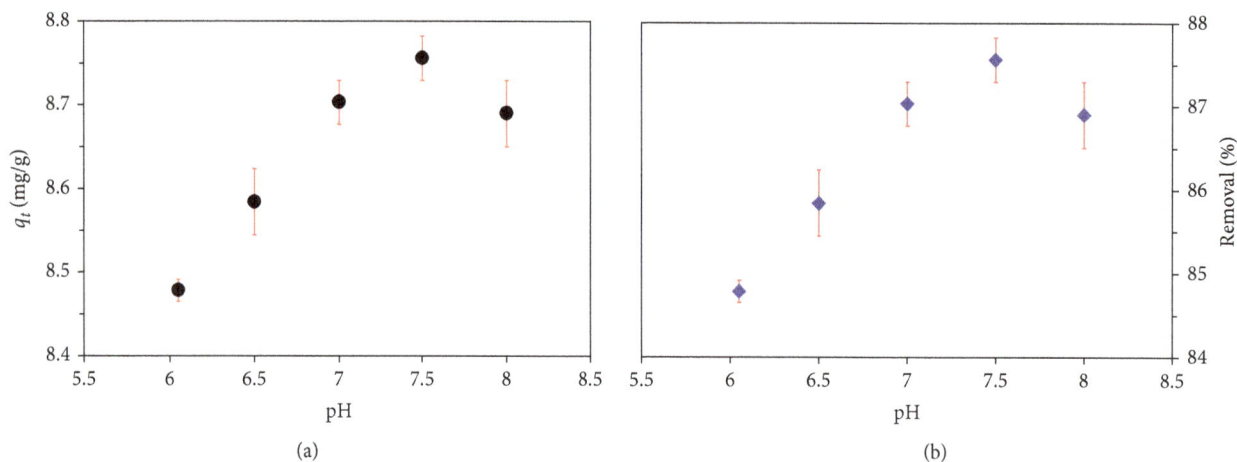

FIGURE 3: Influence of the solution pH on the $NH_4^+$ adsorption capacity and removal efficiency of Al-waste-NaP1. Conditions: contact time = 15 min; adsorbent dose = 5 g/L; Co = 50 mg/L; T = 28 ± 2°C.

adsorption tests was 7.5, where the zeolite mass remains unchanged and the adsorbate would exist mainly in the ionized form, that is, $NH_4^+$. The driving forces for the adsorption process would be described by electrostatic interactions and cation exchange mechanisms. In this sense, the electrostatic attraction would take place between the $NH_4^+$ cations and the zeolitic adsorbent, whose surface is negatively charged in the studied pH range, as shown in the $\zeta$-potential analysis. On the other hand, the adsorbate ($NH_4^+$) and alkali metals (mainly $Na^+$ from the zeolitic framework would be exchanged easily since these cations have similar crystal (1.48 and 0.95 Å for $NH_4^+$ and $Na^+$, resp.) and hydrated radii (3.31 and 3.58 Å for $NH_4^+$ and $Na^+$) [39], thus balancing the total charge of NaP1. The adsorption would involve trapping the $NH_4^+$ cations inside the NaP1 structure, releasing innocuous cations (like $Na^+$) to the aqueous medium:

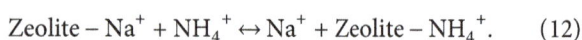

$$Zeolite - Na^+ + NH_4^+ \leftrightarrow Na^+ + Zeolite - NH_4^+. \quad (12)$$

Structurally, the zeolite presents 8-ring pore apertures that would be large enough for the accessibility of certain cations like $NH_4^+$ through the zeolite channel system. The $NH_4^+$ immobilization on NaP1 was studied by the FTIR analysis, comparing the FTIR spectrum of initial Al-waste-NaP1 with that obtained after the uptake of the adsorbate, as shown in Figure 4. Before $NH_4^+$ adsorption, the zeolite shows the assignments of the main absorption bands of NaP1: T-O-T asymmetrical stretching mode (~1000 cm$^{-1}$), T-O-T symmetrical stretching mode (740–680 cm$^{-1}$), external linkage vibration (~607 cm$^{-1}$), and T-O bending mode (430 cm$^{-1}$) of the $TO_4$ tetrahedron [10]. The $NH_4^+$ adsorption process led to very similar absorption bands (represented by dashed lines in Figure 4) to the characteristic vibration modes occurring in the NaP1 framework before adsorption. The results confirmed that the main change was observed at approximately 1400 cm$^{-1}$, attributing to the absorption band (v4 asymmetrical bending mode) of $NH_4^+$ present in the NaP1 structure [40]. Therefore, the cation exchange of $Na^+$ by $NH_4^+$ may be considered as the governing mechanism of the adsorption process, as shown in Figure 5.

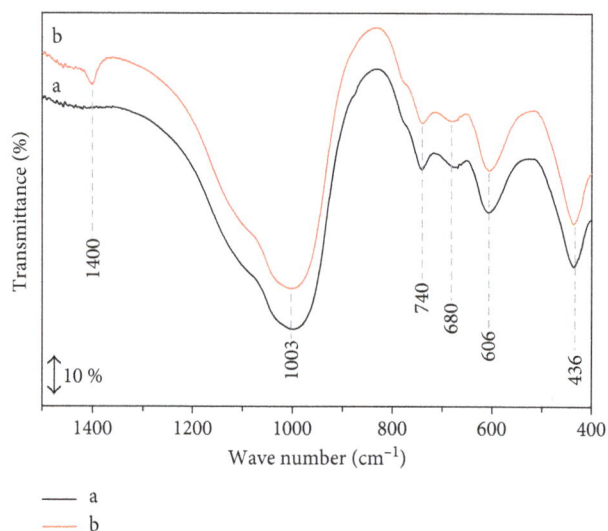

FIGURE 4: FTIR spectra of Al-waste-NaP1 (a) before and (b) after $NH_4^+$ adsorption.

*3.2. Effect of Contact Time: Adsorption Kinetics.* The influence of the contact time on the ammonium removal efficiency was studied from 1 to 360 min, as shown in Figure 6(a). The ammonium adsorption on Al-waste-NaP1 seems to be a very fast process where the equilibrium was reached within the first 5 min. Under the tested operating conditions, the highest removal efficiency of the zeolite was 88%, which was reached at 15 min. The short required contact time would involve a high affinity for the adsorbate, indicating a close electrostatic interaction between the $NH_4^+$ cations and the charged negatively functional groups on the zeolite surface. As the contact time increased, the removal efficiency of Al-waste-NaP1 was almost constant (86.3 and 85.4% for 2 and 30 min) and then slightly decreased from 82 to 75.8% for 60 and 360 min. Some authors found that the $NH_4^+$ adsorption efficiency using natural zeolite was maintained from 30 min to longer contact times (up to 24 h) [41]. It seems to be related to the large number of available active sites on the adsorbent surface and the high adsorbate

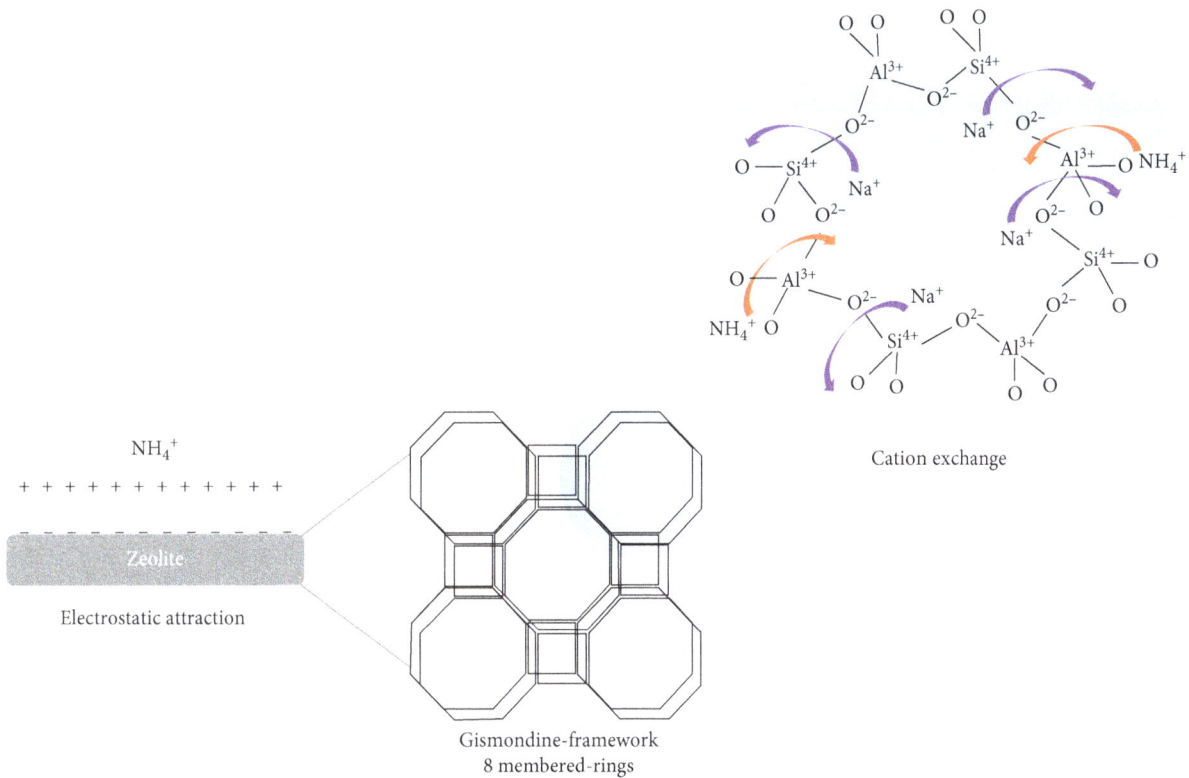

FIGURE 5: Adsorption mechanism through the cation exchange between the extra-framework cations (Na$^+$) of the Al-waste-NaP1 zeolite and the ammonium cations (NH$_4^+$).

concentration gradient at the beginning of the adsorption process, resulting in a fast diffusion and rapid equilibrium. As the adsorbent sites are occupied, the adsorption capacity would decrease significantly. Additionally, the rapid adsorption would take place easily on the adsorbent surface, whereas the slower uptake process would occur inside the pores [42]. In this sense, as a further increase in the contact time had no significant effect on the removal efficiency, the adsorption kinetic of NH$_4^+$ onto Al-waste-NaP1 was only studied between 1 and 60 min. As the kinetic analysis is essential for the process design in water treatment applications, the adsorption kinetic performance was evaluated by applying the pseudo-first-order and pseudo-second-order models, shown in Figure 6(b), using the nonlinear regression method. Apparently, minor differences can be noticed between the plots obtained for the pseudo-second-order and the pseudo-first-order model. Generally, although in most of the reviewed works [42–45], the NH$_4^+$ adsorption kinetic has been evaluated according to linear regression methods, the pseudo-second-order model seems to provide the best results. However, the pseudo-first-order equation is usually more appropriate for the initial stage of adsorption processes (contact time of 20–30 min) not for the whole range [46]. In our case, the adsorption of NH$_4^+$ on Al-waste-NaP1 was significantly rapid, reaching high efficiencies of 80.3 and 86.3% in 1 and 2 min, respectively. Thus, the pseudo-first-order kinetic model seems to describe better the experimental data since this model provided a more accurate correlation, that is, the highest $R^2$ as well as the lowest $\chi^2$,

HYBRID, and RMSE values (Table 1). It is believed that the NH$_4^+$ uptake by NaP1 follows the common transport processes during the adsorption in solid-liquid systems characterized by four steps: (i) the bulk transport, which occurs quickly; (ii) the film diffusion where the adsorbate is transported from the bulk liquid phase to the active sites on the adsorbent external surface (solid phase), taking place slowly; (iii) the intraparticle diffusion where the adsorbate diffuses slowly from the exterior of the adsorbent to the most internal surface (i.e., pores) of the adsorbent; and (iv) the very fast adsorptive attachment [47].

3.3. Effect of Adsorbent Dose. The influence of the adsorbent dose on the removal efficiency and adsorption capacity of NaP1 for the ammonium uptake was studied from 1 to 15 g/L, as shown in Figure 7. The increase of the zeolite dose led to the increase of the removal efficiency from 61 to 92% for 1 and 15 g/L, while the NH$_4^+$ uptake capacity decreased from 30.4 to 3.1 mg/g for 1 and 15 g/L, respectively. The higher the adsorbent mass, the larger the adsorbent surface, and accordingly, the larger the number of adsorption sites on the NaP1 surface is, accelerating the adsorption process. The NH$_4^+$ cations would diffuse from the aqueous medium towards the surface of the adsorbent due to the electrostatic attraction, tending to occupy the adsorption active sites. In this context, the higher the adsorbent dose, the larger the adsorbent sites will be available under the same adsorption conditions (i.e., under the same adsorbate mass gradient).

(a)

○   Experimental
—   First-order
- -   Second-order

(b)

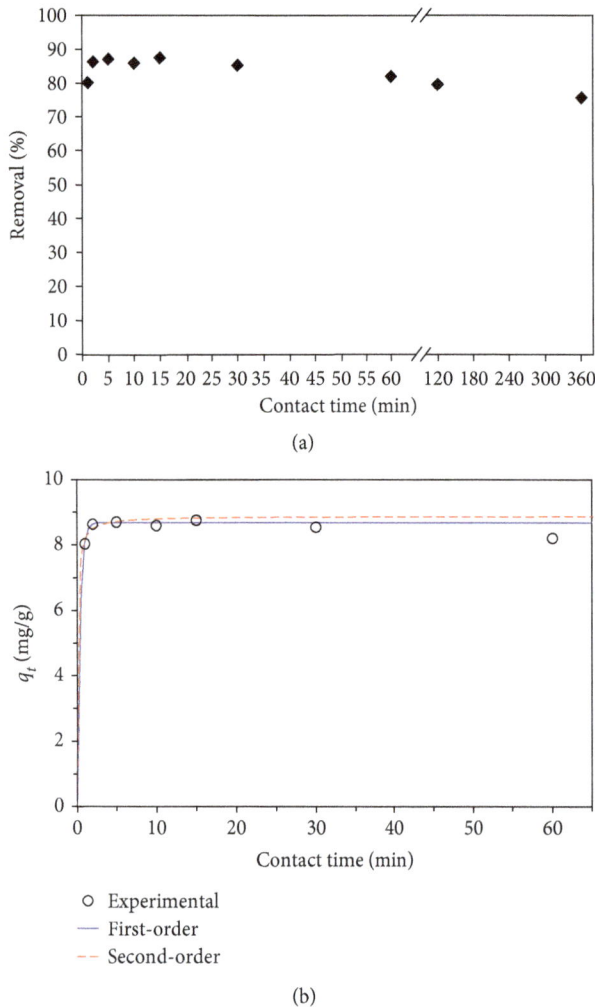

FIGURE 6: (a) Influence of the contact time and (b) kinetic modeling of the $NH_4^+$ adsorption by Al-waste-NaP1. Conditions: pH = 7.5; adsorbent dose = 5 g/L; Co = 50 mg/L; T = 28 ± 2°C.

TABLE 1: Kinetic model parameters and error function values obtained for the $NH_4^+$ adsorption onto Al-waste-NaP1.

| $q_{e, experimental}$ (mg/g) | Pseudo-first-order model | | Pseudo-second-order model | |
|---|---|---|---|---|
|  | $q_{e, model}$ (mg/g) | 8.68 | $q_{e, model}$ (mg/g) | 8.87 |
|  | $k_1$ (1/min) | 2.60 | $k_2$ (g/mg min) | 1.20 |
| 8.75 | $R^2$ | 0.961 | $R^2$ | 0.895 |
|  | $\chi^2$ | 0.022 | $\chi^2$ | 0.035 |
|  | RMSE | 0.196 | RMSE | 0.246 |
|  | HYBRID | 0.460 | HYBRID | 0.720 |

These results are in agreement with values found by other researchers for the adsorption of ammonium onto a geo-polymer-type adsorbent [42]. Both the removal efficiency and adsorption capacity of the zeolite were almost constant from 10 to 15 g/L; thus, the selected adsorbent dose for developing the further adsorption tests was 10 g/L since it would be the lowest adsorbent dose that provides a very high removal efficiency (91%).

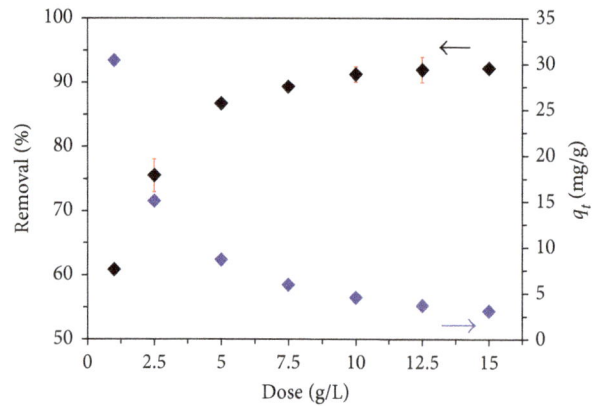

FIGURE 7: Influence of the adsorbent dose on the $NH_4^+$ removal efficiency and uptake capacity of Al-waste-NaP1. Conditions: pH = 7.5; contact time = 15 min; Co = 50 mg/L; T = 28 ± 2°C.

3.4. *Effect of Initial Adsorbate Concentration: Equilibrium Isotherms.* As the initial concentration increased, the amount of ammonium retained by Al-waste-NaP1 also increased gradually up to the adsorbent saturation, which was reached above an initial concentration of approximately 1000 mg/L under the tested conditions. The experimental data indicated that the maximum adsorption capacity of Al-waste-NaP1 was 37.9 mg/g. In general, higher adsorption capacities are obtained from the fit of isotherm models to the experimental data. The applied adsorption isotherms are illustrated in Figure 8. The estimated parameters and error function values for the two-parameter and three-parameter models are shown in Table 2. The Langmuir isotherm was the only two-parameter model that provided the best data tendency in the whole range of initial concentrations. Thus, the Langmuir isotherm would indicate a homogeneous process and monolayer coverage of $NH_4^+$ on the NaP1 surface. On the contrary, the Freundlich isotherm was only suited for the range of low initial concentrations, while the Temkin model was not fit satisfactorily to the experimental data. The relation between the Freundlich exponent ($n = 0.48 < 1$) and the Langmuir separation factor ($R_L = 0.94$ and 0.11 for the lowest and highest initial adsorbate concentration) would indicate that the $NH_4^+$ adsorption by Al-waste-NaP1 was favorable, according to the concave isotherm shape followed by experimental data [23, 24]. The experimental data were described adequately by all the three-parameter models. Such models followed the same isotherm profile, providing the best fits for the experimental data. In particular, the Sips and Toth models provided the highest $R^2$ and the lowest $\chi^2$, HYBRID, and RMSE values. The results show that the maximum amount of adsorbed $NH_4^+$ cation per mass of NaP1 (i.e., $q_{max}$) was of 54.19 and 58.46 mg/g, according to the Sips and Toth isotherms, respectively. The Sips isotherm provided reliable results possibly due to the fact it combines both the Langmuir and Freundlich models, thus covering satisfactorily the whole initial concentration range. Thus, the $NH_4^+$ adsorption using Al-waste-NaP1 could be described by a homogeneous and heterogeneous process. The Redlich–Peterson isotherm is also a combination of both Langmuir and Freundlich

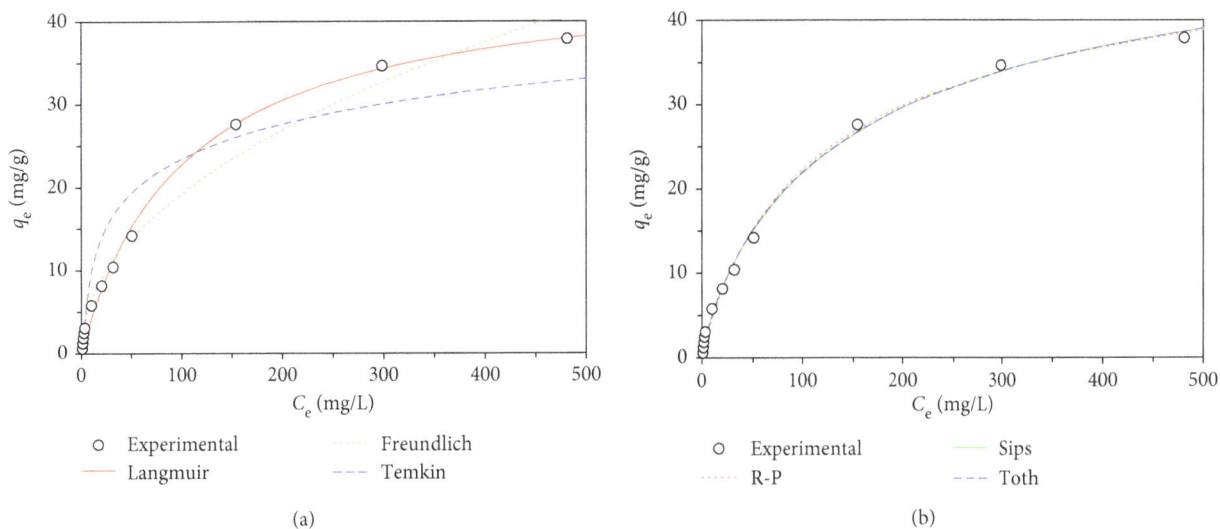

FIGURE 8: Adsorption isotherms for the $NH_4^+$ uptake on Al-waste-NaP1. Conditions: pH = 7.5; contact time = 15 min; adsorbent dose = 10 g/L; and T = 28 ± 2°C. R-P: Redlich–Peterson.

TABLE 2: Isotherm models applied for the $NH_4^+$ adsorption onto Al-waste-NaP1.

| Two-parameter isotherm model | | | Error function | | | |
|---|---|---|---|---|---|---|
| Langmuir | $q_{max}$ (mg/g) | $k_L$ (L/mg) | $R^2$ | $\chi^2$ | RMSE | HYBRID |
| | 46.05 | 0.01 | 0.995 | 3.291 | 0.993 | 25.228 |
| Freundlich | $k_F$ (mg/g)/(mg/L)$^n$ | $n$ | $R^2$ | $\chi^2$ | RMSE | HYBRID |
| | 2.09 | 0.48 | 0.981 | 2.597 | 1.935 | 35.623 |
| Temkin | $A$ (L/g) | $b$ (J/mol) | $R^2$ | $\chi^2$ | RMSE | HYBRID |
| | 0.50 | 413.74 | 0.901 | 11.384 | 4.453 | 179.190 |
| Three-parameter isotherm model | | | | Error function | | |
| Sips | $q_{max}$ (mg/g) | $k_S \times 10^{-3}$ (L/mg) | $n$ | $R^2$ | $\chi^2$ | RMSE | HYBRID |
| | 54.19 | 6.26 | 0.83 | 0.997 | 0.739 | 0.804 | 9.533 |
| Toth | $q_{max}$ (mg/g) | $k_T \times 10^{-2}$ (L/mg) | $n$ | $R^2$ | $\chi^2$ | RMSE | HYBRID |
| | 58.46 | 1.10 | 0.68 | 0.996 | 0.833 | 0.935 | 10.026 |
| Redlich–Peterson | $k_{RP}$ (L/g) | $a_{RP}$ (mg/L)$^{-\beta}$ | $\beta$ | $R^2$ | $\chi^2$ | RMSE | HYBRID |
| | 0.52 | 0.02 | 0.91 | 0.996 | 0.979 | 1.002 | 11.399 |

TABLE 3: Ammonium adsorption capacity for several adsorbents.

| Adsorbent | $q_{max}$ (mg/g) | Isotherm | pH | $T$ (°C) | Time (min) | Dose (g/L) | Reference |
|---|---|---|---|---|---|---|---|
| Natural zeolite | 10.39 | Langmuir | 7 | 25 | 1440 | 32 | Mazloomi and Jalali [48] |
| Geopolymer | 21.07 | Sips | 6 | 22 | 1440 | 5 | Luukkonen et al. [42] |
| Modified biochar | 22.6 | Langmuir | 8–9 | — | 120 | 2 | Vu et al. [6] |
| Natural zeolite | 23.83 | Sips | 6.5 | 25 | 1440 | 3 | Lei et al. [45] |
| Fly-ash zeolite | 37.45 | Langmuir | 8 | 25 | 75 | 4 | Zhang et al. [49] |
| Hydrogel | 42.74 | Langmuir | 6–7 | 30 | 30 | 2 | Zheng et al. [43] |
| Al-waste-NaP1 | 54.19 | Sips | 7.5 | 28 | 15 | 10 | This work |
| Fly ash-zeolite | 95.42 | Langmuir | 7 | 25 | 60 | 2 | Jiang et al. [36] |

models [24]. As the Redlich–Peterson exponent was close to 1 ($\beta = 0.91$), the process would be best described by the Langmuir isotherm instead of Freundlich [23]. Therefore, it is believed that the adsorption of the $NH_4^+$ cation onto this zeolite can be more homogeneous rather than heterogeneous.

### 3.5. Comparison of the $NH_4^+$ Removal Capacity of Al-Waste-NaP1 with Other Adsorbents. The maximum $NH_4^+$ removal

capacity of the studied zeolite was compared with other adsorbent materials (Table 3). Although different operating conditions have been used for the uptake of $NH_4^+$, in general, the results show that the required time to remove the adsorbate by Al-waste-NaP1 was shorter than most of the adsorbents given in Table 3. As can be seen, lower and higher $NH_4^+$ adsorption capacities were found in the literature. The NaP1 adsorbent from Al-waste under effluents recycling showed a high removal capacity, being generally higher than natural zeolites. In addition, it can be considered as a low-cost adsorbent compared to other commercial materials like activated carbons. The studied NaP1 zeolite exhibited adequate sorption properties when it is compared to other sorbent materials obtained from common synthesis processes without recycling of effluents. In this sense, Al-waste-NaP1 showed promising adsorption characteristics, making it a potential adsorbent for the $NH_4^+$ removal from water. It is believed that this zeolite could be also used to remove other contaminants (e.g., heavy metals, radioactive metals, and organic compounds) in aqueous media.

## 4. Conclusions

In this work, the elimination of $NH_4^+$ cation from aqueous solutions using the zeolite NaP1, which was synthesized from a hazardous Al-waste by an eco-friendly process, was studied by means of batch adsorption experiments. The effects of different experimental parameters, including pH, contact time, adsorbent dose, and initial $NH_4^+$ concentration, on the adsorption efficiency and adsorption capacity of Al-waste-NaP1 was studied under ambient conditions. The adsorption kinetic and equilibrium isotherms were also studied by applying the nonlinear method. The results showed that the uptake process of the adsorbate ($NH_4^+$) by the adsorbent (Al-waste-NaP1) was fast leading to removal percentages of 88% in the first 15 min. The $NH_4^+$ removal on the zeolite was better described by the pseudo-first-order kinetic model. The experimental data showed that the highest amount of $NH_4^+$ cations removed by the zeolitic adsorbent was 37.9 mg/g, similar to that found for other sorbent materials. The equilibrium data were better described by the three-parameter isotherm models than the two-parameter equations. In particular, the Sips and Toth isotherms led to maximum capacities of 54.19 and 58.46 mg/g, respectively. Thus, a synergic effect on the environmental protection can be achieved: firstly, the transformation of the hazardous Al-waste into a zeolite can contribute to the end of waste condition, and secondly, the Al-waste-based zeolite can be considered as an alternative adsorbent to other materials used for the treatment of contaminated aqueous effluents.

## Conflicts of Interest

The authors declare that they have no conflicts of interest.

## Acknowledgments

The authors thank MINECO for its financial support (Project CTM2012-34449). R. Sánchez-Hernández thanks MINECO for the grant BES-2013-066269. The authors thank the Research Support Centre of the Geology Faculty, University Complutense of Madrid, for the help given.

## References

[1] J. K. Böhlke, R. L. Smith, and D. N. Miller, "Ammonium transport and reaction in contaminated groundwater: application of isotope tracers and isotope fractionation studies," *Water Resources Research*, vol. 42, no. 5, 2006.

[2] A. Casadellà, P. Kuntke, O. Schaetzle, and K. Loos, "Clinoptilolite-based mixed matrix membranes for the selective recovery of potassium and ammonium," *Water Research*, vol. 90, pp. 62–70, 2016.

[3] A. Urtiaga, I. Ortiz, A. Anglada, D. Mantzavinos, and E. Diamadopoulos, "Kinetic modeling of the electrochemical removal of ammonium and COD from landfill leachates," *Journal of Applied Electrochemistry*, vol. 42, no. 9, pp. 779–786, 2012.

[4] M. Sica, A. Duta, C. Teodosiu, and C. Draghici, "Thermodynamic and kinetic study on ammonium removal from a synthetic water solution using ion exchange resin," *Clean Technologies and Environmental Policy*, vol. 16, no. 2, pp. 351–359, 2014.

[5] L. Megido, B. Suárez-Peña, L. Negral et al., "Relationship between physico-chemical characteristics and potential toxicity of PM10," *Chemosphere*, vol. 162, pp. 73–79, 2016.

[6] T. M. Vu, V. T. Trinh, D. P. Doan et al., "Removing ammonium from water using modified corncob-biochar," *Science of the Total Environment*, vol. 579, pp. 612–619, 2017.

[7] M. Ahmad, A. U. Rajapaksha, J. E. Lim et al., "Biochar as a sorbent for contaminant management in soil and water: a review," *Chemosphere*, vol. 99, pp. 19–33, 2014.

[8] S. Sen Gupta and K. G. Bhattacharyya, "Adsorption of metal ions by clays and inorganic solids," *RSC Advances*, vol. 4, no. 54, pp. 28537–28586, 2014.

[9] L. Lin, Z. Lei, L. Wang et al., "Adsorption mechanisms of high-levels of ammonium onto natural and NaCl-modified zeolites," *Separation and Purification Technology*, vol. 103, pp. 15–20, 2013.

[10] R. Sánchez-Hernández, A. López-Delgado, I. Padilla, R. Galindo, and S. López-Andrés, "One-step synthesis of NaP1, SOD and ANA from a hazardous aluminum solid waste," *Microporous and Mesoporous Materials*, vol. 226, pp. 267–277, 2016.

[11] J. Tian, V. Miller, P. C. Chiu, J. A. Maresca, M. Guo, and P. T. Imhoff, "Nutrient release and ammonium sorption by poultry litter and wood biochars in stormwater treatment," *Science of the Total Environment*, vol. 553, pp. 596–606, 2016.

[12] S. Wang and Y. Peng, "Natural zeolites as effective adsorbents in water and wastewater treatment," *Chemical Engineering Journal*, vol. 156, no. 1, pp. 11–24, 2010.

[13] T. H. Martins, T. S. O. Souza, and E. Foresti, "Ammonium removal from landfill leachate by Clinoptilolite adsorption followed by bioregeneration," *Journal of Environmental Chemical Engineering*, vol. 5, no. 1, pp. 63–68, 2017.

[14] R. Leyva-Ramos, J. E. Monsivais-Rocha, A. Aragon-Piña et al., "Removal of ammonium from aqueous solution by ion exchange on natural and modified chabazite," *Journal of Environmental Management*, vol. 91, no. 12, pp. 2662–2668, 2010.

[15] R. Malekian, J. Abedi-Koupai, S. S. Eslamian, S. F. Mousavi, K. C. Abbaspour, and M. Afyuni, "Ion-exchange process for ammonium removal and release using natural Iranian zeolite," *Applied Clay Science*, vol. 51, no. 3, pp. 323–329, 2011.

[16] N. Koshy and D. N. Singh, "Fly ash zeolites for water treatment applications," *Journal of Environmental Chemical Engineering*, vol. 4, no. 2, pp. 1460–1472, 2016.

[17] R. Sánchez-Hernández, I. Padilla, S. López-Andrés, and A. López-Delgado, "Eco-friendly bench-scale zeolitization of an Al-containing waste into gismondine-type zeolite under effluent recycling," *Journal of Cleaner Production*, vol. 161, pp. 792–802, 2017.

[18] X. You, C. Valderrama, and J. L. Cortina, "Simultaneous recovery of ammonium and phosphate from simulated treated wastewater effluents by activated calcium and magnesium zeolites," *Journal of Chemical Technology and Biotechnology*, vol. 92, no. 9, pp. 2400–2409, 2017.

[19] D. Guaya, C. Valderrama, A. Farran, C. Armijos, and J. L. Cortina, "Simultaneous phosphate and ammonium removal from aqueous solution by a hydrated aluminum oxide modified natural zeolite," *Chemical Engineering Journal*, vol. 271, pp. 204–213, 2015.

[20] M. Uğurlu and M. H. Karaoğlu, "Adsorption of ammonium from an aqueous solution by fly ash and sepiolite: isotherm, kinetic and thermodynamic analysis," *Microporous and Mesoporous Materials*, vol. 139, no. 1–3, pp. 173–178, 2011.

[21] S. Y. Lagergren, "Zur Theorie der sogenannten adsorption gelöster stoffe," *Handlingar*, vol. 24, no. 4, pp. 1–39, 1898.

[22] G. Blanchard, M. Maunaye, and G. Martin, "Removal of heavy metals from waters by means of natural zeolites," *Water Research*, vol. 18, no. 12, pp. 1501–1507, 1984.

[23] H. N. Tran, S.-J. You, A. Hosseini-Bandegharaei, and H.-P. Chao, "Mistakes and inconsistencies regarding adsorption of contaminants from aqueous solutions: a critical review," *Water Research*, vol. 120, pp. 88–116, 2017.

[24] K. Y. Foo and B. H. Hameed, "Insights into the modeling of adsorption isotherm systems," *Chemical Engineering Journal*, vol. 156, no. 1, pp. 2–10, 2010.

[25] H. M. F. Freundlich, "Over the adsorption in solution," *Journal of Physical Chemistry*, vol. 57, pp. 385–471, 1906.

[26] I. Langmuir, "The adsorption of gases on plane surfaces of glass, mica and platinum," *Journal of the American Chemical Society*, vol. 40, no. 9, pp. 1361–1403, 1918.

[27] K. R. Hall, L. C. Eagleton, A. Acrivos, and T. Vermeulen, "Pore- and solid-diffusion kinetics in fixed-bed adsorption under constant-pattern conditions," *Industrial and Engineering Chemistry Fundamentals*, vol. 5, no. 2, pp. 212–223, 1966.

[28] E. Worch, *Adsorption Technology in Water Treatment: Fundamentals, Processes, and Modeling*, De Gruyter, Berlin, Germany, 2012.

[29] M. Temkin and V. Pyzhev, "Kinetics of ammonia synthesis on promoted iron catalysts," *Acta Physiochim*, vol. 12, pp. 217–222, 1940.

[30] O. Redlich and D. L. Peterson, "A useful adsorption isotherm," *Journal of Physical Chemistry*, vol. 63, no. 6, p. 1024, 1959.

[31] R. Sips, "On the structure of a catalyst surface," *Journal of Chemical Physics*, vol. 16, no. 5, pp. 490–495, 1948.

[32] J. Toth, "State equations of the solid-gas interface layers," *Acta Chimica Academiae Scientiarum Hungaricae*, vol. 69, pp. 311–328, 1971.

[33] É. C. Lima, M. A. Adebayo, and F. M. Machado, "Kinetic and equilibrium models of adsorption," in *Carbon Nanomaterials as Adsorbents for Environmental and Biological Applications*, Springer, Berlin, Germany, 2015.

[34] O. Larlus, S. Mintova, and T. Bein, "Environmental syntheses of nanosized zeolites with high yield and monomodal particle size distribution," *Microporous and Mesoporous Materials*, vol. 96, no. 1–3, pp. 405–412, 2006.

[35] L. Y. Zhang, H. Y. Zhang, W. Guo, and Y. L. Tian, "Sorption characteristics and mechanisms of ammonium by coal by-products: slag, honeycomb-cinder and coal gangue," *International Journal of Environmental Science and Technology*, vol. 10, no. 6, pp. 1309–1318, 2013.

[36] Z. Jiang, J. Yang, H. Ma, X. Ma, and J. Yuan, "Synthesis of pure NaA zeolites from coal fly ashes for ammonium removal from aqueous solutions," *Clean Technologies and Environmental Policy*, vol. 18, no. 3, pp. 629–637, 2016.

[37] E. Marañón, M. Ulmanu, Y. Fernández, I. Anger, and L. Castrillón, "Removal of ammonium from aqueous solutions with volcanic tuff," *Journal of Hazardous Materials*, vol. 137, no. 3, pp. 1402–1409, 2006.

[38] Y. Zhao, B. Zhang, X. Zhang, J. Wang, J. Liu, and R. Chen, "Preparation of highly ordered cubic NaA zeolite from halloysite mineral for adsorption of ammonium ions," *Journal of Hazardous Materials*, vol. 178, no. 1–3, pp. 658–664, 2010.

[39] E. R. Nightingale, "Phenomenological theory of ion solvation. effective radii of hydrated ions," *Journal of Physical Chemistry*, vol. 63, no. 9, pp. 1381–1387, 1959.

[40] K. Nakamoto, "Applications in inorganic chemistry," in *Infrared and Raman Spectra of Inorganic and Coordination Compounds*, John Wiley & Sons, Inc., Hoboken, NJ, USA, 2008.

[41] K. Saltalı, A. Sarı, and M. Aydın, "Removal of ammonium ion from aqueous solution by natural Turkish (Yıldızeli) zeolite for environmental quality," *Journal of Hazardous Materials*, vol. 141, no. 1, pp. 258–263, 2007.

[42] T. Luukkonen, M. Sarkkinen, K. Kemppainen, J. Rämö, and U. Lassi, "Metakaolin geopolymer characterization and application for ammonium removal from model solutions and landfill leachate," *Applied Clay Science*, vol. 119, pp. 266–276, 2016.

[43] Y. Zheng, Y. Liu, and A. Wang, "Fast removal of ammonium ion using a hydrogel optimized with response surface methodology," *Chemical Engineering Journal*, vol. 171, no. 3, pp. 1201–1208, 2011.

[44] K. Zare, H. Sadegh, R. Shahryari-ghoshekandi et al., "Equilibrium and kinetic study of ammonium ion adsorption by $Fe_3O_4$ nanoparticles from aqueous solutions," *Journal of Molecular Liquids*, vol. 213, pp. 345–350, 2016.

[45] L. Lei, X. Li, and X. Zhang, "Ammonium removal from aqueous solutions using microwave-treated natural Chinese zeolite," *Separation and Purification Technology*, vol. 58, no. 3, pp. 359–366, 2008.

[46] Y. S. Ho and G. McKay, "A comparison of chemisorption kinetic models applied to pollutant removal on various sorbents," *Process Safety and Environmental Protection*, vol. 76, no. 4, pp. 332–340, 1998.

[47] W. J. Weber and E. H. Smith, "Simulation and design models for adsorption processes," *Environmental Science and Technology*, vol. 21, no. 11, pp. 1040–1050, 1987.

[48] F. Mazloomi and M. Jalali, "Ammonium removal from aqueous solutions by natural Iranian zeolite in the presence of organic acids, cations and anions," *Journal of Environmental Chemical Engineering*, vol. 4, no. 2, pp. 1664–1673, 2016.

[49] M. Zhang, H. Zhang, D. Xu et al., "Removal of ammonium from aqueous solutions using zeolite synthesized from fly ash by a fusion method," *Desalination*, vol. 271, no. 1–3, pp. 111–121, 2011.

# 2

# Novel Synthetic-Based Drilling Fluid through Enzymatic Interesterification of Canola Oil

**Anawe A. L. Paul** ⓘ **and Folayan J. Adewale** ⓘ

*Department of Petroleum Engineering, College of Chemical and Petroleum Engineering, Covenant University, Ota, Nigeria*

Correspondence should be addressed to Anawe A. L. Paul; paul.anawe@covenantuniversity.edu.ng

Academic Editor: Michael Harris

Over the years, the oil industries have avoided aromatic, naphthenic, and paraffinic oils as drilling mud base fluids principally because of their detrimental environmental issues on pelagic and benthic marine ecosystems as a result of their toxicity and nonbiodegradability coupled with the possible deterioration of the oil itself and the rubber parts of the drilling equipment because the aromatic hydrocarbons present in the oil have a tendency to dissolve/damage elastomers present in rubber. Hence, possible insights into how to chemically and/or physically produce synthetic base drilling fluids whose cuttings are nontoxic, readily biodegradable, environmentally friendly, and of nonpetroleum source become imperative. In this study, enzymatic interesterification of canola oil was done with ethanol by using enzyme lipase as catalyst under optimum conditions of temperature and pressure and the physicochemical properties of the produced ester were evaluated and compared with that of diesel and a synthetic hydrocarbon base fluid (SHBF). Results show that the specific gravity, kinematic viscosity, dynamic viscosity, and surface tension of canola oil were reduced by 5.50%, 94.74%, 95.03%, and 9.38%, respectively, upon enzymatic interesterification to conform to standard requirements. Similarly, increased |mud ability to pump fluids and possibility of cold temperature environment can be achieved with the reduction in pour point and cloud point, respectively, of the produced canola oil ester. Finally, the produced ester showed no aromatic content as confirmed from its FTIR analysis which indicates its nontoxicity, biodegradability, and environmental friendliness.

## 1. Introduction

Drilling fluids are complex fluid mixtures which are principally formulated to carry cuttings from beneath the bit, transport them up the annulus and permit their separation at the surface, and prevent the inflow of formation fluids (oil, gas, and water) from the permeable rock that is being penetrated and to form a thin, low-permeability filter cake which seals pores and other openings in formations penetrated by the bit.

The oil industry started with water base fluids but because of formation clays that react, swell, or slough after exposure to water-based mud coupled with the need to penetrate a hole with high temperature, a shift to diesel-based mud began in 1960s [1]. In order to mitigate the environmental and technical issues associated with diesel-based mud, which include high initial cost per barrel, difficulties in detection of gas kick, high cost of lost circulation,

difficulties in keeping the rig clean, and easy deterioration of the rubber part of the drilling equipment; hence, an alternative drilling fluid from nonpetroleum origin becomes imperative.

Diesel is also harmful to the environment particularly marine environment during offshore drilling [2]. OBFs, particularly those made with diesel are more persistent in sediments. Grahl-Nielsen et al. [3] reported little change over five years in the area of sea bottom near a single exploratory well in the North Sea that contained 1,000 to 10,000 mg/kg OBF from the discharge of OBF cuttings.

In many countries that are developing offshore oil and gas resources, cuttings discharge permits require performance of toxicity tests with drilling fluid ingredients and whole drilling fluids [4]. One of the objectives in developing environmentally acceptable alternatives to OBFs was to produce a drilling fluid that would provide the drilling performance of an OBF but would degrade rapidly following discharge to the ocean.

The most important fluid properties that may affect environmental impacts are toxicity and rate of bio-degradation. Environmental impacts that may result from the discharge of drilling fluids and cuttings to the ocean are of two types: effects on water column (pelagic ecosystems) and effects on sea bottom (benthic ecosystems) [5–7].

The Norwegian regulatory authority defines a SBF as a drilling fluid where the base fluid consists of nonwater soluble organic compounds and where neither the base fluid nor the additives are of petroleum origin [8]. These synthetic base fluids may be classified into four general categories, namely, synthetic hydrocarbons, ethers, esters, and acetals.

However, polymerized olefins such as linear alpha olefins (LAOs), poly alpha olefins (PAOs), and internal olefins (IOs) are the most frequently used synthetic hydrocarbons, and they have been reclassified as environmentally unacceptable and are no longer in use because they contain a benzene molecule which is aromatic and nonbiodegradable [9].

Synthetic-based muds (SBMs) were developed to replace OBMs in difficult drilling situation, and though SBMs are expensive than OBMs, they have superior environmental properties that may permit the cuttings to be discharged on-site [10].

Veil et al. [11] describe the environmental benefits of synthetic-based drilling muds to include less waste production from a recyclable product (compared to WBMs); elimination of diesel as a mud base lessens the pollution hazard, improves worker safety through lower toxicity, diminishes irritant properties, and reduces consequent risk (compared to OBMs); increased use of horizontal drilling reduces the areal extent and the environmental impacts of offshore oil and gas operations (compared to WBMs); shortened drilling time results in reduced air emissions from drilling power sources; and improved drilling performance decreases waste-generating incidents, such as pipe stuck in the hole. Mineral oils are less biodegradable than SBF chemicals particularly under anaerobic conditions.

All the SBF base chemicals evaluated by Norman [12] and Steber et al. [13] have measurable aerobic and anaerobic biodegradability. Aerobic and anaerobic biodegradation rates are greatest for ester SBFs, followed by IOs. Meanwhile, Candler et al. [14] ranked both aerobic and anaerobic biodegradability of drilling fluid base chemicals from most to least biodegradable as

$$\text{esters} \gg \text{LAOs} > \text{IOs} \gg \text{PAOs} > \text{mineral oil.} \quad (1)$$

Environmental authorities of the North Sea countries have hypothesized that rapid degradation will minimize the environmental impacts of SBF cuttings discharge and thus speeding ecosystem recovery [15].

## 2. Materials and Methods

### 2.1. Canola Oil.
Canola oil (Canada oil) is processed from the seed of *Brassica napus* and *Brassica rapa* which can be extracted by either solvent extraction method or expeller method. Edible oils and fats composed primarily of triglycerides which are ester of one molecule of glycerol and three molecule of fatty acid. Canola oil analyses show that the triglycerides constitute 94.4 to 99.1% of the total lipid and about 2.5% phospholipids [16]. It has been established that the 18 carbon fatty acids account for about 95% of canola's total fatty acid [17]. It has a larger percentage of oleic acid (about 56%) followed by linoleic acid (26%), linolenic acid (10%), and 4% palmitic acid with only 2% stearic acid [18].

### 2.2. Interesterification Reaction.
From the three main oil modification technologies (fractionation, hydrogenation, and interesterification), interesterification is by far the easiest process to understand and to control [19]. Interesterification process is used to modify the physical properties of the oil or fat blend by rearranging the fatty acid groups within and between the different triglycerides.

Interesterification is the process of exchanging the organic group $R''$ of an ester with the organic group $R'$ of an alcohol. These reactions are often catalyzed by the addition of an acid or base catalyst. The reaction can also be accomplished with the help of enzymes (biocatalysts) particularly lipases:

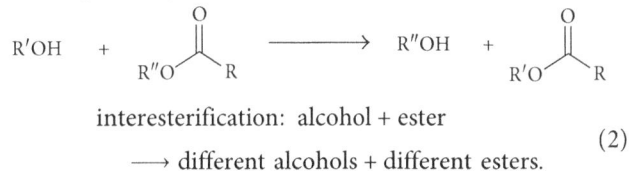

$$\text{interesterification: alcohol + ester}$$
$$\longrightarrow \text{different alcohols + different esters.} \quad (2)$$

Strong acids catalyze the reaction by donating a proton to the carbonyl group, thus making it a more potent electrophile, whereas bases catalyze the reaction by removing a proton from the alcohol, thus making it more nucleophilic. Esters with larger alkoxy groups can be made from methyl or ethyl esters in high purity by heating the mixture of ester, acid/base, and large alcohol and evaporating the small alcohol to drive equilibrium.

### 2.3. Enzymatic Interesterification Procedure.
Six hundred milliliters of chemically degummed canola oil were measured and poured into a 1000 ml PYREX Erlenmeyer flask with rubber stopper. The degummed oil was then heated to a temperature of 60°C in an electric oven. This was followed by the addition of 120 ml of ethanol which was preheated to a temperature of 60°C with the aid of water bath. A 4% concentration of immobilized *Candida antarctica* lipase was then added to the oil-alcohol mixture in the flask, and the flask was then placed in an orbital shaker at 70°C and rotating speed of 300 rpm for 12 hours. After the completion of the reaction time, the reaction mixture was then filtered by vacuum filtration in order to retain the enzyme for subsequent recovery and use.

The resulting filtrate was then taken to a rotary evaporator at 90°C (a temperature higher than the boiling point of ethanol (78.37°C)) in order to evaporate unreacted alcohol. Finally, the reaction mixture was placed in a separating funnel resulting in two distinct layers of upper ethyl ester and a lower layer of monoglyceride, diglyceride, glycerol, and other impurities. The produced ethyl ester was then bleached with 1.5% bleaching clay and deodorized, and its volume was measured in order to deduce its percentage yield:

(a)

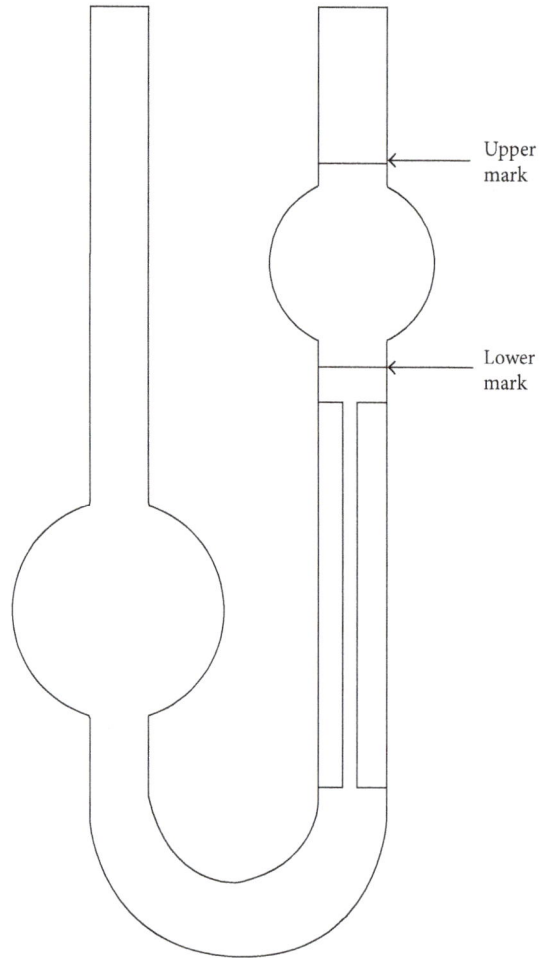

Upper mark

Lower mark

(b)

(c)

(d)

FIGURE 1: Continued.

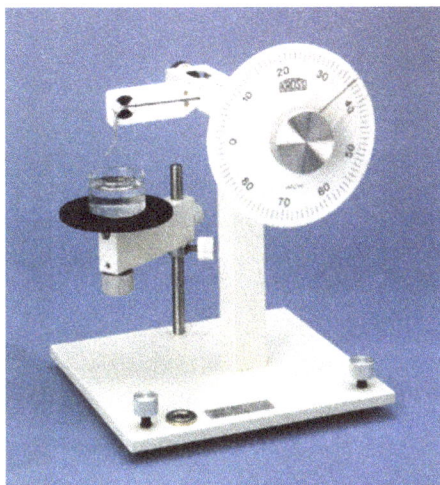

(e)

FIGURE 1: (a) A 50 ml density bottle. (b) Ostwald viscometer. (c) Seta cloud and pour point cryostat [20]. (d) Seta flash point tester [21]. (e) Tensiometer.

$$\% \text{ yield of ester} = \left[\frac{\text{volume of ester produced}}{\text{volume of canola oil used}}\right] * 100,$$

$$\% \text{ yield of ester} = \left[\frac{465}{600}\right] * 100,$$

$$\% \text{ yield of ester} = 77.50\%.$$

(3)

### 2.4. Base Fluid Property Evaluation.
After the synthesis of the ethyl esters, it is imperative to determine the physico-chemical properties of the esters formed and those of the control sample fluids before any mud formulation can occur in order to compare with specified standards to know if there is marked deviation such as physicochemical properties including but not limited to fluid density, kinematic viscosity, dynamic viscosity, cloud point, pour point, flash point, fire point, and surface tension.

### 2.4.1. Specific Gravity Evaluation at Different Temperatures.
Experimental procedure is as follows:

(1) The density bottle was firstly dried to allow for accurate measurement of its mass as shown in Figure 1(a).

(2) The cleaned and dried density bottle was weighed using a digital weighing balance to the nearest 0.01 g, and the weight was recorded.

(3) The density bottle was then filled completely with the fluid sample, and the stopper was inserted into the neck of the bottle resulting in spillage of the fluid outside the density bottle. This is necessary in order to avoid air being entrapped in the density bottle.

(4) The outside of the bottle was carefully dried using a soft tissue paper. The bottle and its content were weighed and the mass recorded.

(5) The liquid sample was poured out of the density bottle and the bottle was rinsed several times with distilled water and dried for subsequent samples weight determination.

(6) Water bath was then used to heat the fluid samples to the required temperatures prior to their weight measurement.

Deductions and calculations are as follows:

$$V_{db} = \text{volume of density bottle}$$

$$= \text{volume of distilled water} = 50 \text{ cm}^3,$$

$$W_{db} = \text{weight of density bottle (g)},$$

$$W_{fs} = \text{weight of fluid sample (g)},$$

$$W_{db+fs} = \text{weight of density bottle} + \text{fluid sample (g)},$$

$$W_{fs} = W_{db+fs} - W_{db},$$

(4)

$$\text{hence, density}\left[\frac{g}{cm^3}\right] = \frac{W_{fs}}{V_{db}},$$

(5)

$$\text{specific gravity} = \frac{\text{density of liquid}}{\text{density of equal volume of water}}.$$

(6)

### 2.4.2. Base Fluid Viscosity Determination

*(1) Method of Determining the Kinematic Viscosity of Base Fluid.* The following steps were used to determine the viscosity of the fluid sample:

(1) The viscometer (Figure 1(b)) was rinsed with ethanol and dried to avoid contaminants.

(2) The lower bulb of the viscometer was filled with the fluid sample to half of its capacity.

(3) Cold air was then blown into the viscometer so as to allow the fluid sample to travel up the viscometer at the other side of the upper mark of the bulb.

(4) The stop watch was switched on at the upper mark oil level, and it was stopped when the oil gets to the lower mark below the upper bulb. The two marks indicate a known volume and the time taken for the level of oil to pass between these marks is proportional to the kinematic viscosity.

(5) The time taken for the oil to travel to the lower bulb mark was recorded in seconds and it was converted to kinematic viscosity (in centistokes) using the viscometer size 200 constant of 0.1 cst/sec.

(6) The above steps were repeated for various oil samples at different temperatures.

Mathematically, the kinematic viscosity is obtained by using the following equation:

$$\mu = \alpha t, \tag{7}$$

where $\mu$ = kinematic viscosity (centistokes), $t$ = effluent time (seconds), and $\alpha$ = viscometer constant (cst/sec).

The dynamic viscosity ($v$) can be obtained by multiplying the kinematic viscosity ($\mu$) with the density of the fluid ($\rho$):

$$v = \mu\rho. \tag{8}$$

### 2.4.3. Cloud and Pour Points of Base Fluid Samples.
The cloud point and pour point were measured by using the seta cloud and pour point cryostat shown in Figure 1(c).

The following procedures were used in the course of the experiment:

(1) One of the three (3) compartments was filled with two liters of butyl glycol.

(2) The glass cups in each of the compartments were filled with different oil samples up to the upper mark.

(3) The glass cups were covered with the thermometer cork, and the thermometer was inserted into the oil sample in the glass cup through the cork.

(4) The outer black insulating gasket and disc were placed on the glassware, and this was placed in the cryostat compartment and the power corresponding to the compartment was switched on.

(5) The glass cups were brought out on periodic intervals to check the cloudiness of the oil samples. The temperature at which the oil sample becomes cloudy is its cloud point.

(6) After recording the cloud point, the thermometer was removed, and the glass cup was inserted into the compartment and the oil samples were allowed to freeze completely.

(7) The thermometer cork was removed, and the glass cup was placed tilted on a flat table with

a thermometer in it. The temperature at which the first drop of oil is formed is the pour point of the oil samples.

### 2.4.4. Flash Point and Fire Point of Base Fluid Samples.
Flash point of a flammable liquid is defined as the lowest temperature at which it can form an ignitable mixture in air while fire point is the temperature at which vapors of the flammable liquid continue to burn after being ignited even after the source of ignition is removed.

The following procedure was used to determine the flash point of each base fluid sample by using the equipment shown in Figure 1(d):

(1) The gas supply was switched on from the gas canister filled with butane gas.

(2) The control valve on the gas canister was adjusted until the pilot jet flame is approximately 12 mm long.

(3) The test jet flame was also adjusted to 4 mm diameter by rotating the pinch valve, and the gas supply was then switched off.

(4) The flash point tester power was then switched on, and the test temperature was set by using the set temperature button.

(5) The tester sample cup was allowed to stabilize at the set temperature, and the syringe was loaded with the base fluid sample and injected into the sample cup through the filler orifice and the syringe was removed.

(6) The gas supply was then switched on, and a fire lighter was used to light and set the pilot and test jet flame at 4 mm.

(7) At the set test temperature, a warning beep sounds.

(8) The shutter was then opened and closed over a period of five seconds.

(9) A flash was detected at the flash point of the sample, and the temperature at which the flash occurred was recoded as the flash point of the fluid sample.

(10) After the flash point, the heating was continued, and the fire point was taken as the temperature at which the application of test flame causes the oil sample to burn for at least five seconds.

(11) The sample cup was allowed to cool to room temperature.

(12) The above steps were repeated for other samples.

### 2.4.5. Surface Tension Determination.
The surface tension of a liquid can be experimentally measured by several methods such as the drop weight method (stallagmometer), Du Nouy ring method, Wilhelmy plate method, and the maximum bubble pressure method. However, the Du Nouy ring method is a rapid, simple, and most widely used method because it does not need to be calibrated using solutions of known surface tension [22].

Du Nouy ring method procedure for surface tension estimation is as follows:

(1) The measurement is performed by an instrument known as tensiometer as shown in Figure 1(e). However, the instrument does not measure surface tension directly but it has an accurate microbalance and a mechanism to vertically move the liquid sample in a glass beaker.

(2) The weight of the circular ring to be immersed into the fluid sample was measured by weighing balance and recorded as $W_r$.

(3) The ring hanging from the hook of the balance was immersed into the fluid sample and then carefully pulled up by lowering the sample vessel.

(4) The force applied on the ring when it pulls through the air-liquid interface was continuously recorded by the microbalance.

(5) The above procedure was repeated for other fluid samples at designated temperature of 25°C.

Hence, the total force required to detach the ring is recorded as $W_{total}$.

Mathematically,

$$W_{total} = W_r + 4\pi R_r \gamma, \tag{9}$$

where $W_{total}$ = the total force needed to detach the ring (Newton); $W_r$ = weight of the circular ring immersed into the fluid sample (Newton); $R_r$ = radius of the ring (meter); and $\gamma$ = surface tension (N/m).

From (4), the surface tension can be expressed as

$$\gamma = \frac{W_{total} - W_r}{4\pi R_r}. \tag{10}$$

Equation (5) provides us with a surface tension estimation that is characterized with about 25% error. Hence, a more accurate estimation is established by introducing a surface tension correction factor $f$ as shown in (6):

$$\gamma = \left[\frac{W_{total} - W_r}{4\pi R_r}\right] f, \tag{11}$$

where $f$ is the surface tension correction factor which was obtained graphically from [23] as variation of $f$ with $R_r^3/V$ and $R_r/r_w$, in which $R_r$ = radius of the ring (meter); $V$ = volume of the liquid raised by the ring during detachment; and $r_w$ = radius of the ring wire.

## 3. Results and Discussion

### 3.1. Base Fluid Properties Characterization

*3.1.1. Base Fluid Specific Gravity.* The results of the specific gravity of the base fluids are presented in Table 1 and Figure 2. The specific gravity values are deduced from the weight of the sample shown in Table 2. From the results, it can be inferred that the EICO base fluid has a lower specific gravity of 0.860 compared with the raw canola oil with the highest specific gravity of 0.910 at ambient temperature. The implication of specific gravity is felt in the amount of weighting material required in mud formulation to achieve the required mud weight standard. In comparison with the

TABLE 1: Base fluids' specific gravity at different temperatures.

| Temp (°C) | Canola oil | EICO | Diesel | SHBF |
|---|---|---|---|---|
| 25 | 0.910 | 0.860 | 0.850 | 0.837 |
| 50 | 0.896 | 0.842 | 0.835 | 0.818 |
| 75 | 0.875 | 0.821 | 0.812 | 0.797 |
| 100 | 0.862 | 0.805 | 0.785 | 0.769 |

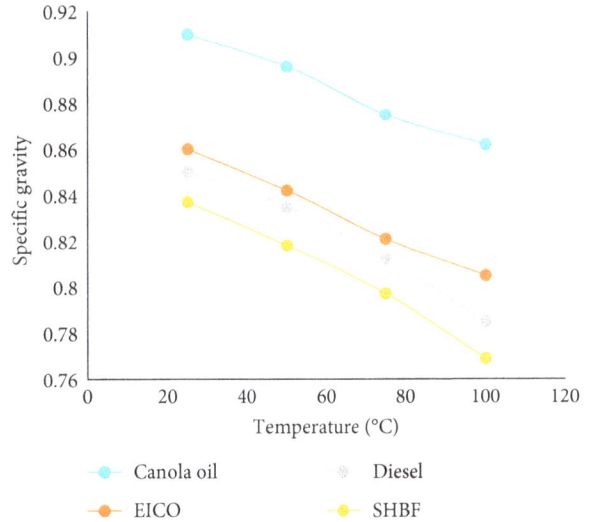

FIGURE 2: Specific gravity variation with temperature.

TABLE 2: Base fluids' weight in grams per 50 ml density bottle at different temperatures.

| Temp (°C) | Canola oil | EICO | Diesel | SHBF |
|---|---|---|---|---|
| 25 | 45.50 | 43.00 | 42.50 | 41.85 |
| 50 | 44.80 | 42.10 | 41.75 | 40.90 |
| 75 | 43.75 | 41.05 | 40.60 | 39.85 |
| 100 | 43.10 | 40.25 | 39.25 | 38.45 |

API standard requirement of 0.76–0.88 for oil-based mud, it can be deduced that the EICO compares favorably with the diesel and synthetic hydrocarbon base fluid (SHBF) standards. Similarly, Hyne [24] proposed that oils with specific gravity of 25–50 API are usually the best for drilling fluid base oil. Hence, the EICO shows a promising result for base oils with specific gravity of 33 API compared with canola oil of 23 API at ambient temperature.

*3.1.2. Base Fluid Viscosity.* The viscosity of a fluid is a measure of its resistance to gradual deformation by shear stress or tensile stress. During the course of evaluation, two types of viscosities were used to characterize the base fluids which are dynamic (shear) viscosity and kinematic viscosity. The dynamic viscosity expresses its resistance to shearing flows, where adjacent layers move parallel to each other with different speeds while kinematic viscosity describes the resistance of a fluid to flow under gravity.

From the effluent time presented in Table 3, it can be seen that it took the canola oil more time to travel down the viscometer which translates to higher kinematic viscosities,

TABLE 3: Ostwald viscometer (size 200) effluent time (seconds) of base fluid samples.

| Temp (°C) | Canola oil | EICO | Diesel | SHBF |
|---|---|---|---|---|
| 25 | 665 | 35 | 32 | 29 |
| 50 | 254 | 27 | 24 | 21 |
| 75 | 132 | 21 | 17 | 13 |
| 100 | 78 | 14 | 11 | 8 |

TABLE 4: Kinematic viscosity (cst) of base fluids at different temperatures.

| Temp (°C) | Canola oil | EICO | Diesel | SHBF |
|---|---|---|---|---|
| 25 | 66.5 | 3.5 | 3.2 | 2.9 |
| 50 | 25.4 | 2.7 | 2.4 | 2.1 |
| 75 | 13.2 | 2.1 | 1.7 | 1.3 |
| 100 | 7.8 | 1.4 | 1.1 | 0.8 |

TABLE 5: Dynamic viscosity (cp) of base fluids at different temperatures.

| Temp (°C) | Canola oil | EICO | Diesel | SHBF |
|---|---|---|---|---|
| 25 | 60.52 | 3.01 | 2.72 | 2.43 |
| 50 | 22.76 | 2.27 | 2.00 | 1.72 |
| 75 | 11.55 | 1.72 | 1.38 | 1.04 |
| 100 | 6.72 | 1.13 | 0.86 | 0.62 |

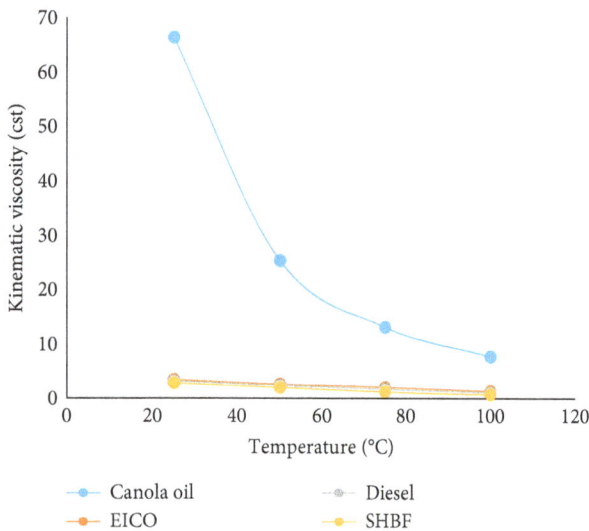

FIGURE 4: Dynamic viscosity variation with temperature.

TABLE 6: Surface tension of fluid samples at 25°C.

| Base oil | Surface tension (N/m) |
|---|---|
| Canola | 0.032 |
| EICO | 0.029 |
| Diesel oil | 0.028 |
| SHBF | 0.026 |

TABLE 7: Cloud and pour points of base fluid samples.

| Base oil sample | Cloud point (°C) | Pour point (°C) |
|---|---|---|
| Canola oil | 8 | 2 |
| EICO | 1 | −4 |
| Diesel | 2 | −1 |
| SHBF | 0 | −3 |

TABLE 8: Flash point and fire point of base fluid samples.

| Base oil sample | Flash point (°C) | Fire point (°C) |
|---|---|---|
| Canola oil | 284 | 315 |
| EICO | 165 | 185 |
| Diesel | 69 | 81 |
| SHBF | 76 | 87 |

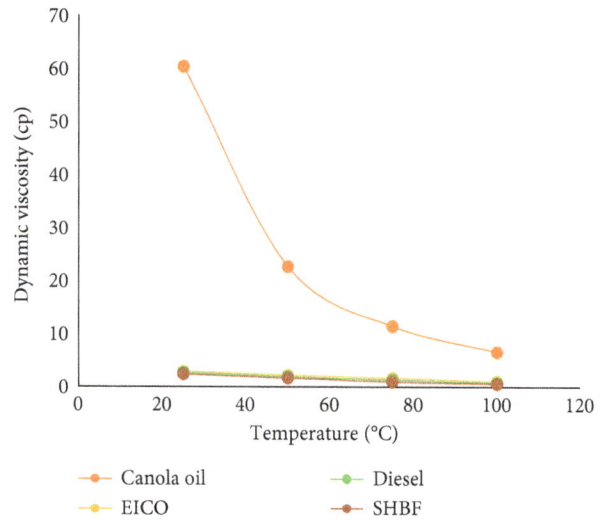

FIGURE 3: Kinematic viscosity variation with temperature.

and the travelling time reduces with temperature. This is partly connected with the high density of the fluid and presence of long chain fatty acid. The EICO took less time to travel down the viscometer just like the diesel and SHBF which are the standards and hence lower kinematic and dynamic viscosities.

Similarly, from the results presented in Tables 4 and 5 and Figures 3 and 4 for kinematic and dynamic viscosities, respectively, it was observed that the canola OBF has the highest kinematic and dynamic viscosity, respectively, at all temperatures. A lower kinematic viscosity is important because it allows the oil-based mud to be formulated at lower oil/water ratios and gives better rheology (lower plastic viscosity) at low mud temperature because too high viscosity increases pump pressure and limits flow properties and thus reduces penetration rates.

Also, the viscosity reduces with the increase in temperature as shown in Figures 3 and 4 because at higher temperature, there is induced movement. Lang et al. [25] and Noureddini et al. [26] also reported that the viscosity of canola and other vegetable oils decreases with an increase in temperature.

*3.1.3. Surface Tension of Base Fluids.* Drilling fluid surface tension measurement is very important in fluid characterization because high surface tensions decrease the ability of

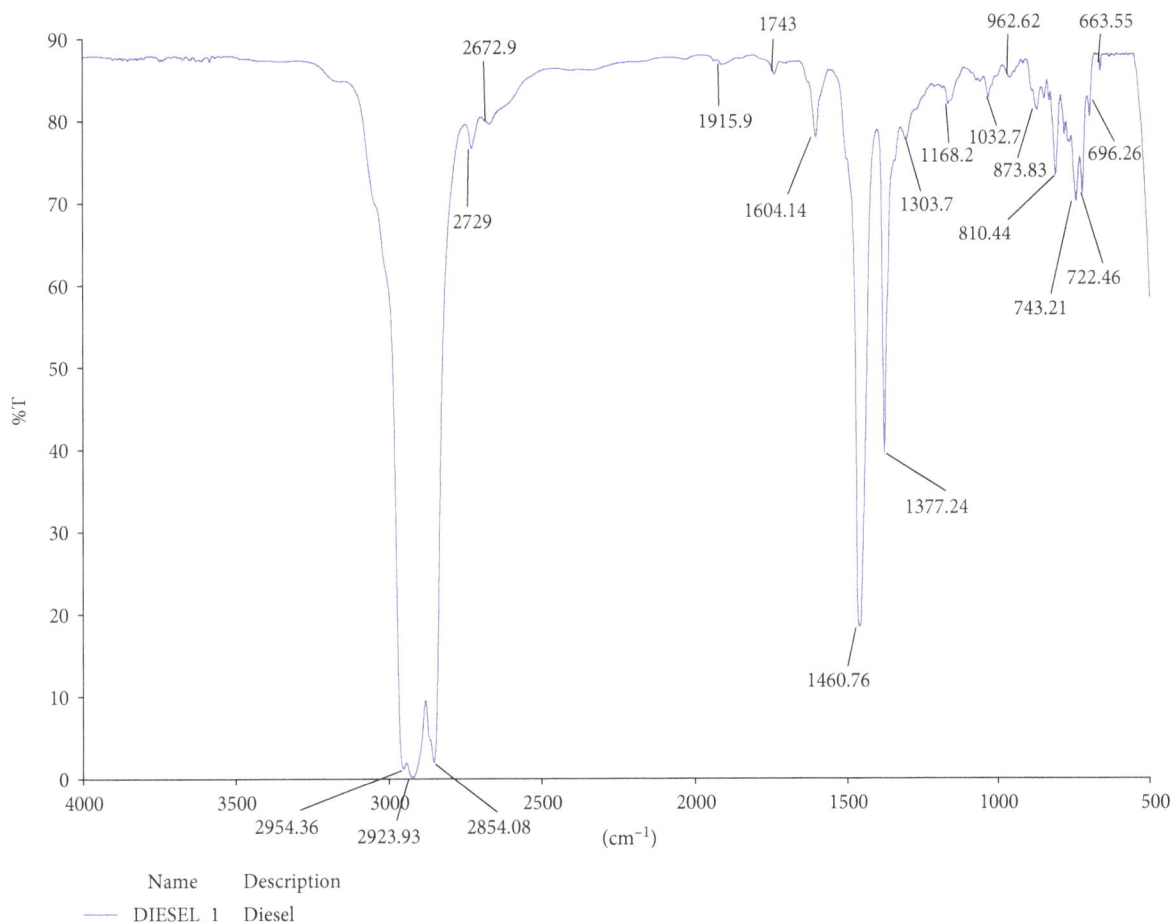

FIGURE 5: Diesel FTIR spectra.

the drilling fluid to pass through a shale shaker screen, particularly fine screens with their small openings [27].

Similarly, the reduced surface tension of EICO base fluid as shown in Table 6 would reduce the amount of surfactant (primary and secondary emulsifiers) required to lower the interfacial tension between oil and water during mud formulation. Bearing in mind that when the oil and water are mixed together mechanically, they separate immediately and the agitation ceases. Hence, lower base oil surface tension will reduce the amount of surfactants required to enable one liquid to form a stable dispersion of fine droplets in the other. Surface tension depends on the nature of the liquid, the surrounding environment, and temperature. The stronger the attractive intermolecular forces between liquids, the larger the surface tension. Surface tension decreases when temperature increases because cohesive forces decrease with an increase in molecular thermal activity.

*3.1.4. Cloud Points and Pour Points of Base Fluids.* The pour point is the lowest temperature at which a liquid will begin to flow while the cloud point is the temperature at which wax crystals begin to form in a liquid as it is cooled.

From Table 7, the canola oil OBF has higher pour points of 2°C which can make the OBMs to suffer from poor screening and excessive pressure, surges in deep water wells, or other operations that are subjected to low temperatures. But upon enzymatic transesterification, the pour point was significantly reduced to −4°C which invariably means that little stresses are needed to be overcome before the base mud begins to flow.

Similarly, knowing the cloud point is important for determining storage stability. Storing formulations at temperatures significantly higher than the cloud point may result in phase separation and instability. Hence, the EICO SBF has low cloud points and hence can be stored under lower temperature conditions.

*3.1.5. Flash Points and Fire Point of Base Fluid Samples.* The flash point is the lowest temperature at which a liquid can form an ignitable mixture in air near the surface of the liquid. The lower the flash point, the easier it is to ignite the material. The fire point of a fuel is the lowest temperature at which the vapor of that fuel will continue to burn for at least 5 seconds after ignition by an open flame. At the flash point, a lower temperature, a substance will ignite briefly, but vapor might not be produced at a rate to sustain the fire. From Table 8, the EICO SBF has higher flash points of 165°C and fire point of 185°C which is good for forming SBMs because it really shows that the fluids can be worked with at higher temperature. According to Johanscvik and Grieve [28], the flash point of oil-based mud must be greater than 100°C as

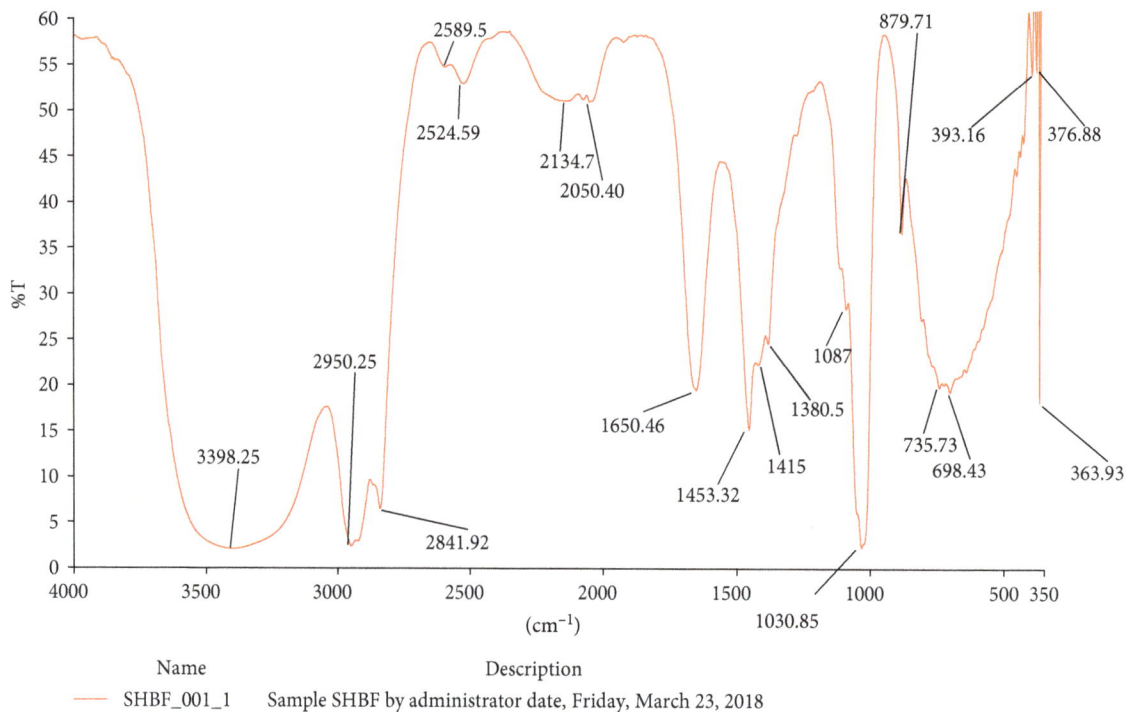

| Name | Description |
| --- | --- |
| —— SHBF_001_1 | Sample SHBF by administrator date, Friday, March 23, 2018 |

FIGURE 6: SHBF FTIR spectra.

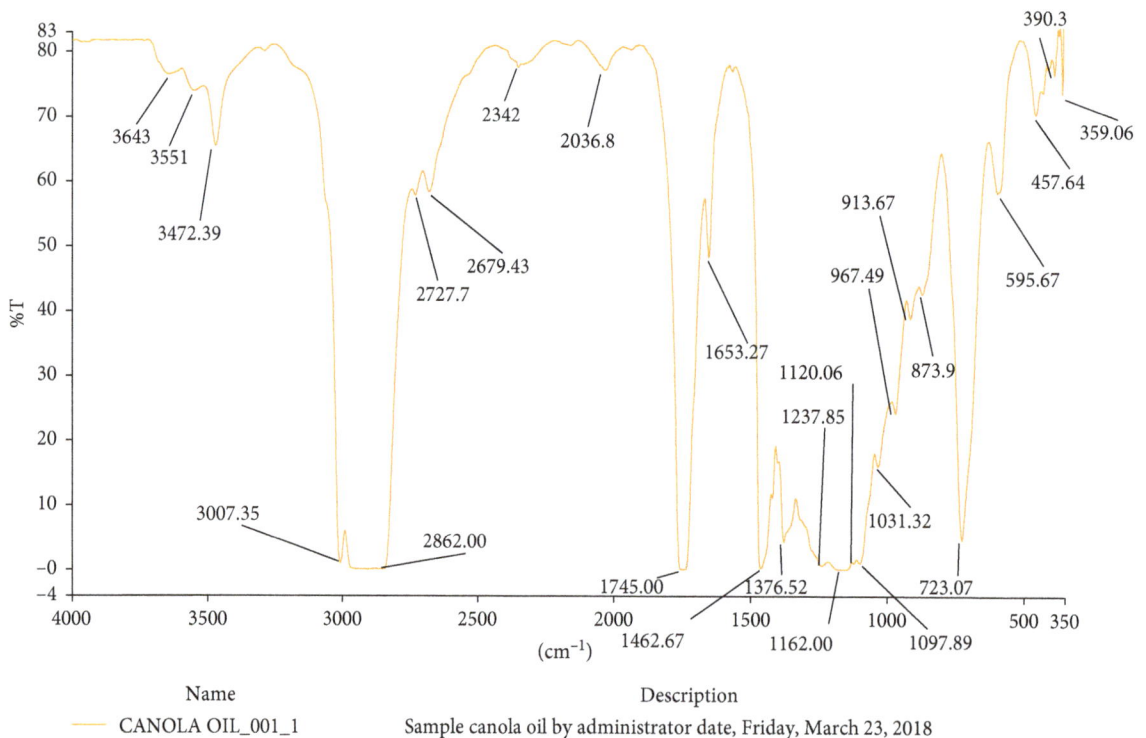

| Name | Description |
| --- | --- |
| —— CANOLA OIL_001_1 | Sample canola oil by administrator date, Friday, March 23, 2018 |

FIGURE 7: Canola oil FTIR spectra.

higher flash point will minimize fire hazard because less hydrocarbon vapor is expected to generate above the mud.

*3.1.6. Degree of Aromaticity.* The relative degree of aromatic hydrocarbon present in a compound can be known through Fourier transform infrared spectroscopy (FTIR) spectra

analysis. From FTIR spectra analysis, a well-defined absorption of one but typically two sets of bands in the region $1615\,cm^{-1}$–$1495\,cm^{-1}$ for aromatic ring stretch and $3130\,cm^{-1}$–$3070\,cm^{-1}$ for aromatic C–H stretch is consistent with aromatic compounds [29]. A careful look at diesel FTIR spectra (Figure 5) and SHBF spectra (Figure 6) showed that

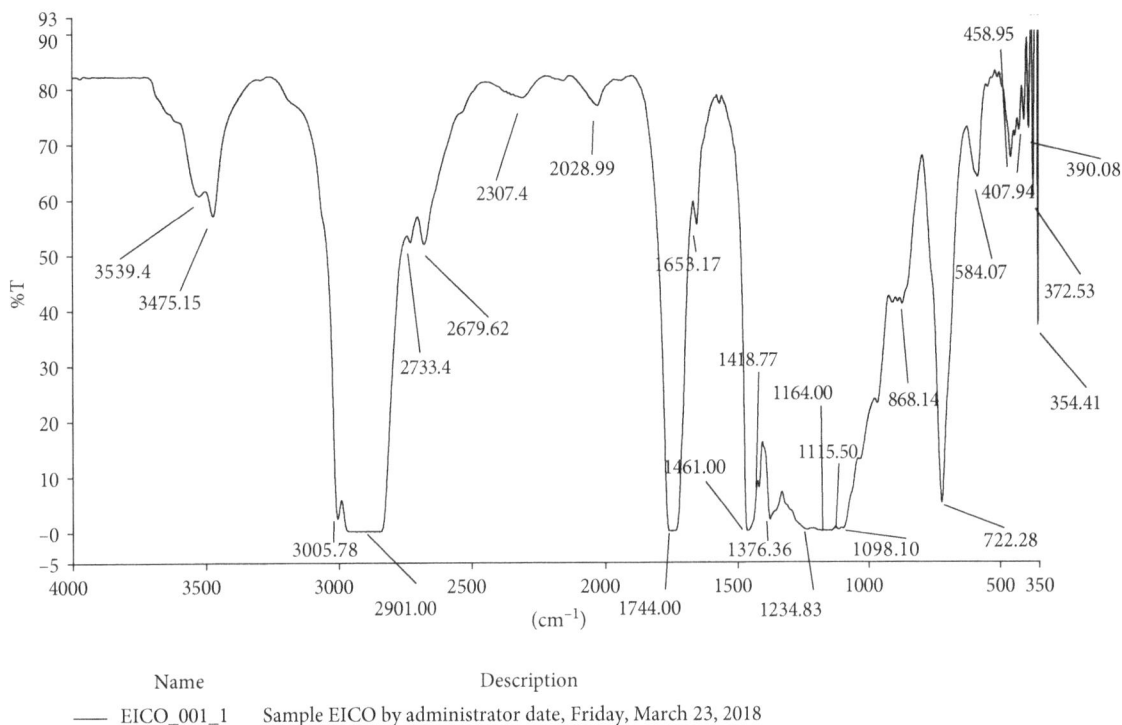

Name | Description
EICO_001_1 | Sample EICO by administrator date, Friday, March 23, 2018

FIGURE 8: EICO FTIR spectra.

an aromatic ring stretch of $1604\,cm^{-1}$–$1460\,cm^{-1}$ and $1650\,cm^{-1}$–$1453\,cm^{-1}$ was found in diesel and SHBF FTIR spectrum, respectively. This aromatic ring stretch is absent in canola oil and the enzymatically interesterified canola oil (EICO) (Figures 7 and 8), respectively. However, field experience as well as laboratory tests has indicated that oils with a high aromatic content are more detrimental to rubber parts in a mud circulating system than those with low aromatic content.

Hence, oils with zero aromatic content are the most desirable for use in drilling fluids in order to minimize damage to rubber equipment on the rig and to reduce death of marine organism when the cuttings are discharged.

## 4. Conclusions

The following conclusions were drawn:

(i) The EICO gave 94.74% reduction in kinematic viscosity to conform to the standard kinematic viscosity requirement of base fluids to be used. For SBMs, a kinematic viscosity of 2.0 to 3.6 cp is recommended according to API standard.

(ii) The cloud point and pour point were considerably reduced in the enzymatically synthesized canola oil, so that an account of possible cold environment operation can be easily accommodated without the SMBs losing their excellent rheological and filtration properties. Because higher pour points can make a SBM to suffer from poor screening and excessive pressure surges in deep water wells or other operations that are subjected to low operation.

(iii) Though the flash point and fire point were decreased after interesterification to a desirable level meaning that the fluids can still be worked with at higher temperatures without possible fair of ignition. Also, its transportation and storage ability will pose no threat.

(iv) A reduction of 5.50% in specific gravity was achieved with the enzymatic interesterification which consequentially helps to formulate SBMs that are of moderate density because loss of circulation may result from excessive pressure due to mud that is too dense or heavy and thus reduces rate of penetration and increase drilling cost.

(v) The produced fluid has no aromatic compound as evaluated by its FTIR Spectra analysis and thus no environmental pollution can arise from its cuttings discharge and no deterioration of the rubber part of drilling equipment can occur during drilling.

## Abbreviations

EICO: Enzymatically interesterified canola oil
FTIR: Fourier transform infrared spectroscopy
OBF: Oil-based fluid
OBM: Oil-based mud
SBF: Synthetic-based fluid
SBM: Synthetic-based mud
SHBF: Synthetic hydrocarbon-based fluid
WBF: Water-based fluid
WBM: Water-based mud.

## Conflicts of Interest

The authors declare that there are no conflicts of interest regarding the publication of this paper.

## Acknowledgments

The authors are very grateful to the Chancellor of Covenant University and the University Management team for their support for Research and Development and the IR unit of University of Ibadan without which this research work would not have seen the light of the day.

## References

[1] PetroWiki, *Drilling Fluids Types*, 2017, http://petrowiki.org/Drilling_fluid_types#cite_ref-r1_1-0.

[2] A. Yassin, A. Kamis, and M. Abdullah, *Palm Oil Diesel as Base Fluid in Formulating Oil based Drilling Fluid*, Society of Petroleum Engineers, Dallas, TX, USA, 1991.

[3] O. Grahl-Nielsen, S. Sporst, C. E. Sjgren, and F. Oreld, *The Five-Year Fate of Sea-Floor Petroleum Hydrocarbons from Discharged Drill Cuttings*, Elsevier Applied Science, London, UK, 1989.

[4] F. V. Jones, C. Hood, and G. Moiseychenko, "International methods of evaluating the discharge of drilling fluids in marine environments," in *Proceedings of the SPE International Conference on Health, Safety and Environment in Oil and Gas Exploration and Production*, p. 18, Caracas, Venezuela, 1998.

[5] National Research Council, *Drilling Discharges in the Marine Environment*, National Academy Press, Washington, DC, USA, 1983.

[6] J. M. Neff, *Biological Effects of Drilling Fluids, Drill Cuttings and Produced Waters. Long-term Effects of Offshore Oil and Gas Development*, Elsevier Applied Science Publishers, London, UK, 1987.

[7] J. B. Hinwood, A. E. Poots, L. R. Dennis et al., "Drilling activities; environmental implications of offshore oil and gas development in Australia," in *Proceedings of the Australian Petroleum Production and Exploration Association*, pp. 123–207, Canberra, Australia, 1994.

[8] Norway, "Synthetic drilling fluids; proposal for a definition of synthetic drilling fluids and experiences on the Norwegian Continental Shelf," in *Proceedings of the Oslo and Paris Conventions for the Prevention of Marine Pollution Working Group on Sea-Based Activities (SEBA)*, Biarritz, France, 1997.

[9] J. E. Friedheim and H. L. Conn, "Second generation synthetic fluids in the North Sea: are they better?," in *Proceedings of the IADC/SPE Drilling Conference*, pp. 215–228, New Orleans, LA, USA, March 1996.

[10] J. A. Veil and J. M. Daly, "Innovative regulatory approach for synthetic based drilling fluids," in *Proceedings of the SPE/EPA Exploration and Production Environment Conference*, Austin, TX, USA, February 1999.

[11] J. A. Veil, C. J. Burke, and D. O. Moses, "Synthetic based muds can improve drilling efficiency without pollution," *Oil and Gas Journal*, vol. 94, no. 10, pp. 49–54, 1996.

[12] M. Norman, "Esters–the only synthetic option for the next millennium?," in *Proceedings of the 5th International [IBC] Conference of Minimizing the Environmental Effects of Drilling Operations*, Aberdeen, UK, p. 13, June 1997.

[13] J. Steber, C. P. Herold, and J. M. Limia, *Solving Fluid Biodegradation in Seabed Drill Cuttings Pile*, Offshore, 1994.

[14] J. E. Candler, S. P. Rabke, and J. J. Leuterma, "Predicting the potential impact of synthetic-based muds with the use of biodegradation studies," in *Proceedings of the 1999 SPE/EPA Exploration and Production Environmental Conference*, Austin, TX, USA, March 1999.

[15] E. A. Vik, B. S. Nesgard, J. D. Berg et al., "Factors affecting methods for biodegradation testing of drilling fluids for marine discharge," in *Proceedings of the SPE International Conference on Health, Safety and Environment*, New Orleans, LA, USA, pp. 697–711, June 1996.

[16] T. K. Mag, "Bleaching theory and practice," in *Proceedings of the Edible Fats and Oils Processing World Conference*, pp. 107–116, Champaign, IL, USA, 1990.

[17] R. G. Ackman, *Canola Fatty Acids-An Ideal Mixture for Health, Nutrition, and Food Use. Canola and Rapeseed*, Springer, Boston, MA, USA, 1990.

[18] F. D. Gunstone, *Vegetable Oils. Industrial Oil and Fat Products*, Vol. 1, John Wiley & Sons, Hoboken, NJ, USA, 6th edition, 2006.

[19] M. Kellens, *Oil Processing Challenges in the 21th Century: Enzymes Key to Quality and Profitability*, Desmet Ballestra Group, Zaventum, Belgium, 2000.

[20] Stanhope-seta, *Seta Cloud and Pour Point Cryostat*, 2017, http://www.stanhope-seta.co.uk/view_pdf.asp?strPdf=93531-7_Cloud_Pour_Point.pdf&strPage=dl.

[21] Stanhope-seta, *Seta Flash Point Tester*, 2017, http://www.stanhope-seta.co.uk/flashpoint- testing.asp.

[22] C. Huh and S. G. Mason, "A rigorous theory of ring tensiometry," *Colloid and Polymer Science*, vol. 253, no. 7, pp. 566–580, 1975.

[23] W. D. Harkins and H. F. Jordan, "A method for the determination of surface and interfacial tension from the maximum pull on a ring," *Journal of the American Chemical Society*, vol. 52, no. 5, pp. 1751–1772, 1930.

[24] N. J. Hyne, *Dictionary of Petroleum Exploration, Drilling and Production*, Pen Well Publishing Company, Tulsa, OK, USA, 1991.

[25] W. Lang, F. W. Sosulski, and S. Sokhansanj, "Modelling the temperature dependence of kinematic viscosity for refined Canola oil," *Journal of American Oil Chemist Society*, vol. 69, no. 10, pp. 1054-1055, 1992.

[26] H. Noureddini, B. C. Teoh, and L. D. Clements, "Densities of vegetable oils and fatty acids," *Journal of American Oil Chemist Society*, vol. 69, no. 10, pp. 1184–1188, 1992.

[27] ASME Shale Shaker Committee, *Drilling Fluids Processing Handbook*, Gulf Professional Publishing, Houston, TX, USA, 2004.

[28] C. Johanscvik and W. R. Grieve, "Oil-based mud reduces bore-hole problems," *Oil and Gas Journal*, vol. 4, p. 47, 1987.

[29] J. Coates, "Interpretation of infrared spectra, a practical approach," in *Encyclopedia of Analytical Chemistry*, R. A. Meyers, Ed., John Wiley & Sons, Chichester, UK, 2000.

# 3

# Recent Development in Ammonia Stripping Process for Industrial Wastewater Treatment

**Lennevey Kinidi ⓘ,[1] Ivy Ai Wei Tan ⓘ,[1] Noraziah Binti Abdul Wahab,[1] Khairul Fikri Bin Tamrin ⓘ,[2] Cirilo Nolasco Hipolito,[1] and Shanti Faridah Salleh[1]**

[1]Department of Chemical Engineering and Energy Sustainability, Faculty of Engineering, Universiti Malaysia Sarawak, 94300 Kota Samarahan, Sarawak, Malaysia
[2]Department of Mechanical and Manufacturing, Faculty of Engineering, Universiti Malaysia Sarawak, 94300 Kota Samarahan, Sarawak, Malaysia

Correspondence should be addressed to Lennevey Kinidi; lennybb93@outlook.com

Academic Editor: Sébastien Déon

It is noteworthy to highlight that ammonia nitrogen contamination in wastewater has been reported to pose a great threat to the environment. This conventional method of remediating ammonia nitrogen contamination in wastewater applies the packed bed tower technology. Nevertheless, this technology appears to pose several application issues. Over the years, researchers have tested various types of ammonia stripping process to overcome the shortcomings of the conventional ammonia stripping technology. Along this line, the present study highlights the recent development of ammonia stripping process for industrial wastewater treatment. In addition, this study reviews ammonia stripping application for varied types of industrial wastewater and several significant operating parameters. Furthermore, this paper discusses some issues related to the conventional ammonia stripper for industrial treatment application. Finally, this study explicates the future prospects of the ammonia stripping method. This review, hence, contributes by enhancing the ammonia stripping treatment efficiency and its application for industrial wastewater treatment.

## 1. Introduction

Human activities appear to be the major contributor to water pollution, for instance, agricultural, industrial, and municipal activities. Nitrogen surplus released into the environment has been proven to cause negative impacts on water qualities, human health, and ecosystems [1]. Nonetheless, a wide range of technologies is available to reduce the release of ammonia nitrogen into the environment, such as ammonia stripping [2], breakpoint chlorination [3], ion exchange [4], electrodialysis [5], and biological nitrification [6]. The ammonia stripping method has several plus points as it is a relatively simple process and cost-effective to remove ammonia in wastewater [7]. Besides, the valuable ammonia stripped from wastewater can be recovered from the stripping process. Due to the stability of this process, the ammonia stripping process has been deemed as an appropriate method in remediating wastewater that contains high concentration of ammonia and toxic compounds [8]. Consequently, ammonia stripping

has emerged as a strong interest research area among researchers and industrial community. As such, numerous lab-scale and pilot-scale studies have been performed, especially for the wide range of industrial wastewater that demands cost-effective remediation.

This paper looks into several emerging issues pertaining to remediating ammonia nitrogen by using the ammonia stripper technique, along with some significant operating parameters. In addition, this paper reviews the recent progress in ammonia stripping method with advanced gas-liquid contactors, as conducted by some researchers. Next, a comparison was made between the advanced liquid contactors and the conventional packed tower. Finally, this paper explores the future prospects of this ammonia stripping process.

## 2. Ammonia Stripping Process

The ammonia stripping process is based on the principle of mass transfer. It is a process, by which wastewater is contacted

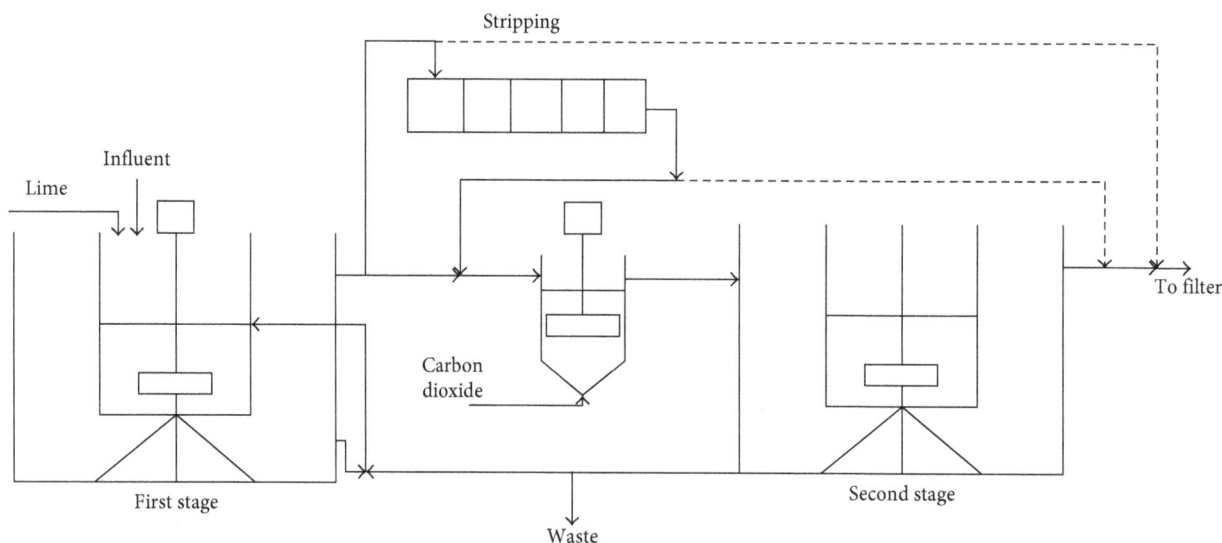

FIGURE 1: Schematic ammonia stripping process for leachate-polluted groundwater [10].

with air to strip the ammonia gas present in the wastewater. The presence of ammonia in wastewater can be found in two forms, namely, ammonium ions and ammonia gas. The relative concentrations of ammonia gas and ammonium ions are subjected to the pH and the temperature of wastewater [9]. The formation of ammonia gas is favored by increasing the pH, which shifts the chemical equilibrium to the right, thus inducing the formation of ammonia gas. Since high pH is required for effective ammonia stripping, lime is used to increase the pH values of wastewater prior to ammonia stripping [9]. In fact, various types of configurations for ammonia stripping process have been applied to remediate the varied types of wastewater containing ammonia nitrogen. For instance, O'Farell et al. conducted a study on nitrogen removal by stripping on a secondary effluent of a municipal wastewater treatment plant [10]. Figure 1 illustrates a schematic diagram of lime precipitation process and ammonia stripping process. Lime is incorporated to hike the pH of the influent prior to stripping, and this is followed by a recarbonation process for neutralization. Aside from raising the wastewater pH, calcium oxide (lime) generates calcium carbonate in the wastewater and serves as a coagulant for hard and particulate matters. Additionally, O'Farell et al. discovered that the ammonia stripping method could remove as much as 90% of ammonia from the secondary effluent [10].

Meanwhile, Raboni et al. investigated the efficiency of the ammonia stripping technique for remediation of groundwater polluted with leachate [11] (Figure 2). In the study, polyelectrolyte, sodium hydroxide, and iron (iii) chloride were added for the coagulation-flocculation and sedimentation processes at pH higher than 11 [11]. The system also comprised of a heater to heat the wastewater at $38°C$ and ammonia recovery via absorption with sulphuric acid. Lastly, the effluent was neutralized after adding sulphuric acid. As a result, they found that the ammonia stripping system for groundwater polluted with leachate displayed removal efficiency of 95.4% with initial ammonia concentration at 199.0 mg/L.

Next, Saracco and Genon investigated the performance of air-stripping system to treat ammonia nitrogen from industrial effluent (Figure 3) [12]. They suggested this route as feasible only if the industrial effluent was characterized by relatively high temperature and ammonia concentration. The stripping process was followed by absorption and crystallization processes. Saracco and Genon concluded that the ammonia stripping and the recovery system, along with its internal air recycle, had been technically feasible and easy to control [12].

## 3. Ammonia Stripping Application in Industrial Wastewater Treatment

To date, ammonia stripping pilot-plants have been employed to treat various types of wastewater containing high concentrations of ammonia and toxic compounds, such as that derived from secondary effluent of municipal wastewater treatment plant [10], animal manure [13], and landfill leachate [14]. Most recently, ammonia stripping was applied to anaerobic-digested effluent as this method offers both economic and environmental advantages. The biogas produced in the anaerobic digestion was used for ammonia removal to prevent inhibition of methanogenesis in the anaerobic reactor [15–17]. Meanwhile, Bonmati and Flotats revealed that no pH modification was required for stripping of ammonia from pig slurry [18]. On the other hand, Limoli et al. investigated ammonia removal from raw manure digestate by employing the turbulent mixing stripping process. They found that the ammonia stripping process via turbulent mixing was indeed feasible for raw manure digestate [19].

Table 1 shows that the ammonia stripping technique is indeed highly efficient in treating wastewater that contains ammonia nitrogen with toxic compounds. Besides, ammonia stripping combined with anaerobic digestion seemed to enhance the performance of anaerobic digestion process, apart from being cost-effective for ammonia removal.

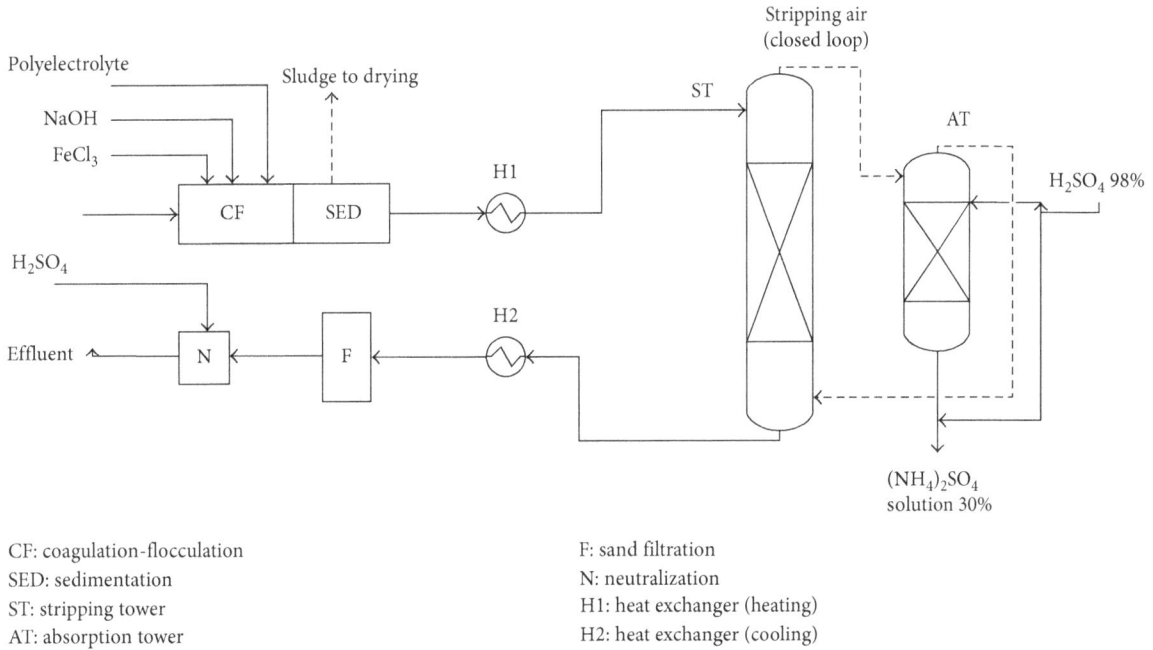

CF: coagulation-flocculation
SED: sedimentation
ST: stripping tower
AT: absorption tower

F: sand filtration
N: neutralization
H1: heat exchanger (heating)
H2: heat exchanger (cooling)

FIGURE 2: Schematic ammonia stripping process for leachate-polluted groundwater [11].

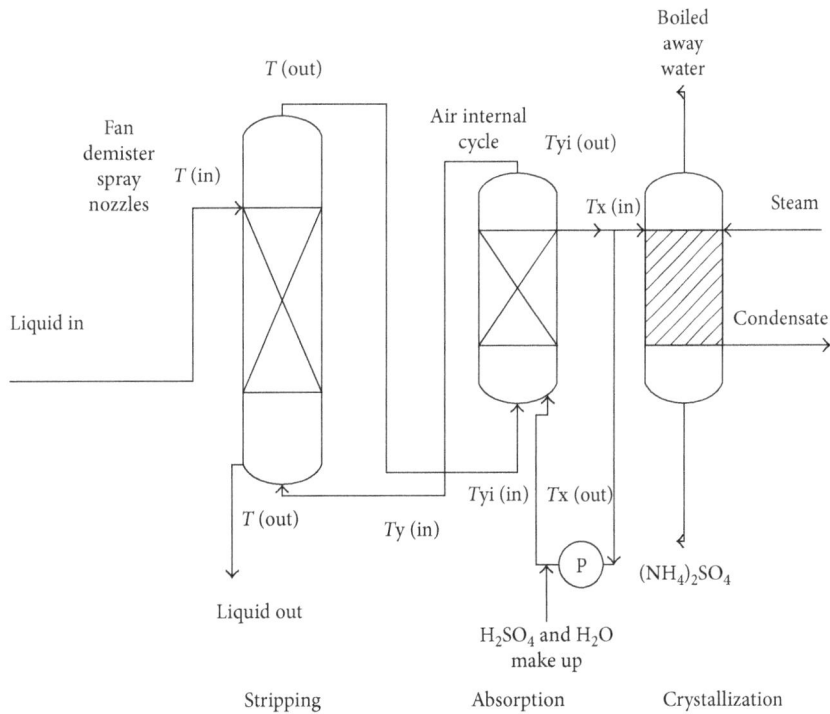

FIGURE 3: Proposed plant for ammonia stripping for industrial effluent application [12].

Nonetheless, Serna-Maza et al. revealed that in-situ ammonia stripping in mesophilic condition was unlikely to have any commercial application for wastes with intermediate total ammonia nitrogen concentrations as only high total ammonia nitrogen concentration stripping had managed to reduce the total ammonia nitrogen concentration below the higher inhibition threshold of approximately $8 \, g \cdot N \cdot L^{-1}$ [29]. Thus, stripping coupled with dilution may offer the best means of controlling total ammoniacal nitrogen concentrations. Next, Collivignarelli et al. found that ammonia stripping without dosage of basificant had been feasible when the initial alkalinity of the leachate was equal to or greater than the acidity of ammonium ions for removal [30]. Notably, this can potentially minimize chemical usage and slash operational cost in removing ammonia nitrogen from leachate. Ammonia stripping also

TABLE 1: Several studies that had looked into the ammonia stripping process with varied types of industrial wastewater.

| Wastewater | Ammonia stripping reactor | Influent ammonia (mg/L) | Ammonia removal percentage (%) | Operating condition | Reference |
|---|---|---|---|---|---|
| Biologically treated blue crab processing wastewater | Packed bed | 2300 | 72 | Liquid loading rate: 25 L/m$^2$/min Temperature: 14°C PH: 12.0 | [20] |
| Petroleum refinery wastewater | Packed bed | 100 | 85 | Airflow rate: 8495.1 L of air/gal of wastewater Temperature: 25°C pH: 10.5 | [21] |
| Landfill leachate | Packed bed | 1213 | 88 | Airflow rate: 4500 L·h$^{-1}$ Temperature: 25°C pH: 11 | [22] |
| Piggery wastewater | Packed bed | 4950 | 80 | Airflow rate: 10 L/min Temperature: 37°C pH: 11 | [23] |
| Fertilizer effluent | Packed bed | 2000 | 99 | Airflow rate: 420 m$^3$·m$^{-2}$·h$^{-1}$ Temperature: 26°C pH: 11 | [24] |
| Raw manure digestate | Mixer | 5000 | 88.7 | Temperature: 23°C pH: 10 | [19] |
| Secondary effluent | Packed bed | 12 | 86.2 | Airflow rate: 100 L/min/sq·m Temperature: 20.6°C pH: 11.7 | [10] |
| Acetylene purification wastewater | Packed bed | 125 | 91% | Airflow rate: 0.5 m$^3$/h Temperature: 60°C pH: 12 | [25] |
| Ammonia-rich soda ash wastewater | Microwave-assisted air stripping | 1350 | 96.3% | pH: 11; time: 5 mins microwave radiation power: 750 W | [26] |
| Sludge liquor from municipal wastewater treatment plants | Ion exchanger loop stripping | 2300 | 84.6% | pH: 10.5 | [27] |
| Swine wastewater | Microwave-assisted air stripping | 2740 | 88.2 | pH: 11 microwave radiation: 700 W | [28] |

appears to be effective and suitable in agriculture due to its simple process and cost-effectiveness in removing ammonia efficiently.

# 4. Process Condition

Numerous studies have highlighted the impacts of varying operational parameters upon the performance of ammonia stripping process. Some important parameters that have been reported to influence the performance of ammonia stripping are temperature, pH, and air to water ratio.

*4.1. Temperature.* Temperature has been proven to have a significant impact on the performance of ammonia stripper. This is because the solubility of ammonia in water is governed by Henry's law. In Henry's law, the constant of gas relies on solute, solvent, and temperature [31]. For example, Campos et al. discovered that the removal of ammonia from landfill leachate at 60°C was relatively significant over a period of 7 hours than at 25°C [32]. Generally, higher efficiency ammonia removal can be obtained at higher temperature. Saracco and Genon also found that the capital

cost of ammonia stripper at a stripping temperature of 80°C was less by half than that at 40°C. Nevertheless, from the economic stance, increment in temperature may lead to a hike in the cost of preheating [12].

*4.2. pH.* Ammonia nitrogen in water exists in equilibrium between the molecular ($NH_3$) and ionic form ($NH_4^+$) according to the following reaction:

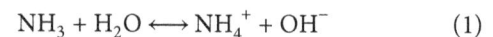

$$NH_3 + H_2O \longleftrightarrow NH_4^+ + OH^- \qquad (1)$$

The distribution between molecular ammonia and ammonium ions in water can be defined by (2) [16] and (3) [18]:

$$[NH_3] = \frac{[NH_3 + NH_4^+]}{1 + [H^+]/K_a} \qquad (2)$$

$$pK_a = 4 \times 10^{-8}T^3 + 9 \times 10^{-5}T^2 + 0.0356T + 10.072, \qquad (3)$$

where $[NH_3]$ is the molecular ammonia concentration, $[NH_3 + NH_4^+]$ is the total ammonia concentration, $[H^+]$ is the hydrogen ion concentration, and $K_a$ is the acid ionization constant. Besides that, $pK_a$ can be expressed in terms of temperature as shown in (3). Higher pH favors the formation of ammonia gas whereas lower pH favors the formation of ammonium ions. Hence, raising the pH level of the wastewater prior ammonia stripping is crucial to favor the formation of molecular ammonia nitrogen for stripping. However, according to Hidalgo et al., excessive rise of pH poses extra cost of lime that is nonfeasible in terms of cost. Hence, an optimum pH is required to strike a balance between process efficiency and economic cost. They found that when the pH exceeded 10.5, the removal efficiency was insignificant because pH no longer the affected the ionization balance between molecular ammonia and ionic ammonium but the cost incurred rose significantly due to the additional lime consumption required to increase the pH levels [33]. Meanwhile, Markou et al. revealed insignificant effect for the types of alkali (potassium hydroxide, sodium hydroxide, and calcium hydroxide) used on the ammonia removal efficiency [34]. However, calcium alkali was preferable due to reduction of solids, heavy metal concentrations, and color of wastewater [35].

### 4.3. Air to Water Ratio.
Air to water ratio is an important parameter that has an impact on the removal rates of ammonia in water. Mass transfer of ammonia into the air is affected by the variance between ammonia concentration level in liquid form and air phase [18]. Lei et al. discovered that the ammonia stripping efficiency of anaerobic effluent was influenced by air/water ratios. The study found that higher ammonia removal rate was achieved after 12 h at an airflow rate of 10 L/min, in comparison to airflow rates at 3 L/min and 5 L/min [16]. Nevertheless, from the engineering stance, Lei et al. concluded that 5 L/min for 1 L of anaerobic effluent should be feasible due to the expensive method of using an airflow rate of 10 L/min for 1 L of wastewater with only 5% increment in removal efficiencies, as compared to airflow rates from 5 L/min until 10 L/min [16]. Next, Campos et al. revealed that the influence of air to water ratio on ammonia stripping performance at higher temperature was less significant as it resulted in ammonia removal greater than 91% at 60°C with an airflow rate between 73 L/h and 120 L/h [32].

## 5. Issues Related to Ammonia Stripper Wastewater in Industrial Treatment Application

The ammonia stripping process has been successfully employed for many types of high-strength ammonia wastewater (Table 1). The method refers to one that is controlled and unaltered by toxic compounds. Nevertheless, the ammonia stripping process has several drawbacks. Among the issues involving the implementation of ammonia stripper to remove ammonia nitrogen in wastewater are fouling problems, sludge production, and release of ammonia gas.

### 5.1. Fouling Problems.
The fouling problems in an ammonia stripper tower are caused by the formation of calcium carbonate scale on the surface of the packing materials. Scale

builds up on the packing materials, thus leading to lower stripping performance [32]. Viotti and Gavasci found that the progressive scaling of the packing reduced stripper efficiency from 98% to 80% after 6 months of operation. The formation of calcium carbonate scale on the packing material is due to the absorption of carbon dioxide from the air stream used for stripping. Moreover, the nature of calcium carbonate varies from soft to hard. Viotti and Gavasci, thus, suggested chemical cleaning to attain higher removal of ammonia from wastewater [36]. The high operation and maintenance cost for air stripping can be attributed to the formation of calcium carbonate scale [37].

### 5.2. Sludge Production.
The stripped effluent of ammonia stripping often fails in meeting the discharge standards. High sludge production and high alkalinity effluent associated with ammonia stripping generate additional treatment cost to this process. However, the calcium carbonate from the ammonia stripper sludge can be recovered. Maree and Zvinowanda, for example, used the flotation technique to recover calcium carbonate from wastewater treatment sludge [38]. As a result, they discovered that floatation technique can potentially recover commercial grade limestone from wastewater sludge [38]. Meanwhile, He et al. assessed the feasibility and performance enhancement for treatment of alkaline-stripped effluent in aerated constructed wetlands [39]. The constructed wetland was relatively simple and was empowered with eco-friendly technology so that it can withstand extreme pH wastewater. He et al. also found that the remediation of alkaline effluent was feasible due to the high buffering capacity of the wetlands [39].

### 5.3. Ammonia Gas.
The ammonia stripping process results in ammonia release into the environment, thus causing additional environmental issues. Ammonia recovery by absorption is generally employed to prevent ammonia gas from being directly released into the environment. Ferraz et al. used sulphuric acid to recover the stripped ammonia from landfill leachate and revealed that 87% of the stripped ammonia was recovered [22]. Next, Zhu et al. discovered that under optimal condition of pH 12, airflow rate of 0.50 m$^3$/h, temperature of 60°C, and stripping time of 120 min; 0.2 mol/L of sulphuric acid can absorb approximately 93% of the ammonia stripped per volume of the acetylene purification wastewater [25]. Meanwhile, Laureni et al. concluded that ammonia stripping coupled with absorption proved to be a feasible option for valorization of nitrogen found in pig slurry. The by-product of this process was ammonium sulphate, which is a marketable product in the agriculture arena as fertilizers [15].

## 6. Advances in Ammonia Stripping Process

Research on ammonia stripping enhancement has continued unabated. Recent development of ammonia removal by ammonia stripping fall into the following: ammonia stripping reactor modifications, membrane contactor, membrane

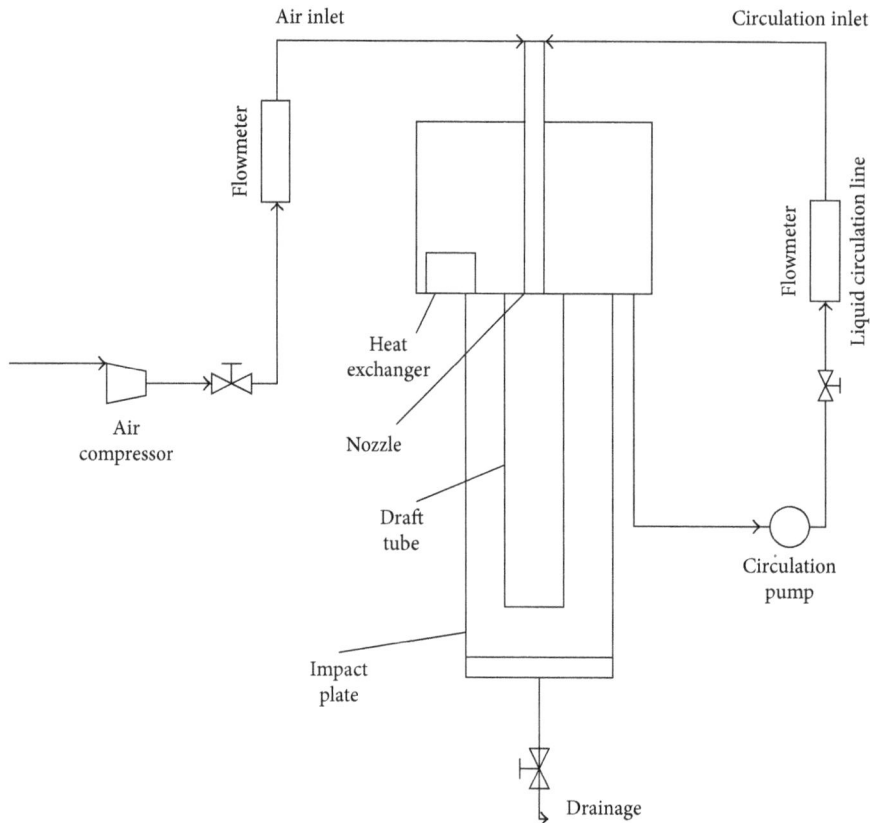

FIGURE 4: Schematic diagram of jet loop reactor [42].

distillation, ion exchange-stripping loop, and microwave-assisted ammonia stripping.

### 6.1. Ammonia Stripping Reactor Modifications.
The construction of a particular ammonia stripping reactor is crucial as it has a strong impact on the whole treatment efficiency and the capital cost. The conventional ammonia stripper reactor employs the packed column technology, in which the packing materials are used to enhance mass transfer between the two phases. The countercurrent-packed tower draws air through its openings at the bottom as the wastewater is pumped to the top of the packed tower. Nonetheless, this process generates carbonate scales on the surface of the packing materials, which can affect ammonia removal efficiencies over time. Apart from that, the average depth of the packed bed tower can range from 6.1 until 7.6 meters, hence consuming a considerably large amount of space. Therefore, some researchers have suggested the use of innovative ammonia stripper reactors as a solution for efficient removal of ammonia. Among the innovative ammonia stripping reactors proposed were rotating packed bed [40], water-sparged aerocyclone reactor [41], and semibatch jet loop reactor [42].

### 6.1.1. Semibatch Jet Loop Reactor.
Removal of ammonia via air stripping in a semibatch jet loop reactor was initiated by Degermenci et al., in which ammonia is removed by a jet loop reactor so as to reduce the construction and operational costs of the conventional ammonia stripping process. It also has a higher mass transfer coefficient and easier adaption from the pilot scale to the industrial-scale [42]. The jet loop reactor was conventionally applied for chemical or biochemical catalyzed reactions [43]. The jet loop reactor offers exceptional mixing performance at relatively low energy consumption for application that involves mass transfer [44].

An overall overview of the jet loop reactor is illustrated in Figure 4. In general, the jet loop reactors were constructed in many designs in terms of apparatus, nozzle dimensions, draft tube, and entry position of the jet stream [45]. The principle of the jet loop reactor is the utilization of the kinetic energy of high-velocity liquid jet to entrain the gas phase, besides producing fine dispersion between the gas and the liquid phases [46].

Degermenci et al. have developed (4) to model the rate of ammonia removal via air-stripping technique in a jet loop reactor [42].

$$-\ln \frac{C_{L,t}}{C_{L,0}} = \frac{K_H Q_G}{V_L} \left[ 1 - e^{-\left( (K_L a S L e)/(Q_G K_H) \right)} \right] t. \quad (4)$$

As a result, the temperature and the gas flow rate exhibited significant impacts on the ammonia removal rate by using the jet loop reactor. Besides, the jet loop appeared to be more effective than the conventional ammonia stripper packed tower. The jet loop reactors can also be used for the conversion processes in treating biochemical wastewater. Farizoglu et al. studied the treatment of cheese whey in a jet loop membrane reactor and achieved 84–94% of chemical

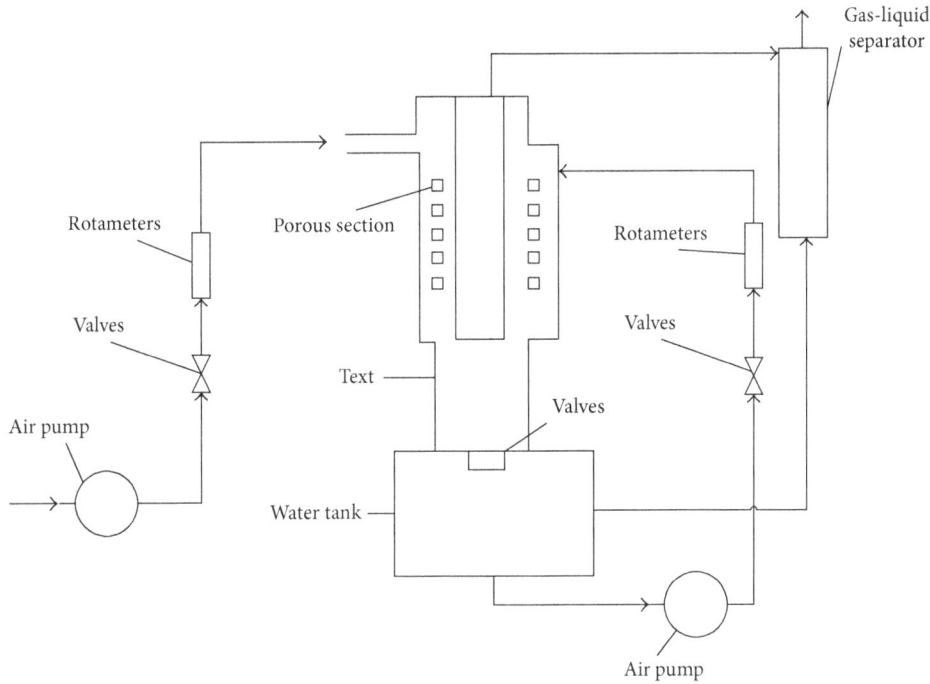

FIGURE 5: Water-sparged aerocyclone reactor configurations.

oxygen demand removal, which possessed the capability to operate at high biomass concentrations [47]. Next, Eusebio et al. investigated the treatment of winery wastewater by using jet loop reactor and found that 80% of COD removal efficiencies had been achieved within 24 hours [48].

*6.1.2. Water-Sparged Aerocyclone Reactor.* Removal of ammonia via water-sparged aerocyclone reactor was first designed by Quan et al. The basic motivation for the innovation was to increase the mass transfer rate and its applicability to treat wastewater with suspended solids [41]. The water-sparged aerocyclone gas-liquid contactor can be used to address two major drawbacks of the conventional packed tower, which are the process performance and the fouling problems in long operations.

The water-sparged aerocyclone reactor is illustrated schematically in Figure 5. The water-sparged reactor is comprised of two concentric right-vertical tubes and a cyclone header on the upper section. Wastewater is pumped into the porous section of the inner tube and sprayed into the centerline of the water-sparged aerocyclone reactor. After that, air is drawn into the aerocyclone at the top header of the inner tube.

Quan et al. adopted (5) developed by Matter-Muller et al. to model the ammonia removal rate via air-stripping technique using the water-sparged aerocyclone [49].

$$-\ln\frac{C_{A,t}}{C_{A,0}} = \frac{H_A Q_G}{V_L}\left[1 - e^{-\left((K_L a V_L)/(Q_G H_A)\right)}\right]t. \quad (5)$$

It was found that the water-sparged aerocyclone removed ammoniacal nitrogen, total phosphorus, and COD from wastewater at 91.0%, 99.2%, and 52.0%, respectively. Due to the promising ammonia removal efficiency by the water-sparged aerocyclone reactor, the structure of the water-sparged aerocylone reactor was improvised by Quan et al. to maximize the mass transfer efficiency of the reactor [50]. Quan et al. also investigated the arrangement and the diameter of the jet holes in water-sparged aerocyclone reactor, thus concluding that the spray holes should be arranged in a square mode with $1.28\,l_c$ of optimum distance between two adjacent spray holes [50].

*6.1.3. Rotating Packed Bed Reactors.* Ammonia removal via air-stripping technique in the rotating packed bed reactor was conducted by Yuan et al. to enhance the high volumetric gas-liquid mass transfer coefficients, as well as to reduce the fouling problem, the equipment size, and the cost incurred, as an attempt to overcome the shortcomings detected in the conventional ammonia stripping technique. The rotating packed bed reactor appeared to be highly efficient in process intensification as it maximized the gas-liquid mass transfer efficiency via strong centrifugal acceleration [51].

In fact, this particular method have been employed in a number of industrial applications, namely, absorption [52], synthesis of biodiesel [53], hydrogen sulfide removal [54], and synthesis of nanoparticles [55].

The rotating packed bed is illustrated schematically in Figure 6. The rotating packed bed consists of a rotating packed bed, gas and influent controls, effluent analyzer, and effluent gas neutralizer [40].

Yuan et al. used (6) to model the ammonia removal rate via air stripping using the rotating packed bed reactor [40]:

$$K_L a = \frac{Q_L}{V_B} \frac{\ln\left[(1 - (1/S))\left(C_{L,in}/C_{L,out}\right) + (1/S)\right]}{1 - (1/s)}. \quad (6)$$

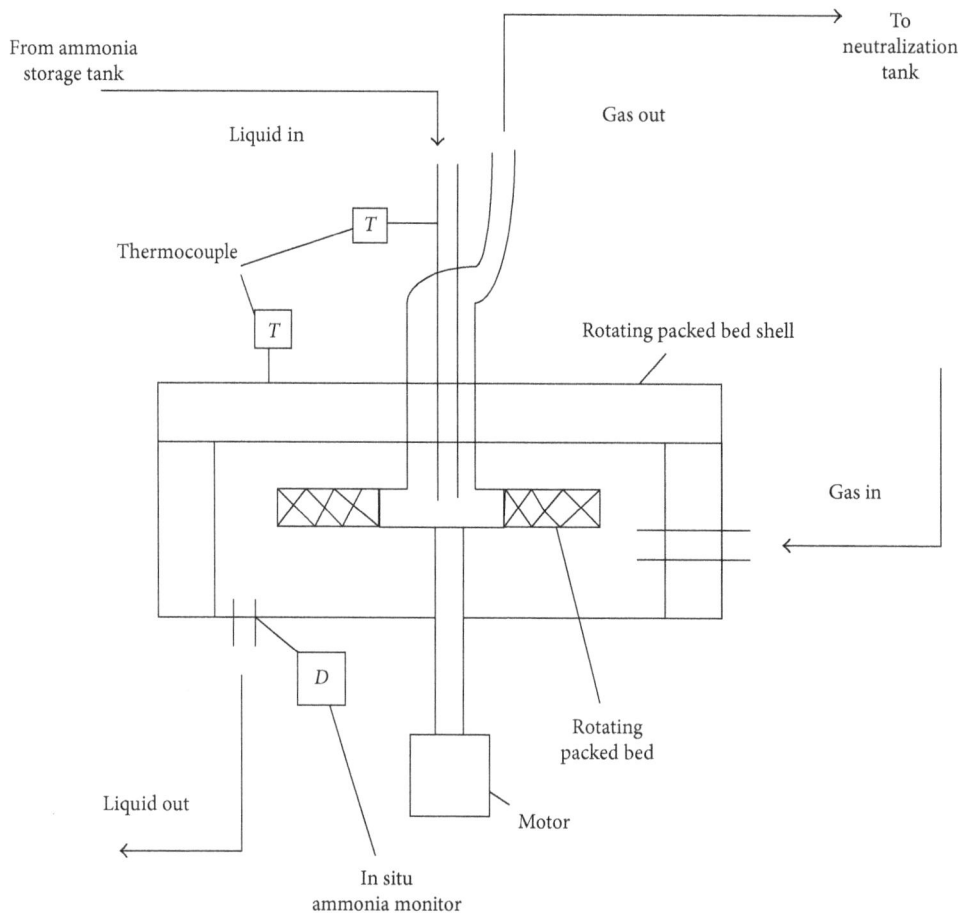

FIGURE 6: Rotating packed bed reactor configuration for ammonia stripping.

It was found that the rotating packed bed displayed higher mass transfer efficiency (12.3–18.41/h), when compared to other conventional and advanced gas-liquid contactors. Nonetheless, information concerning economic feasibility of the operating conditions seemed scarce for packed bed reactors in industrial wastewater treatment [40].

6.2. *Membrane Contactors.* Ammonia stripping by using membrane contactor is another alternative that has lower tendency to fouling and requires no post effluent treatment [56]. Relatively, ammonia stripping by membrane contactor has a higher rate of mass transfer than the conventional ammonia stripping due to its large contact surface area between the wastewater and stripping solution [57]. Semmens et al. have derived (7) to model the ammonia removal rate by ammonia stripping by using membrane contactor [56]:

$$\ln \frac{C_o}{C} = \frac{Q_t}{V}\left(1 - e^{(-kaL/v)}\right). \qquad (7)$$

Ahn et al. have founded that the highest mass transfer coefficient by using PTFE membrane was at $11 \times 10^{-3}$ m/h at the operating condition of 1000 mg/L of ammonia initial concentration with no suspended solids and temperature difference [57]. Hasanoglu et al. investigated the ammonia removal by using flat sheet and hollow fiber membrane

contactors and founded that the circulation configuration solution has a strong impact on the efficiency of the process [58]. Tan et al. studied the ammonia removal by using polyvinylidene fluoride (PVDF) hollow membranes and founded that mass transfer rate is higher at higher feed velocity, but only up to 0.59 m/s [59]. The ammonia stripping by using membrane contactor is illustrated schematically in Figure 7.

6.3. *Membrane Distillation.* In recent years, there has been an increasing research on ammonia removal using membrane distillation. Membrane distillation is driven by the temperature difference across the permeable membrane. It offers prospective recycling and reuse of industrial wastewater and higher process efficiency. Membrane distillation can be grouped into four basic configurations, namely, direct contact membrane distillation [60], vacuum membrane distillation [61], air gap membrane distillation [62], and sweep gas membrane distillation [63]. Liu et al. reported that direct contact membrane distillation process ammonia removal rate was more than 85% at ammonia concentration higher than 400 mg/L, but the removal rate decreased as the ammonia concentration was above 1200 mg/L [60]. El-Bourawi et al. addressed that the most important operating parameters that affect the ammonia removal efficiency of

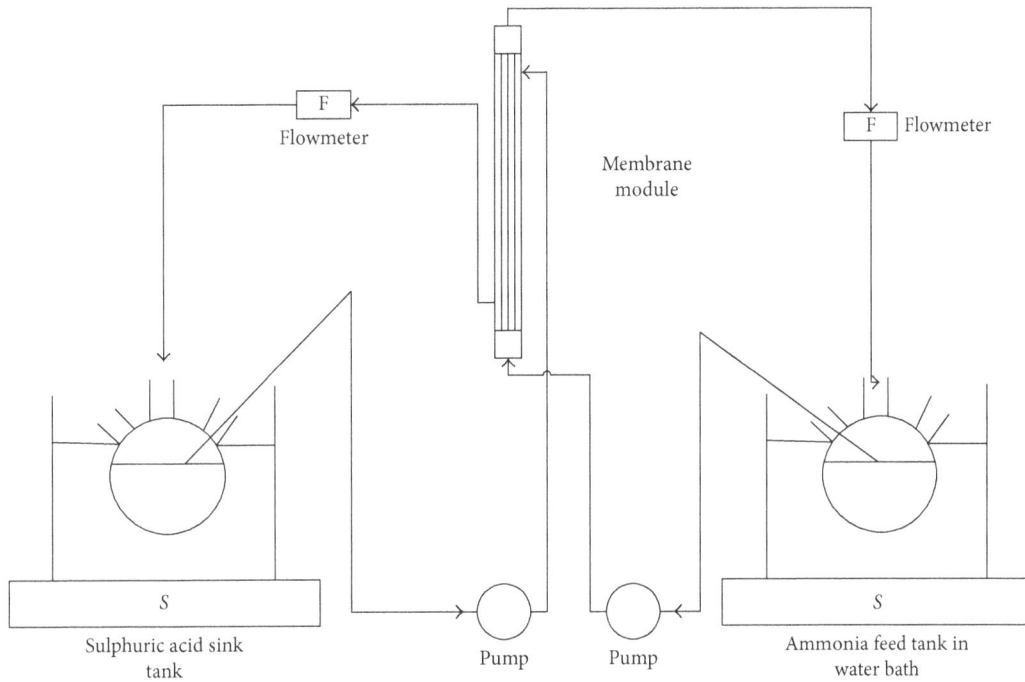

FIGURE 7: Membrane contactor configuration for ammonia stripping [59].

the vacuum membrane distillation are feed temperature, feed flow velocity, and downstream pressure. They reported that ammonia removal efficiencies higher than 90% were achieved by using vacuum distillation [61]. Eykens et al. conducted a lab scale and pilot scale on ammonia stripping by direct contact and air gap membrane distillation. It was founded that air gap membrane distillation has better performance and lower energy requirement than direct contact membrane distillation for larger scale applications [62]. Xie et al. investigated the ammonia removal by sweep gas membrane distillation. The efficiency of the process was affected by the feed temperature, feed flow rate, and gas flow rate. It was founded that sweep gas membrane distillation showed promising result with regards to high-efficiency industrial process water recycling [63]. The removal efficiency of the sweep gas membrane distillation was reported to be up to 97%. Liu et al. reported that the asymmetric PVDF membrane possesses excellent antifouling and sustainable flux in relative with the commercial PTFE (poly-tetrafluoroethylene) membrane [64]. It was founded that the PVDF membrane has less than 8% flux decline in 15 hours continuous operation [64].

*6.4. Ion Exchange Loop Stripping.* Due to the problems associated with operating and maintenance costs of air stripping [65] and the economic viability of the resins used for ion exchange [66], ion exchange and air stripping are combined and called ion exchange loop-stripping. The ion exchange loop stripping offers relatively lower operating and investment cost due to the reduced energy demands and equipment size reductions [67]. The ion exchange loop stripping is illustrated schematically in Figure 8. Ion exchange

loop stripping is made up of the zeolite bed, stripping column, and a scrubber.

Ellersdorfer suggested that sodium hydroxide solution may be an alternative to sodium chloride to reduce chemical consumption [27]. Ellersdorfer has investigated the technical feasibility of ammonium recovery by using ion exchange loop stripping for sludge liquor from municipal wastewater treatment plants and founded that it can be a feasible option for recovering ammonium from sludge liquor wastewater treatment plants at above 900 mg/L [27].

*6.5. Microwave-Assisted Air Stripping.* Reports on microwave radiation that could be used to reduce ammonia nitrogen in wastewater have opened the door for research in the field of ammonia stripping by microwave radiation. Li Lin et al. implemented a pilot-scale study of ammonia removal by using microwave radiation and founded that 80% ammonia removal from coke-plant wastewater can be achieved [26]. Ata et al. carried out studies on optimization of ammonia removal microwave-assisted air stripping and founded that the optimum conditions were at 1800 mg/L of initial concentrations, 7.5 L·min$^{-1}$ of airflow rate, 60°C of temperature, 500 rpm of stirring speed, and 200 W microwave output with 60 minutes of radiation. The removal efficiency of microwave-assisted ammonia stripping was able to achieve 94.2% under optimized conditions [68]. La et al. evaluated the efficiencies of microwave-assisted ammonia removal from swine wastewater. The highest removal efficiency was obtained at 83.1%. Ammonia removal by microwave radiation offers high ammonia removal rate and lower reaction time [28]. However, more research is needed for optimizing the power consumption of this wastewater treatment system.

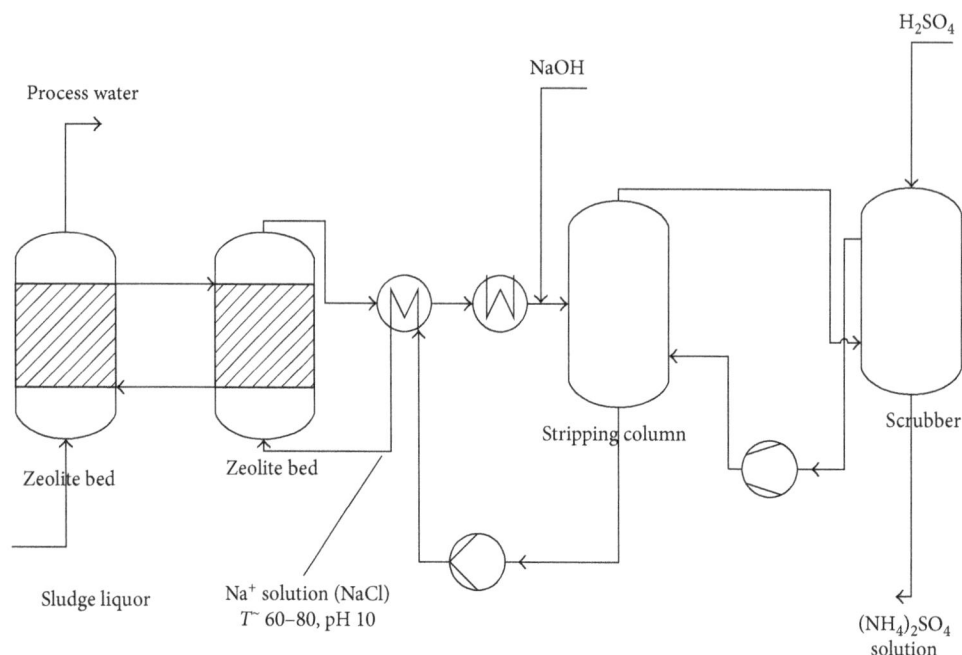

FIGURE 8: Ion exchange loop stripping configurations [27].

## 7. Comparisons between Different Ammonia Stripping Processes

A comparison of various parameters related to the different ammonia stripping processes based on literature was tabulated in Table 2.

Table 2 shows that among the processes evaluated, the packed tower displayed a higher tendency towards fouling, which decreased its efficiency and increased the operational cost of the whole process. Besides, the packed tower also required higher air consumption, when compared to the other ammonia stripper processes. Besides that, the packed tower also requires higher air consumption in relation with other ammonia stripper processes. However, the newer ammonia stripping processes such as, the semibatch jet loop reactor, water-sparged aerocyclone reactor, and rotating packed bed reactor, have lower tendency towards fouling problems. Since the rotating packed bed reactor operates in continuous flow, Yuan et al. suggested that larger rotating packed bed reactor has to be used to ensure higher process efficiency [40]. The water-sparged aerocyclone also offered simultaneous removal of other contaminants, such as total phosphorus and COD. In addition, the conventional packed tower displayed lower tolerance to total suspended solids. Hence, this technique is limited to applications with lower suspended solids present in wastewater. In this case, the ammonia stripping processes via semibatch jet loop reactor, water-sparged aerocyclone reactor, and rotating packed bed reactor are deemed suitable for wastewater that contains higher total suspended solids. It can also be seen that the implementation membrane technologies in ammonia stripping process have been receiving considerable attention in recent years. Separation technologies using membrane incorporated into the ammonia stripping has higher process efficiency and offers prospective wastewater reclamation and reuse [72]. However, the membrane technologies are subjected to membrane fouling which results in a substantial increase in hydraulic resistance [73]. Hence, future research should emphasize on the membrane fouling control and the performance on a larger scale. The microwave-assisted ammonia stripping also showed higher process efficiency at 94.2%. Nonetheless, higher power consumption and running costs posed a serious challenge for the microwave-assisted ammonia stripping process [69].

## 8. Summary of Review and Future Research Perspectives

This review paper has revealed the inherent benefits of the ammonia stripping process, in comparison to the conventional packed tower. Nevertheless, pilot-scale investigation and economic evaluations are required before applying full scale of the ammonia stripping process. Moreover, future researches can specifically focus on the following three aspects.

First, the structure optimization for the each of the ammonia stripping process reactors deserves further research. Since most of the new advanced ammonia stripping reactors were first designed for various types of applications, it is important that these reactors are tailor-made for ammonia stripping processes. One of the most important aspects of ammonia stripping reactor's development refers to the higher air-stripping efficiency at a lower operational cost. Hence, structure optimization can illustrate a detailed design guide for optimized gas-liquid contactors.

Second, more studies are needed to evaluate the capital and operational costs for advanced liquid-gas contactors in ammonia stripping. Since studies regarding these important

TABLE 2: Comparison of the different ammonia stripping processes.

| Ammonia stripping processes | Wastewater volume | Process efficiency (%) | Removal of other contaminant | Suspended solids tolerance | Fouling problem | Stripping time (h) | Airflow rate | Mass transfer coefficient | References |
|---|---|---|---|---|---|---|---|---|---|
| Packed tower | 1000 | 75 | Not available | Low | High | 3.5 | 25 (air to water ratio) | 0.42/h | [69] |
| Semibatch jet loop reactor | 9 L | 97 | Not available | High | Low | 5.8 | 5.6 (air to water ratio) | 0.63/h | [42] |
| Water-sparged aerocyclone | 10 L | 98 | Total P and COD | High | Low | 3.5 | 11.4 (air to water ratio) | 1.2/h | [41] |
| Rotating packed bed reactor | 0.025–0.01 L/min | 64 | Not available | High | Low | 0.0037 | 1800 L/min (continuous flow) | 12.3/h | [40] |
| Membrane contactor | 0.94 L | 99.83 | Not available | Low | High | 10 | Not available | 0.011 m/h | [57] |
| Membrane distillation | 1 L | 98.5 | Not available | Low | High | 4h | Not available | 0.079 | [70] |
| Ion exchange loopstripping | 2 L | 84.6 | Not available | low | High | 2.5 | Not available | Not available | [27] |
| Microwave radiation | 0.75 L | 94.2 | Not Available | High | Not available | 0.0167 | 10 (air to water ratio) | 3.354 | [71] |

aspects are in scarcity, and such information is vital to engineers and decision-makers in-charge of devising new technologies, more evaluations are required to look into the full cost analysis of the advanced gas-liquid contactors so as to determine its economic feasibility for specific wastewater treatment scenario. Additionally, a detailed pilot study on the advanced gas-liquid contactors is also crucial to identify potential hiccups and allay investor concerns.

Third, two of the advanced gas-liquid contactors (rotating packed bed and water-sparged aerocyclone) utilize vortex to induce gas-liquid mass transfer. Hence, there is a possibility of harvesting energy from these water vortexes. As such, it has been proposed that the advanced gas-liquid contactors have to be integrated with water vortex generator. This may be a possible strategy to promote energy self-sufficient ammonia stripping process. Nishi and Inagaki investigated the vortex-type water turbine to generate electricity and discovered its ability in generating electricity by using a low head and a low flow rate using a simple structure [74]. This ammonia stripping reactor liquid, coupled with water vortex generator, seems to be a promising technology for energy self-sufficient wastewater treatment and demands further research.

## 9. Conclusion

Ammonia stripping process is suitable for treating wastewater that contains high concentration of ammonia and toxic compounds with the merits of simpler operation, high efficiency, and excellent treatment stability, thus displaying an exceptional application potential for industrial wastewater treatment. The success of an ammonia stripping process is greatly dependent on temperature, pH, and air to water ratio. As such, the selection of optimized operating parameter is vital for the ammonia stripper to achieve higher efficiency. The different types of ammonia stripping reactors for ammonia stripping are presented in this review article. Its outstanding mass transfer performance and higher total suspended solids tolerance discriminate the conventional packed tower for ammonia stripping method. The following directions are proposed for further research. First, the structure optimization should be done for each of the ammonia stripping processes for higher air-stripping efficiency at a lower operational cost. Secondly, full cost analysis of the advanced ammonia stripper processes is needed to evaluate its economic feasibility for specific wastewater treatment scenario. Lastly, the integration of advanced gas-liquid contactors with vortex power generator for an energy self-sufficient wastewater treatment is proposed.

## Conflicts of Interest

The authors declare that they have no conflicts of interest regarding the publication of this paper.

## References

[1] S. R. M. Kutty, S. N. I. Ngatenah, M. H. Isa, and A. Malakahmad, "Nutrients removal from municipal wastewater treatment plant effluent using *Eichhornia crassipes*," *Engineering and Technology*, vol. 3, no. 12, pp. 826–831, 2009.

[2] V. D. Leite, S. Prasad, W. S. Lopes, J. T. Sousa, and A. J. M. Barros, "Study on ammonia stripping process of leachate from the packed tower," *Journal of Urban and Environmental*, vol. 7, no. 2, pp. 21–222, 2013.

[3] T. A. Pressley, D. F. Bishop, A. P. Pinto, and A. F. Cassel, "Ammonia-nitrogen removal by breakpoint chlorination," 1973, https://nepis.epa.gov/Exe/ZyPDF.cgi/91020N8G.PDF?Dockey=91020N8G.PDF.

[4] Batelle–Northwest Richland, "Wastewater ammonia removal by ion exchange," 1971, https://nepis.epa.gov/Exe/tiff2png.cgi/9100GI2R.PNG?-r+75+g+7+D%3A%5CZYFILES%5CINDEX%20DATA%5C70THRU75%5CTIFF%5C00001708%5C9100GI2R.TIF.

[5] M. Mondor, L. Masse, D. Ippersiel, F. Lamarche, and D. I. Masse, "Use of electrodialysis and reverse osmosis for the recovery and concentration of ammonia from swine manure," *Bioresource Technology*, vol. 99, no. 15, pp. 7363–7368, 2008.

[6] R. G. Rice, C. M. Robson, G. W. G. Miller, J. C. Clark, and W. Kohn, "Biological processes in the treatment of municipal water supplies," 1982, https://nepis.epa.gov/Exe/ZyPDF.cgi/9100LYD1.PDF?Dockey=9100LYD1.PDF.

[7] I. Ozturk, M. Altinbas, I. Koyuncu, and Y. C. Gomec, "Advanced physico-chemical treatment experiences on young municipal landfill leachates," *Waste Management*, vol. 23, no. 5, pp. 441–446, 2003.

[8] USEPA, "Wastewater technology fact sheet: ammonia stripping," 2000, https://nepis.epa.gov/Exe/ZyPDF.cgi/P10099PH.PDF?Dockey=P10099PH.PDF.

[9] L. K. Wang, Y.-T. Hung, and N. K. Shammas, *Advanced Physicochemical Treatment Processes: Handbook of Environmental Engineering*, Vol. 4, The Humana Press Inc., Totowa, NJ, USA, 2006.

[10] T. P. O'Farell, F. P. Frauson, A. F. Cassel, and D. F. Bishop, "Nitrogen removal by ammonia stripping," *Journal of Water Pollution Control Federation*, vol. 44, no. 8, pp. 1527–1535, 1972.

[11] M. Raboni, V. Torretta, O. Viotti, and G. Urbini, "Experimental plant for the chemico-physical treatment of groundwater polluted by MSW leachate, with ammonia recovery," *Revista Ambiente & Agua*, vol. 8, no. 3, pp. 22–32, 2013.

[12] G. Sarraco and G. Genon, "High temperature ammonia stripping and recovery from process liquid wastes," *Journal of Hazardous Materials*, vol. 37, no. 1, pp. 191–206, 1994.

[13] A. Alitalo, A. Kyro, and E. Aura, "Ammonia stripping of biologically treated liquid manure," *Journal of Environmental Quality*, vol. 41, no. 1, pp. 273–20, 2012.

[14] K. C. Cheung, L. M. Chu, and M. H. Wong, "Ammonia stripping as pretreatment for landfill leachate," *Water, Air, and Soil Pollution*, vol. 94, no. 1-2, pp. 209–221, 1995.

[15] M. Laureni, J. Palatsi, M. Llovera, and A. Bonmati, "Influence of pig slurry characteristics on ammonia stripping efficiencies and quality of the recovered ammonium-sulfate solution," *Journal of Chemical Technology and Biotechnology*, vol. 88, no. 9, pp. 1654–1662, 2013.

[16] X. H. Lei, N Sugiura, C. P. Feng, and T. Maekawa, "Pretreatment of anaerobic digestion effluent with ammonia stripping and biogas purification," *Journal of Hazardous Materials*, vol. 145, no. 3, pp. 391–397, 2007.

[17] M. A. Rubia, M. Walker, S. Heaven, C. J. Banks, and R. Borja, "Preliminary trials of in situ ammonia stripping from source segregated domestic food waste digestate using biogas: effect

of temperature and flow rate," *Bioresource Technology*, vol. 101, no. 24, pp. 9486–9492, 2010.

[18] A. Bonmati and X. Flotats, "Air stripping of ammonia from pig slurry: characterisation and feasibility as pre- or post-treatment to mesophilic anaerobic digestion," *Waste Management*, vol. 23, no. 3, pp. 261–272, 2003.

[19] A. Limoli, M. Langone, and G. Andreottola, "Ammonia removal from raw manure digestate by means of a turbulent mixing stripping process," *Journal of Environmental Management*, vol. 176, pp. 1–10, 2016.

[20] G. D. Boardman and P. J. McVeigh, "Use of air stripping technology to remove ammonia from biologically treated blue crab processing wastewater," *Journal of Aquatic Food Product Technology*, vol. 7, no. 4, pp. 81–97, 1998.

[21] B. V. Prather, "Wastewater aeration may be key to more efficient removal of impurities," *Oil and Gas Journal*, vol. 57, pp. 78–89, 1959.

[22] F. M. Ferraz, J. Povinelli, and E. M. Veira, "Ammonia removal from landfill leachate by air stripping and absorption," *Environmental Technology*, vol. 34, no. 15, pp. 2317–2326, 2013.

[23] L. Zhang, Y. W. Lee, and D. Jahng, "Ammonia stripping for enhanced biomethanization of piggery wastewater," *Journal of Hazardous Materials*, vol. 199-200, pp. 36–42, 2012.

[24] V. K. Minocha and A. V. S. P. Rao, "Ammonia removal and recovery from urea fertilizer plant waste," *Environmental Technology Letters*, vol. 9, no. 7, pp. 655–664, 1988.

[25] L. Zhu, D. Dong, X. Hua, Z. Guo, and D. Liang, "Ammonia nitrogen removal from acetylene purification wastewater by air stripping," *Water Science Technology*, vol. 75, no. 11-12, pp. 2538–2545, 2017.

[26] Li Lin, J. Chen, Z. Q. Xu et al., "Removal of ammonia nitrogen in wastewater by microwave radiation: a pilot-scale study," *Journal of Hazardous Materials*, vol. 168, no. 2-3, pp. 862–867, 2009.

[27] M. Ellersdorfer, "The ion-exchange-loop stripping process: ammonium recovery from sludge liquor using NACl-treated clinoptilolite and simultaneous air stripping," *Water Science and Technology*, vol. 77, no. 3, pp. 695–705, 2017.

[28] J. H. La, T. Kim, J. K. Jang, and I. S. Change, "Ammonia nitrogen removal and recovery from swine wastewater by microwave radiation," *Environmental Engineering Research*, vol. 19, no. 4, pp. 381–385, 2014.

[29] A. Serna-Maza, S. Heaven, and C. J. Banks, "In situ biogas stripping of ammonia from a digester using a gas mixing system," *Environmental Technology*, vol. 38, no. 24, pp. 3216–3224, 2017.

[30] C. Collivignarelli, G. Bertanza, M. Baldi, and F. Avezzu, "Ammonia stripping from MSW landfill leachate in bubble reactors: process modelling and optimization," *Waste Management & Research*, vol. 16, no. 5, pp. 455–466, 1998.

[31] K. Kojima, S. Zhang, and T. Hiaki, "Measuring methods of infinite-dilution activity coefficients and a database for systems including water," *Fluid Phase Equilibria*, vol. 131, no. 1-2, pp. 145–179, 1997.

[32] J. C. Campos, A. P. Moura, L. Costa, F. V. Yokoyama, D. F. Arouja, and M. C. Cammarota, "Evaluation of pH, alkalinity and temperature during air stripping process for ammonia removal from landfill leachate," *Journal of Environmental Science and Health*, vol. 48, no. 9, pp. 1105–1113, 2013.

[33] D. Hidalgo, F. Corona, J. M. Martin-Marroquin, J. D. Alamo, and A. Alicia, "Resource recovery from anaerobic digestate: struvite crystallisation versus ammonia stripping," *Desalination and Water Treatment*, vol. 57, no. 6, pp. 2626–2632, 2015.

[34] G. Markou, M. Agriomallou, and D. Georgakakis, "Forced ammonia stripping from livestock wastewater. the influence of some physico-chemical parameters of the wastewater," *Water Science & Technology*, vol. 75, no. 3-4, pp. 686–692, 2016.

[35] S. Gustin and R. Marinsek-Logar, "Effect of pH, temperature and air flow rate on continuous ammonia stripping of the anaerobic digestion effluent," *Process Safety and Environmental Protection*, vol. 89, no. 1, pp. 1–66, 2011.

[36] P. Viotti and R. Gavasci, "Scaling of ammonia stripping towers in the treatment of groundwater polluted by municipal solid waste landfill leachate: study of the causes of scaling and its effects on stripping performance," *Revista Ambient Agua*, vol. 10, no. 2, pp. 241–252, 2015.

[37] C. T. Whitman, G. T. Mehan, G. H. Grubbs et al., "Development document for the proposed effluent limitations guidelines and standards for the meat and poultry products industry point source category," 2002, https://nepis.epa.gov/Exe/ZyPDF.cgi/20002F0Q.PDF?Dockey=20002F0Q.PDF.

[38] J. P. Maree and C. M. Zvinowanda, "Recovery of calcium carbonate from wastewater treatment sludge using a flotation technique," *Journal of Chemical Engineering Process Technology*, vol. 3, no. 2, pp. 1–6, 2012.

[39] K. L. He, S. B. Wu, L. C. Guo, Z. S. Ajmal, H. Z. Luo, and R. J. Dong, "Treatment of alkaline stripped effluent in aerated constructed wetlands: feasibility evaluation and performance enhancement," *Water*, vol. 8, no. 9, pp. 1–11, 2016.

[40] M. H. Yuan, Y. H. Chen, J. Y. Tsai, and C. Y. Chang, "Ammonia removal from ammonia-rich wastewater by air stripping using a rotating packed bed," *Process Safety and Environmental Protection*, vol. 102, pp. 777–785, 2016.

[41] X. J. Quan and Z. L. Cheng, "Mass transfer performance of a water-sparged aerocyclone reactor and its application in wastewater," *Journal of Hazardous Materials*, vol. 170, no. 2-3, pp. 983–938, 2009.

[42] N. Degermenci, O. N. Ata, and E. Yildiz, "Ammonia removal by air stripping in a semi-batch jet loop reactor," *Journal of Industrial and Engineering Chemistry*, vol. 18, no. 1, pp. 399–404, 2012.

[43] A. Behr and M. Becker, "Multiphase catalysis in jetloop-reactors," *Chemical Engineering Transactions*, vol. 17, pp. 141–144, 2009.

[44] H.-J. Warnecke, M. Geisendorfer, and D. C. Hempel, "Mass transfer behaviour of gas-loop reactors," *Acta Biotechnology*, vol. 11, no. 1, pp. 306–311, 1988.

[45] K. H. Tebel and P. Zehner, "Fluid dynamic description of jet loop reactors in multiphase operation," *Chemical Engineering Technology*, vol. 12, no. 1, pp. 274–280, 1989.

[46] C. A. M. C. Dirix and K. Van de Wiele, "Mass transfer in jet loop reactors," *Chemical Engineering Science*, vol. 45, pp. 2333–2340, 1990.

[47] B. Farizoglu, B. Keskinler, E. Yildiz, and A. Nuhoglu, "Cheese whey treatment performance of anaerobic jet loop membrane bioreactor," *Process Biochemistry*, vol. 39, pp. 2283–2291, 2004.

[48] A. Eusobio, M. Petruccioli, M. Lageiro, F. Federici, and J. C. Duarte, "Microbial characterisation of activated sludge in jet-loop bioreactors treating winery wastewaters," *Journal of Industrial Microbiology & Biotechnology*, vol. 31, no. 1, pp. 29–34, 2004.

[49] C. Matter-Muller, W. Gujer, and W. Giger, "Transfer of volatile subtances from water to the atmosphere," *Water Research*, vol. 15, no. 11, pp. 1271–1279, 1981.

[50] X. J. Quan, Z. L. Cheng, F. Xu, F. C. Qiu, D. Li, and Y. P. Yan, "Structural optimization of the porous section in a water-sparged

aerocyclone reactor to enhance the air-stripping efficiency of ammonia," *Journal of Environmental Chemical Engineering*, vol. 2, no. 2, pp. 1199–1206, 2014.

[51] D. P. Rao, A. Bhowal, and P. S. Goswami, "Process intensification in rotating packed beds (HIGEE): an appraisal," *Industrial Engineering Chemical Resource*, vol. 43, no. 4, pp. 1150–1162, 2004.

[52] Y. Sun, T. Nozawa, and S. Furusaki, "Gas holdup and volumetric oxygen transfer coefficient in three-phase fluidized bed reactor," *Journal of Chemical Engineering*, vol. 21, no. 1, pp. 15–20, 1988.

[53] J. T. Xu, C. S. Liu, M. Wang et al., "Rotating packed bed reactor for enzymatic synthesis of biodiesel," *Bioresource Technology*, vol. 224, pp. 292–297, 2017.

[54] K. Guo, J. Wen, Y. Zhao et al., "Optimal packing of a rotating packed bed for H2S removal," *Environmental Science & Technology*, vol. 48, no. 12, pp. 6844–6849, 2014.

[55] J. F. Chen, Y. H. Wang, F. Guo, X. M. Wang, and Z. Chong, "Synthesis of nanoparticles with novel technology: high-gravity reactive precipitation," *Industrial Engineering Chemistry Resource*, vol. 39, no. 4, pp. 948–954, 2000.

[56] M. J. Semmens, D. M. Foster, and E. L. Cussler, "Ammonia removal from water using microporous hollow fibers," *Journal of Membrane Science*, vol. 51, no. 1-2, pp. 127–140, 1990.

[57] Y. T. Ahn, Y. H. Hwang, and H. S. Shin, "Application of PTFE membrane for ammonia removal in membrane contactor," *Water Science and Technology*, vol. 63, no. 12, pp. 2944–2948, 2011.

[58] A. Hasanoglu, J. Romero, B. Perez, and A. Plaza, "Ammonia removal from wastewater streams through membrane contactors: experimental and theoretical analysis of operation parameters and configuration," *Chemical Engineering Journal*, vol. 160, no. 2, pp. 530–537, 2010.

[59] X. Y. Tan, S. P. Tan, W. K. Teo, and K. Li, "Polyvinylidene fluoride (PVDF) hollow fibre membranes for ammonia removal from water," *Journal of Membrane Science*, vol. 271, no. 1-2, pp. 59–68, 2006.

[60] Q. L. Liu, Z. H. Wang, L. W. Chen, and P. P. Wang, "The effect of ammonia initial concentration in membrane distillation process for high ammonia concentration wastewater treatment," in *Proceedings of the International Conference on Consumer Electronics, Communications and Networks (CECNet)*, pp. 1795–1797, Beijing, China, March 2011.

[61] M. S. El-Bourawi, M. Khayet, R. Ma, Z. Ding, Z. Li, and X. Zhang, "Application of vacuum membrane distillation for ammonia removal," *Journal of Membrane Science*, vol. 301, no. 1-2, pp. 200–209, 2007.

[62] L. Eykens, I. Hitsov, K. De Sitter, C. Dotremont, L. Pinoy, and B. V. D. Bruggen, "Direct contact and air gap membrane distillation: differences and similarities between lab and pilot scale," *Desalination*, vol. 422, pp. 91–100, 2017.

[63] Z. L. Xie, T. Duong, M. Hoang, N. Cuong, and B. Bolto, "Ammonia removal by sweep gas membrane distillation," *Water Research*, vol. 43, no. 6, pp. 1693–1699, 2009.

[64] Y. F. Liu, T. H. Xiao, C. H. Bao, J. F. Zhang, and X. Yang, "Performance and fouling study of asymmetric PVDF membrane applied in the concentration of organic fertilizer by direct contact membrane distillation," *Membranes*, vol. 8, no. 1, p. 9, 2018.

[65] A. G. Capodaglio, P. Hlavinek, and M. Raboni, "Physico-chemical technologies for nitrogen removal wastewater: a review," *Revista Ambiente Agua*, vol. 10, no. 3, 2015.

[66] Q. Deng, B. R. Dhar, E. Elbeshbishy, and H. S. Lee, "Ammonium nitrogen removal from the permeates of anaerobic membrane bioreactors: economic regeneration of exhausted zeolite," *Environmental Technology*, vol. 35, no. 16, pp. 2008–2017, 2014.

[67] M. Ellersdorfer, "Recovery of ammonia from liquid digestate for NOx removal," in *Proceedings of the 11th DepoTech Conference*, vol. 835, pp. 367–372, Leoben, Austria, November 2012.

[68] O. N. Ata, A. Kanca, Z. N. Demir, and V. Yigit, "Optimization of ammonia removal from aqueous solution by microwave-assisted air stripping," *Water Air Soil Pollition*, vol. 228, no. 11, pp. 448–458, 2017.

[69] L. Le, H. W. Wang, and H. H. Lu, "Nitrogen removal using air stripping tower in urban wastewater treatment plant," *China Wastewater*, vol. 22, pp. 92–99, 2006.

[70] Q. Xia, Y. B. Yun, J. J. Chen, D. Qu, and C. L. Li, "Treatment of ammonia nitrogen wastewater by membrane distillation PVDF membrane," *Desalination and Water Treatment*, vol. 61, pp. 126–135, 2016.

[71] O. N. Ata, K. Aygun, H. Okur, and A. Kanca, "Determination of ammonia removal from aqueous solution and volumetric mass transfer coefficient by microave-assisted air stripping," *International Journal of Science and Technology*, vol. 13, no. 10, pp. 2459–2466, 2016.

[72] S. H. You, D. H. Tseng, and G. L. Guo, "A case study on the wastewater reclamation and reuse in the semiconductor industry," *Resources, Conservation and Recycling*, vol. 32, no. 1, pp. 73–81, 2001.

[73] M. Mulder, *Basic Principal of Membrane Technology*, Kluwer Academic Publisher, Norwell, MA, USA, 2nd edition, 1996.

[74] Y. Nishi and T. Inagaki, "Performance and flow field of gravitation vortex type water turbine," *International Journal of Rotating Machinery*, vol. 2017, Article ID 2610508, 11 pages, 2017.

# 4

# *Moringa oleifera* Lam. and its Potential Association with Aluminium Sulphate in the Process of Coagulation/Flocculation and Sedimentation of Surface Water

Karina Cardoso Valverde ⓘ,[1] Priscila Ferri Coldebella ⓘ,[1] Marcela Fernandes Silva ⓘ,[1] Letícia Nishi,[1] Milene Carvalho Bongiovani ⓘ,[2] and Rosângela Bergamasco ⓘ[1]

[1]*Departamento de Engenharia Química, Universidade Estadual de Maringá, Av. Colombo 5790, Bloco D-90, 87020-900 Maringá, PR, Brazil*
[2]*Universidade Federal do Mato Grosso, Av. Alexandre Ferronato, 1200, 78557-267 Sinop, MT, Brazil*

Correspondence should be addressed to Karina Cardoso Valverde; karinacvalverde@gmail.com

Academic Editor: Julio Sánchez Poblete

The present study aims to optimize the operational conditions in surface water coagulation/flocculation and sedimentation step, besides evaluating the association between seeds of *Moringa oleifera* Lam. (*M. oleifera*) and the synthetic coagulant aluminium sulphate for surface water treatment. The assays were performed in Jar Test using surface water from Pirapó River basin, Maringá, PR. It was observed that the operational conditions affect the coagulation/flocculation and sedimentation process efficiency. Optimal operational conditions for coagulants association are as follows: rapid mixing velocity (RMV) of 105 rpm, rapid mixing times (RMT) of 1 min, slow mixing velocity (SMV) of 30 rpm, slow mixing times (SMT) of 15 min, and sedimentation time (ST) of 15 min; this enables an improvement in the process, contributing to a reduction in synthetic coagulant aluminium sulphate demand of up to 30%, combined with an increase in *M. oleifera* dosage, not affecting the coagulation/flocculation and sedimentation process efficiency, considering the water pH range between 7 and 9.

## 1. Introduction

Currently, millions of people are exposed to dangerous levels of chemical pollutants and biological contaminants in drinking water due to the inadequate handling of urban population and industrial or agricultural wastewaters [1]. Thus, it is necessary to carry out physical, chemical, and/or microbiological processes in order to remove the impurities present in water and coagulants have been widely used in this way in conventional processes of water treatment [2].

Aluminium sulphate ($Al_2(SO_4)_3 \cdot 18H_2O$) stands out as the most used synthetic coagulant in Brazil when it comes to water treatment of public supplies as a result of its high efficiency in suspended solids removal and low cost [3]. However, its effect is strongly dependent on pH, especially in the range from 5.5 to 8 [4], and at the end of the treatment there is the possibility of a high concentration of residual aluminium remaining in water [5], which can be associated

with the acceleration of degenerative processes of Alzheimer's disease [6–8]. There is also the problem of the reaction between aluminium and the natural alkalinity present in water, which leads to a pH reduction [9]. Therefore, it is interesting to propose alternatives to reduce the quantity of the synthetic coagulant, such as the utilization of natural coagulants in water treatment.

Natural coagulants are biodegradable and present low toxicity and low levels of residual sludge production [5, 10], besides being considered health-friendly [6, 11].

Seeds of *Moringa oleifera* Lam. (*M. oleifera*) stand out as a promising natural coagulant [7, 12–15] by being considered as a low-cost and safe alternative, besides having attested efficiency in water treatment [16].

Amagloh and Benang [17] assure that when seed powder of *M. oleifera* is added to turbid water, the proteins present in this coagulant produce positive charges by means of electrostatic attractions with negatively charged particles [18], such

TABLE 1: Operational conditions of coagulation/flocculation and sedimentation process.

| ASSAY | 1 | 2 | 3 | 4 | 5 | 6 | 7 | 8 | 9 | 10 | 11 | 12 | 13 | 14 | 15 | 16 | 17 | 18 | 19 | 20 | 21 | 22 | 23 | 24 | 25 | 26 | 27 |
|---|---|---|---|---|---|---|---|---|---|---|---|---|---|---|---|---|---|---|---|---|---|---|---|---|---|---|---|
| RMV (rpm) | 100 | 105 | 110 | 100 | 105 | 110 | 100 | 105 | 110 | 100 | 105 | 110 | 100 | 105 | 110 | 100 | 105 | 110 | 100 | 105 | 110 | 100 | 105 | 110 | 100 | 105 | 110 |
| RMT (min) | 1 | 1 | 1 | 2 | 2 | 2 | 3 | 3 | 3 | 1 | 1 | 1 | 2 | 2 | 2 | 3 | 3 | 3 | 1 | 1 | 1 | 2 | 2 | 2 | 3 | 3 | 3 |
| SMV (rpm) | 15 | 15 | 15 | 15 | 15 | 15 | 15 | 15 | 15 | 30 | 30 | 30 | 30 | 30 | 30 | 30 | 30 | 30 | 45 | 45 | 45 | 45 | 45 | 45 | 45 | 45 | 45 |
| SMT (min) | 15 | 15 | 15 | 15 | 15 | 15 | 15 | 15 | 15 | 15 | 15 | 15 | 15 | 15 | 15 | 15 | 15 | 15 | 15 | 15 | 15 | 15 | 15 | 15 | 15 | 15 | 15 |

SMV: slow mixing velocity; RMV: rapid mixing velocity; SMT: slow mixing times; RMT: rapid mixing times.

as mud, clay, bacteria, and other toxic particles present in water. The flocculation process occurs when the proteins get bound to the negative charges of the particles, producing flakes and bringing together the impurities present in the water. However, the coagulation mechanism is not yet well defined among researchers.

Investigations into the behavior of seeds of *M. oleifera* in conjunction with aluminium salts presented promising results [11, 13, 19], attesting that *M. oleifera* is a feasible coagulant as a partial replacement of synthetic coagulants [16]. Valverde et al. [4] observed that the association of seeds of *M. oleifera* and aluminium sulphate brings about an increase in apparent color and turbidity removal efficiency of surface water. Dalen et al. [20] mention that the association of coagulants can improve water sanitation in developing countries.

Considering the operational conditions in the coagulation/flocculation process, studies carried out by Cordeiro Cardoso et al. [21] demonstrated that rapid mix, slow mix, and sedimentation do influence color and turbidity removal during the process performed with seeds of *M. oleifera*. Since these parameters affect the global efficiency of impurity removal in surface water, they do not have to be simply adopted, but operational conditions from laboratory studies must be established for the water treatability [22].

Pritchard et al. [15] assert that rapid mixing of a few seconds is important after the addition of a coagulant so as to assure a uniform dispersion and also increase the opportunity of contact between the particles [18]. In this way, in order to ensure coagulation efficiency, the occurrence of a uniform and intense mixture of the coagulant in water is necessary so that the contact probability between coagulant and particles is excellent before the completion of the reactions.

According to Vijayaraghavan et al. [7], there is a shortage of comprehensive studies that compare natural and synthetic coagulants association efficiency. In this way, the general objective of this work is to optimize the operational conditions of coagulation/flocculation and sedimentation step, besides evaluating the efficiency of natural coagulant seeds of *M. oleifera* and synthetic coagulant aluminium sulphate association in surface water treatment, by means of the alteration of coagulants dosage and coagulation pH, using coagulation diagram tools.

## 2. Materials and Methods

Surface water collected at the Water Treatment Plant (Sanepar) of the city, coming from Pirapó River basin, Maringá city, Paraná state, Brazil, was characterized by means of the following quality parameters: apparent color and compounds with $UV_{254\,nm}$ absorption (DR 5000 Hach spectrophotometer), turbidity (2100P Hach turbidimeter), total dissolved solids (TDS), and pH (Thermo Scientific Orion VSTAR92 Versa Star pH meter).

The coagulation/flocculation assays were carried out in Jar Test, Nova Ética, 218/LDB0 of six jars, with rotation regulator of mixer shafts, in duplicate, in 700 mL recipients of surface water.

*2.1. Coagulants Preparation.* For the preparation of synthetic coagulant standard solution, 1 g of aluminium sulphate hydrate (Vetec) was dissolved in distilled water and the volume was brought to 100 mL in order to obtain a 1% $w\cdot v^{-1}$ solution.

In order to obtain *M. oleifera* powdered coagulant, 15 g of seeds ordered from Aracaju, SE, were manually peeled, ground in a blender (NL-41 Mondial), and dried in a forced air buffer (Digital Timer SX CR/42) at 40°C until constant weight was observed [17].

*2.2. Operational Conditions of Coagulation/Flocculation and Sedimentation Process.* In this step, coagulant dosages added during the assays were adapted from values cited in the literature, namely, 25 mg·L$^{-1}$ for aluminium sulphate [23, 24] and 50 mg·L$^{-1}$ for seeds of *M. oleifera* [25].

Operational conditions of coagulation/flocculation and sedimentation process used in the association of seeds of *M. oleifera* and aluminium sulphate coagulants are presented on Table 1. It is worth mentioning that these values are based on optimal operational conditions of seeds of *M. oleifera*, earlier studied by Madrona et al. [10] and Cordeiro Cardoso et al. [21], and of aluminium sulphate, obtained through information at Sanepar, thus adopting real operational conditions used at the Water Treatment Plant in Maringá city.

*2.3. Coagulation Diagrams.* Coagulation diagrams developed on 3DField 3.5.3.0. software were created from coagulation pH variation and seeds of *M. oleifera* and aluminium sulphate association dosage by evaluating the removal efficiency of apparent color, turbidity, and compounds with $UV_{254\,nm}$ absorption.

Surface water pH used in the assays was adjusted in the range between 4 and 10 with 0.1 mol·L$^{-1}$ and 1 mol·L$^{-1}$ sodium hydroxide (NaOH) and 0.1 mol·L$^{-1}$ and 1 mol·L$^{-1}$ hydrochloric acid (HCl).

For this step of the work, coagulants dosages specified in Table 2 were used, according to adaptations from the methodology proposed by Nwaiwu and Bello [19].

TABLE 2: Coagulants dosage in association.

| Point | % coagulant $(Al_2(SO_4)_3 \cdot 18H_2O/$ M. oleifera) | Coagulant dosage (mg·L$^{-1}$) | |
|---|---|---|---|
| | | $Al_2(SO_4)_3 \cdot 18H_2O$ | Seeds of M. oleifera |
| 1 | 0%/100% | 0 | 50 |
| 2 | 10%/90% | 2.5 | 45 |
| 3 | 20%/80% | 5 | 40 |
| 4 | 30%/70% | 7.5 | 35 |
| 5 | 40%/60% | 10 | 30 |
| 6 | 50%/50% | 12.5 | 25 |
| 7 | 60%/40% | 15 | 20 |
| 8 | 70%/30% | 17.5 | 15 |
| 9 | 80%/20% | 20 | 10 |
| 10 | 90%/10% | 22.5 | 5 |
| 11 | 100%/0% | 25 | 0 |

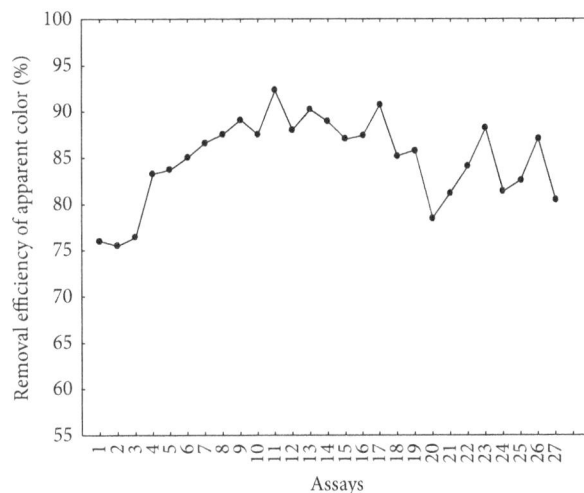

FIGURE 1: Removal efficiency of M. oleifera and aluminium sulphate for apparent color.

*2.4. Coagulation/Flocculation and Sedimentation Process Evaluation.* After completing the coagulation/flocculation and sedimentation process, 15 mL of treated water was collected for results evaluation, which was carried out on the basis of percentage reduction in apparent color, turbidity, and compounds with $UV_{254\,nm}$ absorption.

*2.5. Statistical Analysis.* For the purpose of comparing the results obtained in the coagulation/flocculation and sedimentation assays, Analysis of Variance (ANOVA) and multiple comparison of means, or Tukey's Test, were carried out, using confidence interval of 95%, for significant $p$ values < 0.05, in order to verify significant differences by means of SISVAR version 5.3 statistical program [26].

In order to assess the removal efficiency of quality parameters after the water clarification process, a 27 × 4 factorial delineation was used, whose factors were as follows: assays (27 variations for slow and rapid mixing velocity (SMV and RMV) and slow and rapid mixing times (SMT and RMT) and sedimentation times (four ST), with two replications [18].

# 3. Results and Discussions

Table 3 presents the characterization of surface water.

*3.1. Operational Conditions: Optimization Step of the Coagulation/Flocculation and Sedimentation Process.* By evaluating the results obtained by means of Tukey's Test, it is possible to observe that there was no significant statistical difference concerning operational conditions presented in Table 4.

Earlier studies [2, 4] have shown results of apparent color and turbidity removal with 60 min and 120 min sedimentation time (ST), using saline and aqueous extract of M. oleifera, respectively, as the sole coagulant in the coagulation/flocculation and sedimentation process. The elapsed time required for sedimentation when using M. oleifera powdered seed as the sole coagulant is higher than that of M. oleifera and aluminium sulphate association [8, 19].

Bongiovani et al. [27] confirm that the utilization of a polymer along with seeds of M. oleifera improves removal efficiency of quality parameters and it increases the size of flocs, thus diminishing significantly the overall sedimentation time.

Several results are presented in Table 4. However, ST lies in the range from 30 min to 60 min. Since 15 min is the standard elapsed time used at Water Treatment Plants [27], optimal operational conditions for seeds of M. oleifera and aluminium sulphate association are as follows: rapid mixing velocity (RMV) of 105 rpm, rapid mixing time (RMT) of 1 min, slow mixing velocity (SMV) of 30 rpm, slow mixing time (SMT) of 15 min, and sedimentation time (ST) of 15 min.

Figures 1, 2, and 3 present assay results for 15 min ST concerning efficiency removal of apparent color, turbidity, and compounds with $UV_{254\,nm}$ absorption.

By evaluating the results obtained for the three quality parameters studied, one can conclude that the optimal operational conditions for seeds of M. oleifera and aluminium sulphate association are: RMV of 105 rpm, RMT of 1 min, SMV of 30 rpm, SMT of 15 min, and ST of 15 min.

*3.2. Coagulation Diagrams Construction Step.* Coagulation diagrams are constructed from removal efficiency of the quality parameters evaluated, considering that the best results are identified by dark-colored regions.

Figures 4, 5, and 6 show coagulation diagrams that evaluate removal efficiency of apparent color, turbidity, and compounds with $UV_{254\,nm}$ absorption, respectively.

As for coagulants association, in a general way, it is possible to observe in the coagulation diagrams that as the aluminium sulphate dosage increases, the coagulation effect is enhanced.

Valverde et al. [4] ascertained that the association of natural and synthetic coagulants enables an increase in the removal efficiency of apparent color and turbidity of the water under scrutiny, thus corroborating the study of Dalen et al. [20], which suggests that aluminium sulphate and powdered M. oleifera present better synergic characteristics than

TABLE 3: Characterization of surface water.

| Quality parameter | Apparent color | Turbidity | $UV_{254 nm}$ | TDS | pH |
|---|---|---|---|---|---|
| Surface water | 406 uH | 73.9 NTU | $0.285 \, cm^{-1}$ | $115.33 \, mg \cdot L^{-1}$ | 7.817 |

TABLE 4: Values of removal efficiency for the best operating conditions in association of seeds of *M. oleifera* and aluminium sulphate.

| Parameters | | | | | Removal efficiency (%) | | |
|---|---|---|---|---|---|---|---|
| RMV (rpm) | RMT (min) | SMV (rpm) | SMT (min) | ST (min) | Apparent color | Turbidity | $UV_{254 nm}$ |
| 105 | 1 | 30 | 15 | 15–45 | 92.3–93.6 | 91.0–92.0 | 75.5–76.3 |
| 110 | 1 | 30 | 15 | 45 | 90.9 | 90.0 | 72.4 |
| 100 | 2 | 30 | 15 | 30–45 | 91.2–92.5 | 92.2–92.9 | 74.3–75.2 |
| 105 | 3 | 30 | 15 | 45–60 | 93.8–95.4 | 92.8 – 94.1 | 77.2 |
| 110 | 3 | 30 | 15 | 60 | 92.9 | 92.6 | 79.0 |
| 110 | 1 | 45 | 15 | 60 | 92.8 | 91.2 | 76.1 |
| 105 | 2 | 45 | 15 | 45–60 | 93.1–93.3 | 92.9–93.3 | 75.3–75.8 |
| 100 | 3 | 45 | 15 | 45 | 91.0 | 91.0 | 74.0 |
| 105 | 3 | 45 | 15 | 45–60 | 92.9–93.2 | 92.9–93.0 | 78.6 |

SMV: slow mixing velocity; RMV: rapid mixing velocity; SMT: slow mixing times; RMT: rapid mixing times; ST: sedimentation times.

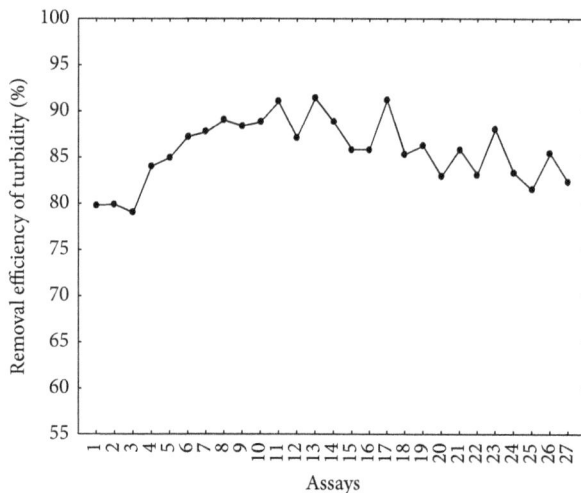

FIGURE 2: Removal efficiency of *M. oleifera* and aluminium sulphate for turbidity.

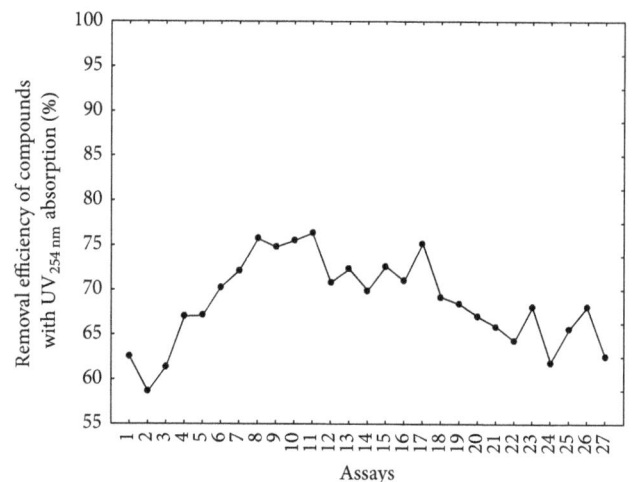

FIGURE 3: Removal efficiency of *M. oleifera* and aluminium sulphate for compounds with $UV_{254 nm}$ absorption.

the ones described in the utilization of *M. oleifera* as the sole coagulant. This occurrence can be evidenced in the coagulation diagrams presented.

Considering the water pH range between 7 and 9, it can be asserted that coagulation addition with dosages higher than $17.5 \, mg \cdot L^{-1}$ aluminium sulphate/$15 \, mg \cdot L^{-1}$ seeds of *M. oleifera* (point 8) would be interesting for the attainment of removal efficiency of about 75.0% for color and turbidity and 70.0% for compounds with $UV_{254 nm}$ absorption.

It is possible to achieve a 30% reduction in the consumption of synthetic coagulant aluminium sulphate combined with an increase in *M. oleifera* dosage without affecting the efficiency of the process. Therefore, research in this area must continue in order to obtain water that meets the required potability standard.

## 4. Conclusions

It was observed that the operational conditions affect the efficiency of the coagulation/flocculation and sedimentation process. Coagulant aluminium sulphate, applied in combination with seeds of *M. oleifera*, enables an improvement in

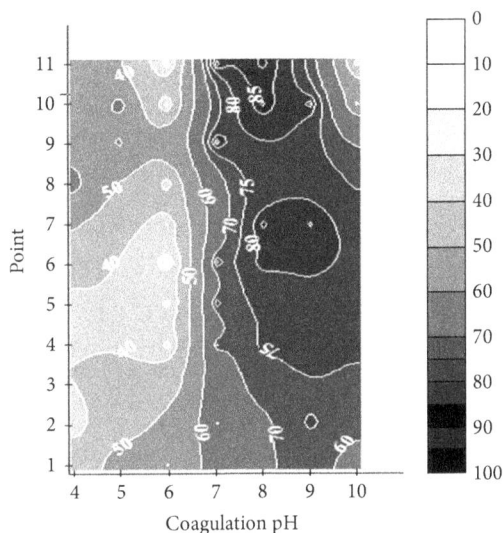

FIGURE 4: Coagulation diagrams with curves of apparent color removal in percentage (%).

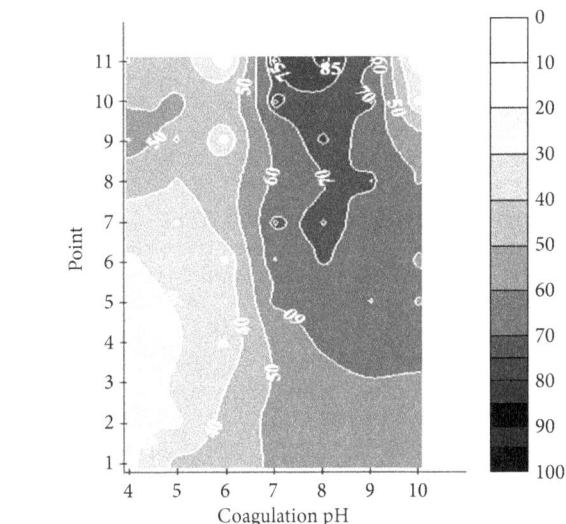

FIGURE 6: Coagulation diagrams with curves of compounds with $UV_{254\,nm}$ absorption removal in percentage (%).

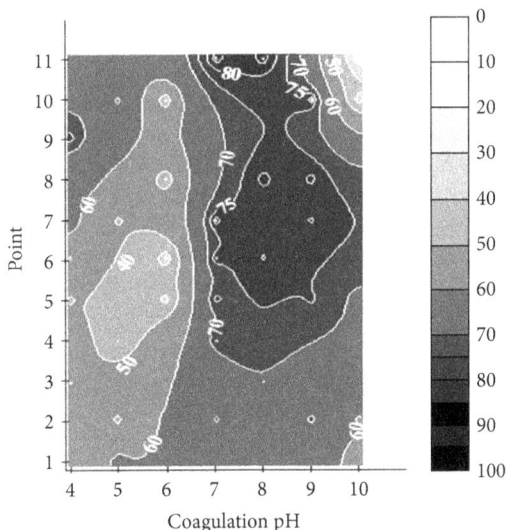

FIGURE 5: Coagulation diagrams with curves of turbidity removal in percentage (%).

## Conflicts of Interest

The authors declare that there are no conflicts of interest.

## Acknowledgments

The authors would like to acknowledge Universidade Federal de Sergipe (UFS) for seeds of *M. oleifera* seeds and Sanepar for samples of surface water. They would also like to acknowledge CAPES for financial support in the form of the scholarships received.

sedimentation time of the overall process, so that 15 min are enough. Optimal operational conditions for coagulants association are as follows: RMV of 105 rpm, RMT of 1 min, SMV of 30 rpm, and SMT of 15 min.

For the water pH range between 7 and 9, the association of natural and synthetic coagulants with dosages higher than $17.5\,mg{\cdot}L^{-1}$ aluminium sulphate/$15\,mg{\cdot}L^{-1}$ seeds of *M. oleifera* presents synergic characteristics, and removal efficiency of about 75.0% for color and turbidity and of 70.0% for compounds with $UV_{254\,nm}$ absorption is obtained.

The use of seeds of *M. oleifera* contributes to a reduction of up to 30% in the required demand of aluminium sulphate, without affecting the efficiency of coagulation/flocculation and sedimentation process, besides being an interesting environmental option nowadays.

## References

[1] WHO, World Health Organization, Water Quality and Health Strategy: 2013-2020, 2013.

[2] G. S. Madrona, G. B. Serpelloni, A. M. Salcedo Vieira, L. Nishi, K. C. Cardoso, and R. Bergamasco, "Study of the effect of Saline solution on the extraction of the Moringa oleifera seed's active component for water treatment," *Water, Air, & Soil Pollution*, vol. 211, no. 1-4, pp. 409–415, 2010.

[3] P. A. Lo Monaco, A. T. Matos, I. C. Ribeiro, F. d. Nascimento, and A. P. Sarmento, "Utilização de Extrato de Sementes de Moringa como Agente Coagulante no Tratamento de Água para Abastecimento e Águas Residuárias," *Ambi-Agua*, vol. 5, no. 3, pp. 222–231, 2010.

[4] K. C. Valverde, L. C. K. Moraes, M. C. Bongiovani, F. P. Camacho, and R. Bergamasco, "Coagulation diagram using the Moringa oleifera Lam and the aluminium sulphate, aiming the removal of color and turbidity of water," *Acta Scientiarum - Technology*, vol. 35, no. 3, pp. 485–489, 2013.

[5] S. Kawamura, "Effectiveness of natural polyelectrolytes in water treatment," *Journal - American Water Works Association*, vol. 83, no. 10, pp. 88–91, 1991.

[6] T. Okuda, A. U. Baes, W. Nishijima, and M. Okada, "Improvement of extraction method of coagulation active components from Moringa oleifera seed," *Water Research*, vol. 33, no. 15, pp. 3373–3378, 1999.

[7] G. Vijayaraghavan, T. Sivakumar, and A. Vimal Kumar, "Application of plant based coagulants for waste water treatment," *International Journal of Advanced Engineering Research and Studies*, vol. 1, no. 1, pp. 88–92, 2011.

[8] K. Ravikumar and A. K. Sheeja, "Water clarification using Moringa oleifera seed coagulant," in *Proceedings of the 2012 International Conference on Green Technologies, ICGT 2012*, pp. 64–70, India, December 2012.

[9] M. R. Gidde and A. R. Bhalerao, "Optimisation of Physical Parameters of Coagulation-Flocculation Process in Water Treatment, JERAD," *Journal of Environmental Research and Development*, vol. 6, no. 1, pp. 99–110, 2011.

[10] G. S. Madrona, I. G. Branco, V. J. Seolin, B. D. A. A. Filho, M. R. Fagundes-Klen, and R. Bergamasco, "Evaluation of extracts of moringa oleifera lam seeds obtained with nacl and their effects on water treatment," *Acta Scientiarum - Technology*, vol. 34, no. 3, pp. 289–293, 2012.

[11] J. R. Rodríguez-Núñez, D. I. Sánchez-Machado, J. López-Cervantes, J. A. Núñez-Gastélum, R. G. Sánchez-Duarte, and M. A. Correa-Murrieta, "Moringa oleifera Seed Extract in the Clarification of Surface Waters," *International Journal of Engineering Pedagogy*, vol. 2, no. 11, pp. 17–21, 2012.

[12] A. Ndabigengesere and K. S. Narasiah, "Influence of operating parameters on turbidity removal by coagulation with moringa oleifera seeds," *Environmental Technology (United Kingdom)*, vol. 17, no. 10, pp. 1103–1112, 1996.

[13] J. K. Abaliwano, K. A. Ghebremichael, and G. L. Amy, "Application of the purified *Moringa oleifera* coagulant for surface water treatment, watermill work," *WaterMill Work. Paper Serie*, vol. 5, pp. 1–19, 2008.

[14] E. N. Ali, S. A. Muyibi, H. M. Salleh, M. Z. Alam, and M. R. Salleh, "Production of Natural Coagulant from Moringa Oleifera Seed for Application in Treatment of Low Turbidity Water," *Journal of Water Resource and Protection*, vol. 02, no. 03, pp. 259–266, 2010.

[15] M. Pritchard, T. Craven, T. Mkandawire, A. S. Edmondson, and J. G. O'Neill, "A comparison between Moringa oleifera and chemical coagulants in the purification of drinking water - An alternative sustainable solution for developing countries," *Physics and Chemistry of the Earth*, vol. 35, no. 13-14, pp. 798–805, 2010.

[16] M. Awad, H. Wang, and F. Li, "Preliminary study on combined use of *Moringa* seeds extract and PAC for water treatment," *Research Journal of Recent Sciences*, vol. 2, no. 8, pp. 52–55, 2013.

[17] F. K. Amagloh and A. Benang, "Effectiveness of *Moringa oleifera* seed as coagulant for water purification," *African Journal of Agricultural Research*, vol. 4, no. 2, pp. 119–123, 2009.

[18] K. Cardoso Valverde, P. Ferri Coldebella, and R. Bergamasco, "Otimização das condições de operação no processo de clarificação de água superficial por meio da associação dos coagulantes moringa oleifera lam e cloreto férrico," *Periódico Eletrônico Fórum Ambiental da Alta Paulista*, vol. 9, no. 11, pp. 46–54, 2013.

[19] N. E. Nwaiwu and A. A. Bello, "Effect of *Moringa oleifera*-alum Ratios on Surface Water Treatment in North East Nigeria," *Research Journal of Applied Sciences, Engineering and Technology*, vol. 3, no. 6, pp. 505–512, 2011.

[20] M. Dalen, J. Pam, A. Izang, and R. Ekele, "Synergy Between Moringa oleifera Seed Powder And Alum In The Purification of Domestic Water," *The Scientific World Journal*, vol. 4, no. 4, 2010.

[21] K. Cordeiro Cardoso, R. Bergamasco, E. Sala Cossich, and L. C. Konradt Moraes, "Otimização dos tempos de mistura e decantação no processo de coagulação/floculação da água bruta por meio da *Moringa oleifera* Lam," *Acta Scientiarum. Technology*, vol. 30, no. 2, 2008.

[22] M. de Julio, I. Volski, D. A. Fioravante, O. Selhorst Filho, and F. I. Oroski, "Avaliação da Viabilidade Técnica do Emprego da Tecnologia de Filtração Direta no Tratamento da Água Bruta Afluente à ETA do Município de Ponta Grossa, PR," *Ciência & Engenharia*, vol. 18, no. 1, pp. 21–30, 2009.

[23] A. Baghvand, A. D. Zand, N. Mehrdadi, and A. Karbassi, "Optimizing coagulation process for low to high turbidity waters using aluminum and iron salts," *American Journal of Environmental Sciences*, vol. 6, no. 5, pp. 442–448, 2010.

[24] H. F. Makki, A. F. Al-Alawy, N. N. Abdul-Razaq, and M. A. Mohammed, "Using Aluminum Refuse as a Coagulant in the Coagulation and Flocculation Processes," *Iraqi Journal of Chemical and Petroleum Engineering*, vol. 11, no. 3, pp. 15–22, 2010.

[25] R. Joshua and V. Vasu, "Characteristics of Stored Rain Water and its Treatment Technology Using Moringa seeds," *International Journal of Life science and Pharma Research*, vol. 2, no. 1, pp. 154–175, 2013.

[26] D. F. Ferreira, "Sisvar: a computer statistical analysis system," *Ciência e Agrotecnologia*, vol. 35, no. 6, pp. 1039–1042, 2011.

[27] M. C. Bongiovani, F. P. Camacho, L. Nishi et al., "Improvement of the coagulation/flocculation process using a combination of Moringa oleifera Lam with anionic polymer in water treatment," *Environmental Technology (United Kingdom)*, vol. 35, no. 17, pp. 2227–2236, 2014.

# 5

# Effect of Doping Metals on the Structure of PEO Coatings on Titanium

**Nykolay D. Sakhnenko** ⓘ, **Maryna V. Ved'** ⓘ, **and Ann V. Karakurkchi** ⓘ

*National Technical University "Kharkiv Polytechnical Institute", Kyrpychova St. 2, Kharkiv 61002, Ukraine*

Correspondence should be addressed to Ann V. Karakurkchi; anyutikukr@gmail.com

Academic Editor: Eric Guibal

The structure and properties of the oxide films formed on titanium alloys by means of plasma-electrolytic oxidizing in alkali electrolytes based on pyrophosphates, borates, or acetates of alkali metals with the addition of dopants' oxides or oxoanions of varying composition have been studied. Anodic polarization in the spark discharge (microarc) mode at application of interelectrode potential 90 to 160 V has been used to obtain mixed-oxide systems $TiO_x \cdot WO_y$, $TiO_x \cdot MoO_y$, $TiO_x \cdot ZrO_2$, and $TiO_x \cdot V_2O_5$. The possibility to obtain the oxide layers containing the alloying elements by the modification of the composition of electrolytes has been stated. The chemical and phase composition as well as the topography, the microstructure, and the grain size of the formed layers depend on the applied current, interelectrode voltage, and the layer chemical composition. The effect of formed films composition on the resistance of titanium to corrosion has been discussed. Catalytic activity of mixed-oxide systems was determined in the model reaction of methyl orange dye MO photodestruction.

## 1. Introduction

Intensive economic activity and increase in the production capacities of different sectors of the industry lead to the growth of the pollution in air and water basins by toxic substances with different nature and chemical stabilities. Given this, the organization of removal of natural and technogenic contaminators from the air and aqueous medium is impossible without effective and accessible catalyst application [1]. Catalytic materials based on titanium oxide found application in the heterogeneous catalysis and especially photocatalysis [2–4]. The introduction of additional components into the composition of oxide layers makes it possible to improve the functional properties and activity of the catalyst and the scope of its application [5, 6].

Better technological forms of a catalyst are the thin-film coatings, formed directly on the metal substrate by the method of plasma-electrolytic oxidizing (PEO) [7]. The anodic oxidation in the microarc mode allows formation of different types of titanium oxides and incorporation of electrolyte components into the oxide layers [8, 9]. Such mixed metal-oxide systems differ with high adhesion to metal substrate and have developed the surface. On the surface relief protrusions of coatings, there are a significant number of active catalytic centers. They are oxides of doped metals, differing in physical and chemical properties and affinity for oxygen [7, 10].

The use of electrolytes of different compositions and varying conditions for titanium alloys PEO treatment allows obtaining coatings doped with transition [11, 12], refractory [13, 14], rare metals, and dispersed oxides [15, 16]. So, the studies targeted at improvement of the techniques used for the formation of mixed-oxide coatings on titanium alloys are of great interest.

At the same time, issues of managing the PEO coatings morphology and selection of dopants and their influence on the composition and properties of the formed oxide layers remain unresolved. A successful solution of these problems creates the prerequisites for the formation of coatings that possess an improved resource, a proper chemical resistance to the aggressive media action, prescribed morphology, and catalytic activity [5, 17, 18].

TABLE 1: Electrolytes for PEO.

| Electrolyte number | 1 | 2 | 3 | 4 |
|---|---|---|---|---|
| Electrolyte components | $K_4P_2O_7$ $MoO_3$ $Na_4B_2O_7$ | $K_4P_2O_7$ $V_2O_5$ | $K_4P_2O_7$ $Na_2WO_4$ $Na_4B_2O_7$ | $K_4P_2O_7$ $ZrO_2$ $Na_3Cit$ |

TABLE 2: PEO parameters.

| Electrolyte number | 1 | 2 | 3 | 4 |
|---|---|---|---|---|
| Current density, $i$ (A/dm$^2$) | | 1.0–5.0 | | |
| Sparking voltage, $U_s$ (V) | 90–110 | 100–120 | 140–160 | 110–120 |
| Maximum voltage, $U_{max}$ (V) | 180–190 | 190–200 | 210–220 | 230–240 |

This work presents certain results of investigation on the features of oxide coatings forming on the surface of titanium alloys using plasma-electrolytic oxidizing in alkali complex electrolytes containing oxides or oxoanions of doping metals and studying the composition and morphology of the synthesized mixed-oxide systems.

## 2. Materials and Methods

*2.1. Materials.* The formation of mixed-oxide systems $TiO_x·MO_y$ (M = Mo, W, V, and Zr) was carried out on the titanium alloy VT1-0 by plasma-electrolytic oxidizing.

Electrochemical treatment performed in a thermostatic cell in aqueous electrolyte solutions based on pyrophosphates, borates, and acetates of alkali metals with the addition of dopants' oxides or oxoanions is shown in Table 1. Working solutions for the research were prepared using certified reagents of the grade "chemically pure" and distilled water.

A preliminary treatment of the specimens included the mechanical purification from process impurities, degreasing in 0.2–0.3 M NaOH, etching in the mixture of 0.1–0.3 M HF and 0.3–0.9 M $HNO_3$, and the distilled water flushing.

*2.2. Synthesis Methods.* The oxidation was carried on in the galvanostatic mode using the stabilized DC source B5-50 (Ukraine). The process was carried out at current densities of 1.0–5.0 A/dm$^2$ and total voltage up to 250 V (Table 2) in a thermostatic cell with vigorous stirring of the electrolyte and flow-through circulation cooling. The processing time was 30–90 minutes.

*2.3. Methods of the Study.* The specimen surface was studied using the scanning electron microscope ZEISS EVO 40XVP. The surface pattern was obtained by recording secondary electrons and scanning the surface with the electron beam that enabled the investigation of surface morphology with a high resolution and contrast range. The images were processed using the SmartSEM software environment. Chemical composition of the surface was defined by the analysis of characteristic X-ray spectrum that was recorded by the INCA Energy 350 electron probe microanalysis integrated into the system of the SEM. The X-ray was excited by the radiation of specimens with the electron beam of 15 kV.

In addition, the element composition of coatings was defined using the X-ray fluorescent method and portable all-purpose commercial X-ray spectrometer "SPRUT"; a relative standard deviation was in the range of $10^{-3}$ to $10^{-2}$.

The surface roughness of coatings was defined using the contact method and the scanning probe microscope AFM

NT-206 (Microtestmashine Co, the Republic of Belarus). The specimens were scanned using the CSC-37 probe (cantilever B, the lateral resolution of 3 nm) at least at the three points on the surface in different sections of the sample for data averaging. A size of crystallite grains and a level of the roughness of surface coatings were defined using the obtained 2D and 3D surface charts and the cross section of the surface.

A corrosion behavior of the titanium alloys with oxide coatings was studied by the electrode impedance spectroscopy method using the automatic AC bridge P-5083 at the fixed frequencies of 1 and 10 kHz and in the frequency range of 20–1·10$^5$ Hz [19] in the solution of 0.1 M NaCl. The measurements taken in compliance with the series circuit were realized using auxiliary electrodes, in particular a co-axially arranged platinum grid or coplanar plates made of stainless steel.

The catalytic activity of conversion coatings on titanium was determined in the model photodestruction reaction of methyl orange dye MO in a thermostatic reactor at a temperature of 25°C and constant mixing. The reactor was filled with a solution with a dye concentration of $2 \times 10^{-2}$ g/dm$^3$, and the catalyst was placed and then kept in the dark for 24 hours to establish adsorption equilibrium. The irradiation was carried out with a fluorescent lamp of DeLux EBT-01, giving a soft ultraviolet, and oxygen or hydrogen peroxide was used as an oxidizing agent. The MO concentration was determined at the same intervals by the calorimetric method. The processing of the measurement results and the determination of the degree of photodegradation, as well as the apparent reaction rate constant k′, were carried out according to the developed algorithm [20].

## 3. Results and Discussion

Previous studies have shown that obtaining oxide-metallic coatings with a high-content of dopants was possible in the process of one-stage plasma-electrolytic oxidation of titanium [12, 21] and aluminum alloys [22] when using complex electrolytes.

The synthesized oxide materials did not require further treatment and possessed a broad range of functional properties, including catalytic activity in heterogeneous redox reactions. This approach to catalytic coating formation on the titanium alloy surface underlies the present research as a working hypothesis.

It was established that voltage chronograms (Figure 1) in studied electrolytes (Table 2) have a similar form, but they differ somewhat from the classical one [17]. The growth of

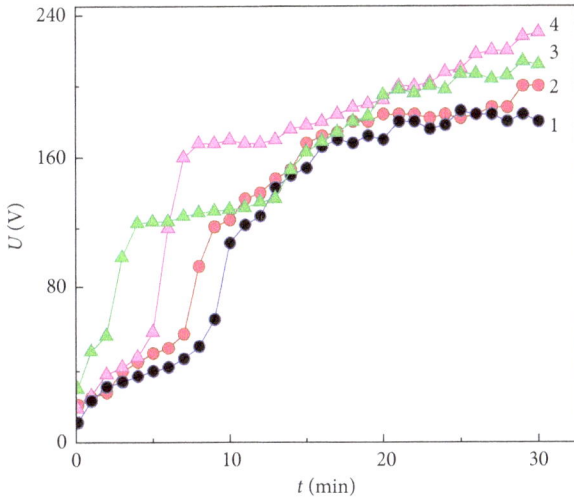

FIGURE 1: Chronograms of the forming voltage at PEO with current density 2 A/dm² for oxide coatings. 1: $TiO_x \cdot MoO_y$; 2: $TiO_x \cdot V_2O_5$; 3: $TiO_x \cdot WO_y$; 4: $TiO_x \cdot ZrO_2$.

TABLE 3: Specific electrical resistance and thermal stability of oxides [23].

| Metal | Oxide | Specific electrical resistance $\rho$ ($\Omega \cdot$cm) at 293 K | Thermal stability |
|---|---|---|---|
| Mo | $MoO_2$ | $8.8 \cdot 10^{-5}$ metallic conductivity | $T > 2100$ $3MoO_2 \rightarrow Mo + 2MoO_3$ |
| | $MoO_3$ | $10^4$–$10^8$ semiconductor | — |
| W | $WO_2$ | $3 \cdot 10^{-3}$ $p$-type semiconductor | $T > 2150$ $3WO_2 \rightarrow W + 2WO_3$ |
| | $WO_3$ | $10^3$–$10^5$ semiconductor | At a temperature above 1098 K sublimes |
| V | $VO_2$ | $8 \cdot 10^{-4}$ $n$-type semiconductor | — |
| | $V_2O_5$ | $4 \cdot 10^{-3}$ $n$-type semiconductor | 1000–1550 K $2V_2O_5 \rightarrow 4VO_2 + O_2$ $(V_6O_{13})$ |
| Zr | $ZrO_2$ | $10^{13}$ dielectric | — |

the forming voltage at the initial stage of the process is extremely slow. On the voltage chronograms, a gently sloping plot is observed. Obviously, such characters of $U, t$ dependencies is associated with the competition of the direct process of phase titanium oxide formation and reverse one—its chemical dissolution. Upon reaching a voltage of 50–55 V, all dependences exhibit a sharp almost linear rise, corresponding to the formation of an oxide layer.

The voltage growth again significantly slows with the onset of film breakdown, and on the $U, t$ dependencies, we can see plateaus. It should be noted that both the level of first-limiting voltage $U_1$ and time interval for its stabilization depend on the nature of the dopant. Obviously, reactions of coating defects formation and healing, as well as electrolyte components incorporation in the oxide layer, balance on these plateau sites of dependences. The voltage $U_1$ for the systems $TiO_x \cdot (Mo\ V\ W)O_y$ is at the level 110–120 V and for $TiO_x \cdot ZrO_2$ is higher–160 V.

The microarc region is characterized by a considerable number of oscillations with a general trend of increasing the voltage to 200–220 V. This is obviously due to the active inclusion of electrolyte components in the coating composition, which forms compounds with different oxidation degrees and, accordingly, different resistivity, and thermal stability (Table 3).

For all oxide coatings, the dependences of the voltage change rate are similar in shape (Figure 2). In the prespark area, they are characterized by a slight decrease with a minimum in the range of 45–50 V. It is related to the balancing of the processes of formation and dissolution of titania [12].

With an increase in voltage, there is a rapid increase in the rate of voltage change, reflecting the predominance of the oxide layer formation and growth over its dissolution. A further decrease in $dU/dt$ is evidently a consequence of the breakdown of the oxide film, as well as of the accompanying processes of electrolyte components incorporation and their high-temperature transformations in the sparking zone.

At the same time, the microarc region is characterized by process instability and the appearance of $dU/dt$ oscillations. It is associated with competition and a stochastic distribution of processes breakdown, healing for a complex oxide system.

The data of micro-X-ray spectral analysis and the scanning electron microscopy show inclusion of electrolyte components remelt into $TiO_2$ coatings. The surface morphology strongly depends on the content of alloying elements in mixed-oxide coatings. Figure 3 depends to a larger extent on the nature of the dopant compound. Uniform low-porosity oxide coatings with a doping component content $\omega$ of (% by weight) Mo: 3.0, V: 4.0, and Zr: 4.0 were formed in solutions based on dispersed metal oxides.

Analysis of the topography and components surface distribution for coatings $TiO_x \cdot WO_y$ obtained by PEO in electrolyte solutions based on tungstate of different concentrations allows us to conclude, that picks of the clusters are enriched in tungsten (Figure 4(a)), and the valleys matrix is characterized by maximal content of titanium (Figure 4(b)). The coatings have a tubular microporous structure, a characteristic of titanium nanotubes, and the surface is covered with toroidal rings.

It should be noted that varying the concentration of sodium tungstate in solution and increasing the current density make it possible to obtain coatings of different dopant contents, which also differ in relief (Figure 5).

With the increase in the tungsten content in the coating, large clusters disappear, and a more perfect microglobular structure with rounded grains of similar diameter (up to 1 $\mu$m), as well as toroidal structures, are formed. When the content of tungsten in oxides is increased by more than 50% by weight (10 at.%) (Figure 6(a)), the surface is covered by a network of microcracks due to high internal stresses associated with the concentration of the refractory metal and with high applied current density (Figure 6(b)).

Proceeding from the specifics of the synthesized materials application as catalytically active components of the complex systems at elevated temperatures and the effect of various fields, analysis of the composition and morphology

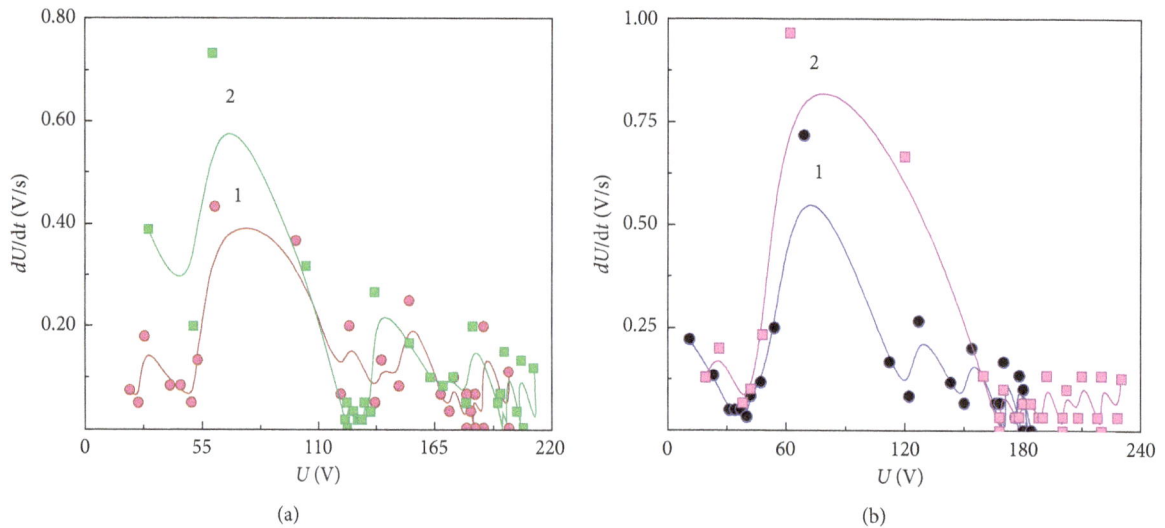

FIGURE 2: Rate of voltage change when forming coatings: (a) 1: $TiO_x \cdot V_2O_5$; 2: $TiO_x \cdot WO_y$. (b) 1: $TiO_x \cdot MoO_y$; 2: $TiO_x \cdot ZrO_2$.

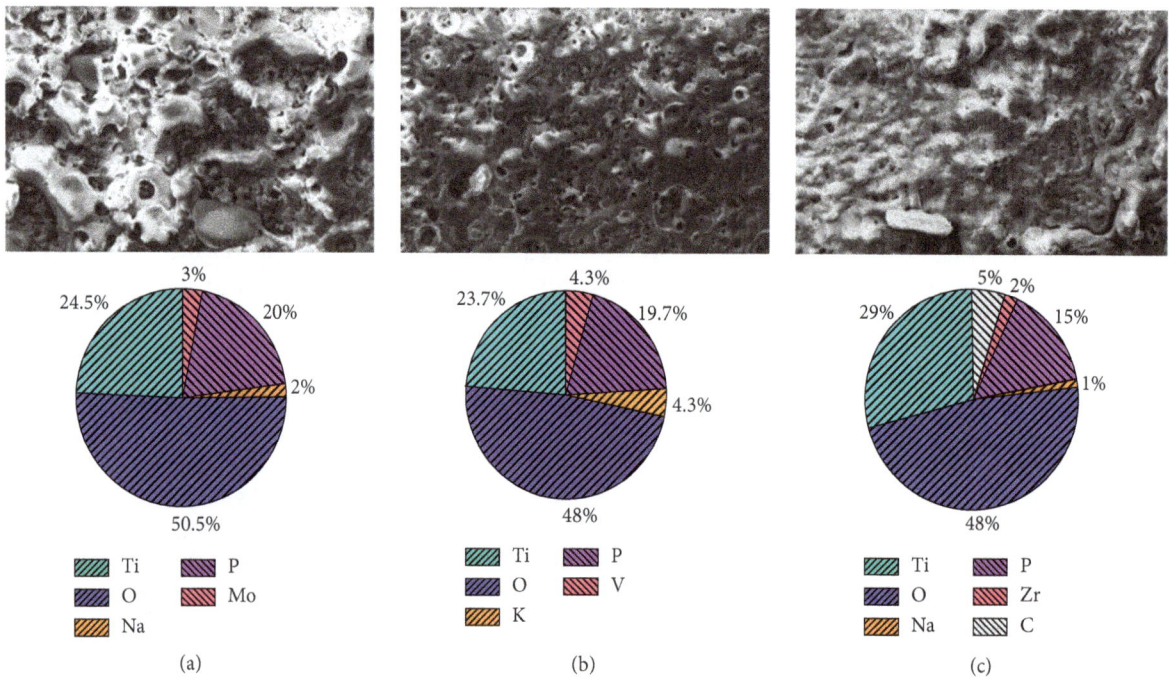

FIGURE 3: Surface microphotographs (magnification ×500) and composition (% by weight) of mixed-PEO coatings: (a) $TiO_x \cdot MoO_y$; (b) $TiO_x \cdot V_2O_5$; (c) $TiO_x \cdot ZrO_2$.

of the surface after thermal treatment at 450°C for 6 hours was very informative. As seen from Figure 7, heat treatment leads to the enlargement of the grains dimensions and an increase in the surface roughness, as well as a decrease in the oxygen content.

So, the quantitative composition of the coatings, as well as the surface cluster nature, creates the prerequisites for high catalytic activity of mixed oxides [24], which can be increased by additional heat treatment.

The roughness is an indicator of surface quality and depends on the material processing method. So, it can be regarded as an additional indicator of the surface

development during oxidation [21, 25]. Estimating the topography of the titanium alloy VT1-0 surface oxidized in potassium diphosphate, we can conclude the coatings to be unevenly rough (Figure 8).

Analysis of the cross section of grain between markers 1 and 2 (Figure 8(b)) indicates that the grain size varies between 400 and 500 nm, and the height of the parabolic picks of the relief is 100–400 nm (Figure 8(c)).

The topography of the $TiO_2 \cdot ZrO_2$ system surface (Figure 9(a)) differs significantly from the previous one: it has a fine-crystalline structure with the maximum degree of development in the series of materials considered.

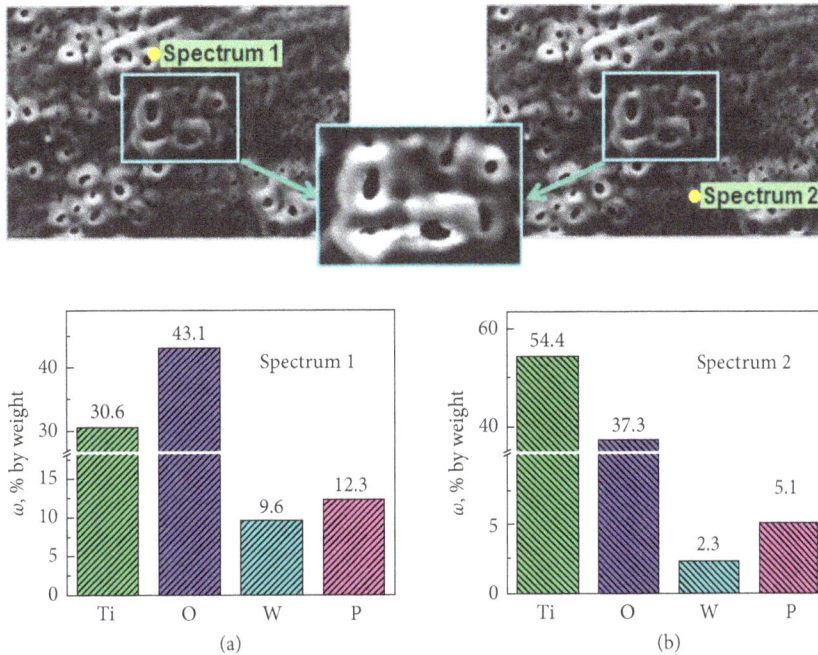

FIGURE 4: Surface morphology (×2000) and the elemental composition of $TiO_x \cdot WO_y$ oxide coatings: (a) at picks of clusters; (b) at valleys.

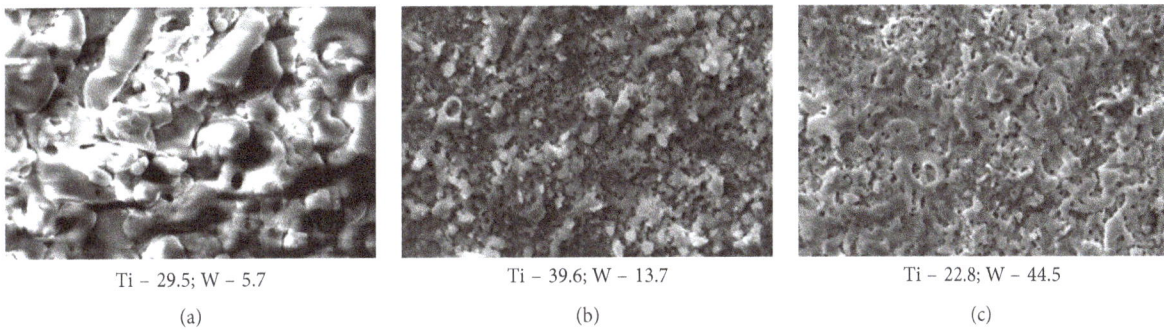

Ti – 29.5; W – 5.7      Ti – 39.6; W – 13.7      Ti – 22.8; W – 44.5

(a)          (b)          (c)

FIGURE 5: Surface morphology and the elemental composition (% by weight) of the $TiO_x \cdot WO_y$ oxide PEO coatings, obtained at different current densities ($i$, A/dm$^2$): (a) 1.5; (b) 2.5; (c) 4.0. Magnification ×2000.

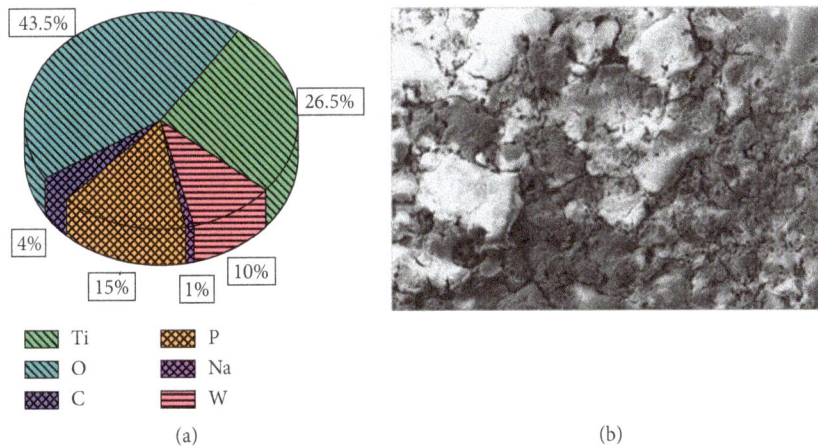

FIGURE 6: Oxide coatings $TiO_x \cdot WO_y$, obtained at $i = 5.0$ A/dm$^2$: (a) elemental composition, at.%; (b) surface morphology. Magnification ×200.

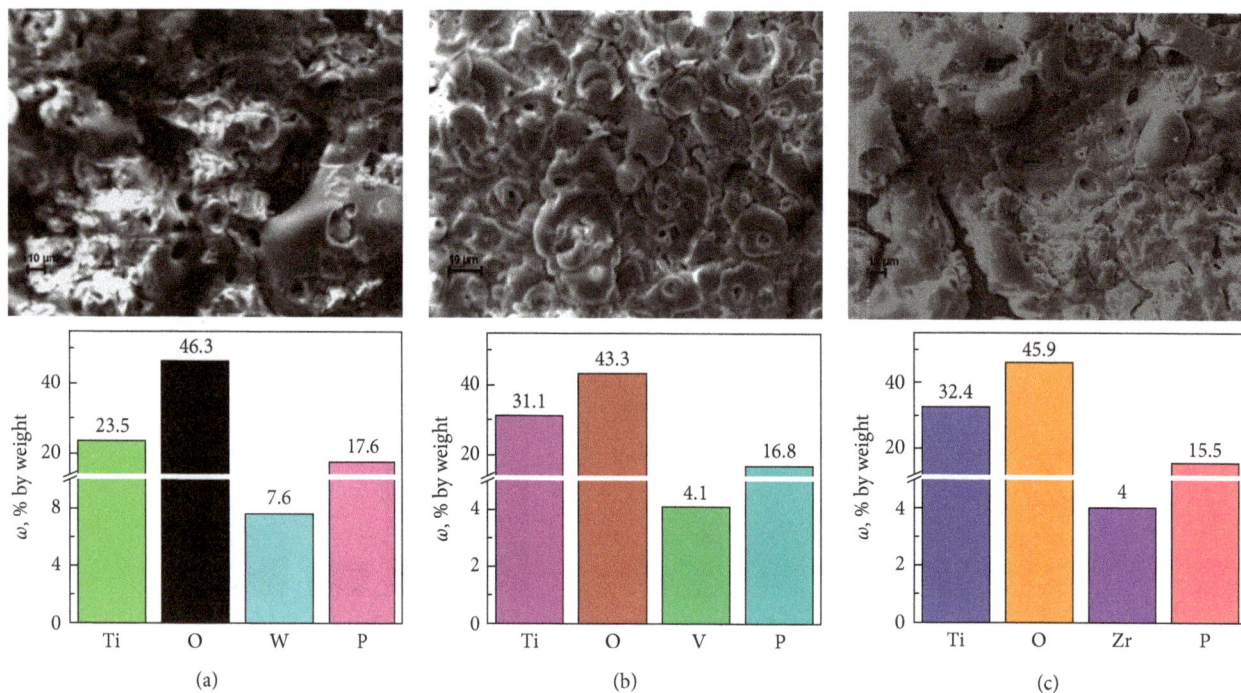

FIGURE 7: Surface morphology and elemental composition of mixed-oxide coatings TiO$_x$·MO$_y$ after heat treatment: (a) TiO$_x$·WO$_y$; (b) TiO$_x$·VO$_y$; (c) TiO$_x$·ZrO$_y$. Magnification ×500.

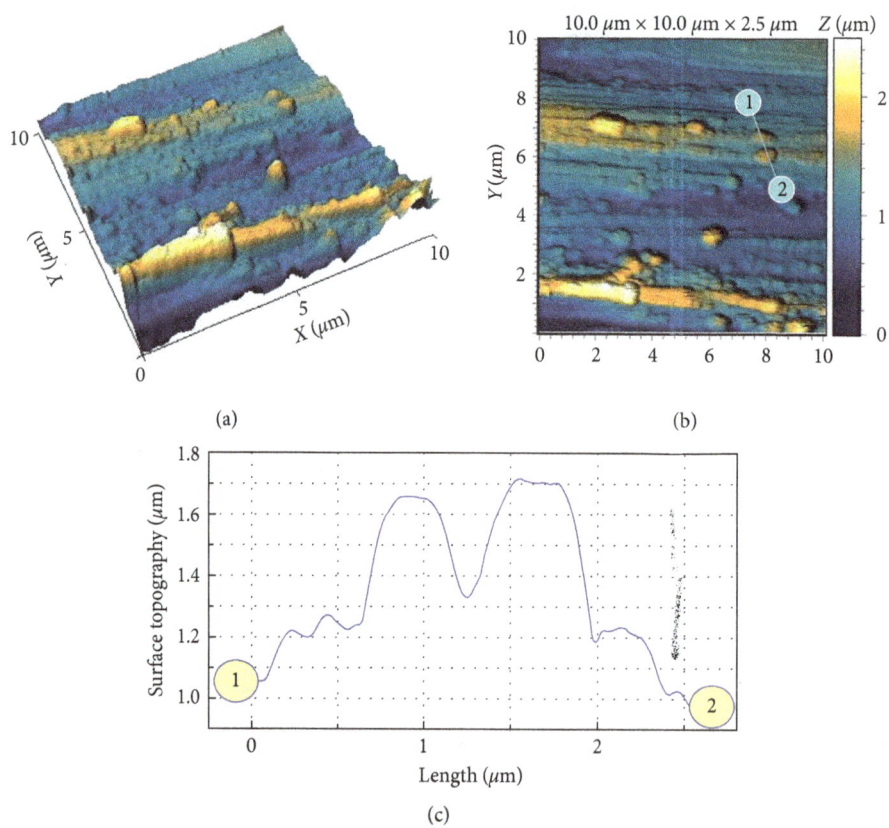

FIGURE 8: 3D (a) and 2D maps (b) of the surface and cross-sectional profile between markers 1 and 2 (c) for Ti│TiO$_2$ coating, obtained from 1 M K$_4$P$_2$O$_7$, $i = 1.5$ A/dm$^2$, $U = 60$ V, $t = 25$°C. Scanning area AFM $10 \times 10$ μm.

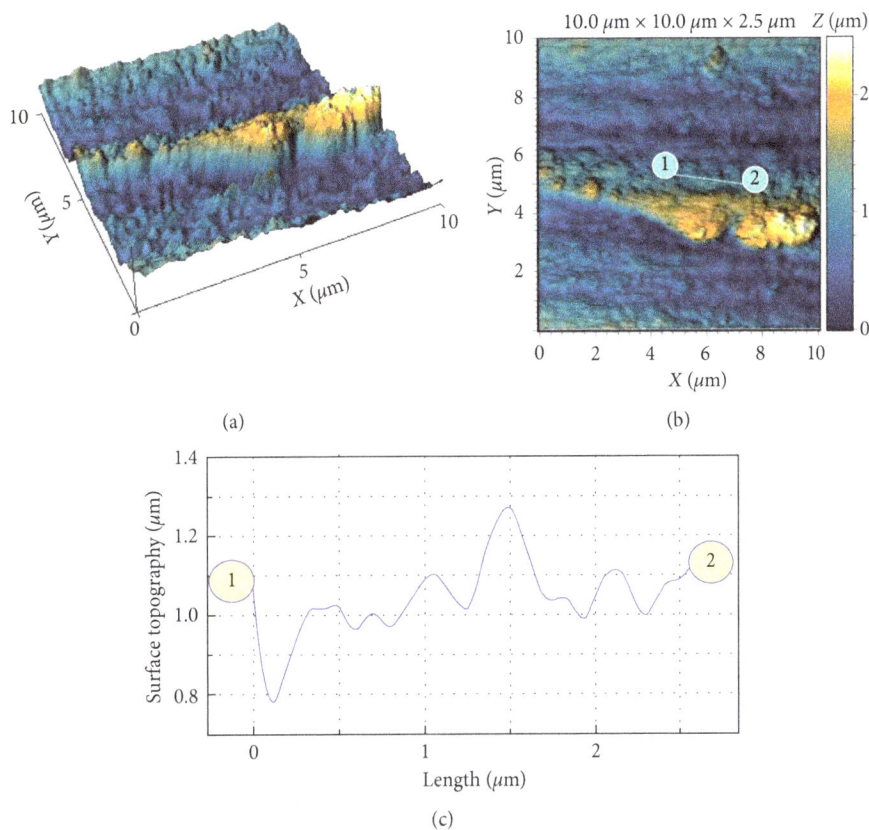

FIGURE 9: 3D (a) and 2D maps (b) of the surface and cross-sectional profile between markers 1 and 2 (c) for coating Ti│TiO$_2$·ZrO$_2$. Scanning area AFM $10 \times 10\,\mu$m. Content of Zr 2.1% by weight.

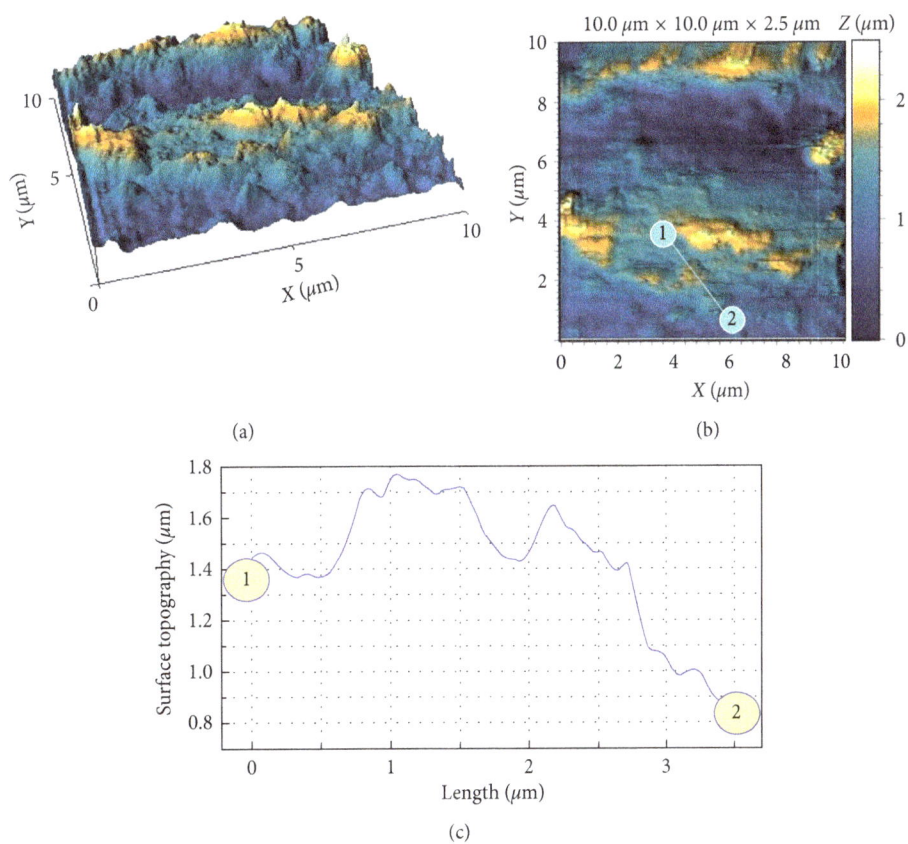

FIGURE 10: 3D (a) and 2D maps (b) of the surface and cross-sectional profile between markers 1 and 2 (c) for Ti│TiO$_2$·ZrO$_2$ coating after calcination. Scanning area AFM $10 \times 10\,\mu$m. Content of Zr 2.1% by weight.

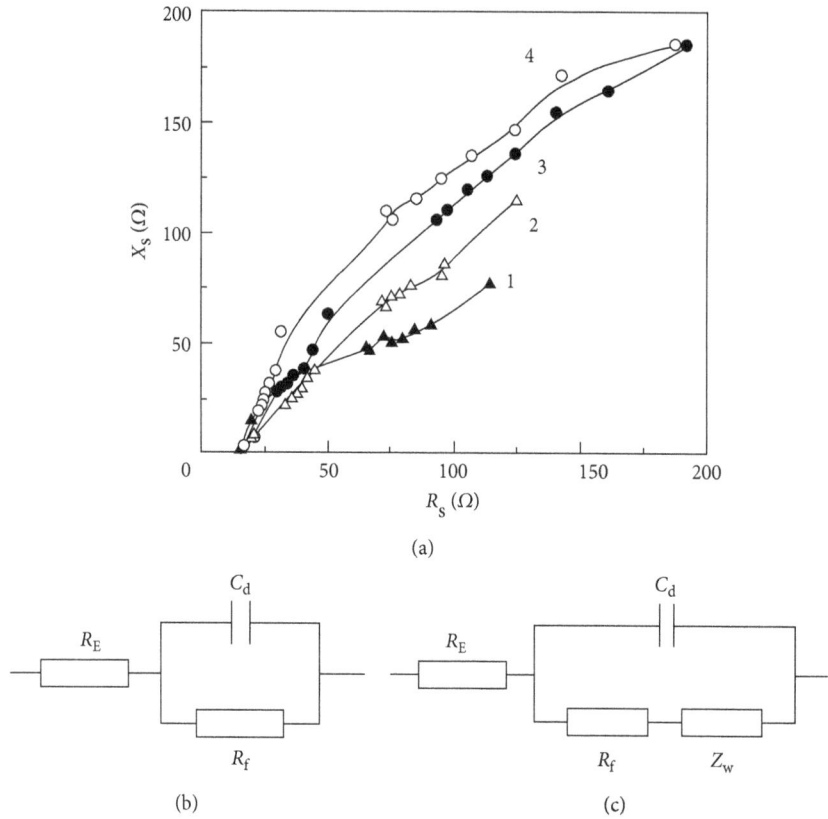

FIGURE 11: Nyquist plots (a) of mixed-oxide systems: 1: $TiO_x \cdot WO_y$; 2: $TiO_x \cdot MoO_y$; 3: $TiO_x \cdot V_2O_5$; 4: $TiO_x \cdot ZrO_2$ and the corresponding equivalent substitution schemes for systems $TiO_x \cdot WO_y$ and $TiO_x \cdot MoO_y$ (b) and $TiO_y \cdot V_2O_5$ and $TiO_x \cdot ZrO_2$ (c).

The spread in the sizes of grains agglomerates is 150–300 nm and varies in height from 100 to 500 nm (Figure 9(b)). The cross-sectional profile of the $TiO_2 \cdot ZrO_2$ between markers 1–2 is characterized by a variety of grain shapes with a predominance of pointed crystals (Figure 9(c)). Heat treatment of mixed-oxide coatings also changes the topography of the surface (Figures 10(a) and 10(b)). The relief acquires a globular shape and becomes more uniform, the agglomerates sizes are reduced to 140–200 nm, and the grain sizes remain within 80–100 nm (Figure 10(c)). The spread of the picks and valleys along the height also decreases in comparison with the unheated material. The uniformly developed surface is one of the factors ensuring an increase in the catalytic activity of the oxide system after calcinations.

Thus, doping of titania with zirconium oxides leads to a change in the morphology of the surface and an increase in its specific area. Thermal treatment of mixed-oxide systems helps to reduce the size of agglomerates and grains and provides a uniformly developed surface. The combination of these factors is the methodological basis for increasing the catalytic activity of materials, especially in the application to photocatalysis [5, 10].

Corrosion tests were carried out for mixed systems $TiO_x \cdot V_2O_5$ and $TiO_x \cdot ZrO_2$. Nyquist plots constructed from the results of electrode impedance measurements (Figure 11, curves 3, 4) are fragments of semicircles, reflecting the kinetic control of the corrosion process.

TABLE 4: Corrosion parameters of samples with coatings: $TiO_x$ oxides of metals in solution 0.1 M NaCl.

| Coating | Corrosion rate | | Open circut potential, $E_{cor}$ (V) |
|---|---|---|---|
| | $i_{cor} \cdot 10^6$ (A/cm$^2$) | $k_h \cdot 10^4$ (mm/year) | |
| $TiO_x \cdot WO_y$ | 1.509 | 1.97 | −0.17 |
| $TiO_x \cdot MoO_y$ | 0.755 | 0.98 | −0.122 |
| $TiO_x \cdot V_2O_5$ | 0.377 | 0.49 | −0.104 |
| $TiO_x \cdot ZrO_2$ | 0.279 | 0.36 | −0.08 |

TABLE 5: Characteristics of the photocatalytic activity of oxide systems obtained at $i = 1.5$ A/dm$^2$.

| Coating composition | Electrolyte | Degree of destruction (%) | $k' \cdot 10^2$ (min$^{-1}$) |
|---|---|---|---|
| Ti \| $TiO_2$ | 0.5 M $H_2SO_4$ | 28.00 | 0.88 |
| Zr \| $ZrO_2$ | | 24.10 | 0.8 |
| | 1 M $K_4P_2O_7$ | 18.00 | 0.66 |
| Ti \| $TiO_x \cdot ZrO_2$ | 0.5 M $H_2SO_4$; 0.1 M $ZrO_2$ | 59.00 | 1.69 |
| | 1 M $K_4P_2O_7$; 0.1 M $ZrO_2$ | 53.92 | 1.45 |
| Ti \| $TiO_x \cdot V_2O_5$ | 1 M $K_4P_2O_7$; 0.1 M $V_2O_5$ | 68.90 | 1.89 |

The impedance plots of the samples with $TiO_x \cdot WO_y$ and $TiO_x \cdot MoO_y$ coatings containing metal oxides in the intermediate oxidation state (Figure 11(a), curves 1 and 2, resp.) consist of two sections: a semicircular fragment and a straight line with an inclination angle close to 45°. This form of dependence indicates a mixed diffusion-kinetic control of the corrosion process.

In Figures 11(b) and 11(c), equivalent substitution schemes for the considered mixed-oxide systems are presented. Scheme (Figure 11(b)) includes elements such as $R_E$, electrolyte resistance, and parameters of Faraday reactions on the other surface parts are visualized by the elements $C_d$ and $R_f$. The other scheme (Figure 11(c)) differs by diffusion Warburg impedance $Z_W$. The values of the current $i_{cor}$ and depth $k_h$ of the corrosion rate indicators, as well as the corrosion potential of $E_{cor}$, indicate a sufficiently high-corrosion resistance of the systems under study and allow the materials to be classified as very stable (Table 4). The coatings containing zirconium oxide have the highest protective properties. The results obtained are quite natural, since it is zirconium oxide which increases the resistance to pitting corrosion.

Thus, the method of plasma-electrolytic oxidizing of titanium allows synthesizing mixed coatings containing doped-metals oxides, differing in composition, porosity, surface roughness, and corrosion resistance in aggressive media [26].

Tests of the catalytic activity of mixed-oxide systems show that the MO dye oxidation rate and degree of photodestruction at Ti | TiO$_2$·ZrO$_2$ catalyst increases by 10% and at Ti | TiO$_2$·V$_2$O$_5$ by 30% relative to the pure titania coatings (Table 5) [26].

## 4. Conclusions

(1) The method of plasma-electrolytic oxidizing allows synthesizing uniform microporous conversion and composite coatings with a developed surface containing mixed-oxide systems TiO$_x$ with WO$_y$, MoO$_y$, ZrO$_2$, and V$_2$O$_5$. A distinctive feature of the synthesis in its technical implementation is the variability, due to the possibility of forming such coatings in electrolytes both on the basis of dopant oxoanions and their dispersed oxides.

(2) The influence of electrolyte nature and PEO parameters on the mixed-oxide composition, surface morphology and roughness, and consequently, functional properties is shown.

(3) Corrosion resistance of oxide systems increase in the range TiO$_x$·ZrO$_2$ > TiO$_x$·V$_2$O$_5$ > TiO$_x$·MoO$_y$ > TiO$_x$·WO$_y$.

(4) Analysis of the catalytic activity of mixed oxides and composite coatings in the model reaction of methyl orange oxidation showed not only their high efficiency but also the synergism. The established values of the quantitative characteristics of the photocatalytic degradation of a number of organic compounds provide grounds for asserting the prospects of using the systems studied to neutralize toxic reagents, in particular, such as formaldehyde, phenol, and their derivatives.

## Conflicts of Interest

The authors certify that they have no conflicts of interest.

## Acknowledgments

The authors acknowledge the Karpenko Physico-Mechanical Institute of the NAS of Ukraine for providing all the support during the study period.

## References

[1] S. L. Suib, *New and Future Developments in Catalysis: Catalysis for Remediation and Environmental Concerns*, Elsevier, New York, NY, USA, 2013.

[2] S. Bagheri, N. Muhd Julkapli, and S. Bee Abd Hamid, "Titanium dioxide as a catalyst support in heterogeneous catalysis," *The Scientific World Journal*, vol. 2014, Article ID 727496, 21 pages, 2014.

[3] A. Fujishima, K. Hashimoto, and T. Watanabe, *TiO$_2$ Photocatalysis: Fundamentals and Applications*, BKC, Tokyo, Japan, 1999.

[4] L. Lin, Y. Chai, B. Zhao et al., "Photocatalytic oxidation for degradation of VOCs," *Open Journal of Inorganic Chemistry*, vol. 3, no. 1, pp. 14–25, 2013.

[5] V. S. Rudnev, I. V. Lukiyanchuk, M. S. Vasilyeva, M. A. Medkov, M. V. Adigamova, and V. I. Sergienko, "Aluminum- and titanium-supported plasma electrolytic multicomponent coatings with magnetic, catalytic, biocide or biocompatible properties," *Surface and Coatings Technology*, vol. 307, pp. 1219–1235, 2016.

[6] H. Liu, L. Yu, W. Chen, and Y. Li, "The progress of TiO$_2$ nanocrystals doped with rare earth ions," *Journal of Nanomaterials*, vol. 2012, Article ID 235879, 9 pages, 2012.

[7] I. V. Lukiyanchuk, V. S. Rudnev, and L. M. Tyrina, "Plasma electrolytic oxide layers as promising systems for catalysis," *Surface and Coatings Technology*, vol. 307, pp. 1183–1193, 2016.

[8] X. H. Wu, Q. Wei, X. B. Ding, W. D. He, and Z. H. Jiang, "Dopant influence on the photo-catalytic activity of TiO$_2$ films prepared by micro-plasma oxidation method," *Journal of Molecular Catalysis A: Chemical*, vol. 268, no. 1-2, pp. 257–263, 2007.

[9] V. S. Rudnev, "Multiphase anodic layers and prospects of their application," *Protection of Metals*, vol. 44, no. 3, pp. 263–272, 2008.

[10] V. D. Binas, D. Kotzias, and G. Kiriakidis, "Modified TiO$_2$ based photocatalysts for improved air and health quality," *Journal of Materiomics*, vol. 3, no. 1, pp. 3–16, 2017.

[11] M. S. Vasilyeva, V. S. Rudnev, A. Y. Ustinov, I. A. Korotenko, E. B. Modin, and O. V. Voitenko, "Cobalt-containing oxide layers on titanium, their composition, morphology, and catalytic activity in CO oxidation," *Applied Surface Science*, vol. 257, no. 4, pp. 1239–1246, 2010.

[12] L. Yan, T. Lihong, T. Xinyu, L. Xin, and C. Xiaobo, "Synthesis, properties, and applications of black titanium dioxide nanomaterials," *Science Bulletin*, vol. 62, no. 6, pp. 431–441, 2017.

[13] M. R. Bayati, R. Molaei, A. Z. Moshfegh, and F. Golestani-Fard, "A strategy for single-step elaboration of V$_2$O$_5$-grafted TiO$_2$ nanostructured photocatalysts with evenly distributed pores," *Journal of Alloys and Compounds*, vol. 509, no. 21, pp. 6236–6241, 2011.

[14] S. Stojadinović, N. Radić, R. Vasilić et al., "Photocatalytic properties of $TiO_2/WO_3$ coatings formed by plasma electrolytic oxidation of titanium in 12-tungstosilicic acid," *Applied Catalysis B: Environmental*, vol. 126, pp. 334–341, 2012.

[15] S. Di, Y. Guo, H. Lv, J. Yu, and Z. Li, "Microstructure and properties of rare earth $CeO_2$-doped $TiO_2$ nanostructured composite coatings through micro-arc oxidation," *Ceramics International*, vol. 41, no. 5, pp. 6178–6186, 2015.

[16] M. Babaei, C. Dehghanian, P. Taheri, and M. Babaei, "Effect of duty cycle and electrolyte additive on photocatalytic performance of $TiO_2$-$ZrO_2$ composite layers prepared on CP Ti by micro arc oxidation method," *Surface and Coatings Technology*, vol. 307, pp. 554–564, 2016.

[17] N. D. Sakhnenko, M. V. Ved, and A. V. Karakurkchi, "Nanoscale oxide PEO coatings forming from diphosphate electrolytes," in *Nanophysics, Nanomaterials, Interface Studies, and Applications*, O. Fesenko and L. Yatsenko, Eds., pp. 159–184, Springer International Publishing, Basel, Switzerland, 2017.

[18] Y. Zhang, W. Fan, H. Q. Du, and Y. W. Zhao, "Microstructure and wearing properties of PEO coatings: effects of $Al_2O_3$ and $TiO_2$," *Surface Review and Letters* vol. 25, no. 8, article 1850102, 2017.

[19] M. V. Ved', N. D. Sakhnenko, and K. V. Nikiforov, "Stability control of adhesional interaction in a protective coating/metal system," *Journal of Adhesion Science and Technology*, vol. 12, no. 2, pp. 175–183, 1998.

[20] V. V. Bykanova, N. D. Sakhnenko, and M. V. Ved', "Synthesis and photocatalytic activity of coatings based on the $Ti_xZn_yO_z$ system," *Surface Engineering and Applied Electrochemistry*, vol. 51, no. 3, pp. 276–282, 2015.

[21] M. V. Ved', N. D. Sakhnenko, A. V. Karakurkchi, M. V. Mayba, and A. V. Galak, "Synthesis and functional properties of mixed titanium and cobalt oxides," *Advanced Functional Materials*, vol. 24, no. 4, pp. 534–540, 2017.

[22] G. S. Yar-Mukhamedova, M. V. Ved, A. V. Karakurkchi, and N. D. Sakhnenko, "Mixed alumina and cobalt containing plasma electrolytic oxide coatings," *IOP Conference Series: Materials Science and Engineering*, vol. 213, p. 012020, 2017.

[23] J. Rumble, *Handbook of Chemistry and Physics*, CRC, Boca Raton, FL, USA, 98th edition, 2017.

[24] A. E. R. Friedemann, K. Thiel, U. Haßlinger, M. Ritterd, M. Gesing, and P. Plagemann, "Investigations into the structure of PEO-layers for understanding of layer formation," *Applied Surface Science*, vol. 443, pp. 467–474, 2018.

[25] A. Karakurkchi, M. Sakhnenko, M. Ved, A. Galak, and S. Petrukhin, "Application of oxide-metallic catalysts on valve metals for ecological catalysis," *Eastern-European Journal of Enterprise Technologies*, vol. 5, no. 10, pp. 12–18, 2017.

[26] N. Sakhnenko, M. Ved, and A. Karakurkchi, "Morphology and properties of coatings obtained by plasma-electrolytic oxidation of titanium alloys in pyrophosphate electrolytes alloys," *Protection of Metals and Physical Chemistry of Surfaces*, vol. 53, no. 6, pp. 1082–1090, 2017.

# 6

# Coupling Solvent Extraction Units to Cyclic Adsorption Units

**Mariana Busto ⓘ,[1] Enrique Eduardo Tarifa ⓘ,[2] and Carlos Román Vera ⓘ[1]**

[1]*Institute of Research on Catalysis and Petrochemistry, INCAPE, FIQ-UNL, CONICET, Collecting Ring, National Road 168 Km 0, El Pozo, 3000 Santa Fe, Argentina*
[2]*Faculty of Chemical Engineering, Universidad Nacional de Jujuy, CONICET, Ítalo Palanca No. 10, San Salvador de Jujuy, Argentina*

Correspondence should be addressed to Carlos Román Vera; cvera@fiq.unl.edu.ar

Academic Editor: Jose C. Merchuk

The possibility of regenerating the solvent of extraction units by cyclic adsorption was analyzed. This combination seems convenient when extraction is performed with a high solvent-to-impurity ratio, making other choices of solvent regeneration, typically distillation, unattractive. To our knowledge, the proposed regeneration scheme has not been considered before in the open literature. Basic relations were developed for continuous and discontinuous extraction/adsorption combinations. One example, deacidification of plant oil with alcohol, was studied in detail using separate experiments for measuring process parameters and simulation for predicting performance at different conditions. An activated carbon adsorbent was regenerated by thermal swing, making cyclic operation possible. When extracting the acid with methanol in a spray column, feed = 4 L min$^{-1}$, solvent = 80 L min$^{-1}$, feed impurity level 140 mmol L$^{-1}$, and extract concentration 7.6 mmol L$^{-1}$, the raffinate reaches a purity of 1.2 mmol L$^{-1}$, the solvent being regenerated cyclically in the adsorber (364 kg) to an average of 0.7 mmol L$^{-1}$. Regeneration of the solvent by cyclic adsorption had a low heat duty. Values of 174 kJ per litre of solvent compared well with the high values for vaporization of the whole extract phase (1011 kJ L$^{-1}$).

## 1. Introduction

Liquid-liquid extraction (LLE) is an important technique of separation used in many applications of the chemical process industry. Distillation, the workhorse of separation processes, is based on boiling point differences; LLE instead is based on different relative solubilities of solutes in two immiscible, or partially miscible, liquids. Extraction is typically used in cases in which distillation is not cost-effective or directly not possible at all. This is the case when azeotropes are formed, or when volatility differences between components are too low, or when heat-sensitive materials are present that could decompose at the high temperatures of distillation. Also, if the component to be recovered has a very high boiling point or is present in very small concentrations, distillation is not cost-effective.

One of the most important steps in the design of LLE is the choice of the solvent, which must meet several criteria in order to achieve a maximum transfer rate: (i) a high solubility for the solute and low solubility for the feed/raffinate; (ii) a density difference with the carrier higher than 0.15 g cm$^{-3}$; (iii) a medium surface tension (5–30 dyne cm$^{-1}$); (iv) high resistance to thermal degradation when thermal regeneration is used; (v) a high boiling point and low viscosity, for ease of handling. It is readily apparent that not all criteria can be met and that a careful screening is needed to choose the best solvent from a given set.

One aspect not always conveniently stressed in LLE is that of solvent regeneration. This must be easy and energy-efficient. As a consequence, when the solvent is being chosen, it must be decided how is to be regenerated. Since most solvents are regenerated by distillation, aspects to be analyzed are selectivity, solute distribution, and volatility. Solvents that display high selectivity usually have low solute distribution coefficients. If they also have lower volatility than the impurity, the impurity can be recovered as a distillate. However, if the solvent has a lower boiling point than the impurity, then the solvent should be distilled off for regeneration. If the impurity distribution coefficient is low, then a high solvent-to-feed ratio is needed. High solvent recycle rates and high solvent regeneration rates are hence needed. For these systems, separation by distillation might not be an option, and other principle could be chosen. An option that has not received attention in the scientific literature or the industrial practice is that of solvent regeneration by adsorption.

The possibilities are analyzed in this work of a process using liquid-liquid extraction for removing impurities from a feed, and adsorption for solvent regenerating the solvent. This combination has not been previously discussed in the open literature. Main features from the kinetic, thermodynamic, and process point of view are considered and discussed; and the main parameters for designing such a process are written. Equations are revised for batch and continuous units involving local and global interphase mass transfer coefficients, and the range of practical values of these parameters for these two operations is discussed.

One example involving experimental work is used as proof-of-concept, deacidification of vegetable oils by extraction with alcohol, coupled to the cyclic adsorption of carboxylic acids from alcoholic solutions in an activated carbon packed column. The liquid-liquid extraction (LLE) of free fatty acids (FFA) from some plant oils has received some attention lately, because fatty acids can be recovered easily for further processing, the yield of neutral oil being maximum [1–3]. These are advantages of LLE over caustic refining in which oil is lost because of reaction with the caustic (saponification) and by emulsification, FFAs being converted to a difficult to handle soap-stock [4]. It has also advantages over physical refining (removal of acids by distillation at 250°C) that produces undesirable changes in color and can also produce the degradation of valuable nutraceuticals and antioxidants [5–7].

Applications of an extraction/adsorption process for refining plant oils could be useful in the biodiesel industry, the food industry, and the industry of biodegradable technical oils (lubricants, dielectric transformer oil, etc.).

## 2. Materials and Methods

The general procedure was as follows: (i) liquid-liquid thermodynamic equilibrium data was got for the solvent-feed-impurity system in the form of partition coefficients; (ii) solid-liquid thermodynamic equilibrium isotherms were obtained for the solvent-adsorbent-impurity system; (iii) kinetic parameters for extraction were obtained, in the form of global average $aK_L$ values (min$^{-1}$) for a column and stirred tank extractor; (iv) adsorption kinetic parameters for adsorption were obtained, in the form of global average linear driving force parameter ($K_{LDF}$) values (min$^{-1}$); (v) tests of adsorbent regeneration by thermal swing were made, measuring the relevant parameters; (vi) simulations were run for continuous and discontinuous units, varying process conditions.

*2.1. Materials.* Edible sunflower oil and oleic acid (Sigma–Aldrich 99% grade) were used as a source of triglyceride and fatty acid, respectively. Acidified solutions of plant oil of variable concentration were obtained by dissolving weighed amounts of oleic acid in sunflower oil. The solvent used, methanol, was supplied by Biopack (Buenos Aires, Argentina). The chemical purities were higher than 99%. All compounds were used without further purification. Activated carbon (Filtrasorb, Calgon Carbon) was used in this study. The carbon was conditioned upon receiving by boiling in deionized water for 1 hour, then drying in an oven at 110°C for 24 hours. The activated carbon had a BET specific surface area of 972 m$^2$ g$^{-1}$, a total pore volume of 0.68 mL g$^{-1}$, and a bulk density of 0.502 g mL$^{-1}$.

*2.2. Liquid Extraction Equilibrium.* The feed-impurity-solvent system was sunflower oil-oleic acid-methanol. The oleic acid was distributed between the sunflower oil (oil phase) and methanol (alcohol phase). The alcohol and oil phases were mostly immiscible. Experimental LLE data were obtained in a stirred tank reactor. This had an AISI 304 stainless steel vessel with 100 ml total volume, 40 mm of diameter, and 80 mm of length and a magnetic coupling between the motor and the stirrer. The tank was heated with a tubular furnace and the temperature was controlled with a Novus N1100 controller. The amounts of each component for preparing the solutions were determined by weighing on an analytical balance (Model Shimadzu AUW220D Dual Range Balance, 0.0001 g precision). The mixtures were vigorously stirred for 4 h and then left to rest for at least 12 h. This led to the formation of two clear and transparent phases, with a well-defined interface that were sampled for analysis.

The oleic acid concentration was determined by potentiometric titration (AOCS Method Ca 5a-40) with a microburet. The amount of methanol in the oil phase was determined by weighing the liquid before and after evaporating the solution (80°C, 300 mmHg vacuum). The amount of oil in the methanol phase was determined from a mass balance of the previous components. The analysis was repeated at least three times, and the average of these readings was taken as the liquid phase composition.

*2.3. Extraction Kinetics.* Values of the average mass transfer coefficient on the solvent side $aK_{MeOH}$ were calculated from extraction tests in two kinds of extractors: a spray column and a laboratory stirred tank reactor. Coefficients for the column were obtained from single drop experiments using the methodology of Azizi et al. [24].

In the stirred tank tests, the technique of Schindler and Treybal [18] was followed. A stirred tank was used that had the same flange and stirrer as the extraction tests. The internal volume, diameter, and length were also the same as in Section 2.2. The only difference was that the tank had two additional connections for continuous operation. The flowrates of solvent and feed were controlled with peristaltic pumps. The oleic acid concentration in the raffinate and extract phases were determined by titration after adequate settling and formation of two distinct separate phases.

*2.4. Adsorption Equilibrium.* Adsorption isotherms were measured in a continuously stirred tank batch reactor. The method chosen was that of solid addition in which different

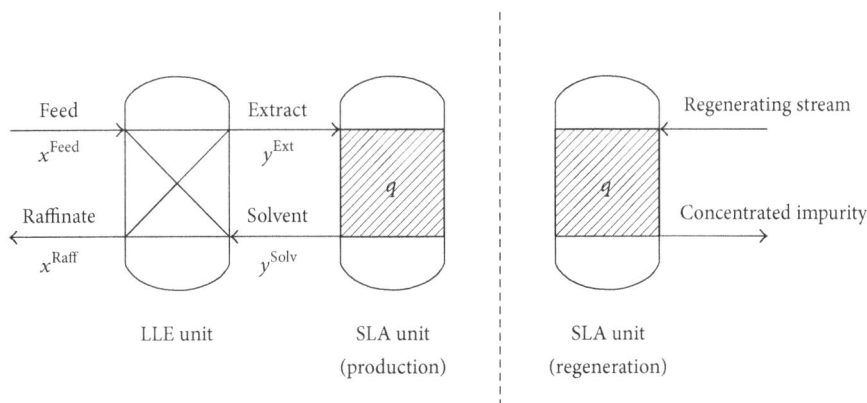

FIGURE 1: Scheme of liquid-liquid extraction (LLE) unit coupled to solid-liquid adsorption (SLA) unit.

amounts of adsorbent in powder form (about 200 meshes) were allowed to reach equilibrium. The stirring rate was kept at 1600 rpm, and the temperature was kept at 30 and 40°C. The acidity of the methanol solution was determined by potentiometric titration using the average of two measurements. This technique had an average error of 0.69%. Concentration of oleic acid in the solid was determined by a mass balance.

*2.5. Adsorption Kinetics.* Kinetics of adsorption over the activated carbon in pellet form were measured in a packed bed column with recycle. The mass of the bed was 1 g, and the flowrate of the extract (solvent with dissolved oleic acid) was 7 L h$^{-1}$. The carbon particle size was 35–60 meshes. The mass of the liquid phase was 40 g and the test lasted 2 h. Samples were taken periodically and oleic acid content of the liquid phase was measured by titration of the acidity as indicated above.

*2.6. Settling Tests.* Tests of settling rates were made for the sunflower oil-methanol system by vigorously stirring mixtures of varying solvent-to-oil ratio, then being allowed to rest at three different temperatures, 25, 40, and 50°C. The time was recorded when two distinctive phases were formed and no oil remained in suspension in the upper phase.

## 3. Results and Discussion

*3.1. Theoretical Analysis.* A scheme of the proposed combination of operations is included in Figure 1. The LLE unit can be a batch or continuous mixer/settler unit, a countercurrent, or cocurrent contact column. The solid-liquid adsorption unit (SLA) can be a bleaching stirred tank or a packed adsorbent column. The latter seems better suited for the proposed combination because it allows an easy separation of the solvent and an easy regeneration of the adsorbent.

The successful matching of the SLA and LLE units seems to relay on the adequate design of the equipment and the choice of solvent and adsorbent. The *feed* (with impurity concentration $x^{\text{Feed}}$) is mixed with the *solvent* (with impurity concentration $y^{\text{Solv}}$) and leaves the contactor, with a lower concentration of impurity, as the *raffinate* stream

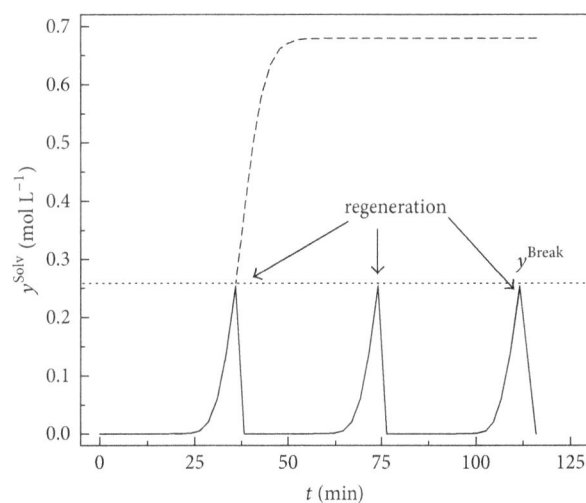

FIGURE 2: SLA unit operated to saturation (dashed line) or in cyclic mode with intermediate regeneration (solid line). Impurity concentration in the fluid phase at the adsorption column outlet, $y^{\text{Solv}}$, as a function of time. Continuous operation. $y^{\text{break}}$ = value of $y^{\text{Solv}}$ at regeneration time.

(of impurity concentration $x^{\text{Raff}}$). The *extract* (with impurity concentration $y^{\text{Ext}}$) that leaves the unit must then be fed to the adsorbent column. This column is operated in production mode until the impurity concentration in the exit reaches a limit value ($y^{\text{break}}$). This is called the breakthrough point. At this point, the feed is stopped, and the column is put into regeneration. The regeneration step can be typically of the thermal type, the column being flushed with a hot fluid to desorb the impurity. For example, a stream of hot solvent can be used, the volume of solvent for regeneration being conveniently small. Figure 2 shows a plot of the concentration of the impurity at the SLA unit exit. The dashed line corresponds to the outlet concentration for the case in which the column is operated to saturation with no intermediate regeneration.

It must be noted that when the column becomes saturated the exit concentration becomes equal to $y^{\text{Ext}}$, the

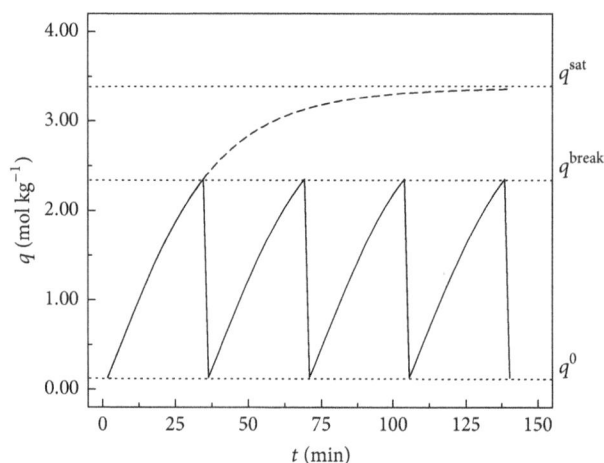

FIGURE 3: Concentration of impurity in the solid phase in an adsorbent column with cyclic regeneration.

concentration at the inlet. The concentration of impurity in the regenerated column is usually not zero, because total regeneration can be rather costly and not practically necessary. As seen in Figure 2, the column also yields a regenerated solvent with variable purity. However, this is not a problem provided $y^{break}$ is conveniently low.

A similar kind of plot can be seen in Figure 3 but this time the average concentration of the impurity in the solid phase is plotted as a function of time. Again, it can be seen that the column is cycled at a $q^{break}$ point different to the saturation value. This is because at saturation mass transfer kinetics become too slow due to the decrease in the driving force.

While columns have both a space and time-dependent concentration profiles, other liquid-liquid contactors do not. Stirred tanks with adsorbent in suspension or short packed columns operated with a high liquid recycle have a practically uniform concentration with respect to the spatial coordinate and vary only as a function of time.

The adequate design of the LLE and the SLA units should meet some obvious criteria: (i) the period of operation/regeneration of the column should not be too short, and a minimum value should be specified, for example, 10–40 min; (ii) the adsorbent should have an adequate capacity and adequate adsorption kinetics, with an adequate utilization of the total surface at the point of breakpoint, for example, 40–70% saturation. The latter is usually a problem for most adsorbents because the internal surface area is very high for materials with small pores, and the mass transfer intrapellet resistance limits the access to the inner pore volume.

A comparison of the mathematical expression for liquid-liquid and solid-liquid equilibrium is necessary for understanding the nature of both phenomena. The same can be said for the kinetic expressions for mass transfer between the two phases, either liquid-liquid or liquid-solid. Rather than working with general expressions, the expressions will be written for the practical example: the system of extraction

of oleic acid from acidic sunflower oil with methanol and adsorption of oleic acid from methanol over activated carbon. For simplification, the solvent and the feed are supposed to be practically immiscible and that Nernst law is always valid. There is also no reaction involved. For the column equations, plug flow of the individual phases is assumed.

Equation (1) in Table 1 is an example of the isotherm equation for a solid-liquid-adsorbate system in equilibrium and depicts the equilibrium concentration of the impurity on the solid as a function of the concentration in the liquid phase. The function used is that of the Langmuir isotherm. Equation (2) is Freundlich isotherm. Equation (3) is the definition of the partition coefficient for the impurity between the raffinate and extract phases, according to the Nernst law. The coefficient $m$ is a complex function of fluid thermodynamic properties. Nernst law is deduced for low concentrations of solute but can be applied to solutions of higher concentration, though its validity is reduced to a narrower range.

Equations (4)–(6) are the equations for the flux densities (moles per unit area and time) through the liquid-solid interface, while (7)–(9) are the equations corresponding to the transfer to the liquid-liquid interphase. Equations (7)–(9) correspond to the double film model, while (4)–(6) correspond to transfer due to Fickian diffusion on the porous solid side and film diffusion on the liquid side. In (7)–(9), the underlying hypotheses of the double film model apply, that is, the liquid phases are separated by an interface and one film in each phase adheres to this interface. The mass transfer takes place exclusively in this double stagnant film by a molecular diffusion mechanism. In the bulk of each phase the concentration of the impurity is uniform due to perfect mixing.

Equations (10)–(12) and (13)–(16) of Table 1 correspond to flux equations in terms of driving forces and overall mass transfer coefficients. The former are the differences between equilibrium and actual values of concentration at any point in time. In the case of adsorption, the definition of global mass transfer coefficient resembles that of the linear driving force model, $K_{LDF}$, and hence it will be used as such. The LDF model was first proposed by Glueckauf and Coates [25] as an approximation to mass transfer phenomena in adsorption processes in the gas phase but has been found useful to model adsorption in packed beds because it is simple and consistent both analytically and physically [26]. Several authors have inspected the nature of $K_{LDF}$. Ruthven and Farooq [27] considered that it is composed of two contributions, related to the intrapellet mass transfer resistance ($R_D$) and the film mass transfer resistance ($R_f$), the explicit formulation being that of (17). While $R_f$ depends on the diffusivity of the impurity in the fluid phase, $R_D$ depends on the diffusivity of the impurity inside the porous matrix of the solid adsorbate. Hence, in most cases, and particularly in adsorption in liquid phase, $R_D$ is the highest resistance and $R_f$ can be neglected.

Equations (18) and (19) depict the relations between the local liquid-liquid mass transfer coefficients and the overall coefficients. The latter can be expressed in terms of driving forces in the raffinate or extract side leading to two different coefficients. For the local coefficients, depending on which phase is continuous and which is disperse, different

TABLE 1: Relation between SLA and LLE equilibrium and kinetic coefficients. Application to the extraction of oleic acid with methanol from sunflower oil and the adsorption of oleic acid from methanol over a solid adsorbent (activated carbon).

| Concept | Adsorption | | L-L extraction | |
|---|---|---|---|---|
| Equilibrium distribution between the two phases | $q_{OA} = \dfrac{\alpha L_{OA}\, y}{1 + L_{OA} y + L_{Oil} y' + L_{MeOH} y''}$ | (1) | | |
| | $q_{OA} = K_F y^{1/n}$ | (2) | | |
| | $q_{OA} = q$ = concentration of impurity in the solid phase, $mol\,kg^{-1}$ | | | |
| | $y$ = concentration of impurity in the liquid phase, $mol\,L^{-1}$ | | $y = mx$ | (3) |
| | $y'$ = concentration of oil in the liquid phase, $mol\,L^{-1}$ | | $x$ = concentration of impurity (oleic acid) in the raffinate phase (sunflower oil) $mol\,L^{-1}$ | |
| | $y''$ = concentration of methanol in the liquid phase, $mol\,L^{-1}$ | | $y$ = concentration of impurity (oleic acid) in the extract phase (methanol), $mol\,L^{-1}$ | |
| | $L_{OA}$ = Langmuir constant for OA adsorption, $L\,mol^{-1}$ | | | |
| | $\alpha$ = saturation capacity for OA over the adsorbent, $mol\,kg^{-1}$ | | | |
| | $K_F, n$ = Freundlich constants for specific adsorbent and adsorbate | | | |
| | $H_{OA} = L_{OA}\alpha$ = Henry's constant for adsorption of OA | | | |
| Relation between flux densities and interfacial gradients | $N_{OA} = k_f\left(y - y_{surf}\right)$ | (4) | | |
| | $q_{surf} = Q\left(y_{surf}\right)$ | (5) | | |
| | $N_{OA} = D\rho_p\left(\dfrac{\partial q}{\partial r}\right)_{surf}$ | (6) | | |
| | $N_{OA}$ = flux across the film surrounding the particle (eq. (4)) | | $N_{OA} = k_{MeOH}\left(y_{int} - y\right)$ | (7) |
| | $N_{OA}$ = flux due to diffusion (eq. (6)). Both fluxes are equal at steady-state | | $N_{OA} = k_{Oil}\left(x - x_{int}\right)$ | (8) |
| | $k_f$ = film coefficient | | $y_{int} = mx_{int}$ | (9) |
| | $q_{surf}$ = surface concentration of adsorbate | | $k_{MeOH}$ = film coeff., MeOH side | |
| | $y_{surf}$ = concentration of impurity on the surface of the adsorbent | | $k_{Oil}$ = film coeff., oil side | |
| | $Q$ = function that gives the value of $q$ from the value of concentrations in the liquid phase (eq. (1)) | | $N_{OA}$ = impurity molar flux, molecules per unit time and area | |
| | $\rho_p$ = bulk density of the adsorbent particle | | int = interface | |
| | $D$ = net diffusivity of the adsorbate inside the adsorbent particle | | | |
| Relation between fluxes and driving forces | $\dfrac{\partial q_{av}}{\partial t} = k_e\left(q_{surf} - q_{av}\right)$ | (10) | $N_{OA} = K_{MeOH}\left(y_{eq} - y\right)$ | (13) |
| | $\dfrac{\partial q_{av}}{\partial t} = K_{LDF}\left(q_{eq} - q_{av}\right)$ | (11) | $N_{OA} = K_{Oil}\left(x - x_{eq}\right)$ | (14) |
| | $q_{eq} = Q\left(y\right)$ | (12) | $y_{eq} = mx_{eq}$ | (15) |
| | $q_{av}$ = average adsorbate concentration in the adsorbent particle | | $x_{eq} = \dfrac{y}{m}$ | (16) |
| | $q_{surf}$ = surface concentration | | $K_{MeOH}$ = overall transfer coefficient | |
| | $k_e$ = effective film coefficient for intrapellet diffusion | | $K_{Oil}$ = overall transfer coefficient | |
| | $q_{eq}$ = equilibrium adsorbate concentration for $y$ (eq. (1)) | | eq = equilibrium | |
| | $K_{LDF}$ = linear driving force mass transfer coefficient for adsorption | | | |

TABLE 1: Continued.

| Concept | Adsorption | | L-L extraction | |
|---|---|---|---|---|
| Mass transfer coefficients | | | $\dfrac{1}{K_{\text{MeOH}}} = \dfrac{1}{k_{\text{MeOH}}} + \dfrac{m}{k_{\text{Oil}}}$ | (18) |
| | | | $\dfrac{1}{K_{\text{Oil}}} = \dfrac{1}{k_{\text{Oil}}} + \dfrac{1}{mk_{\text{MeOH}}} = \dfrac{1}{mK_{\text{MeOH}}}$ | (19) |
| | | | $Sh_c = 0.725 S_c^{0.42} \mathfrak{Re}_c^{0.57} (1 - \phi)$ | (20) |
| | | | $Sh_c = \dfrac{k_c d_{32}}{D_c}$ | (21) |
| | $\dfrac{1}{K_{\text{LDF}}} = R_f + R_D = \dfrac{r_p}{3k_f} + \dfrac{r_p^2}{15\varepsilon D}$   (17) | | $\mathfrak{Re}_c = \dfrac{\rho_c d_{32} v_{\text{slip}}}{\mu_c}$ | (22) |
| | $\varepsilon$ = porosity of the adsorbent particle | | $Sc_c = \dfrac{\mu_c}{D_c}$ | (23) |
| | $r_p$ = radius of the adsorbent particle | | $k_d = 0.023 v_{\text{slip}} \, Sc_c^{-0.5}$ | (24) |
| | $k_f$ = film transfer coefficient. | | $\phi = \dfrac{V_d}{V_d + V_c}$ | (25) |
| | $R_f$ = film transfer resistance | | $\phi$ = hold-up of the disperse phase (oil) | |
| | $R_D$ = intrapellet diffusion resistance | | $d_{32}$ = average Sauter diameter | |
| | | | $v_{\text{slip}}$ = slip velocity between phases | |
| | | | $\mu_c$ = viscosity of the continuous phase | |
| | | | $V_d$ = volume of the disperse phase | |
| | | | $V_c$ = volume of the continuous phase | |
| | | | $k_c$ = mass transfer coefficient, continuous phase | |
| | | | $k_d$ = mass transfer coefficient, disperse phase | |
| Mass balance: batch unit perfectly mixed | $V\dfrac{dy}{dt} = WK_{\text{LDF}} \left( q_{\text{eq}} - q_{\text{av}} \right)$   (26) | | $\dfrac{dy}{dt} = \dfrac{aK_{\text{MeOH}}}{1 - \phi} \left( y_{\text{eq}} - y \right)$ | (30) |
| | $y = y^0, \ q = q^0, \ t = 0$   (27) | | $\dfrac{dx}{dt} = \dfrac{aK_{\text{Oil}}}{\phi} \left( x - x_{\text{eq}} \right)$ | (31) |
| | $\dfrac{dy}{dt} = K'_{\text{LDF}} \left( y - y_{\text{eq}} \right)$   (28) | | $x = x^0, \ y = y^0, \ t = 0$ | |
| | $q_{\text{av}} = Q \left( y_{\text{eq}} \right)$   (29) | | $a$ = interfacial area per unit volume of whole liquid phase | |
| | $V$ = volume of adsorbent | | | |
| | $W$ = weight of adsorbent | | | |
| Mass balance: continuous contact tower equations | $\dfrac{\partial y}{\partial t} - D_L \dfrac{\partial^2 y}{\partial z^2} + \dfrac{\partial (uy)}{\partial z} + \dfrac{1 - \varepsilon_B}{\varepsilon_B} \rho_p \dfrac{\partial q}{\partial t} = 0$   (32) | | | |
| | $y(0, t) = y^0$   (33) | | | |
| | $\dfrac{\partial y}{\partial z} = 0, \ z = L$   (34) | | $z = \displaystyle\int_{x_1}^{x_2} \dfrac{v_{\text{oil}} dx}{a_{\text{av}} K_{\text{Oil}} \left( x - x_{\text{eq}} \right)}$ | (36) |
| | $y(z, 0) = 0$   (35) | | $z = \displaystyle\int_{y_1}^{y_2} \dfrac{v_{\text{MeOH}} dy}{a_{\text{av}} K_{\text{MeOH}} \left( y_{\text{eq}} - y \right)}$ | (37) |
| | $\varepsilon_B$ = bed porosity | | $v$ = superficial velocity | |
| | $y$ = fluid phase concentration of the impurity | | $z$ = axial coordinate, height of the column | |
| | $q$ = solid phase concentration of the impurity | | $a_{\text{av}}$ = average interfacial area per unit volume of the contactor vessel | |
| | $u$ = interstitial velocity = $v/\varepsilon_B$ | | | |
| | $\rho_p$ = particle density | | | |
| | $D_L$ = diffusivity of the adsorbate in fluid. | | | |
| | $z$ = axial coordinate | | | |
| | $L$ = height of the column | | | |

correlations will be applied to each side, though both are usually of the form of an equation of the Sherwood number as a function of the Reynolds number, the Schmidt number, and the holdup of the disperse phase. Example equations for calculating $k_c$ (local coefficient for the continuous phase) and $k_d$ (coefficient for the disperse phase) for column continuous contactors have been written in (20)–(25) according to the suggestions of Koncsag and Barbulescu [28].

Discontinuous operation in perfectly mixed units shows more similarities for both operations as it can be seen in (26) to (31) in Table 1. This is also a problem easily handled mathematically and corresponds to the operation of stirred tank extraction units and stirred tank adsorption units. Packed columns operated with a high recycle ratio can also be considered as perfectly mixed stirred tanks. In (30) and (31), $a$ is the interfacial area per unit equipment volume. Alternatively, the balance in the liquid phase can be written in terms of a $K'_{LDF}$ coefficient, as in (28) and (29).

Equations (32) to (37) correspond to the differential equations, border, and initial conditions that express the mass balance for the adsorbate species during the movement along the packed bed and diffusion inside the porous adsorbent. The full model takes into account the backmixing in the axial direction by considering a Fickian diffusion term with axial diffusivity $D_L$. These equations must be coupled to a local equation like (26) that describes the law of variation of $q$ as a function of the process variables.

Equations (36) and (37) are the integrated forms of the local mass balance. They are simpler to handle than the previous one for adsorption. However, this is a simplified view, and more sophisticated models are needed to reflect phenomena of emulsion formation and collapse, carryover, flooding, drop coalescence and breakage, and so forth.

The interfacial area during extraction is a function of the drop size. The drop size is bigger at higher values of holdup of the disperse phase and at bigger values of the stirring power (in stirred tanks). However, the dependence is soft.

The inspection of (26) to (29) and (30) and (31), and their comparison with (32) and (33), shows that adsorption columns can never work in a true steady state like liquid-liquid extraction units do. To describe their operation, a solution as a function of time and space must be found. This is because the solid phase is fixed while the fluid phase is flowing continuously. As shown in Figure 2 when the breakthrough condition of the column is reached, the operation of the adsorber must be stopped and regeneration must be performed. This is different from the extraction column in which a continuous steady state can always be established between the two flowing liquid phases.

For both adsorption and extraction, the throughput for any separation unit is mainly given by intrinsic parameters such as the kinetic mass transfer parameters, the thermodynamic constants for the L-L and S-L equilibrium, the parameters describing the interface, and the total volume of the phases. For any given choice of contacting device and set of process conditions, that is, temperature, pressure, and liquid phase flowrates, the volume of the unit would be the result of a design procedure for a given desired rate of extraction and adsorption, because the process conditions

will dictate the values of the intrinsic parameters. Therefore, it is of interest to list the range of values of the most important intrinsic parameters involved in the design of adsorption and extractors. This is done in Table 2.

Equivalent coefficients have been placed in the same row. In the case of the interfacial area for adsorption, all available surface area, external and intrapellet, has been included. It must be noted that, due to diffusional resistance, not all surface is readily available. However, this is taken into account when calculating the intrapellet mass transfer resistance. Since, for mesoporous and microporous adsorbents, the inner surface is much higher than external one, $a$ is practically the intrapellet area divided by the pellet volume. It is apparent from this comparison that the S-L interfacial area is much higher than the L-L for most adsorbents and L-L contactors.

Inspection of the last row of Table 2 yields the most important insight. If $K'_{LDF}$ and $aK_L$ values are compared, this is a comparison of parameters with similar driving forces, it can be seen that in global terms adsorption is slower than extraction under most conditions, especially for the case of extraction in stirred tanks or static mixers. More similar values are obtained when we match adsorption with a low energy extraction operation, for example, in a spray column. For the coupling of both units however what it must be similar is the uptake of impurity per unit time and this is a function also of the driving force. In this sense, slow adsorption kinetics can be compensated by high solid affinities (high $H$, $q_{eq}$ values), while fast extraction kinetics could be inhibited by low impurity solubilities (low $m$ values). All these considerations will have a better insight once the examples are discussed in detail.

*3.2. Example of Plant Oil Deacidification.* In this example, sunflower oil is first extracted with methanol in order to remove the impurities, that is, oleic acid. The solvent is then regenerated by adsorption of oleic acid from the methanol solution. The adsorbent is in turn regenerated by a thermal swing. In order to obtain sound values of the parameters that describe the phenomena, separate experiments for determining the liquid-liquid and liquid-solid thermodynamic and kinetic parameters were performed.

*3.2.1. Determination of the Partition Coefficient for Oleic Acid.* For the system oleic acid-methanol-sunflower oil, values of $m$ were obtained from plots of $y$ as a function of $x$ at three different temperatures (Figure 4). $m$ was found to be equal to 0.875 at 30°C, 0.922 at 40°C, and 1.125 at 50°C. $m$ was calculated as the ratio of the concentration of oleic acid in the alcohol phase (free of oil) to the concentration of oleic acid in sunflower oil (free of methanol). The concentration of oleic acid in either phase was really a little lower due to dissolution of methanol in the oil phase and dissolution of oil in the alcohol phase. In this sense, the higher solubility of oleic acid in methanol at higher temperatures is also accompanied by a higher solubility of the oil, and hence there must be a balance when choosing the right temperature of operation, because the relative purity of the extract or the yield of raffinate can

TABLE 2: Typical values and ranges of intrinsic parameters for adsorption and extraction. AC = activated carbon.

| Adsorption parameter | System | Range | Extraction parameter | System | Range |
|---|---|---|---|---|---|
| $H$, adim. $(mol\ kg^{-1})/(mol\ L^{-1})$ | Oleic acid-biodiesel-silica [8] | 20–50 | $m$, adim. $(mol\ L^{-1})/(mol\ L^{-1})$ | Toluene-Acetone-Water [15] | 0.967 |
| | p-aminobenzoic acid-water- resin [9] | 20 | | Oleic acid-cannoli oil-methanol [16] | 0.705 |
| | Naphthalene-water-carbon [10] | 1 | | Methylene chloride-ethanol-water [17] | 0.133 |
| | Sulfur compounds-diesel-AC [11] | 0.6 | | Sunflower oil-oleic-acid-methanol (this work) | 0.7415 |
| | Thiophene-naphtha-NaY [12] | 2.1 | | | |
| | Naphthenic acid-toluene-clay [13] | 2.3 | | | |
| | dichlorobenzene-methanol-graphene [14] | 598 | | | |
| | Oleic acid-methanol-AC (this work) | 33 | | | |
| $q^{max}$, mol kg$^{-1}$ | Oleic acid over silica [8] | 4.96 | $y^{max}$, mol L$^{-1}$ | Toluene-acetone-water [15] | 0.498 |
| | p-aminobenzoic acid-water- resin [9] | 2.372 | | Oleic acid-canola oil-methanol [16] | 0.31 |
| | Naphthalene-water-carbon [10] | 0.272 | | Methylene chloride-ethanol-water [17] | 0.0225 |
| | sulfur compounds-diesel-AC [11] | 0.003 | | Sunflower oil-oleic-acid-methanol (this work) | 0.104 |
| | thiophene-naphtha-NaY [12] | 2.713 | | | |
| | Oleic acid-methanol-AC (this work) | 2.98 | | | |
| $a$, cm$^2$ cm$^{-3}$* | Silica [8] | $10^4$–$10^5$ | $a$, cm$^2$ cm$^{-3}$** | Stirred tank, Re > 10000 [18] | 10–250 |
| | Resin [9] | $10^7$ | | Continuous stirred tank [19] | 3.2–130 |
| | NaY [12] | $10^6$–$10^7$ | | Packed bed [20] | 1.5–3 |
| | Norit [11] | $10^6$ | | Slug flow [21] | 20–40 |
| | AC [10] | $10^7$ | | Static mixer [14] | 30–300 |
| | AC (this work) | $10^7$ | | Spray column (this work) | 3 |
| $K'_{LDF}$, min$^{-1}$*** | Oleic acid-biodiesel-silica [8] | 0.0365 | $aK_L$, min$^{-1}$**** | Water-kerosene-acetic acid, column [22] | 0.054 |
| | Naphthalene-water-carbon [10] | 0.044 | | Toluene-acetic acid-water, column [23] | 0.049 |
| | p-aminobenzoic acid-water- resin [9] | 0.060 | | Sunflower oil-oleic acid-methanol, stirred tank (this work) | 0.75 |
| | thiophene-naphtha-NaY [12] | 0.167 | | | |
| | sulfur compounds-diesel-AC [11] | 0.004 | | | |
| | Oleic acid-methanol-AC (this work) | 0.066 | | | |

*Per unit volume of packed bed; ** per unit volume of extractor, coefficients for the continuous phase (usually the solvent for high solvent-to-feed systems); *** Calculation of $K'_{LDF}$ is done with equations (26) and (29) of Table 1. ****$K_L$ = overall mass transfer coefficient, L-L extraction, continuous phase side ($K_{MeOH}$ in example of this work). Calculation of $K_L$ is done with equations (18) and (19) of Table 1.

TABLE 3: Experimental values of drop size ($d$), terminal velocity ($V$) and mass transfer coefficients (continuous phase side) ($aK_L$) at three different temperatures. Single drop tests, $\phi = 0.15$.

| $T$, °C | $d$, mm | $a$, cm$^2$ cm$^{-3}$ | $V$, cm s$^{-1}$ | $aK_L$, min$^{-1}$ |
|---|---|---|---|---|
| 30 | 3.19 | 18.8 | 6.21 | 0.842 |
| 40 | 3.13 | 19.2 | 4.93 | 1.190 |
| 50 | 3.13 | 19.2 | 3.13 | 1.610 |

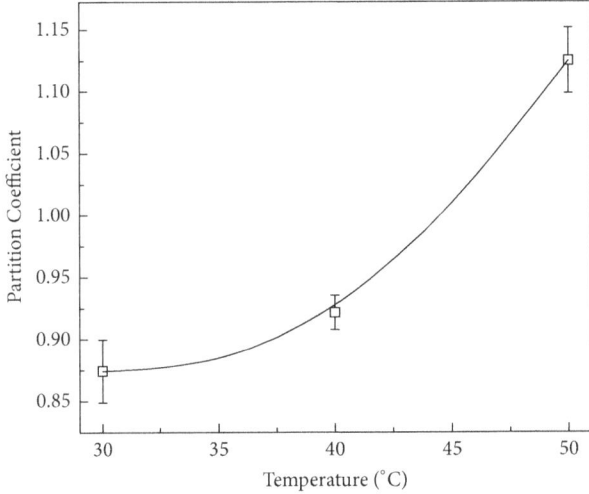

FIGURE 4: Plot of the partition coefficient as a function of the temperature of the experiment.

be an issue. The linearity of the $y$ versus $x$ plots was very good with $r^2$ values of about 0.997.

### 3.2.2. Determination of Mass Transfer Coefficient of Oleic Acid between Methanol and Sunflower Oil.
Mass transfer coefficients varied widely depending on the type of contact equipment used. Results are presented in Table 3 for spray column single drops experiments. Average diameter values were calculated with Sauter's formula (see (38)). Interfacial area is calculated with (39). In the case of the column, increasing temperature values (from 30 to 50°C) increased the overall mass transfer coefficient by almost a factor of 2. This is possibly related to the decrease in viscosity and a significant increase of the Reynolds number (Re).

$$d_{32} = \sum \frac{n_i d_i^3}{n_i d_i^2}, \tag{38}$$

$$a = \frac{6\phi}{d_{32}}. \tag{39}$$

The values obtained compared fairly well with others reported. Sankarshana et al. [29] found values of $aK_L$ (continuous phase side, feed/solvent = 0.2–1) equal to 0.02–0.06 min$^{-1}$ for packed columns with random and ordered packing, while working with a system of acetic acid-ethyl acetate-water system under countercurrent mode. Nosratinia et al. [30] found $aK_L$ values of 0.12–0.84 min$^{-1}$ in a spray column with jet injection of the disperse phase.

Geankoplis and Hixson [31] found values of overall $aK_L$ (water, continuous phase side) of 0.07–1.2 min$^{-1}$ at varying disperse phase flow rates, in a ferric chloride-isopropyl ether-HCl$_{(aq)}$ system, in a spray tower.

In spray towers for liquid-liquid extraction, the Sauter diameter is a function of the disperse phase holdup and fluid dynamic conditions. Salimi-Khorshidi et al. [32] found that $d_{32}$ varied within 2.5–4 mm when varying $\phi = 0.1$–0.6 and flowrates, a volcano plot being found for $d_{32}$ as a function of Re or $\phi$.

Considerations for the scale-up of $aK_L$ coefficients from single drop measurements to full-scale drop swarms should be discussed. Hughmark [33] studied comprehensive data sets, with $\phi = 0.006$–0.2, and early found that, for ratios of the continuous to disperse phase viscosity less than one, the mass transfer coefficients (in the form of Sh$_c$ or $k_c$) for the continuous phase of drop swarms were the same as for single drops, while, for viscosity ratios greater than one, the multiple drop coefficients were somewhat smaller. Hughmark fitted his data with Ruby and Elgin $k_c$ equation [34]. In this system, the coefficient $aK_L$ is thus a function of the impeller Reynolds and also directly proportional to $\phi$.

The value of the mass transfer coefficient for the stirred tank experiment was 0.75 min$^{-1}$ using an experimental setup similar to that of Schindler and Treybal [18] and using a holdup of disperse phase of 0.5. These authors early correlated the mass transfer coefficients for stirred tanks studying the mass transfer between two liquid phases in an agitated baffled vessel and found that mass transfer coefficient increased with impeller Reynolds number and disperse phase holdup. Baffling roughly increased $aK_L$ by 1.5 times. Average volumetric $aK_L$ values for the continuous phase ranged within 3–25 min$^{-1}$ for values of the impeller Reynolds number of 20000–60000. The dependence of $aK_L$ on impeller Reynolds number was strong, being roughly proportional to Re$^{2-2.5}$, while the dependence on holdup of the disperse phase was weaker, being proportional to about $\phi^{0.9-1}$. They also found that $k_c$ was almost insensitive to variations in the holdup. These trends can be easily rationalized by considering that, for stirred tanks the Sauter diameter, as in (38), is imposed by the impeller Reynolds, while the interfacial area per unit volume of disperse phase corresponds to the value given by (39). For this reason, values of $aK_L$ for the simulations will be extrapolated from experimental data at similar stirring conditions by scaling with the value of $\phi$.

### 3.2.3. Adsorption Properties.
For the oleic acid-methanol-sunflower oil system, an adsorbent of activated carbon was chosen because of its good performance in preliminary screening tests. This is a fairly novel application for carboxylic acid adsorption since, in the literature, silica, silicates, clays, and zeolites have usually been employed [35–37] while reports on the use of carbons are concentrated on decontamination of water [38, 39].

Results of adsorption of oleic acid in sunflower oil overactivated carbon are plotted in Figure 5. The curves correspond to a virgin activated carbon. The last value in the abscissae axis is 0.10 mol L$^{-1}$; therefore the plotted results

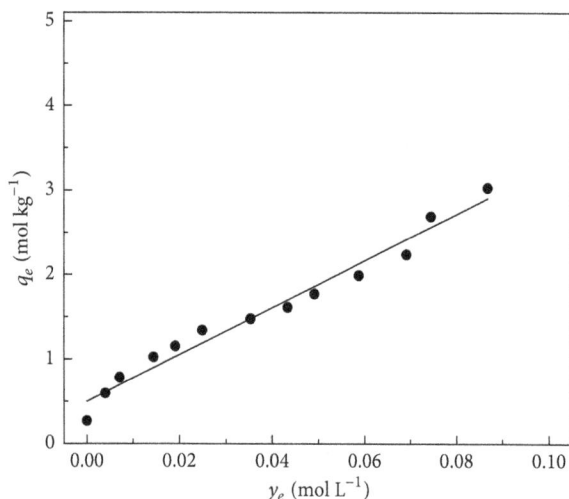

FIGURE 5: Adsorption of oleic acid from methanol at room temperature (30°C) and overactivated carbon (Calgon Carbon Filtrasorb 200). Correlation coefficient for Henry's constant (slope of the line), $r^2 = 0.97$.

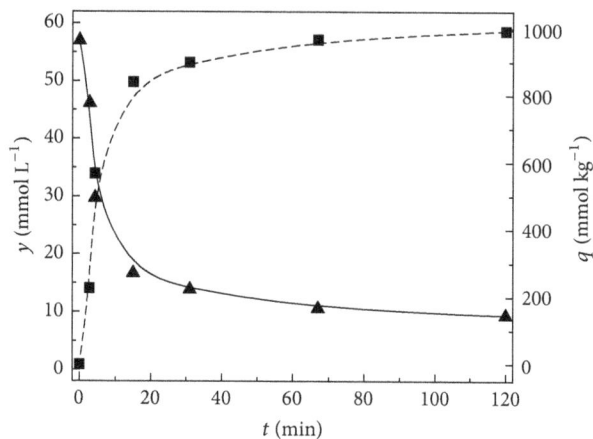

FIGURE 6: Concentration of oleic acid in both the liquid phase ($y$, ▲) and solid phase ($q$, ■) as a function of time. Packed column with fast recycle, 30°C. Granular activated carbon, average pellet diameter, 0.4 mm.

correspond to mildly acidic sunflower oil and sunflower oil of low acid content. The curve is better interpreted by the Freundlich model (Table 1, equation (2)), with $K_F = 8.34$ and $n = 1.66$, with an $r^2 = 0.971$. For simplicity of the treatment however, the data can be fitted with Henry's linear isotherm yielding $H = 33$ (L kg$^{-1}$).

Another isotherm was taken at 40°C in order to calculate the heat of adsorption. This isotherm had an $H$ value of 16 (L kg$^{-1}$). Applying the integrated form of van't Hoff equation and considering that the heat of adsorption was not a function of temperature, the heat of adsorption was estimated as 57 kJ mol$^{-1}$. This value compares well with other found in the literature. Li et al. [40] found that adsorption of phenol on resin from aqueous solutions had a heat of adsorption of about 38 kJ mol$^{-1}$. Chiou and Li [41] found a heat of adsorption of 52.9 kJ mol$^{-1}$ for reactive dye in aqueous solution on chemical cross-linked chitosan beads. Ilgen and Dulger [42] measured a value of about 34 kJ mol$^{-1}$ for the adsorption of oleic acid from sunflower oil over zeolite 13x.

Adsorption tests were also made in a packed bed column with fast recycle. In this column the axial concentration gradient was negligible and the behavior was similar to a stirred tank with perfect mixing. The results were fitted with the simple model of the linear driving force model, in the form of (26). The results for one of such tests are plotted in Figure 6. The calculated value for $K_{LDF}$ from the experiment is 0.066 min$^{-1}$.

*3.2.4. Settling Times for Phase Separation in a Gravity Decanter.* The results of the experiments to measure the settling time as a function of the volumetric methanol-to-oil ratio and the temperature are included in Figure 7. At 25°C for methanol-to-oil ratios lower than 1 no complete phase separation could be achieved even after 1 day, some oil

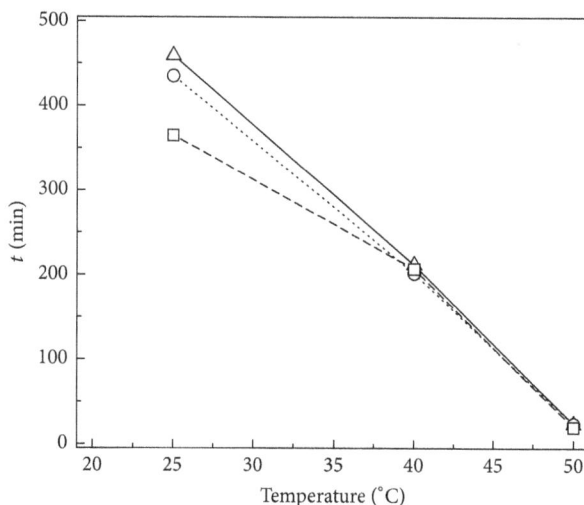

FIGURE 7: Experimental settling time as a function of the temperature. Tank with no internals and no coalescing aid. Disperse phase (oil) holdup: 0.5 (△), 0.25 (○), and 0.2 (□).

remaining disperse in the methanol phase as indicated by the opacity of the upper phase.

Complete separation was achieved for methanol-to-oil ratios equal to or higher than 1. The general trends were that, at high temperatures, for example, 50°C, the settling time was independent of the methanol-to-oil ratio, while at lower temperatures higher methanol-to-oil ratios lowered the settling time. For a continuous operation of a decanter with a settling time of one hour, a temperature of about 50°C is needed.

*3.2.5. Simulation*

*(1) Simulation of a Continuous Extraction/Adsorption Process.* The layout of a process using an extraction column coupled to a set of twin adsorption columns is depicted in Figure 8.

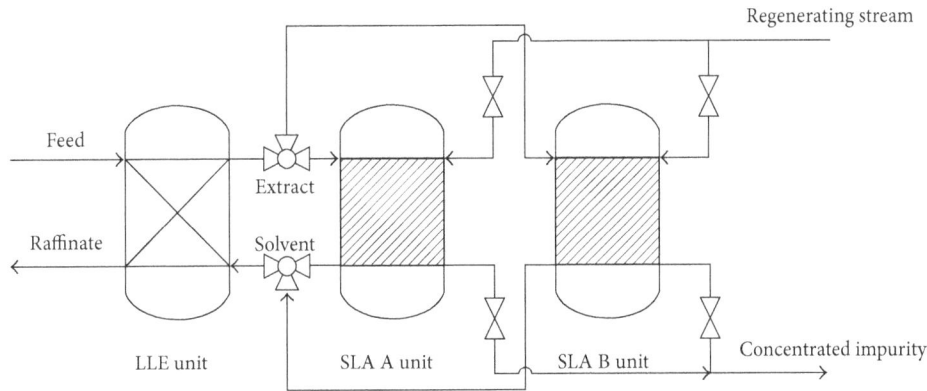

FIGURE 8: Flowsheet of extraction column coupled to a set of twin adsorption columns. $W^{Ads} = 364\,kg$, $V^{Ext} = 0.15\,m^3$.

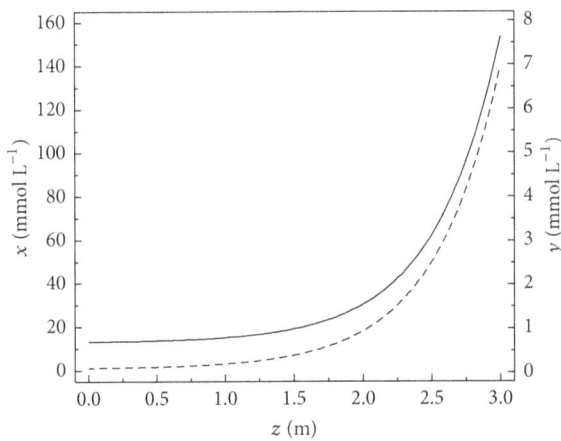

FIGURE 9: Plot of $x$ (dashed line) and $y$ (solid line) as a function of the position in the countercurrent extraction column. $Q^{Feed} = 4\,L\,min^{-1}$, $Q^{Solv} = 80\,L\,min^{-1}$, $x^{Feed} = 0.140\,mol\,L^{-1}$, $y^{Solv} = 0.7\,mmol\,L^{-1}$.

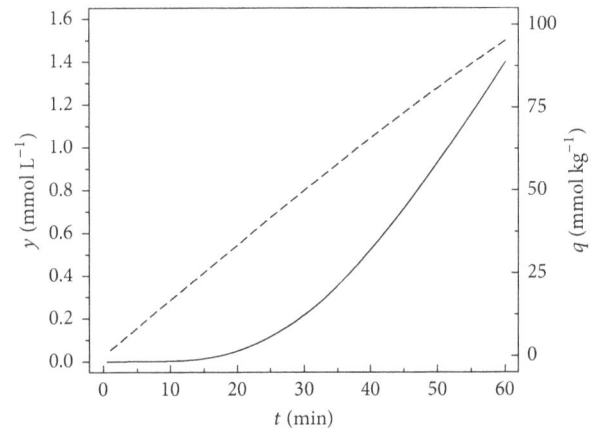

FIGURE 10: Adsorption column. Plot of $y$ (solid line) and $q$ (dashed line) as a function of time. $W^{Ads} = 364\,kg$, $Q^{Ext} = 80\,L\,min^{-1}$, $y^{Ext} = 7.64\,mmol\,L^{-1}$, $q_0 = 0\,mol\,kg^{-1}$.

In this layout the operation of the extraction column is continuous, and the equations describing the relation between the concentration in any phase and time are (32)–(35) in the case of the adsorption column. The equations describing the exchange in the case of extraction are (36) and (37). The solution of this system of equations for an example of extraction of acidic sunflower oil with methanol is depicted from Figures 9–11.

The holdup of the disperse phase was calculated from the experimentally measured characteristic velocity of the drop ($v_K$), and the values of the flowrates of the feed and solvent, by means of the equation of Gayler (40). Gayler proposed that for many different types of columns the following equation held [43]:

$$\frac{u_d}{\phi} + \frac{u_c}{1 - \phi} = u_k\left(1 - \phi\right), \qquad (40)$$

where $u_c$ and $u_d$ are the superficial velocities of the continuous and dispersed phases, respectively, and $u_k$, the characteristic velocity, is the mean relative velocity of droplets extrapolated to zero flowrate and can be identified with

the terminal velocity of a single drop in the equipment concerned. Equation (40) was numerically solved, giving $\phi = 0.044$.

The equations of the column extraction unit were solved analytically in order to avoid the problem of solving the two-point boundary value problem imposed by the countercurrent flow. Equations (36) and (37) were solved by obtaining the eigenvalues and eigenvectors of the matrix of derivatives and by considering that the solution eigenfunction was $y = k\exp(\lambda z)$. One eigenvalue was found to be zero, so both $x$ and $y$ had the form $g = A + B\exp(\lambda z)$, where $\lambda$ is a function of $m$, $v^{Feed}$, $v^{Solv}$, and $aK_L$; and $A$ and $B$ are function of the previous parameters and also the initial conditions.

The impurity concentration in the feed, 140 mmol L$^{-1}$, is equivalent to about 4.4% acidity, which should be reduced to about 30 mmol L$^{-1}$ in order to be suitable as a feed for the biodiesel alkali-catalyzed process. For some applications, the maximum acidity is even lower. For insulating oils, ASTM D3487 establishes a maximum acidity of 0.03 mg KOH g$^{-1}$, that is, 0.5 mmol L$^{-1}$. This is near the concentration value of the raffinate in Figure 9, 1.1 mmol L$^{-1}$. Anyway, a special limit for insulating oils of plant origin is established in ASTM

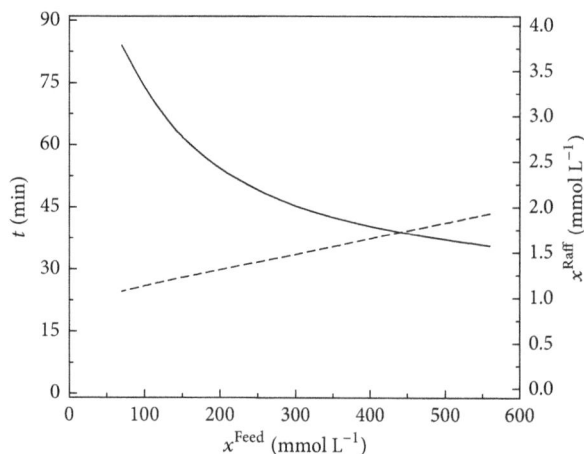

FIGURE 11: Adsorption unit cycle time (solid line) and raffinate concentration (dashed line) as a function of the initial concentration of the impurity in the oil phase ($x^{Feed}$), for a required residual concentration at the outlet of the adsorption column ($y^{break}$) of 1.40 mmol L$^{-1}$.

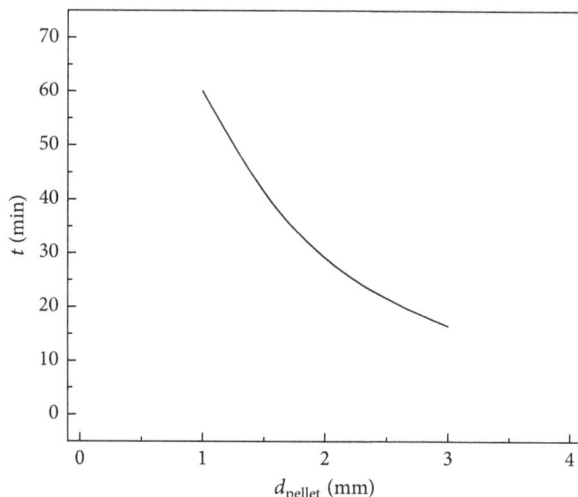

FIGURE 12: Adsorption unit cycle time as a function of the pellet diameter, for a required residual concentration at the outlet of the adsorption column ($y^{break}$) of 1.40 mmol L$^{-1}$ and impurity in the oil phase ($x^{Feed}$) equal to 140 mmol L$^{-1}$.

D6871, 0.6 mg KOH g$^{-1}$, that is, 10 mmol L$^{-1}$, nine times the final value in the raffinate of the example. 0.6 mg KOH g$^{-1}$ is also the limit established for refined edible oil for human consumption (FAO CODEX-STAN 210, 1999) and therefore the extraction step is suitable for this application also, though it must be noted that ethanol would be more appropriate than methanol for a better compliance with health restrictions.

For simulating the adsorption column the set of equations (32)–(35) for the adsorption column was solved after analyzing the underlying hypotheses. The axial dispersion sometimes produces the broadening of the adsorption front due to the contribution of both molecular diffusion and dispersion caused by fluid flow [44]. The impact of the axial dispersion is assessed by the Peclet number (Pe) small values indicating backmixing is important. According to Carberry [45], for Pe values much greater than 100, the flow can be considered plug flow type. Values of molecular diffusivity of oleic acid in methanol were calculated with Wilke-Chang equation. Axial diffusivity was calculated with the correlation of Wakao and Funazkri [46]. The Pe value was found to be equal to 3300.

For the case of linear isotherm, a solution to ((32)–(35)) in the form $(q, y) = f(z,t)$ can be got by using the "quasi-log normal distribution" (Q-LND) [47, 48]. In this case, it is assumed that the quasi-log normal probability density function can be used to represent the impulse response of the system. It has been demonstrated that the analytical solution and the Q-LND approximate solution are similar for a wide range of the model parameters and that deviations appear only for very low values of the residence time. This solution is used to plot the breakthrough curve of Figure 10.

In Figure 10, the concentration of the extract is 7.64 mmol L$^{-1}$ and must be reduced to 0.7 mmol L$^{-1}$. The concentration of impurity in the solvent at the column outlet is nonlinear function of time, and the initial concentration is about zero. An outlet concentration of 1.4 mmol L$^{-1}$

corresponds to an average concentration of solvent somewhat lower than 0.7 mmol L$^{-1}$. 1.4 mmol L$^{-1}$ could then be safely considered the column breakthrough condition ($y^{break}$), the cycling time for the packed bed being of 1 hour. At this time, the average load of impurity on the adsorbent is 95 mmol kg$^{-1}$. The way of removing this load to regenerate the bed will be dealt later in detail.

$t^{break}$ will be a function of $x^{feed}$ and the desired purity of the solvent, for a given fixed operation condition of the extraction unit. This is illustrated in Figure 11. The adsorbent column must be maintained in operation until the breakthrough occurs. As expected $t^{break}$ decreases with higher concentration of impurity in the feed but the curve is enough soft to allow handling varying impurity concentration in the 100–600 mmol L$^{-1}$ range with $t^{break}$ in the 30–60 min range. The concentration in the raffinate varies from 1 to 2 mmol L$^{-1}$, which can be considered negligible.

The simulation runs of Figures 9, 10, and 11 were made with a pellet size of 1 mm. For a bigger 12 × 40 meshes granular carbon, a maximum size of 1.5 mm can be found. For pelletized carbon, sizes of up to 3 mm are common. Particle size has a great influence on the cycling time, because the intrapellet diffusion mass transfer resistance $R_D$ is proportional to the square of the pellet radius. This is clear in the plot of Figure 12. The time of operation must be reduced from 1 h to about 15 min when the pellet size is increased from 1 to 3 mm.

*(2) Regeneration.* An assessment of the regeneration of the solvent by evaporation/distillation and adsorption should be made. Distillation is the most common method but it requires a relatively high amount of energy. Adsorption was demonstrated to be a feasible regeneration method but it needs energy for desorbing the impurity from the adsorbent bed. A comparison of the amount of heat involved

TABLE 4: Amounts of energy involved in the regeneration of the extract (in kJ per litre extract).

| Distillation | | Adsorption | |
|---|---|---|---|
| Step | Energy | Step | Energy |
| Heating to boiling point | 63 | Adsorption (exothermal) | 0 |
| Solvent evaporation | 948 | Heating 1 bed volumes of methanol to 100°C | 22 |
| | | Desorption (endothermal) by elution of hot methanol | <1 |
| | | Heating to boiling point, evaporation of solvent of eluting stream | 151 |
| Total heat duty | 1011 | | 174 |

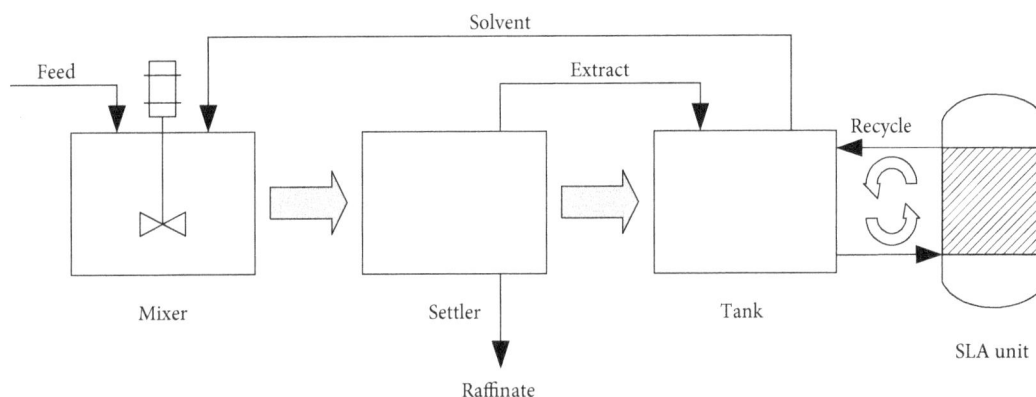

FIGURE 13: Flowsheet of a sequential discontinuous combination of extraction tank and adsorption column with fast recycle. Dimensions and process parameters: $W^{Ads} = 1334$ kg, $V^{Feed} = 0.24$ m$^3$, $V^{Solv} = 4.8$ m$^3$, $x^{Feed} = 140$ mmol L$^{-1}$, $T = 303$ K.

in each case is done in Table 4 for the case of Figures 9 and 10. The information of these figures indicated that the regenerated solvent had an average concentration of oleic acid of 0.7 mmol L$^{-1}$ and that the average concentration on the solid at breakthrough was about 95 mmol kg$^{-1}$. The amount of solvent regenerated at $t^{break}$ was 4800 L. If this volume had been regenerated by distillation the amount of heat would have been 1000 kJ L$^{-1}$. This is considering that all the methanol is evaporated at 100% efficiency and that there are no schemes for heat recovery. In the case of the regeneration by adsorption, the heat is consumed in the heating of the eluting volume for the thermal swing, heating this same stream to the boiling point and evaporating the solvent in it to recover the free oleic acid. A temperature of regeneration of 100°C and an elution volume equal to 1 bed volume was chosen. These values permit achieving a residual concentration of impurity in the solid lower than 5% of the original load before regeneration. With these parameters, the heat duty of the regeneration by adsorption amounts to about 151 kJ L$^{-1}$, just a 15% of the classical regeneration. The regeneration temperature demands running the regeneration step with a little overpressure of 2.3 bar due to the high vapor pressure of methanol. In general terms, it was deduced that temperature of regeneration is the most influential variable, the elution volume having lower impact on the residual concentration of impurity in the solid. For simplicity of the involved calculation, regeneration will be assumed to be complete in what follows.

Time for regeneration was found experimentally. Residual oleic acid on the solid did not vary for time spans for regeneration higher than 2.5 min.

Some other authors have used only flushing with solvent in order to regenerate the adsorbent. Yori et al. [49] removed glycerol from biodiesel by adsorption over a silica column and regenerated the bed by flushing with a small amount of methanol. In their case the great affinity of methanol for the adsorbed impurity (glycerol) was the crucial factor for regenerating the bed. In the studied case, a thermal swing is needed to help desorption.

*(3) Simulation of a Batch Extraction/Adsorption Process.* For this simulation, extraction tanks were chosen with a volume equal to that processed by the equipment of Figure 8 for 1 h operation (see Figure 13). This enables a comparison of performance and equipment requirements for similar throughput. In order to use completely discontinuous units, a column with fast recycle was programmed that obeys (30) and (31). Choosing a stirred tank with adsorbent suspended in the liquid would have yielded the same operation equations. However packed columns make regeneration easier. In order to have a short residence time and work as a perfectly mixed stirred tank, the recycling flow rate must be made fast enough. Reducing the residence time to a fraction of a minute makes this possible. For not making pressure drop an issue at this flow conditions the L/D of the column should be low.

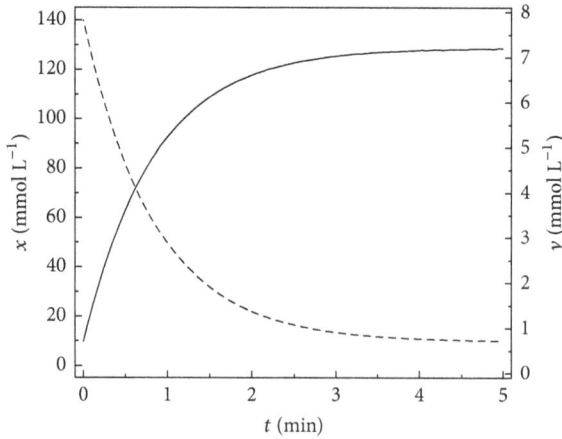

FIGURE 14: Plot of $x$ (dashed line) and $y$ (solid line) as a function of time in the batch extraction unit. $W^{Ads} = 1334$ kg, $V^{Feed} = 0.24$ m$^3$, $V^{Solv} = 4.8$ m$^3$, $x^{Feed} = 140$ mmol L$^{-1}$, $T = 30°$C.

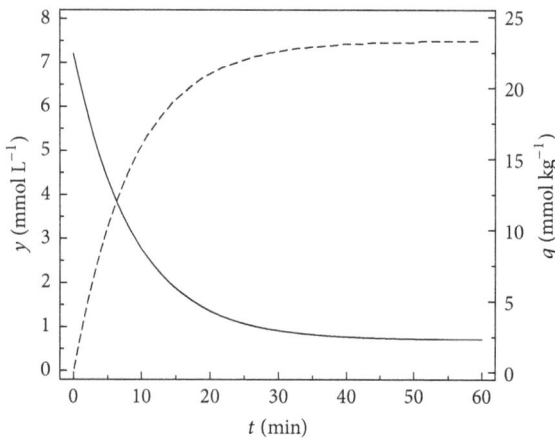

FIGURE 15: Plot of $y$ (solid line) and $q$ (dashed line) as a function of time in the adsorption column with recycle. $W^{Ads} = 1334$ kg, $V^{Feed} = 240$ L, $V^{Solv} = 1800$ L, $y^{Raff} = 7.2$ mmol L$^{-1}$, $T = 30°$C.

Results of the simulation are given in Figure 14. The concentration of the impurity can be reduced from 140 to about 10 mmol L$^{-1}$. This is worse than the final value of the countercurrent column, 1.1 mmol L$^{-1}$, and is a consequence of the unfavorable behavior of perfectly mixed systems with equilibrium restrictions. Two stages would be necessary for achieving the final raffinate concentration of Figure 9.

The results of Figure 15 bring similar conclusions as in the case of the batch extraction unit. The performance of the column with fast recycle is worse than that seen for the column with once-through flow. Although the extract to be refined has a similar concentration of impurity (7.2 mmol L$^{-1}$), an outlet solvent concentration of 0.7 mmol$^{-1}$ like in Figure 10 can only be achieved by increasing the mass of adsorbent 3.6 times. This is also explained by the unfavorable behavior of perfectly mixed systems with equilibrium restrictions.

The comparison of the saturation time values for both the extraction and adsorption column shows that the extraction tank has a saturation time of about 5 min and the packed

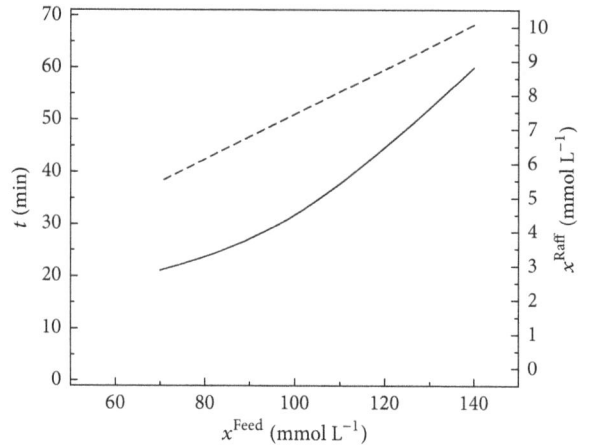

FIGURE 16: Adsorption time (solid line) as a function of the initial concentration of the impurity in the oil phase ($x^{Feed}$), for a required residual concentration at the outlet of the adsorption column ($y^{break}$) of 0.70 mmol L$^{-1}$. Resulting $x^{Raff}$ as a function of $x^{Feed}$ (dashed line). $W^{Ads} = 1334$ kg.

bed about 40 min. This is the result obtained for a volume of 4800 L liquid phase in the extractor and a volume of about 2600 L of packed bed. It can be deduced that the throughput of each unit per unit volume is different, about 200 L min$^{-1}$ per L of process vessel for the extractor and 0.05 for the adsorber. These values are not too different as it could be expected from the slow kinetics of adsorption and could be the results of a compensation with a high available surface area for the chosen adsorbent.

A comparison of all characteristic times must also include the settling time of the decanter. At 50°C this is about 1 h, more similar to the saturation time of the adsorber. However, the packed bed should be operated at a conveniently low temperature, like 30°C, to have a favorable adsorbing capacity. At 30°C the settling time increases to about 3 h, making this step the slowest of the process.

A plot of the necessary minimum adsorption time as a function of the feed impurity concentration is included in Figure 16. The same trends of Figure 11 are seen. The concentration of the raffinate increases when the concentration of impurity in the feed is increased. Keeping the raffinate at a constant composition can only be achieved by increasing the solvent-to-feed ratio in the extraction unit. The operation time for the adsorption column increases when the concentration of impurity in the feed increases. This is due to the higher concentration of impurity that demands more time to be removed.

## 4. Conclusions

The deacidification of sunflower oil by extraction with methanol and regeneration of the solvent by adsorption on activated carbon were tried. The method is considered useful for the food and biodiesel industries as a means of economically achieving low impurity levels in raffinate streams in systems with a high solvent-to-feed ratio.

Equations for the design of extractors and packed columns, both in continuous and discontinuous mode, were developed. Suitable equations for design were written from general principles, highlighting similarities in formulation for driving forces, mass transfer rates, thermodynamic and kinetic parameters. A comparison of ranges of thermodynamic and kinetic parameters seems to indicate that the matching of extraction and adsorption units needs to accommodate the fast kinetics of extraction with the relatively slow kinetics of adsorption, and the relatively high affinity of adsorbents per unit mass with the low capacity of most solvents per unit volume. In this sense a solution is the cyclic operation of a column adsorber of adequate size.

Deacidification by extraction with methanol can be carried out with relatively high efficiency due to a fairly high value of the partition coefficient for oleic acid between the polar and organic phase, with values of $m = 0.74$ a $0.93$ in the range of temperatures 30 a $50°C$. Extraction in a countercurrent column with a solvent-to-feed ratio of 20 allows reduction of the acidity of an oil with 4.4% acidity to a final value of 0.04%, adequate for its use as a feedstock for the production of biodiesel with the alkali-catalyzed process, its use as edible oil for human consumption, or its use as dielectric biodegradable oil. Extraction permits a maximum yield of oil and the recovery of the fatty acid impurity.

Typical operation of a stirred tank extraction unit yielded a value of the global coefficient for mass transfer $aK_L$ of $0.75\,min^{-1}$ (methanol side). Fast kinetics of extraction permitted the operation of a stirred tank extraction unit with a saturation time of about 5 min. This was faster than the characteristic time for saturation of a batch adsorption column that had a value of about 1 h. However, a totally discontinuous process needs also of a decanter that had big settling times, making it the slowest step of the process.

The extraction/adsorption combination seems well suited for extraction operations with a high solvent-to-feed ratio in which solvent regeneration by distillation becomes prohibitive due to a lower vapor pressure of the impurity to be removed/recovered, thus demanding evaporation of large amounts of solvent. It was demonstrated, with the example of sunflower oil deacidification, that, for a solvent-to-feed ratio of 20, the heat duty of a distillation-based solvent regeneration could be as large as 1011 kJ per litre of solvent. A regeneration process based on adsorption needs of a heat duty much lower, of about 174 kJ per litre solvent, the heat duty mainly being related to the thermal swing of the packed bed.

Applications for extraction with a high solvent-to-feed ratio for which an extraction/adsorption combination would be convenient could be those using a solvent with high selectivity but low affinity for the impurity, or "polishing" operations with a low driving force due to the high dilution of the impurity in the feed.

The extraction/adsorption system is amenable for both continuous and discontinuous operation. However, the continuous operation has a higher efficiency due to the intrinsic advantages of plug flow as compared to perfect mixing, for systems in thermodynamic equilibrium is a limitation.

For adsorption the main mass transfer resistance is intrapellet diffusion. In this sense small adsorbent particles improve the turnover of the process and increase the percentage of utilization of the adsorbent volume.

## Conflicts of Interest

The authors declare that there are no conflicts of interest regarding the publication of this paper.

## Acknowledgments

This work was performed with the funding of CONICET (PIP Grants 11220130100457CO and 11420110100235CO), Universidad Nacional del Litoral (CAI + D Grant 50420150100074LI), and Universidad Nacional de Jujuy (SeCTER 08/D138).

## References

[1] M. S. Cuevas, C. E. C. Rodrigues, and A. J. A. Meirelles, "Effect of solvent hydration and temperature in the deacidification process of sunflower oil using ethanol," *Journal of Food Engineering*, vol. 95, no. 2, pp. 291–297, 2009.

[2] C. E. C. Rodrigues, C. B. Gonçalves, E. Batista, and A. J. A. Meirelles, "Deacidification of Vegetable Oils by Solvent Extraction," *Recent Patents on Engineering*, vol. 1, no. 1, pp. 95–102, 2007.

[3] S. Turkay and H. Civelekoglu, "Deacidification of sulfur olive oil. I. Single-stage liquid-liquid extraction of miscella with ethyl alcohol," *Journal of the American Oil Chemists' Society*, vol. 68, no. 2, pp. 83–86, 1991.

[4] R. A. Carr, "Degumming and refining practices in the U.S.," *Journal of the American Oil Chemists' Society*, vol. 53, no. 6, pp. 347–352, 1976.

[5] V. Kale, S. P. R. Katikaneni, and M. Cheryan, "Deacidifying rice bran oil by solvent extraction and membrane technology," *Journal of the American Oil Chemists' Society*, vol. 76, no. 6, pp. 723–727, 1999.

[6] S. S. Koseoglu, "Membrane technology for edible oil refining," *Oils & Fats International*, vol. 5, pp. 16–21, 1991.

[7] N. Othman, Z. A. Manan, S. R. Wan Alwi, and M. R. Sarmidi, "A review of extraction technology for carotenoids and vitamin e recovery from palm oil," *Journal of Applied Sciences*, vol. 10, no. 12, pp. 1187–1191, 2010.

[8] C. Vera, M. Busto, J. Yori et al., "Adsorption in Biodiesel Refining - A Review," in *Biodiesel - Feedstocks and Processing Technologies*, Intech Open Access Publishers, 2011.

[9] H. Zhang, A. Li, J. Sun, and P. Li, "Adsorption of amphoteric aromatic compounds by hyper-cross-linked resins with amino groups and sulfonic groups," *Chemical Engineering Journal*, vol. 217, pp. 354–362, 2013.

[10] M. Yuan, S. Tong, S. Zhao, and C. Q. Jia, "Adsorption of polycyclic aromatic hydrocarbons from water using petroleum coke-derived porous carbon," *Journal of Hazardous Materials*, vol. 181, no. 1-3, pp. 1115–1120, 2010.

[11] M. Muzic, K. Sertic-Bionda, Z. Gomzi, S. Podolski, and S. Telen, "Study of diesel fuel desulfurization by adsorption," *Chemical Engineering Research and Design*, vol. 88, no. 4, pp. 487–495, 2010.

[12] M. L. M. Oliveira, A. A. L. Miranda, C. M. B. M. Barbosa, C. L. Cavalcante Jr., D. C. S. Azevedo, and E. Rodriguez-Castellon, "Adsorption of thiophene and toluene on NaY zeolites exchanged with Ag(I), Ni(II) and Zn(II)," *Fuel*, vol. 88, no. 10, pp. 1885–1892, 2009.

[13] L. Zou, B. Han, H. Yan, K. L. Kasperski, Y. Xu, and L. G. Hepler, "Enthalpy of adsorption and isotherms for adsorption of naphthenic acid onto clays," *Journal of Colloid and Interface Science*, vol. 190, no. 2, pp. 472–475, 1997.

[14] F. Wang, J. J.-H. Haftka, T. L. Sinnige, J. L. M. Hermens, and W. Chen, "Adsorption of polar, nonpolar, and substituted aromatics to colloidal graphene oxide nanoparticles," *Environmental Pollution*, vol. 186, pp. 226–233, 2014.

[15] J. Saien and S. Daliri, "Mass transfer coefficient in liquid-liquid extraction and the influence of aqueous phase pH," *Industrial & Engineering Chemistry Research*, vol. 47, no. 1, pp. 171–175, 2008.

[16] E. Batista, S. Monnerat, K. Kato, L. Stragevitch, and A. J. A. Meirelles, "Liquid-liquid equilibrium for systems of canola oil, oleic acid, and short-chain alcohols," *Journal of Chemical & Engineering Data*, vol. 44, no. 6, pp. 1360–1364, 1999.

[17] C. L. Munson and C. J. Klng, "Factors Influencing Solvent Selection for Extraction of Ethanol from Aqueous Solutions," *Industrial & Engineering Chemistry Process Design and Development*, vol. 23, no. 1, pp. 109–115, 1984.

[18] H. D. Schindler and R. E. Treybal, "Continuous-phase mass-transfer coefficients for liquid extraction in agitated vessels," *AIChE Journal*, vol. 14, no. 5, pp. 790–798, 1968.

[19] B. Schuur, W. J. Jansma, J. G. M. Winkelman, and H. J. Heeres, "Determination of the interfacial area of a continuous integrated mixer/separator (CINC) using a chemical reaction method," *Chemical Engineering and Processing: Process Intensification*, vol. 47, no. 9-10, pp. 1484–1491, 2008.

[20] R. P. Verma and M. M. Sharma, "Mass transfer in packed liquid-liquid extraction columns," *Chemical Engineering Science*, vol. 30, no. 3, pp. 279–292, 1975.

[21] B. Xu, W. Cai, X. Liu, and X. Zhang, "Mass transfer behavior of liquid-liquid slug flow in circular cross-section microchannel," *Chemical Engineering Research and Design*, vol. 91, no. 7, pp. 1203–1211, 2013.

[22] S.-S. Ashrafmansouri and M. Nasr Esfahany, "The influence of silica nanoparticles on hydrodynamics and mass transfer in spray liquid-liquid extraction column," *Separation and Purification Technology*, vol. 151, Article ID 12450, pp. 74–81, 2015.

[23] H. Bahmanyar and M. J. Slater, "Studies of drop break-up in liquid-liquid systems in a rotating disc contactor. Part I: Conditions of no mass transfer," *Chemical Engineering & Technology*, vol. 14, no. 2, pp. 79–89, 1991.

[24] Z. Azizi, A. Rahbar, and H. Bahmanyar, "Investigation of packing effect on mass transfer coefficient in a single drop liquid extraction column," *Iranian Journal of Chemical Engineering*, vol. 7, no. 4, pp. 3–11, 2010.

[25] E. Glueckauf and J. I. Coates, "Theory of chromatography. Part IV. The influence of incomplete equilibrium on the front boundary of chromatograms and on the effectiveness of separation," *Journal of the Chemical Society (Resumed)*, pp. 1315–1321, 1947.

[26] M. Otero, C. A. Grande, and A. E. Rodrigues, "Adsorption of salicylic acid onto polymeric adsorbents and activated charcoal," *Reactive and Functional Polymers*, vol. 60, no. 1-3, pp. 203–213, 2004.

[27] D. M. Ruthven and S. Farooq, "Air separation by pressure swing adsorption," *Gas Separation and Purification*, vol. 4, no. 3, pp. 141–148, 1990.

[28] C. I. Koncsag and A. Barbulescu, "Liquid-Liquid Extraction with and without a Chemical Reaction," in *Mass Transfer in Multiphase Systems and its Applications*, Chapter 10, InTech Open Access Publishers, 2011.

[29] T. Sankarshana, S. Ilaiah, and U. Virendra, *Liquid-Liquid Extraction: Comparison in Micro and Macro Systems*, AIChE Annual Meeting, 2013.

[30] F. Nosratinia, M. R. Omidkhah, D. Bastani, and A. A. Saifkordi, "Investigation of mass transfer coefficient under jetting conditions in a liquid-liquid extraction system," *Iranian Journal of Chemistry and Chemical Engineering*, vol. 29, no. 1, pp. 1–12, 2010.

[31] C. J. Geankoplis and A. N. Hixson, "Mass Transfer Coefficients in an Extraction Spray Tower," *Industrial & Engineering Chemistry*, vol. 42, no. 6, pp. 1141–1151, 1950.

[32] A. Salimi-Khorshidi, H. Abolghasemi, A. Khakpay, Z. Kheirjooy, and M. Esmaieli, "Spray and packed liquid-liquid extraction columns: Drop size and dispersed phase mass transfer," *Asia-Pacific Journal of Chemical Engineering*, vol. 8, no. 6, pp. 940–949, 2013.

[33] G. A. Hughmark, "Liquid-liquid spray column drop size, holdup, and continuous phase mass transfer," *Industrial & Engineering Chemistry Fundamentals*, vol. 6, no. 3, pp. 408–413, 1967.

[34] C. L. Ruby and J. C. Elgin, "Mass transfer between liquid drops and a continuous phase in a countercurrent fluidized system: liquidliquid extraction in a spray tower," *Chemical Engineering Progress Symposium*, vol. 51, pp. 17–29, 1955.

[35] S. Kang and B. Xing, "Adsorption of dicarboxylic acids by clay minerals as examined by in situ ATR-FTIR and ex situ DRIFT," *Langmuir*, vol. 23, no. 13, pp. 7024–7031, 2007.

[36] C. D. Hatch, R. V. Gough, and M. A. Tblbert, "Heterogeneous uptake of the Cl to C4 organic acids on a swelling clay mineral," *Atmospheric Chemistry and Physics*, vol. 7, no. 16, pp. 4445–4458, 2007.

[37] N. Kanazawa, K. Urano, N. Kokado, and Y. Urushigawa, "Exchange characteristics of monocarboxylic acids and monosulfonic acids onto anion-exchange resins," *Journal of Colloid and Interface Science*, vol. 271, no. 1, pp. 20–27, 2004.

[38] E. Suescún-Mathieu, A. Bautista-Carrizosa, R. Sierra, L. Giraldo, and J. C. Moreno-Piraján, "Carboxylic acid recovery from aqueous solutions by activated carbon produced from sugarcane bagasse," *Adsorption*, vol. 20, no. 8, pp. 935–943, 2014.

[39] Y. S. Aşçı and I. M. Hasdemir, "Removal of Some Carboxylic Acids from Aqueous Solutions by Hydrogels," *Journal Chemical & Engineerind data*, vol. 53, pp. 2351–2355, 2008.

[40] A. Li, Q. Zhang, G. Zhang, J. Chen, Z. Fei, and F. Liu, "Adsorption of phenolic compounds from aqueous solutions by a water-compatible hypercrosslinked polymeric adsorbent," *Chemosphere*, vol. 47, no. 9, pp. 981–989, 2002.

[41] M. S. Chiou and H. Y. Li, "Adsorption behavior of reactive dye in aqueous solution on chemical cross-linked chitosan beads," *Chemosphere*, vol. 50, no. 8, pp. 1095–1105, 2003.

[42] O. Ilgen and H. S. Dulger, "Removal of oleic acid from sunflower oil on zeolite 13X: Kinetics, equilibrium and thermodynamic studies," *Industrial Crops and Products*, vol. 81, pp. 66–71, 2016.

[43] R. Gayler, N. W. Roberts, and H. R. Pratt, "liquid-liquid extraction: Part IV. A further study of hold-up in packed columns," *Chemical Engineering Research and Design*, vol. 31, pp. 57–68, 1953.

[44] D. M. Ruthven, *Principles of Adsorption and Adsorption Processes*, John Wiley & Sons Inc, New York, NY, USA, 1984.

[45] J. J. Carberry, *Chemical and Catalytic Reaction Engineering*, Mc-Graw Hill, New York, NY, USA, 1976.

[46] N. Wakao and T. Funazkri, "Effect of fluid dispersion coefficients on particle-to-fluid mass transfer coefficients in packed beds. Correlation of sherwood numbers," *Chemical Engineering Science*, vol. 33, no. 10, pp. 1375–1384, 1978.

[47] M. E. Sigrist, H. R. Beldomenico, E. E. Tarifa, C. L. Pieck, and C. R. Vera, "Modelling diffusion and adsorption of As species in Fe/GAC adsorbent beds," *Journal of Chemical Technology and Biotechnology*, vol. 86, no. 10, pp. 1256–1264, 2011.

[48] G.-H. Xiu, T. Nitta, P. Li, and G. Jin, "Breakthrough Curves for Fixed-Bed Adsorbers: Quasi-Lognormal Distribution Approximation," *AIChE Journal*, vol. 43, no. 4, pp. 979–985, 1997.

[49] J. C. Yori, S. A. D'Ippolito, C. L. Pieck, and C. R. Vera, "Deglycerolization of biodiesel streams by adsorption over silica beds," *Energy & Fuels*, vol. 21, no. 1, pp. 347–353, 2007.

# 7

# One-Step Ultrafiltration Process for Separation and Purification of a Keratinolytic Protease Produced with Feather Meal

**Juliana R. Machado,[1] Emanuelle E. Severo,[1] Janaina M. G. de Oliveira ⓘ,[1] Joana da C. Ores ⓘ,[1] Adriano Brandelli,[2] and Susana J. Kalil ⓘ[1]**

[1]*Escola de Química e Alimentos, Universidade Federal do Rio Grande, P.O. Box 474, 96203900 Rio Grande, RS, Brazil*
[2]*Laboratório de Bioquímica e Microbiologia Aplicada, Universidade Federal do Rio Grande do Sul, 91501-970 Porto Alegre, RS, Brazil*

Correspondence should be addressed to Susana J. Kalil; dqmsjk@furg.br

Academic Editor: Donald L. Feke

A purification technique to obtain keratinolytic proteases produced by *Bacillus* sp. P45 in a medium containing chicken feather meal as substrate is presented. The experiments were carried out in a dead-end ultrafiltration unit, and the influence of the membrane cutoff, pH of enzymatic extract, and operating pressure on the purification of keratinase were studied. The one-step ultrafiltration process with the membrane molecular mass cutoff of 10 kDa at pH 8.0 and operating pressure of 0.147 MPa showed an enzyme recovery of 87.8% and a 4.1-fold purification factor. It is showed that ultrafiltration could be potentially used in the purification of keratinases.

## 1. Introduction

Brazil is a major producer of poultry meat globally, generating tons of organic waste such as viscera, feet, bones, feathers, and blood that are not exploited for human consumption. These organic by-products are mostly used to prepare animal feed or soil fertilizers though methods of incineration in the disposal of such waste are still applied in some locations [1]. Chicken feathers represent about 10% of the waste disposed of by the poultry industry. Feathers are essentially composed of keratin (about 90% w/w), a protein of difficult degradation due to the presence of strong chemical bonds in the polypeptide chain as disulfide bridges, hydrogen, and hydrophobic interactions that hinder their rapid degradation in the environment [2–4].

The recovery of keratin poses a great challenge to the poultry industry. An alternative to recycle these keratinous materials is the bioconversion into products with higher added value by specific microorganisms producing keratinolytic proteases. These proteases, named keratinases, are often serine or metalloproteases capable of degrading keratinous wastes [5, 6]. Alkaline keratinases are produced by several bacterial species, including *Bacillus licheniformis* [7–9], *Kocuria rosea* [10], *Streptomyces* sp. [11], and even fungi such as *Aspergillus niger* [12]. A keratinase-producing bacterium, *Bacillus* sp. P45, was isolated from the intestine of the Amazon basin fish *Piaractus mesopotamicus* [4]. *Bacillus* sp. P45 efficiently degraded feather keratin during submerged cultivations, producing extracellular keratinolytic enzymes.

This type of enzyme has gained biotechnological interest for use in the fertilizer, detergents, and cosmetic industries, and also for the potential use of keratinous residues to produce biohydrogen and biogas [1, 3, 13]. Furthermore, keratinolytic proteases can be applied to obtain animal feed rich in amino acids. The enzymatic hydrolysis preserves essential amino acids such as methionine, lysine, and tryptophan and avoids the formation of nonnutritive amino acids such as lanthionine and lysinoalanine [14]. New applications have also been developed such as prion degradation for prevention of mad cow disease [3], biodegradable plastic manufacturing, and keratin peptide production [14, 15].

The feasibility of an industrial application of keratinases resides initially in obtaining the enzyme from a viable source, such as poultry feathers, and its purification using a simple protocol. Several techniques have been studied to purify microbial keratinases, including ammonium sulfate precipitation, solvent precipitation, ultrafiltration, ion exchange chromatography, gel filtration chromatography, hydrophobic interaction chromatography, and hydroxyapatite chromatography [5, 7–9, 11, 16]. However, those studies employ purification processes with sequences of different techniques in order to obtain highly purified enzyme preparations needed to characterize the enzyme. In those protocols, the membrane separation has been applied only as a concentration step. For industrial applications, high degrees of purity are often not required. Thus, it is necessary to study industrially applicable techniques to purify microbial keratinases, reducing the costs of this process. In addition to scale-up problems, which limit protein production levels, the traditional techniques such as chromatography require complex instrumentation support to run efficiently and yield low throughput of product at an extremely high cost. Hence, a separation technique that can provide high productivity and purity at the same time at low process cost would certainly be beneficial to the biotechnology industry [17].

Ultrafiltration has been widely used for protein concentration and separation because of the lower complexity compared to the previously mentioned purification techniques [18]. The major advantage of the ultrafiltration processes over a conventional bioseparation processes is the high product throughput. However, in spite of widespread use of ultrafiltration in processes such as diafiltration and concentration, the potential for its use in protein fractionation has not been exploited in the biotechnology industry [17]. Nevertheless, there are studies that show that ultrafiltration can be applied to purification of enzymes, obtaining high yields and product purity at the same time [19–21].

This study provides a potential one-step purification process for obtaining a microbial keratinase. Furthermore, this process adds value to the most problematic by-product of the poultry industry. In the present research, the use of an industrially applicable technique for the purification of the keratinase obtained from *Bacillus* sp. P45 using feathers was studied. The enzyme was concentrated and purified by ultrafiltration. This approach allows obtaining a purified keratinase in a single step.

## 2. Materials and Methods

*2.1. Microorganism and Inoculum Preparation.* *Bacillus* sp. P45 (GenBank accession number AY962474) was maintained at 4°C on brain heart infusion (BHI) agar plates. For inoculum preparation, *Bacillus* sp. P45 was inoculated on BHI plates and incubated at 30°C for 24 h. The cultures were gently scraped from the agar surface, added to a sterile NaCl solution (8.5 g/L), and mixed until a homogeneous suspension with an optical density of 0.5 at 600 nm was obtained [22].

TABLE 1: Experimental design $2^3$ in coded levels (real values).

| Trial | MWCO (kDa) | pH | Pressure (MPa) |
|---|---|---|---|
| 1 | −1 (10) | −1 (7.0) | −1 (0.147) |
| 2 | +1 (30) | −1 (7.0) | −1 (0.147) |
| 3 | −1 (10) | +1 (8.0) | −1 (0.147) |
| 4 | +1 (30) | +1 (8.0) | −1 (0.147) |
| 5 | −1 (10) | −1 (7.0) | +1 (0.245) |
| 6 | +1 (30) | −1 (7.0) | +1 (0.245) |
| 7 | −1 (10) | +1 (8.0) | +1 (0.245) |
| 8 | +1 (30) | +1 (8.0) | +1 (0.245) |

MWCO: molecular weight cutoff.

*2.2. Submerged Cultivation.* The enzyme was produced by submerged cultivation as described by Daroit et al. [22] using the culture medium composed of (g/L) feather meal (50) and $NH_4Cl$ (5.25) prepared in a mineral medium (NaCl (0.5), $K_2HPO_4$ (0.3), and $KH_2PO_4$ (0.4)). In Erlenmeyer flasks (250 mL) containing 50 mL of medium, the initial pH was adjusted to 7.0 before sterilization by autoclaving at 121°C for 15 min. Cultures were initiated with 1% (v/v) inoculum. The growing conditions were 30°C and 125 rpm for 48 h. At the end of the cultivation, the supernatant was separated by centrifugation (5.000 ×g for 20 min), obtaining the crude enzyme extract.

*2.3. Ultrafiltration (UF).* Experiments were conducted in a dead-end ultrafiltration unit with a working volume of 160 mL stirred by a magnetic bar suspended down to 5 mm of the membrane. The module was equipped with a regenerated cellulose membrane (Millipore) having a total filtering area of 19.63 cm$^2$. A new membrane was used for each experiment, and two different sizes were used with molecular weight cutoff of 10 kDa and 30 kDa. The system was pressurized with compressed nitrogen, and the temperature was kept at 15°C to avoid enzyme denaturation. Prior to the ultrafiltration process with crude enzyme extract, a flux of water was passed through the membrane. A volume of 40 mL of crude enzyme extract was added, and the process was stopped when the volumetric concentration factor reached the value of 4 [21]. The enzyme activity and the protein content of the feed (input crude extract), retentate, and permeate were assayed at the end of each experiment.

*2.4. Concentration and Purification of Keratinase by Ultrafiltration.* The influence of the operating pressure, the pH of the enzyme extract, and the molecular weight cutoff (MWCO) during the ultrafiltration process was studied by a 2$^3$ experimental design, totaling eight experimental trials carried out in duplicate. The responses evaluated were enzyme recovery and purification factor. The statistical analysis of the experimental design was performed using analysis of variance (ANOVA) with a confidence level of 95%. Statistica 5.0 software (StatSoft Inc., USA) was used for the regression and graphical analysis of the data obtained by the experimental design. Table 1 presents the matrix of experimental design.

The efficiency of the process of concentration and purification of the enzyme keratinase by ultrafiltration was evaluated through the enzyme recovery and purification factor. The enzyme recovery (%R) was obtained by the ratio between the total activity in the retentate and the total activity in the feed. The purification factor (PF) was calculated by dividing the specific activity of the enzyme in the retentate (U/mg) by the specific activity of the enzyme extract used in the feed (U/mg).

*2.5. Enzyme Assay.* Keratinase activity was monitored using the soluble substrate azocasein (Sigma, Saint Louis, USA) as described by Daroit et al. [4]. One unit (U) of protease activity was defined as the amount of enzyme that caused an increase of 0.1 absorbance unit at the defined assay conditions.

*2.6. Total Protein Determination.* Protein was determined by the method of Lowry et al. [23], using bovine serum albumin (BSA) as the standard.

## 3. Results and Discussion

*3.1. Study of Flux with Variation of pH, Operating Pressure, and MWCO in the Ultrafiltration Process.* During the trials, the flux behavior in the process of concentration and purification of keratinase was evaluated (Figures 1 and 2). In the ultrafiltration at pH 7.0 (Figure 1), it is possible to observe that the flux variation of trial 2 was 76.4 to 25 $L/m^2 \cdot h$ with a reduction of 67.5% and an operating time of 0.6 h. Trials 5 and 6 showed a flux reduction of 65.1 and 64.4%, respectively. Trial 1 (10 kDa and 0.147 MPa) showed the smallest flux variation from 38.2 to 19.5 $L/m^2 \cdot h$ with a reduction of 48.9% and an operating time of 0.73 h.

In Figure 2, the flux profiles at pH 8.0 show that the membrane cutoff and the operating pressure affect the permeate flux more than the pH—the change for pH 8.0 in trials 3, 4, 7, and 8 has no significant influence on the flux permeate, compared with the same conditions at pH 7.0. Trial 3 (10 kDa, 0.147 MPa, and pH 8.0) had the lowest percentage decrease of the flux with a value of 46.1% (39 to 21 $L/m^2 \cdot h$), almost the same as trial 1 (48.9%), which is performed with the same MWCO and operating pressure but with pH 7.0.

When the effect of the membrane cutoff on the permeate flux was analyzed, it was observed that, with the 30 kDa membrane, the reduction of the flux is larger than with the 10 kDa membrane at low pressures (0.147 MPa). This may be attributed to the fact that the extract contains peptides, amino acids, and other proteolytic enzymes with molecular mass below 30 kDa, which facilitates their initial transport through the membrane, thereby obtaining a higher initial flux of the process. However, due to the concentration polarization and fouling phenomena, the flow rate drops to values as low as those obtained with 10 kDa.

The flux reductions caused by the fouling and concentration polarization have long been recognized as major problems in the protein ultrafiltration. It could be seen that, at the beginning of the process, there is a rapid decline of the flux. After this initial period, there is a gradual decline of the

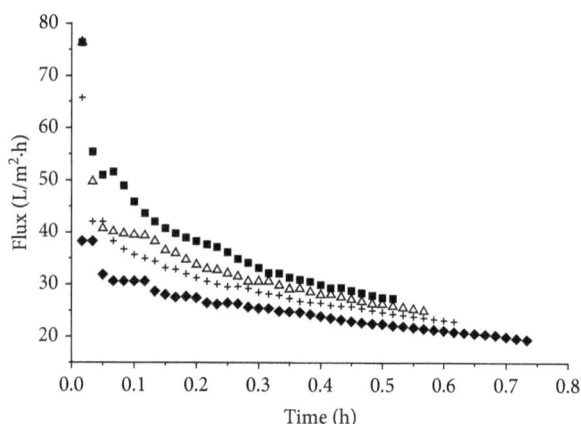

FIGURE 1: The flux permeate during the ultrafiltration process at pH 7.0: trial 1 (◆): 10 kDa and 0.147 MPa; trial 2 (△): 30 kDa and 0.147 MPa; trial 5 (+): 10 kDa and 0.245 MPa; trial 6 (◁): 30 kDa and 0.245 MPa.

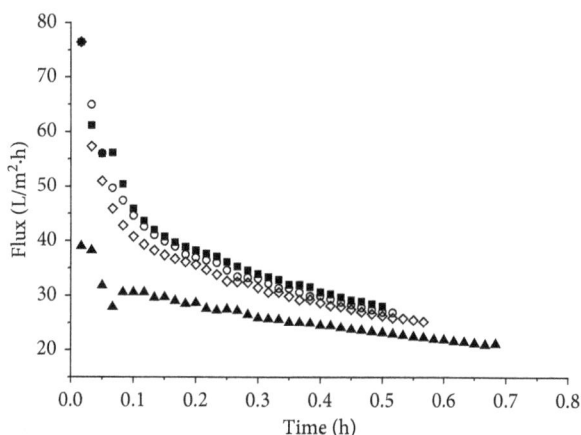

FIGURE 2: The flux permeate during the ultrafiltration process at pH 8.0: trial 3 (▲): 10 kDa and 0.147 MPa; trial 4 (◊): 30 kDa and 0.147 MPa; trial 7 (○): 10 kDa and 0.245 MPa; trial 8 (■): 30 kDa and 0.245 MPa.

flux that occurs due to the formation of incrustations of proteins into the membrane surface forming the effect of concentration polarization and the fouling phenomena. The proteins that are retained by the membrane can form a gel layer on the membrane surface, which acts as a second dynamic membrane, increasing protein retention. Other authors who used the ultrafiltration process for concentration and separation of proteins [17, 21, 24] observed the same behavior. During the process of concentration and purification of a pretreated protease from the tuna spleen extract, a decline of 59% was obtained in the flux [25], which is close to values obtained in this work.

However, at higher pressures (0.245 MPa), the difference on the flux reduction between the 10 kDa and 30 kDa membranes is less pronounced. In fact, higher operating pressures provided higher initial flow rates in the process. This higher pressure may have decreased the effect of concentration polarization; thus, the process with the 10 kDa membrane achieved initial fluxes close to the experiments with the 30 kDa membrane.

TABLE 2: Responses of $2^3$ experimental design.

| Trial | %R predicted | %R experimental | Relative standard (%) | PF predicted | PF experimental | Relative standard (%) |
|---|---|---|---|---|---|---|
| 1 | 80.5 | 79.3 | −1.5 | 3.8 | 3.4 | −11.2 |
| 2 | 40.1 | 41.3 | 3.0 | 2.3 | 2.1 | −11.4 |
| 3 | 86.7 | 87.8 | 1.3 | 3.8 | 4.1 | 7.8 |
| 4 | 46.2 | 44.6 | −3.7 | 2.3 | 1.9 | −23.2 |
| 5 | 82.9 | 77.3 | −7.3 | 3.7 | 3.6 | −2.2 |
| 6 | 62.7 | 62.2 | −0.9 | 3.4 | 3.5 | 1.7 |
| 7 | 82.9 | 82.4 | −0.7 | 3.7 | 3.7 | 0.5 |
| 8 | 68.9 | 69.4 | 0.7 | 3.4 | 3.4 | −1.2 |

$n = 2$; %R: enzyme recovery; PF: purification factor.

TABLE 3: Analysis of variance for enzyme recovery and purification factor.

| | Factor | Sum of square | Degrees of freedom | Mean squares | $F_{calculated}$ | $F_{tabulated}$ |
|---|---|---|---|---|---|---|
| | Regression | 4175.8 | 4 | 1043.9 | 300.8 | 3.4 |
| %R ($R = 0.99$) | Residue | 38.2 | 11 | 3.5 | | |
| | Total corrected | 4213.9 | 15 | | | |
| | Regression | 5.3 | 3 | 1.8 | 9.00 | 3.5 |
| PF ($R = 0.84$) | Residue | 2.2 | 12 | 0.2 | | |
| | Total corrected | 7.5 | 15 | | | |

*3.2. Concentration and Purification of Keratinase by Ultrafiltration.* Table 2 shows the values of the parameters evaluated during keratinase purification and the relative standard deviations with the predicted and experimental values obtained in the trials. The recovery of the enzyme keratinase was in the range of 41.3 to 87.8% and a purification factor between 1.9- and 4.1-fold. It was possible to verify that the increase of the membrane cutoff from 10 to 30 kDa decreases the recovery of enzyme and consequently the purification factor. Trial 3, with a membrane cutoff of 10 kDa, pH 8.0, and pressure of 0.147 MPa, provided the highest enzyme recovery (87.8%) and purification factor (4.1-fold). The membrane cutoff, the operating pressure, and the interaction between these two variables had significant effect on the two responses, enzyme recovery and purification factor; the pH, on the other hand, had a significant effect on the purification factor only. ANOVA was carried out using Fisher's statistical test (Table 3) for the validation of the empirical model obtained for the PF and recovery of the enzyme. Correlation coefficients ($R$) of 0.99 and 0.84 were obtained for the %R and PF, respectively. In addition to this, the $F_{calculated}$ value was 3 times higher than the $F_{tabulated}$ value for the PF and 90 times higher for the %R, showing that the model fitted the data satisfactorily and was considered predictive for both responses. Equations (1) and (2) represent empirical models codified for recovery (%R) and purification factor (FP), respectively.

$$\%R = 68.11 - 13.62 * MWCO + 3.08 * pH$$
$$+ 4.735 * P + 6.60 * MWCO * P, \quad (1)$$

$$PF = 3.31 - 0.42 * MWCO + 0.25 * P + 0.3 * MWCO * P, \quad (2)$$

where %R is the enzyme recovery, PF is the purification factor, MWCO is the molecular mass cutoff, and $P$ is the operating pressure.

Figures 3 and 4 show the contour plots obtained from the empirical models for better understanding of the interaction of the variables pH, pressure, and MWCO and to obtain the best conditions for keratinase concentration and purification.

Analyzing the interaction of pH and pressure on enzyme recovery, it is possible to verify that when a 10 kDa membrane was employed with higher values of pH (between 7.5 and 8.0) and pressure between 0.196 and 0.147 MPa, the enzyme recovery increased (up to 80%), as shown in Figure 3. When a 30 kDa membrane was used, the effect of pressure on recovery values was even more pronounced. This is probably attributed to the formation of fouling that favors the retention of the enzyme.

With respect to the purification factor, the pH of the enzyme extract did not affect this response significantly, not generating the contour curve. The effect of pH is more related to the enzyme activity—keratinases usually have a great stability in neutral and alkaline pH [2, 14]. Only the MWCO and operating pressure had an influence on the purification factor (Figure 4). It was observed that higher purification factor values were obtained using lower MWCO values with operation pressures between 0.147 and 0.196 MPa.

Evaluating the influence of the MWCO, a negative effect was observed for both responses (recovery and purification factor). In other words, when the smallest MWCO is used (10 kDa), the highest values of recovery and purification factor are obtained. An MWCO of 30 kDa decreased the enzyme recovery by 27.2%. This can be attributed to the molecular mass of the enzyme under study. According to Daroit et al. [26], the molecular mass of keratinase from *Bacillus* sp. P45 is approximately 26 kDa, and thus, higher MWCO favors the passage of the enzyme to permeate, decreasing the recovery of the enzyme.

Despite the enzyme having a molecular mass of 26 kDa, the 30 kDa membrane was tested because it was expected that the concentration polarization and fouling phenomena could favor the retention of the enzyme due to formation of

(a)

(b)

(c)

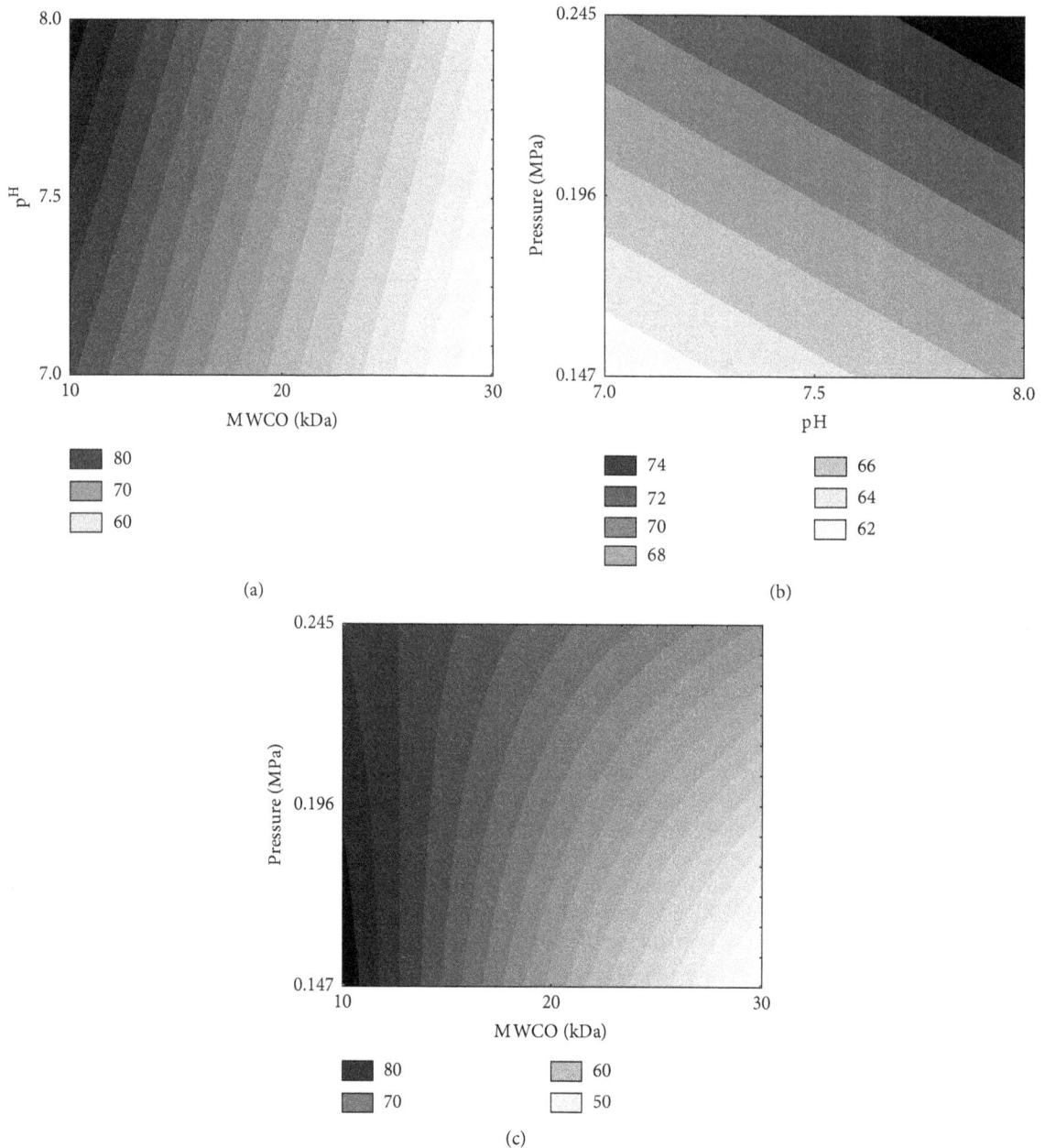

FIGURE 3: Contour curves for the recovery of enzyme as a function of (a) pH and MWCO, (b) pH and the operating pressure, and (c) MWCO and operating pressure.

the gel layer without major losses in the flux since it is a membrane with a larger pore diameter. Actually, a higher flow was observed, but the retention of the keratinase was much less efficient compared to the 10 kDa membrane. Therefore, the enzyme permeated the membrane of 30 kDa resulting in a smaller recovery and purification factor.

In relation to the pressure, it can be verified that lower pressures and MWCO provide values of higher purification factors. Furthermore, membranes of lower molecular mass cutoff with high values of pH and low operation pressures provide higher recovery values. Analyzing the flux behavior, at higher pressures (0.245 MPa), the reduction of the flux was more pronounced in the best

condition (10 kDa membrane and pH 8) compared with lower pressures (0.147 MPa).

Ultrafiltration has been applied in the purification of keratinases only intended to concentrate the enzyme for the next purification step [7, 8, 27–29]. However, it is observed that the values found in the cited works are lower than those found in this study that use just one step. Radha and Gunasekaran [27] produced a keratinase by a recombinant *Bacillus megaterium*. During the purification process, the ultrafiltration step (MWCO of 10 kDa) achieved a purification factor of 2.3-fold with a recovery of 73.5%. Lin et al. [7] purified and characterized a keratinase isolated from a feather-degrading culture medium inoculated *with Bacillus licheniformis* PWD-1.

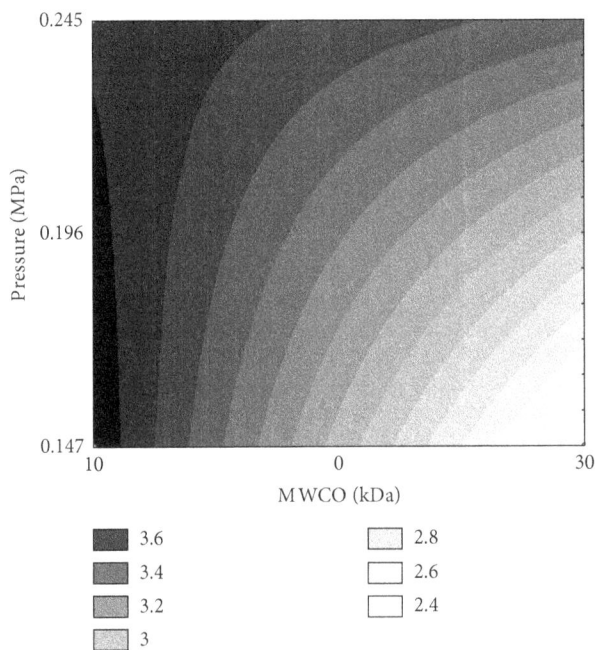

FIGURE 4: Contour curve for purification factor as a function of operating pressure and MWCO.

The ultrafiltration step (MWCO of 10 kDa) achieved a purification factor of 1.8-fold and recovery of 69.3%. In the study of Allpress et al. [28], the extracellular keratinase produced by *Lysobacter* NCIMB 9497 was purified for further characterization. In the ultrafiltration step (MWCO of 10 kDa), a purification factor of 2-fold was obtained.

Cheng et al. [8] and Suntornsuk et al. [29] also used a membrane MWCO of 10 kDa for keratinase concentration. Cheng et al. [8] characterized the keratinase produced by *Bacillus licheniformis* PWD-l produced with feather powder. Prior to the purification process, the enzyme was concentrated using a spiral cartridge concentrator with a MWCO membrane of 10 kDa. Suntornsuk et al. [29] isolated and determined the properties of a keratinase produced by a thermotolerant feather-degrading bacterial strain from Thai soil. For its purification, the enzyme was firstly concentrated by a 10 kDa MWCO membrane.

In this study, an ultrafiltration process was developed in the purification of keratinase from *Bacillus* sp. P45 in a single step. A system with an operating pressure of 0.147 MPa, membrane MWCO of 10 kDa, and pH of 8.0 provided a 4-fold purification factor without major recovery losses. Thus, it is possible to reduce the purification steps through an optimized ultrafiltration method for purification of this enzyme. The ease of operation and high efficiency makes the ultrafiltration process an interesting alternative in the purification of keratinases, which can be used in the degradation of resistant materials in effluents or even for applications in industry cleaning products like detergents. Besides, the enzyme has potential importance for production of protein hydrolysates [3].

## 4. Conclusions

The ultrafiltration system formed by the membrane with the molecular weight cutoff of 10 kDa, an operating pressure of 0.147 MPa, and a pH of 8.0 provided a recovery of 87.8% and 4.1-fold purification factor of the enzyme keratinase from *Bacillus* sp. P45. The ultrafiltration process is positioned as a potential alternative to be used for the industrial concentration and purification of keratinase from *Bacillus* sp. P45 in just one step. Furthermore, this process adds value to the most problematic by-product of the poultry industry, chicken feathers.

## Conflicts of Interest

The authors declare that there are no conflicts of interest regarding the publication of this paper.

## Acknowledgments

This work was supported by the Coordenação de Aperfeiçoamento de Pessoal de Nível Superior (CAPES) and the Conselho Nacional de Desenvolvimento Científico e Tecnológico (CNPq) through research scholarships.

## References

[1] A. Lasekan, F. Abu Bakar, and D. Hashim, "Potential of chicken by-products as sources of useful biological resources," *Waste Management*, vol. 33, no. 3, pp. 552–565, 2013.

[2] B. Bockle, B. Galunsky, and R. Muller, "Characterization of a keratinolytic serine proteinase from *streptomyces pactum* dsm 40530," *Applied and Environmental Microbiology*, vol. 61, no. 10, pp. 3705–3710, 1995.

[3] A. Brandelli, "Bacterial keratinases: useful enzymes for bioprocessing agroindustrial wastes and beyond," *Food and Bioprocess Technology*, vol. 1, no. 2, pp. 105–116, 2008.

[4] D. J. Daroit, A. P. F. Correa, and A. Brandelli, "Keratinolytic potential of a novel *bacillus* sp. P45 isolated from the Amazon basin fish Piaractus mesopotamicus," *International Biodeterioration and Biodegradation*, vol. 63, no. 3, pp. 358–363, 2009.

[5] C. Bernal, J. Cairo, and N. Coello, "Purification and characterization of a novel exocellular keratinase from kocuria rosea," *Enzyme and Microbial Technology*, vol. 38, no. 1-2, pp. 49–54, 2006.

[6] R. C. S. Thys and A. Brandelli, "Purification and properties of a keratinolytic metalloprotease from *microbacterium* sp.," *Journal of Applied Microbiology*, vol. 101, no. 6, pp. 1259–1268, 2006.

[7] X. A. Lin, C. G. Lee, E. S. Casale, and J. C. H. Shih, "Purification and characterization of a keratinase from a feather-degrading bacillus licheniformis strain," *Applied and Environmental Microbiology*, vol. 58, no. 10, pp. 3271–3275, 1992.

[8] S. W. Cheng, H. M. Hu, S. W. Shen, H. Takagi, M. Asano, and Y. C. Tsai, "Production and characterization of keratinase of

a feather-degrading *bacillus licheniformis* pwd-1," *Bioscience, Biotechnology, and Biochemistry*, vol. 59, no. 12, pp. 2239–2243, 1995.

[9]  E. Tiwary and R. Gupta, "Medium optimization for a novel 58 kDa dimeric keratinase from *bacillus licheniformis* er-15: biochemical characterization and application in feather degradation and dehairing of hides," *Bioresource Technology*, vol. 101, no. 15, pp. 6103–6110, 2010.

[10]  C. Bernal, L. Vidal, E. Valdivieso, and N. Coello, "Keratinolytic activity of kocuria rosea," *World Journal of Microbiology and Biotechnology*, vol. 19, no. 3, pp. 255–261, 2003.

[11]  F. H. Xie, Y. P. Chao, X. Q. Yang et al., "Purification and characterization of four keratinases produced by *streptomyces* sp. strain 16 in native human foot skin medium," *Bioresource Technology*, vol. 101, no. 1, pp. 344–350, 2010.

[12]  A. M. Farag and M. A. Hassan, "Purification, characterization and immobilization of a keratinase from Aspergillus oryzae," *Enzyme and Microbial Technology*, vol. 34, no. 2, pp. 85–93, 2004.

[13]  G. Forgacs, S. Alinezhad, A. Mirabdollah, E. Feuk-Lagerstedt, and I. S. Horvath, "Biological treatment of chicken feather waste for improved biogas production," *Journal of Environmental Sciences*, vol. 23, no. 10, pp. 1747–1753, 2011.

[14]  R. Gupta and P. Ramnani, "Microbial keratinases and their prospective applications: an overview," *Applied Microbiology and Biotechnology*, vol. 70, no. 1, pp. 21–33, 2006.

[15]  A. Brandelli, D. J. Daroit, and A. Riffel, "Biochemical features of microbial keratinases and their production and applications," *Applied Microbiology and Biotechnology*, vol. 85, no. 6, pp. 1735–1750, 2010.

[16]  J. M. Kuo, J. I. Yang, W. M. Chen et al., "Purification and characterization of a thermostable keratinase from *meiothermus* sp. 140," *International Biodeterioration and Biodegradation*, vol. 70, pp. 111–116, 2012.

[17]  R. Ghosh and Z. F. Cui, "Protein purification by ultrafiltration with pre-treated membrane," *Journal of Membrane Science*, vol. 167, no. 1, pp. 47–53, 2000.

[18]  R. van Reis and A. Zydney, "Bioprocess membrane technology," *Journal of Membrane Science*, vol. 297, no. 1-2, pp. 16–50, 2007.

[19]  E. Nakkeeran, K. S. Venkatesh, R. Subramanian, and S. U. Kumar, "Purification of *aspergillus carbonarius* polygalacturonase using polymeric membranes," *Journal of Chemical Technology and Biotechnology*, vol. 83, no. 7, pp. 957–964, 2008.

[20]  J. B. Wang, J. G. Liu, J. R. Lu, and Z. F. Cui, "Isolation and purification of superoxide dismutase from garlic using two-stage ultrafiltration," *Journal of Membrane Science*, vol. 352, no. 1-2, pp. 231–238, 2010.

[21]  S. Golunski, V. Astolfi, N. Carniel et al., "Ethanol precipitation and ultrafiltration of inulinases from kluyveromyces marxianus," *Separation and Purification Technology*, vol. 78, no. 3, pp. 261–265, 2011.

[22]  D. J. Daroit, A. P. F. Correa, and A. Brandelli, "Production of keratinolytic proteases through bioconversion of feather meal by the Amazonian bacterium *bacillus* sp. P45," *International Biodeterioration and Biodegradation*, vol. 65, no. 1, pp. 45–51, 2011.

[23]  O. H. Lowry, N. J. Rosebrough, A. L. Farr, and R. J. Randall, "Protein measurement with the folin phenol reagent," *Journal of Biological Chemistry*, vol. 193, pp. 265–275, 1951.

[24]  D. M. Kanani and R. Ghosh, "A constant flux based mathematical model for predicting permeate flux decline in constant pressure protein ultrafiltration," *Journal of Membrane Science*, vol. 290, no. 1-2, pp. 207–215, 2007.

[25]  Z. Li, A. H-Kittikun, and W. Youravong, "Purification of protease from pre-treated tuna spleen extract by ultrafiltration: an altered operational mode involving critical flux condition and diafiltration," *Separation and Purification Technology*, vol. 66, no. 2, pp. 368–374, 2009.

[26]  D. J. Daroit, A. P. F. Correa, J. Segalin, and A. Brandelli, "Characterization of a keratinolytic protease produced by the feather-degrading Amazonian bacterium *bacillus* sp. P45," *Biocatalysis and Biotransformation*, vol. 28, no. 5-6, pp. 370–379, 2010.

[27]  S. Radha and P. Gunasekaran, "Purification and characterization of keratinase from recombinant *pichia* and *bacillus* strains," *Protein Expression and Purification*, vol. 64, no. 1, pp. 24–31, 2009.

[28]  J. D. Allpress, G. Mountain, and P. C. Gowland, "Production, purification and characterization of an extracellular keratinase from *Lysobacter* ncimb 9497," *Letters in Applied Microbiology*, vol. 34, no. 5, pp. 337–342, 2002.

[29]  W. Suntornsuk, J. Tongjun, P. Onnim et al., "Purification and characterisation of keratinase from a thermotolerant feather-degrading bacterium," *World Journal of Microbiology and Biotechnology*, vol. 21, no. 6-7, pp. 1111–1117, 2005.

# 8

# Droplet Impact Phenomena in Fluidized Bed Coating Process with a Wurster Insert

**T. Havaić, A.-M. Đumbir, M. Gretić, G. Matijašić ⓘ, and K. Žižek**

*Faculty of Chemical Engineering and Technology, University of Zagreb, Zagreb, Croatia*

Correspondence should be addressed to G. Matijašić; gmatijas@fkit.hr

Academic Editor: Michael Harris

The aim of this study is to determine favorable process conditions for the coating of *placebo* tablets. Tablets made of micro-crystalline cellulose are coated with hydroxypropyl cellulose polymer and Advantia™ Prime polymeric mixture film in lab-scale fluid-bed environment with a Wurster tube. In order to determine favorable process conditions (concentration, Wurster tube position, inlet air temperature, and atomization pressure), evaluation factors expressing process efficiency were calculated. Stereomicroscopy analysis provided good results with respect to the coating layer adherence and consistency. Results showed that the increased number of the coating cycles contributes to the desired featureless film morphology, when sufficiently high temperature and pressure are applied, thus resulting in high intra- and intertablet uniformity. Additionally, this paper analyzes the coating process from a mechanistic perspective of the underlying phenomena occurring on a tablet surface. Provided diagrams can help efficiently in detecting proper conditions that will result in coated tablets with strictly defined aimed properties. Process and formulation properties synergically result in a preferential occurrence of a deposition mechanism for all experiments conducted. Moreover, collision is found as a prevalent impact regime for the coating process studied. Finally, our intention here is to correlate hydrodynamic conditions and droplet breakup occurrence with a droplet diameter.

## 1. Introduction

Coating process is widely used in many diverse industries. Numerous process variables affect fluidized bed coating performance, thus influencing the product properties [1]. Complexity of the process can be overwhelmed by studying the process in detail using a multiscale approach [2]. High coating quality and higher process efficiency can be accomplished if the right process settings are strictly applied [3].

Coating of pharmaceuticals serves to enhance therapeutic, technological, and marketing properties of particular drugs. Among other common dosage forms, tablets emerge as the most popular solid oral dosage form. Coated tablets can provide optimal drug release profiles, helping enhance therapeutic performance, moisture protection, and protection from the gastric environment. Coatings are also convenient for incorporating another drug to avoid chemical incompatibilities or formula adjuvant.

Pharmaceutical industry emphasizes significance of coatings, being an important factor in releasing an active ingredient. Consequently, there is a growing interest in carrying out more efficient coating processes [4]. A great deal of practical problems arises with the proper selection of formulation and process properties. A good selection of process conditions is essential to avoid coalescence and breakage of the particle, as well as undesirable film formation mechanism, resulting in poor coating qualities.

Assorted purposes of coating specify the experimental program. The objective is to avoid tablet sticking on the equipment wall and prevent agglomeration, thus providing homogeneous coating thickness distribution on the tablet surface. It should be stressed that certain combinations of process properties lead to unevenly distributed coating thickness and irregular morphology. This results in a wide tablet-to-tablet coating variation and consequently may result in undesired variation in drug-release profiles [5], causing the process to be quite inefficient, therefore highly uneconomical [6].

The major advantage of the fluidized-bed process over drum coating is high level of film uniformity, since the film quality changes in line with the process conditions adjustment. Coating needs to be uniformly distributed on the tablet surface, while maintaining good mechanical properties. Certain alterations in process conditions (temperature, atomization air flow, solution concentration, etc.) strikingly affect the average size of the coating solution droplet. Droplet properties (size and velocity) directly affect the impact regime, hence on the film morphology and its final consistence.

The film coating process produces thin polymer-based coating with layer thickness varying from 5 to $50\,\mu m$, depending on the coating solution properties and the number of coating cycles applied. In the case of film coating, tablet weight increases only 2–5%, that is much less to those applied to sugar coating where weight increase is up to 50%. Weight reduction directly affects the tablet size and transport-related costs [7].

The main ingredient of coating solution used in the film coating process is a polymer. In most cases, it comes to cellulose esters, acrylic polymers, and copolymers. Being the key element, the polymer properties substantially influence the features of the coating layer. It is essential for excellent mechanical properties to coincide with good solubility, viscosity, and permeability. In order to meet the end-use requirement, more than one polymer component are combined [8].

Together with aforementioned specifications, coating needs to be compact, glassy, opaque, and affixed firmly. Extra ingredients are frequently added to ensure aimed coating performance.

Currently, industrial coating process optimization relies mostly on empirically determined conclusions. Furthermore, common coating quality factors are found to be complex to relate to process conditions alteration. For this reason, calculation-based adjustment of process settings in many cases gives poor result [6]. Another way is to conduct experimental research trying to reach desired product appearance.

Fluidized bed coating process variables being solid particles velocity, coating period, coating solution concentration and flow rate, viscosity, and droplet size directly affect applied film properties. Accurate and thorough interpretation of results can give a hint to the needed changes in process modeling. This way the optimization process can be accelerated and hence more affordable.

For successful process evolution, it is essential to detect the effect of all process and formulation properties on the underlying mechanisms controlling the process and therewith on the final process outcome. Regarding this, there is a strong need to carefully consider the coating process as a complex synergy of many events occurring in the coating process volume. These events correspond to three underlying phenomena. They are fluidization, spraying, and drying. We have to apply an adequate proper fluidization flow pattern for tablets, to reach spray droplets uniformly and to dry spray droplets at the suitable kinetics. These phenomena act simultaneously and thus competitively create the final coating performance.

With process and material properties, we are highly able to adjust the events on each tablet's surface in a manner we strive.

Accordingly, the main goal of this research is to comprehensively investigate the effect of many process variables on coating layer formation. In addition, we seek to understand droplet formation, to reveal its size and to provide an explanation of its spreading on a tablet surface. Factors that greatly influence the droplet characteristics are properties of atomizer (nozzle type and spray width), coating solution properties (viscosity, density, and surface tension), flow rates (air pressure), and temperature inside the coating unit [9]. Air stream plays a role in fluidization of particles, as well as evaporation of the excess solvent. Sudden evaporation leads to increase in viscosity. The progression of the droplet spreading over the tablet surface becomes limited, and coalescence may take over.

Additionally, there is a possibility that droplet will overdry before it hits the particle if the air is too hot. This problem is emphasized in the case of organic solvent appliance or when polymeric solution rapidly changes as the dry matter content increases.

*1.1. Droplet Impact Regime.* Among the most influential factors in coating layer formation, one can find droplet size. Many papers addressed the problem of understanding the effect of droplet formation and impact conditions on wetting dynamics [9–12]. However, a little attention has been paid to the exploration of other present phenomena such as drying, droplet-particle and droplet-droplet collisions, and so on.

Droplet deposition is considered to be a successful impact outcome. Adherence efficiency depends on momentum, angle of incidence, solution properties, and characteristics of the surface. Particles and droplets ratio indicates the dominant impact regime [9].

Calculation of impact efficacy can be based on relation of space surrounding the critical trajectory and projected tablet surface area, yet Cheng and Turton [13] pointed out unpredictable "shading" effect. Further papers appraise this remark, still leaving out other important details (turbulence, droplet shape, surface features, etc.) without proven contribution to the impact outcome, which must exist.

Dimensionless numbers, namely, Reynolds (Re), Ohnesorge (Oh), and Weber (We), have proven to be useful in explanation of the droplet formation process. They are highly able to nondimensionalise droplet breakup phenomenon and impact mechanism occurrence. Relationships of dimensionless numbers strongly define the droplet-surface impact model as well as coating mechanisms:

$$\mathrm{Re} = \frac{v_d \cdot d \cdot \rho}{\mu}, \tag{1}$$

$$\mathrm{Oh} = \frac{\sqrt{\mathrm{We}}}{\mathrm{Re}} = \frac{\mu}{\sqrt{\rho \cdot d \cdot \gamma_{liq}}}, \tag{2}$$

$$\mathrm{We} = \frac{\rho \cdot d \cdot v_d^2}{\gamma_{liq}}. \tag{3}$$

The lack of Weber number is that it does not take into account viscosity. There are some upgrades of the Weber dimensionless number. These are Ohnesorge (Oh) and Laplace (La) numbers which include liquid viscosity. Toviakka [9] has found that high impact velocity of low-viscous tiny droplets stimulates the spreading. Accordingly, Pasandideh-Ferd et al. [14] have developed maximum spreading factor ($\xi_{max}$) expression:

$$\xi_{max} = \sqrt{\frac{We + 12}{3\left(1 - \cos\theta_a\right) + 4\left(We/\sqrt{Re}\right)}}. \tag{4}$$

The denominator comprises elements of restriction to spreading (solidification by drying, surface tension, and viscous losses). In view of the fact that droplet retraction is not included in this calculation, it only partially explains the issue of thin film formation.

Maximum spreading factor is the ratio of maximum droplet diameter after the impact on the solid surface when the droplet starts spreading and the mean diameter of the droplet before the impact.

Simple model by Asai et al. [12] excludes the contact angle. Meanwhile, it seems to work well in experiments with micron sized droplets:

$$\xi_{max} = 1 + 0.48 \cdot We^{0.5} \exp\left(-1.48 \cdot We^{0.22}Oh^{0.21}\right). \tag{5}$$

Another model, established by Roisman [15], likewise contains no variables. They noted that impact phenomena calculation based on energy balance is inaccurate and suggested mass and momentum balance of spread droplet and pin line. From the equation of the spreading factor through the conservation law of mass, the maximum spreading factor can be expressed:

$$\xi_{max} = 0.87 \cdot Re^{1/5} - 0.40 \cdot Re^{2/5}We^{-1/2}. \tag{6}$$

The impact regime map was adapted from Khoufech et al. [16]. The critical Ohnesorge number is considered as a boundary between the deposition and splashing impact model.

### 1.2. Droplet Size.

The droplet size is estimated by the following empirical expression:

$$D_{50} = 604.53 \cdot \frac{\gamma_{liq}^{0.41} \cdot \mu^{0.32}}{\left(v_{REL}^2 \cdot \rho_{air}\right)^{0.57} \cdot A_G^{0.36} \cdot \rho_{sol}^{0.16}}$$

$$+ 330.71 \cdot \left[\left(\frac{\mu}{\gamma_{liq} \cdot \rho_{sol}}\right)^{0.17} \cdot \left(\frac{1}{v_{REL}^{0.54}}\right)\left(\frac{\dot{M}_{sol}}{\dot{M}_{air}}\right)^{0.5}\right]. \tag{7}$$

Nevertheless, estimation of the droplet size according to Waltzel's model for Sauter mean diameter gives more accurate results. Unlike other available empirical equations, this correlation takes the spray air pressure into account (8). Hede et al. [17] have done an exhaustive research study on the results of different correlations. They concluded that Waltzel's model gives substantial results on solution spraying, while other models are far more adjusted for suspension spray characterization:

$$D_{32} = d_0 \cdot 0.35 \cdot \left[\frac{\Delta p_{air} \cdot d_0}{\gamma_{liq}\left(1 + \left(\dot{M}_{sol}/\dot{M}_{air}\right)\right)^2}\right]^{-0.4} \cdot (1 + 2.5 \cdot Oh). \tag{8}$$

### 1.3. Efficacy Assessment of Coating Process: Estimation of Parameters

1.3.1. Intratablet Uniformity. Intratablet coating uniformity is a term used to describe the variation of coating thickness on a single tablet's surface. It is given as minimal coating thickness and the span of coating thickness distribution. Obtained film morphology is determined by the micrographic analysis. Coated tablets are usually sorted into predefined quality category, depending on the coating uniformity. Categories are afterwards ranked by the numeric quality score order. More recently, Raman spectroscopy [18, 19], near-infrared spectroscopy [20], terahertz pulsed imaging [21], and optical coherence tomography [22] have been used to determine the intensity of the colored coatings.

1.3.2. Intertablet Uniformity. Evaluation of the coating uniformity among the tablets in every single experiment is the most challenging part of coating quality assessment. Coefficient of variation (CV) is defined as the ratio of the standard deviation to the mean:

$$CV = \frac{\sigma}{\overline{m}_c}, \tag{9}$$

$$\sigma = \sqrt{\frac{1}{N-1} \sum_{i=1}^{N}\left(m_{c,i} - \overline{m}_c\right)^2}. \tag{10}$$

## 2. Experimental

2.1. Tablets. Tablet cores used in experiments are round-shaped and are made of microcrystalline cellulose (MCC; Avicel® PH Microcrystalline Cellulose, FMC BioPolymer). Characterization was carried out on twenty randomly selected tablets. Averaged tablet's properties are presented in Table 1. The Erweka TBH 30 device was used to determine tablets' dimensions and hardness. Tablets were inserted in a star-shaped feeder and then automatically transported to the test station. At the test station, each tablet is first vertically pressed by the piston to measure thickness. Next, the diametrical compression test is performed on the same tablet. The tablet is pressed diametrically by the horizontal piston to measure the diameter. The force is then increased until the tablet breaks which is recorded as the amount of force (in Newtons) needed to break the tablet, and it corresponds to the hardness value.

From the features measured (Table 1), equivalent diameters (surface area and volume) were calculated and presented in Table 2. Waddell's sphericity factor was calculated as well according to the following equation:

TABLE 1: Measured characteristics of the tablets—average values for 20 tablets.

| Weight (mg) | $75.2 \pm 5.2$ |
|---|---|
| Thickness (mm) | $3.09 \pm 0.02$ |
| Diameter (mm) | $5.01 \pm 0.02$ |
| Hardness (N) | $62.00 \pm 6.22$ |

TABLE 2: Calculated characteristics of tablet cores.

| Equivalent volume diameter, $d_V$ (mm) | $4.88 \pm 0.01$ |
|---|---|
| Equivalent surface diameter, $d_S$ (mm) | $5.29 \pm 0.01$ |
| Waddell's sphericity factor, $\psi_{Wa}$ | $0.85 \pm 0.00$ |
| Tablet density, $\rho_p$ (kg·m$^{-3}$) | $1276.3 \pm 85.6$ |

$$\psi_{Wa} = \left(\frac{d_V}{d_S}\right)^2. \tag{11}$$

Assessment of uncoated tablets friability was carried out using double drum tablet friability tester (J. Engelsmann, AG, Germany). The drums have an inside diameter of 287 mm and are 38 mm in depth. The friability test was performed according to European Pharmacopoeia [23]. Tablets are weighed before the test and after being exposed to mechanical abrasion. The percent of the mass loss of the tablet cores was calculated according to the following equation:

$$F = \frac{m_1 - m_2}{m_1} \cdot 100. \tag{12}$$

Obtained friability was 0.16%. Results showed acceptable friability (must not exceed 1%) for the tablets used according to European Pharmacopoeia [23].

Surface morphology of coated tablets was examined using the stereomicroscope SZX 16 (Olympus, Japan) at magnifications of 20x, 25x, 60x, and 64x. Tablet thickness, hardness, and diameter were determined with the Erweka TBH 30 tester (Erweka GmbH, Heusenstamm, Germany). Testing was conducted on 20 randomly selected tablets. Hardness, diameter, and thickness (Table 1) are expressed as the mean value of all measurements including error bars expressed as standard deviation of the measured values.

## 2.2. Coating Solutions.

The coating film is obtained from aqueous solution of polymer powder mixture. Advantia Prime is a commercially available powder mixture. It contains hydroxypropyl methylcellulose (HPMC), hydroxypropyl cellulose (HPC), ethyl cellulose (EC), titanium dioxide, talc, iron oxide, silicon dioxide, polysorbate 80, and polyethylene glycol (PEG). The aforementioned combination of ingredients promotes water solubility and ensures regulated release of the active pharmaceutical ingredient (API).

Advantia Prime properties are shown in Table 3.

Determination of rheological behavior and viscosity of the coating solution was conducted using a rotational viscometer DV III+ (Brookfield Engineering Laboratories, Inc., USA). Measurements were performed with SC4-21/SC4-13R spindle/chamber combination, and data were analyzed using

TABLE 3: Rheological properties of the Advantia Prime coating solution.

| Mass fraction of Advantia Prime | 0.02 | 0.05 | 0.10 |
|---|---|---|---|
| Density (kg·m$^{-3}$) | 1008 | 1023 | 1036 |
| Viscosity (mPa·s) | 1.71 | 15.60 | 87.80 |
| Surface tension (mN·m$^{-1}$) | 46.1 | 44.6 | 41.8 |

Rheocalc software package (Rheocalc). Rheological diagrams showed a Newtonian behavior for all tested coating solutions with a constant viscosity during shearing.

The surface tension of coating solutions was measured according to the pendant drop method using a goniometer *DataPhysics Contact Angle System OCA 20*. All measurements were conducted at 23°C. The surface tension was calculated from the geometry of the drop and density of the liquid (Table 3). While the droplet is suspended in the air, gravity and the surface tension act on it. The curvature of the drop at the equilibrium state is exactly mathematically defined by the Young–Laplace equation:

$$\Delta p = \frac{2 \cdot \gamma_{liq}}{R}. \tag{13}$$

## 2.3. Process Equipment.

The coating process was performed in a laboratory fluidized-bed unit UniGlatt (Glatt GmbH, Binzen, Germany) equipped with a Wurster insert (Figure 1).

The cone-shaped process chamber owns a small glass window for observing the fluidization occurrence. Two-fluid nozzle is positioned under the bottom end of the Wurster tube (bottom spray mode) fixed in the center of the chamber. Coating solution and air stream contact takes place outside the nozzle. The position of the nozzle is fixed. The opening of the nozzle determines the spray angles and the droplet size. The opening is defined by the position of a small ring on the top of the nozzle. The upper side of the chamber contains a filter that prevents particles leaving the chamber. There is a plate on the bottom that enables compressor air distribution and saves the tablet cores from collapse.

### 2.3.1. Atomization Process: Two-Fluid Nozzle.

Simplified models of two-fluid nozzle with the external mixing zone imply both fluids flow axially. Liquid runs through the central pipe of the atomizer, while air stream fills the outer pipe around it. Contact between the phases takes place in a mixing zone ($L_{mix}$) outside the nozzle orifice (Figure 2). Air stream accelerates the liquid jet that exits the nozzle. Therefore, the same principle is used to establish desired relative velocity of any given solution droplets, hence defining the size of droplets in a spray [17].

When the liquid flow rate is considerably lower than air flow rate, droplets adopt the air velocity. In the case of increased liquid flow rate, droplets acquire certain velocity to the end of the mixing zone as follows:

$$v_d = \frac{v_{air}}{\left(1 + \left(\dot{M}_{sol}/\dot{M}_{air}\right)\right)}. \tag{14}$$

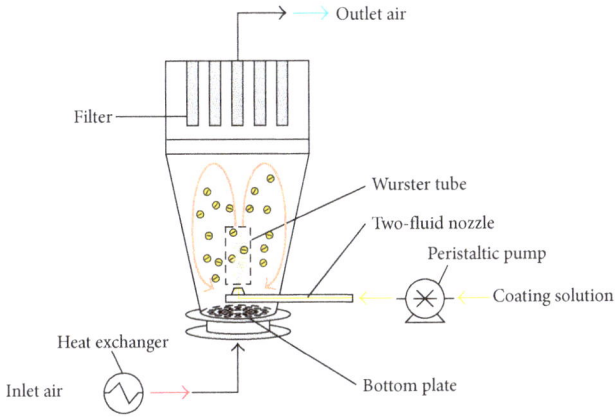

FIGURE 1: Schematic of the fluidized bed with a Wurster insert.

Air velocity was calculated according to the following equation:

$$v_{air} = \frac{\dot{V}_{air}}{A_G} = \frac{\dot{V}_{air}}{(\pi/4) \cdot \left(d_G^2 - d_{0V}^2\right)}. \quad (15)$$

The additional benefit of the atomization process is the rapid increase of the droplet surface, thus providing fast evaporation. Theoretically, maximum atomization efficiency can be reached. The tiniest droplets will be generated when $(\dot{M}_{sol}/\dot{M}_{air}) \to 0$ [13].

*2.3.2. Coating Process Implementation.* Two steps of the process were applied in order to perform coating successfully. Firstly, heating is activated in the process unit containing weighed portion of tablets. When the outlet air temperature appears constant, stationary conditions in the coating chamber are accomplished. The first dose of the coating solution is now being sprayed onto tablets. Transported through the peristaltic pump, solution is brought to the two-fluid nozzle. Coating solution is dosed in several cycles, each one lasting for 30 seconds and being followed by the drying period in the hot air stream for another 30 seconds.

Minimum fluidization velocity is calculated using empirical equation (16), which is valid for particles larger than $100\,\mu m$ [4, 24]:

$$v_{mf} = \frac{\mu}{\rho_{air} \cdot d_p} \left[ (11135.7 + 0.0408 \cdot Ar)^{1/2} - 33.7 \right], \quad (16)$$

$$Ar = \frac{\rho_{air} \cdot d_p^3 \cdot \left(\rho_p - \rho_{air}\right) \cdot g}{\mu^2}. \quad (17)$$

However, estimated values of the minimal fluidization velocities have been insufficient for tablet fluidization in the described experimental setting. Real values are therefore determined visually. The process conditions are shown in Table 4.

Table 5 summarizes combinations of temperature and atomization pressure as well as designation of experiments.

## 3. Results and Discussion

*3.1. Coefficient of Variation (CV) and Content of Poorly Coated Tablets (PCT).* A previous work presented by Matijašić et al. [25] was used as preliminary research to establish optimal process conditions. Table 6 summarizes the values of intertablet uniformity in terms of coefficient of variation (CV) and the content of poorly coated tablets (PCT) from previous research at different process conditions. All experiments were performed in 6 coating/drying cycles and at atomization air pressure of 1 bar, while other process conditions are shown in Table 6. Coefficient of variation is determined over a population of 100 tablets coated in every experiment, randomly selected out of approximately 700 tablets in a batch. Tablets were weighed before and after coating, and CV values were calculated from (10). Preliminary research showed the relationship between the droplet Reynolds number (1) and coefficient of variation (CV) (9). CV decreases with the increase of the droplet Reynolds number. Additionally, diluted solutions have lower viscosity which resulted in smaller droplets that are characterized with high Re values. Low viscosity makes those solutions more suitable for atomization occurring in the chamber. Although the differences in CV values are negligible, the more reasonable explanation arises from the droplet-surface impact study explained in Section 3.3 Furthermore, the content of poorly coated tablets was calculated. Each tablet in the batch of 700 tablets was visually analyzed. Tablets that were unevenly coated all over the surface were separated as bad-coated ones. Their mass content in the batch represents the value of PCT. The results (Table 6) led to a general conclusion that higher temperature ensures successful drying of the coating film and preventing the sticking of the tablets which could result in uneven coating. PCT content was generally lowered when temperature was raised from 50°C to 60°C. Different concentrations of Advantia Prime coating solutions were examined, 2, 5, and 10 wt.%. PCT content was increased with the increase of the mass flow rate and concentration of coating solution. As can be seen, highly viscous coating solution (10 wt.%) applied at low temperature led to poor coating on almost half a batch of tablets (PCT = 49.7%). Considering the latter observations, the aim of this research is to reach the lowest possible CV and PCT content by multiplying cycles of the coating process, rather than increasing the concentration of coating solution. This assumption was loosely based on the fact that more spraying and drying cycles will induce better coverage of the surface and will also decrease imperfections on film morphology.

Considering all preliminary results, further research was performed at a low mass flow rate and concentration of coating solution, increased temperature of drying air, different atomization pressures, and increased number of coating/drying cycles (Table 4).

The results of present research are shown in Table 7.

The results showed minor deviations in CV values that were altogether slightly higher than in the preliminary experiments (Table 6). Detected deviations in the CV quality parameter might be the result of tablet-to-tablet sheltering in

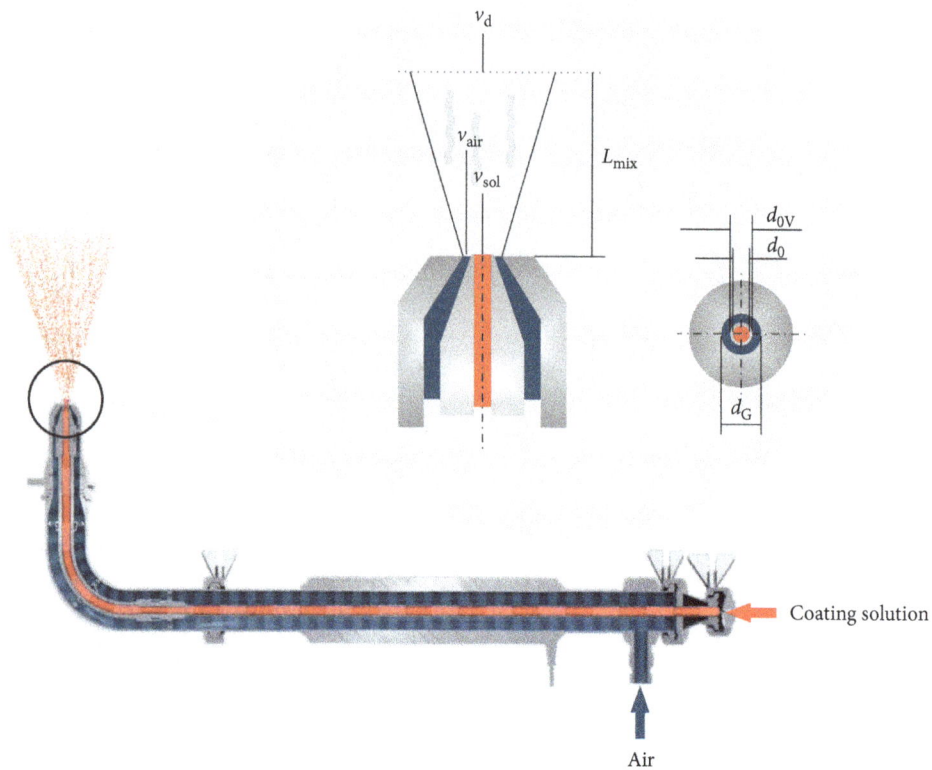

FIGURE 2: Schematic of the external mixing two-fluid nozzle.

TABLE 4: Process conditions during coating in fluid bed.

| | |
|---|---|
| Mass flow rate of coating solution | $7.85 \, \text{g} \cdot \text{min}^{-1}$ |
| Concentration of coating solution | 2 wt.% |
| Tablet mass | $70 \pm 0.1 \, \text{g}$ |
| Distance of the Wurster partition from the perforated plate | 2 cm |
| Nozzle diameter | 0.8 mm |
| Atomization pressure | 1, 1.4, 2 bar |
| Atomization air flow rate | $5.82, 7.03, 8.76 \times 10^{-4} \, \text{m}^3 \cdot \text{s}^{-1}$ |
| Inlet air temperature | 60°C, 67°C |
| Coating period | 30 s |
| Drying period | 30 s |

TABLE 5: Experiments and their process conditions.

| Experiment | A | B | C | D | E |
|---|---|---|---|---|---|
| Temperature (°C) | 60 | 67 | 67 | 67 | 67 |
| Atomization pressure (bar) | 1.0 | 1.0 | 1.2 | 2.0 | 1.4 |

the coating region due to the significant number of tablets passing through the spray zone at a given time [5]. Furthermore, the drawback of this method is the fact that coated tablet weight, when it is coated properly, often equals the tablet weight in the case when coating layer adheres poorly or unevenly covers the surface. So, the mass of the coated tablets will not differ significantly from the median, resulting with low CV values. However, the quality of the coating film of such tablets is not acceptable. The mass of coated tablets may also vary due to the wear of the tablets in the fluidized

bed. Moreover, compared to preliminary experiments that generally have lower CV values, it was concluded that CV is not the reliable quality parameter.

For that reason, poorly coated tablets were separated from the batches, and their content (PCT) was calculated. As explained, PCT content represents the mass fraction of tablets in the batch that were unevenly coated all over the entire surface (Table 7). Preliminary results showed lower PCT percentage at higher drying temperature (60.0°C). Therefore experiments were performed at 60.0°C and 67°C. Generally, it has been shown that the method of trial and error, when in line with a theoretical framework and process capability, may provide considerable process improvement. Increase in the mass of coating film was obtained due to the low solution concentration (2 wt.%) and increased number of coating/drying cycles. Low viscosity of the solution facilitated atomization process, while multiple coating cycles provided a more compact coating film. Higher temperature is beneficial in prevention of overwetting and sticking defects. Wet tablets preferentially stick to each other as well as to the walls or elements of the coating chamber (Figure 3).

Further experiments are directed primarily to the additional increase in atomization pressure and temperature, in order to obtain smaller droplets that will dry out faster. Accelerated evaporation is crucial for good adherence of new layers. Smaller droplets will dry faster ensuring dry surface of the tablet, thus prepared for the new coating cycle. This will prevent overwetting of the surface and sticking of the tablets which will result in a lower PCT content. It might enhance

TABLE 6: Preliminary results [25].

| Temperature (°C) | Mass flow rate of coating solution (g·min$^{-1}$) | Concentration of coating solution (wt.%) | Poorly coated tablets content (PCT) (%) | Droplet Reynolds number | Intertablet uniformity (CV) |
|---|---|---|---|---|---|
| 50 | 7.85 | 2 | 0.9 | 1672 | 0.0116 |
| 50 | 9.90 | 2 | 9.9 | 1647 | 0.0126 |
| 50 | 7.85 | 5 | 14.0 | 345 | 0.0150 |
| 50 | 9.90 | 5 | 34.0 | 338 | 0.0170 |
| 50 | 7.85 | 10 | 49.7 | 103 | 0.0170 |
| 60 | 7.85 | 2 | 2.3 | 1672 | 0.0106 |
| 60 | 9.90 | 2 | 3.6 | 1647 | 0.0132 |
| 60 | 7.85 | 5 | 4.0 | 345 | 0.0146 |
| 60 | 9.90 | 5 | 13.1 | 338 | 0.0166 |

TABLE 7: CV and PCT content values for run process conditions.

| Process conditions | # | Number of cycles | CV | PCT content (%) |
|---|---|---|---|---|
| A: 60°C; 1 bar | A1 | 10 | 0.0236 | 3.6 |
| | A2 | 15 | 0.0273 | 20.9 |
| | A3 | 20 | 0.0254 | 18.2 |
| B: 67°C; 1 bar | B1 | 6 | 0.0269 | 0.0 |
| | B2 | 10 | 0.0273 | 0.0 |
| | B3 | 15 | 0.0257 | 3.7 |
| | B4 | 20 | 0.0237 | 5.6 |
| | B5 | 25 | 0.0255 | 6.0 |
| | B6 | 30 | 0.0250 | 2.2 |
| C: 67°C; 1.2 bar | C1 | 6 | 0.0258 | — |
| | C2 | 10 | 0.0266 | 0.1 |
| | C3 | 15 | 0.0228 | 0.0 |
| | C4 | 20 | 0.0257 | 0.0 |
| | C5 | 25 | 0.0241 | 1.5 |
| | C6 | 30 | 0.0241 | 3.3 |
| D: 67°C; 2.0 bar | D1 | 20 | 0.0228 | — |
| | D1-II | 20 | 0.0248 | — |
| E: 67°C; 1.4 bar | E1 | 25 | 0.0221 | 1.1 |

intratablet coating uniformity as well, since increased number of faster and smaller droplets will now spread more evenly and create a thinner film.

PCT content is significantly lowered when air temperature is increased to 67°C (Table 7). The high value of air pressure (Table 7; D1 and D1-II) and low number of coating/drying cycles (Table 7; C1) resulted in thin film coats that were hardly visible which caused minor difficulties related to intertablet uniformity perceptibility. Consequently, PCT content could not be determined for those experiments.

Coatings obtained in B and C set of experiments are much more uniform than those in A series. Most of the poorly coated tablets in A series refer to the tablets that stuck on the internal surface of the Wurster tube. It is important to emphasize that no tablets stuck to the Wurster tube in experiments B2, D1-II, or C1, C3, and C4. Altogether, results show that a set of experiments performed in C series are sufficient for running an effective and efficient coating process using atomization pressure of 1.2 bar and temperature of 67°C producing the low content of poorly coated tablets.

3.2. Stereomicroscopy. Stereomicroscopy is potentially useful for intra- and intertablet uniformity assessment. Initially, the idea was to gather the stereomicrographs of all coated tablets in order to establish the 1–10 grade scale and carry out the coating quality evaluation. However, the aggravating issue here is that there are more than 700 tablets in each batch. It is practically impossible to analyze every single one of them, compare them, and to rate the quality for each coating.

Anyway, a worthwhile stereomicroscopic analysis has been carried out. Several tablets from each experiment were meticulously examined. Stereomicroscopy analysis revealed various "volcano craters" (Figure 4), "bitten off," and berry-like structures spread all over the tablet surfaces, whereas on the well-coated tablets, there are several minor or almost intangible differences that scarcely lend themselves to description. However, flat and even surface (Figure 5) cannot be precisely characterized by the stereomicroscope, thus being another reason to seek for a more sophisticated method.

3.3. Droplet-Surface Impact Regime. Impact regime determination is based on dimensionless numbers as this gave rise to a solid connection between process conditions and the morphology of the coating layer.

Figure 6 represents illustration of coating mechanisms through relationship of Ohnesorge and Reynolds number. A well-known Mundo et al. deposition diagram [26] gives empirical critical curve that represents the transition in deposition-splashing mechanism where critical constant ($K_{crit}$) is defined as the ratio of Ohnesorge and Reynolds number ($Oh/Re^{-1.25}$). It can be seen that our results fall in the area of deposition mechanism for 5 experimental conditions given in Table 5. However, only 4 points were obtained because experiments A and B were performed at same atomization pressure which gave similar values of Reynolds and Ohnesorge numbers. Reynolds number was calculated according to (1), Ohnesorge number from (2), while droplet velocity and its size were calculated using (15) and (8). When the droplet hits the dry and solid surface, it can rebound from it, deposit on the surface, or splash with secondary drops due to more energetic impact. If the droplet deposits on the surface, it will form a liquid film [26] which is a preferable mechanism for film coating.

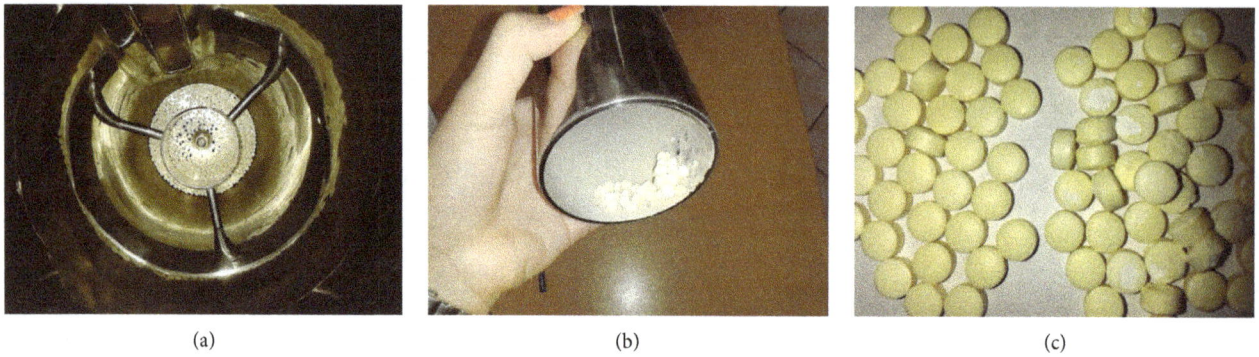

(a)                                    (b)                                    (c)

FIGURE 3: Photographs of the coating chamber filled with tablets, tablets stuck on Wurster tube, and coated tablets.

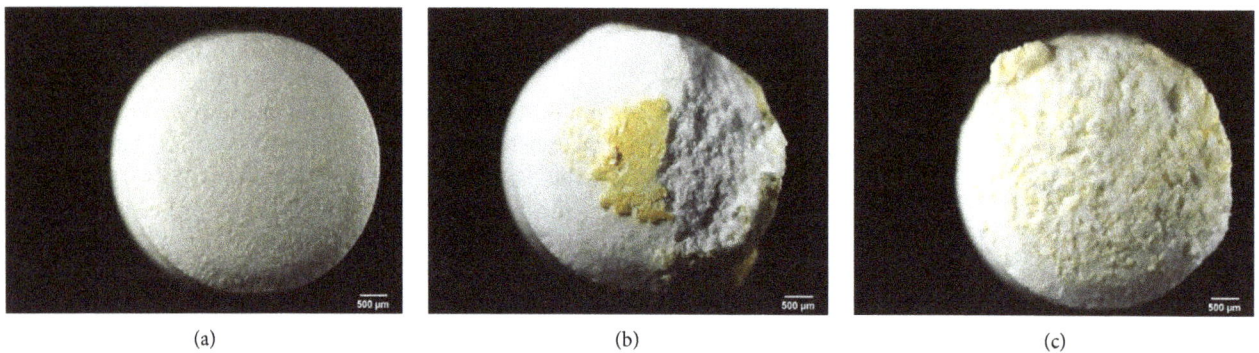

(a)                                    (b)                                    (c)

FIGURE 4: Stereomicrographs: experiment A1—well-coated tablet versus badly-coated tablet (both sides).

FIGURE 5: Stereomicrograph of well-coated tablet from experiment A3 compared to C2.

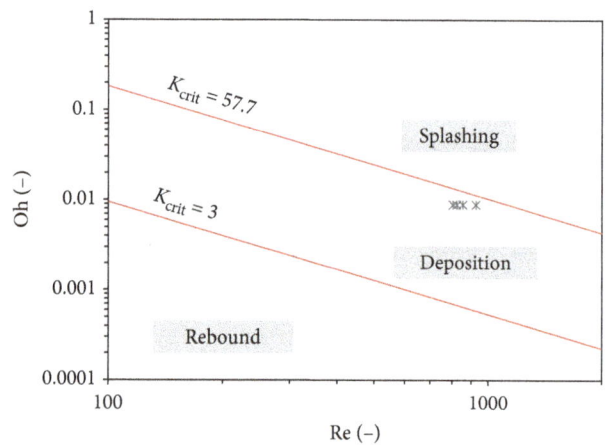

FIGURE 6: Impact mechanism diagram [26] of experiments A–E.

Figure 7 illustrates the impact regime map as defined by Schiaffino and Sonin [27]. Four impact regimes are distinguished: collision for inviscid and viscous fluids as well as the capillary-driven region for both types of fluids. When all four regimes are divided by assuming Ohnesorge, Weber, and Reynolds number equals to unity, transition lines could be plotted [27–29]. According to obtained results, experimental data fall within the region of high Weber number in the area of inviscid fluid and collision as a prevailing mechanism. According to 5 experimental conditions given in Table 5, only 4 points were obtained because experiments A and B were performed at same atomization pressure, resulting with same droplet velocity which gave similar values of Weber number.

Inviscid, impact-driven region assumes that droplet spreading occurs in a short period where the flow is driven by the dynamic pressure of impact [27]. Due to intense collision (splashing) of droplets, there is a risk of absorption of coating solution through the pores in the tablet structure. However, tablets were made by pressing, and the pores are not exposed which resulted in formation of surface coating film.

Khoufech et al. [16] have described droplet adherence to the surface analyzed through We–Oh relation. They have experimented with CMC (carboxymethyl cellulose) which

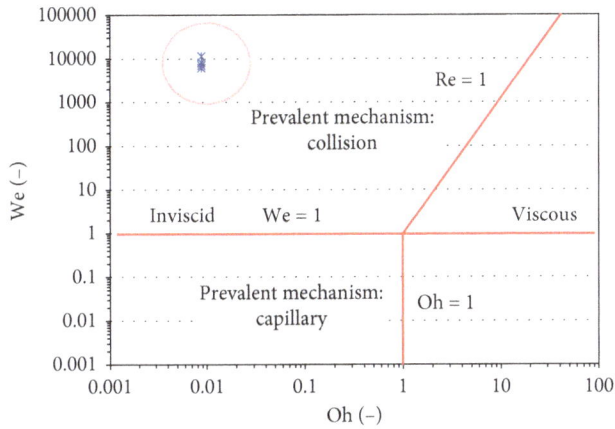

FIGURE 7: Impact regime diagram [27] of experiments A–E.

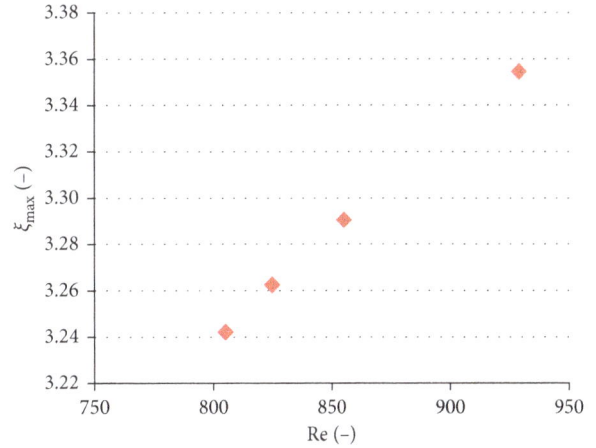

(a)

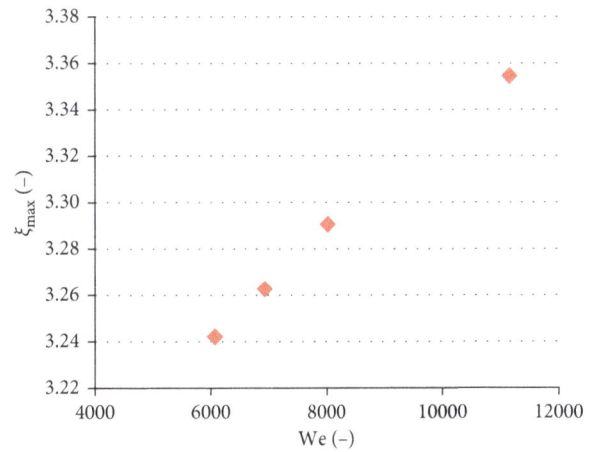

(b)

FIGURE 8: Diagrams presenting the relation between the maximum spreading factor and dimensionless Re and We numbers in experiments A–E.

is widely used for pharmaceutical production and coatings as well. It has valuable properties in API release and can be used in polymer blends with HPC. Assuming they are both water soluble and share few other properties (viscoelasticity and shear-thinning), experiments are comparable. As their theory suggests, step out of the area of collision and splashing dominance is reached only with 5% polymer concentration in coating formulation. Unfortunately, we have encountered technical issues due to significantly increased viscosity of 5% solution of Advantia Prime (15.6 mPa·s).

The impact regime can be successfully adjusted by droplet-surface adherence investigation. At high Weber number values, spreading is driven by impact velocity. Weber number, defined as the ratio of kinetic and surface energy, decreases when droplet size lowers (higher We → higher $\beta_{max}$). Smaller droplets move faster, and according to Toivakka [9], high velocity implies higher collision energy, therefore enhanced spreading over the surface. Figure 8 illustrates the influence of Weber and Reynolds numbers on the maximum spreading factor, ratio of maximum droplet spreading diameter, and droplet diameter before collision calculated according to Roisman [15], (6). Results showed that the increase in Reynolds and Weber numbers would lead to the increase in the maximum spreading factor due to the intense impact of the coating droplet and high kinetic energy. This is consistent with results obtained by Bolleddula et al. [28] and Asai et al. [12].

## 4. Conclusions

Following our intention to perform the coating process more efficiently, many experimental runs have been provided for placebo tablets in a lab-scale fluid-bed environment with a Wurster tube.

Using CV and PCT content with coating layer adherence and its consistency as process quality metrics, favorable process conditions have been detected.

Increasing number of coating cycles is found to significantly contribute to better film morphology in the case of sufficiently high temperatures.

This paper reveals a mechanistic perspective on the complex phenomena occurring on a tablet surface during coating process. Impact regime and impact mechanism diagrams are successfully defined for the system studied. Provided diagrams suggest efficient process conditions for the process with good quality metrics.

Deposition is found to be a preferential mechanism for conditions studied. Furthermore, collision is addressed as a prevalent impact regime.

Droplet breakup occurrence in terms of droplet diameter was efficiently correlated with hydrodynamic conditions.

## Notations

| | |
|---|---|
| $A_G$: | Cross section of nozzle ($m^2$) |
| Ar: | Archimedes number |
| CV: | Coefficient of variation |
| $D$: | Droplet mean diameter before impact (m) |
| $d_G$: | Outer diameter of nozzle (air flow) (m) |
| $D_{max}$: | Maximum droplet diameter after impact (m) |
| $d_{0V}$: | Inner diameter of nozzle (liquid flow) (m) |

$d_0$: Nozzle diameter (m)
$d_p$: Particle (droplet) size (m)
$d_S$: Equivalent surface diameter of tablet (m)
$d_V$: Equivalent volume diameter of tablet (m)
$D_{32}$: Sauter's droplet diameter (m)
$D_{50}$: Droplet median diameter (m)
$F$: Friability (%)
$K_{crit}$: Critical constant $(\text{Oh}/\text{Re}^{-1.25})$
$\dot{M}_{sol}$: Mass flow rate of solution $(\text{kg}\cdot\text{s}^{-1})$
$\dot{M}_{air}$: Mass flow rate of air $(\text{kg}\cdot\text{s}^{-1})$
$\overline{m}_c$: Mean mass of coated tablets (kg)
$m_{c,i}$: Mass of one coated tablet (kg)
$m_1$: Mass of tablets before the friability test (kg)
$m_2$: Mass of tablets after the friability test (kg)
$N$: Number of analyzed tablets (11)
Oh: Ohnesorge number
$R$: Capillary diameter (m)
Re: Reynolds number
$t$: Time (s)
$\dot{V}_{air}$: Volume flow rate of air $(\text{m}^3\cdot\text{s}^{-1})$
$v_d$: Droplet velocity $(\text{m}\cdot\text{s}^{-1})$
$v_{mf}$: Minimal fluidization velocity $(\text{m}\cdot\text{s}^{-1})$
$v_{REL}$: Relative velocity $(\text{m}\cdot\text{s}^{-1})$
We: Weber number
$\gamma_{liq}$: Liquid surface tension $(\text{N}\cdot\text{m}^{-1})$
$\Delta p$: Pressure drop (Pa)
$\Delta p_{air}$: Atomization pressure (bar)
$\xi_{max}$: Maximum spreading factor
$\mu$: Fluid viscosity $(\text{Pa}\cdot\text{s})$
$\rho$: Fluid density $(\text{kg}\cdot\text{m}^{-3})$
$\rho_{air}$: Air density $(\text{kg}\cdot\text{m}^{-3})$
$\rho_{sol}$: Solution density $(\text{kg}\cdot\text{m}^{-3})$
$\rho_p$: Particle (tablet) density $(\text{kg}\cdot\text{m}^{-3})$
$\sigma$: Standard deviation of mass (kg)
$\psi_{Wa}$: Wadell's sphericity factor
API: Active pharmaceutical ingredient
CV: Coefficient of variation (intertablet uniformity)
CMC: Carboxymethyl cellulose
EC: Ethyl cellulose
HPC: Hydroxypropyl cellulose
HPMC: Hydroxypropyl methylcellulose
MCC: Microcrystalline cellulose
PEG: Polyethylene glycol
PCT: Poorly coated tablets content.

## Conflicts of Interest

The authors declare that they have no conflicts of interest.

## References

[1] M. Benković and I. Bauman, "Oblaganje čestica u prehrambenoj industriji," *Croatian Journal of Food Technology, Biotechnology and Nutrition*, vol. 6, pp. 13–24, 2011.

[2] B. Guignon, A. Duquenoy, and E. D. Dumoulin, "Fluid bed encapsulation of particles: principles and practice," *Drying Technology*, vol. 20, no. 2, pp. 419–447, 2002.

[3] J. Pranjić, "Oblaganje u fluidiziranom sloju Wurster tehno-

logijom," M.S. thesis, University of Zagreb, Zagreb, Croatia, 2013.

[4] E. Teunou and D. Poncelet, "Batch and continuous fluid bed coating–review and state of the art," *Journal of Food Engineering*, vol. 53, no. 4, pp. 325–340, 2002.

[5] S. Shelukar, J. Ho, J. Zega et al., "Identification and characterization on factors controlling tablet coating uniformity in a Wurster coating process," *Powder Technology*, vol. 110, no. 1-2, pp. 29–36, 2000.

[6] T. Havaić, "Fenomeni sudara kapi u procesu oblaganja Wurster tehnologijom," M.S. thesis, University of Zagreb, Zagreb, Croatia, 2016.

[7] S. Rujivipat, "Novel formulation and processing aspects for compression-coated tablets and for the compression of polymer-coated multiparticulates," Dissertation zur Erlangung des akademischen Grades des Doktors der Naturwissenschaften, Fachbereich Biologie, Chemie, Pharmazie der Freien Universität Berlin, Berlin, Germany, 2010.

[8] S. R. L. Werner, J. R. Jones, A. H. J. Paterson, R. H. Archer, and D. L. Pearce, "Air-suspension coating in the food industry: Part II—micro-level process approach," *Powder Technology*, vol. 17, no. 1, pp. 34–45, 2007.

[9] M. Toivakka, "Numerical investigation of droplet impact spreading in spray coating of paper," in *Proceedings of 2003 TAPPI 8th Advanced Coating Fundamentals Symposium*, Chicago, IL, USA, May 2003.

[10] A. Khoufech, M. Benali, and K. Saleh, "Influence of liquid formulation and impact conditions on the wetting of hydrophobic surfaces by aqueous polymeric solutions," *Chemical Engineering Research and Design*, vol. 110, pp. 233–244, 2016.

[11] D. Izbassarov and M. Muradoglu, "Effects of viscoelasticity on drop impact and spreading on a solid surface," *Physical Review Fluids*, vol. 1, no. 2, article 023302, 2016.

[12] A. Asai, M. Shioya, S. Hirasawa, and T. Okazaki, "Impact of an ink drop on a paper," *Journal of Imaging Science and Technology*, vol. 37, pp. 205–207, 1993.

[13] X. X. Cheng and R. Turton, "The prediction of variability occurring in fluidized bed coating equipment. II. The role of nonuniform particle coverage as particles pass through the spray zone," *Pharmaceutical Development and Technology*, vol. 5, no. 3, pp. 323–332, 2000.

[14] M. Pasandideh-Fard, Y. M. Qiao, S. Chandra, and J. Mostaghimi, "Capillary effects during droplet impact on a solid surface," *Physics of Fluids*, vol. 8, no. 3, pp. 650–659, 1996.

[15] I. Roisman, "Inertia dominated drop collisions. II. An analytical solution of the Navier–Stokes equations for a spreading viscous film," *Physics of Fluids*, vol. 21, no. 5, article 052104, 2009.

[16] A. Khoufech, M. Benali, and K. Saleh, "Influence of liquid formulation and impact conditions on the coating of hydrophobic surfaces," *Powder Technology*, vol. 270, pp. 599–611, 2015.

[17] P. D. Hede, P. Bach, and A. D. Jensen, "Two-fluid spray atomisation and pneumatic nozzles for fluid bed coating/agglomeration purposes: a review," *Chemical Engineering Science*, vol. 63, no. 14, pp. 3821–3842, 2008.

[18] S. Romero-Torres, J. D. Pérez-Ramos, K. R. Morris, and E. R. Grant, "Raman spectroscopy for tablet coating thickness quantification and coating characterization in the presence of strong fluorescent interference," *Journal of Pharmaceutical and Biomedical Analysis*, vol. 41, no. 3, pp. 811–819, 2006.

[19] J. F. Kauffman, M. Dellibovi, and C. R. Cunningham, "Raman

spectroscopy of coated pharmaceutical tablets and physical models for multivariate calibration to tablet coating thickness," *Journal of Pharmaceutical and Biomedical Analysis*, vol. 43, no. 1, pp. 39–48, 2007.

[20] A. Ariyasu, Y. Hattori, and M. Otsuka, "Non-destructive prediction of enteric coating layer thickness and drug dissolution rate by near-infrared spectroscopy and X-ray computed tomography," *International Journal of Pharmaceutics*, vol. 525, no. 1, pp. 282–290, 2017.

[21] M. Haaser, K. C. Gordon, C. J. Strachan, and T. Rades, "Terahertz pulsed imaging as an advanced characterisation tool for film coatings–a review," *International Journal of Pharmaceutics*, vol. 457, no. 2, pp. 510–520, 2013.

[22] Y. Dong, H. Lin, V. Abolghasemi, L. Gan, J. A. Zeitler, and Y.-C. Shen, "Investigating intra-tablet coating uniformity with spectral-domain optical coherence tomography," *Journal of Pharmaceutical Sciences*, vol. 106, no. 2, pp. 546–553, 2017.

[23] European Pharmacopoeia, *EDQM, European Pharmacopoeia 01/2010:20907 Friability of Uncoated Tablets*, Council of Europe, Strasbourg, France, 8th edition, 2013.

[24] G. Toschkoff and J. G. Khinast, "Mathematical modeling of the coating process," *International Journal of Pharmaceutics*, vol. 457, no. 2, pp. 407–422, 2013.

[25] G. Matijašić, K. Žižek, M. Gretić, A.-M. Đumbir, T. Havaić, and M. Bokulić, "Fluid-bed coating with Wurster insert," in *Proceedings of 5th Croatian Meeting of Chemist and Chemical Engineers, Book of Abstract*, Zagreb, Croatia, 2015.

[26] C. Mundo, M. Sommerfeld, and C. Tropea, "Droplet-wall collisions: experimental studies of the deformation and breakup process," *International Journal of Multiphase Flow*, vol. 21, no. 2, pp. 151–173, 1995.

[27] S. Schiaffino and A. A. Sonin, "Molten droplet deposition and solidification at low Weber," *Physics of Fluids*, vol. 9, no. 11, pp. 3172–3187, 1997.

[28] D. A. Bolleddula, A. Berchielli, and A. Aliseda, "Impact of a heterogeneous liquid droplet on a dry surface: application to the pharmaceutical industry," *Advances in Colloid and Interface Science*, vol. 159, no. 2, pp. 144–159, 2010.

[29] T. Lim, S. Han, J. Chung, J. T. Chung, S. Ko, and C. P. Grigoropoulos, "Experimental study on spreading and evaporation of inkjet printed pico-liter droplet on a heated substrate," *International Journal of Heat and Mass Transfer*, vol. 52, no. 1-2, pp. 431–441, 2009.

# 9

# Energy and Productivity Yield Assessment of a Traditional Furnace for Noncentrifugal Brown Sugar (*Panela*) Production

**Luis F. Gutiérrez-Mosquera** ⓘ,[1] **Sebastián Arias-Giraldo** ⓘ,[2] **and Adela M. Ceballos-Peñaloza** ⓘ[2]

[1]*Department of Engineering, Food and Agribusiness Research Group, Universidad de Caldas, Calle 65 No. 26–10, Manizales, Colombia*
[2]*Food and Agribusiness Research Group, Universidad de Caldas, Calle 65 No. 26–10, Manizales, Colombia*

Correspondence should be addressed to Luis F. Gutiérrez-Mosquera; fernando.gutierrez@ucaldas.edu.co

Academic Editor: Junwu Wang

Noncentrifugal brown sugar (called *panela* in Colombia) is a natural sweetener obtained from the extraction, purification, and concentration of sugarcane juices. In this work, energy and productivity yield of a traditional furnace for *panela* production were evaluated, considering five performance indices. Experimental productions were developed in a pilot plant facility, analyzing furnace gas emissions of furnace and bagasse properties. Mass, energy, and exergy balances were performed. The following indices were obtained from the experimental runs: energy efficiency $12.726 \pm 1.091\%$, exergy efficiency $9.013 \pm 0.710\%$, energy losses through chimney $72.293 \pm 11.507\%$, yield $0.144 \pm 0.021\,kg_{panela}/kg_{bagasse}$, productivity $7.450 \pm 0.520\,kg_{panela}/h$, and bagasse consumption $1.258 \pm 0.139\,kg_{bagasse\,consumed}/kg_{bagasse\,produced}$. It was found that these outcomes were strongly influenced by excess air and gas circulation velocity through the furnace, which affects the combustion rate and heat transfer between the gases and the juices. Finally, it was concluded that the traditional scheme is inefficient and requires various critical operational adjustments, such as combustion chamber, chimney draft control, and heat exchangers design.

## 1. Introduction

Noncentrifugal brown sugar, called *jaggery* in India, *panela* in Colombia, and *rapadura* in Brazil, is a natural food obtained by extraction and concentration of sugarcane juices (*Saccharum officinarum*). Worldwide, it is used as a sweetener or as a ready-to-consumer product, highly valued for its appreciable energy supply and contribution to the food security. The main component of *panela* is sucrose, although glucose, fructose; vitamins A, C, D, E, and B; and minerals such as calcium, iron, potassium and zinc also stand out [1, 2]. In the world, Colombia has the highest consumption of noncentrifugal brown sugar per capita (22 kg/year) and is the second highest international producer with a 12% global market share. Approximately 350,000 families work in the Colombian noncentrifugal brown sugar sector, which produces more than 1,330,000 tons of *panela* annually in 236 municipalities [3, 4].

The production of *panela* from sugarcane is performed in locations called sugar mills (*trapiche* in Spanish), through ancestral and traditional methods. To obtain noncentrifugal brown sugar, the sugarcane juices are extracted using a mill, to be subsequently filtered, purified, and clarified. When the contaminants have been removed, it proceeds to evaporation and concentration of sugarcane juices using a series of metal receptacles, heat exchangers, or pans. The residual bagasse of the milling is used as solid fuel material. The determination of appropriate heating time is made empirically. Finally, the concentrate taken from the pans is beaten and molded, and then the noncentrifugal brown sugar is packed as the final product [5].

The technological system, in which thermal energy transfer is carried out between the combustion gases and the juices, in order to reach dissolved solids concentration between 88–94° Brix, is called traditional furnace (Figure 1). The traditional furnace is composed of the bagasse feed zone,

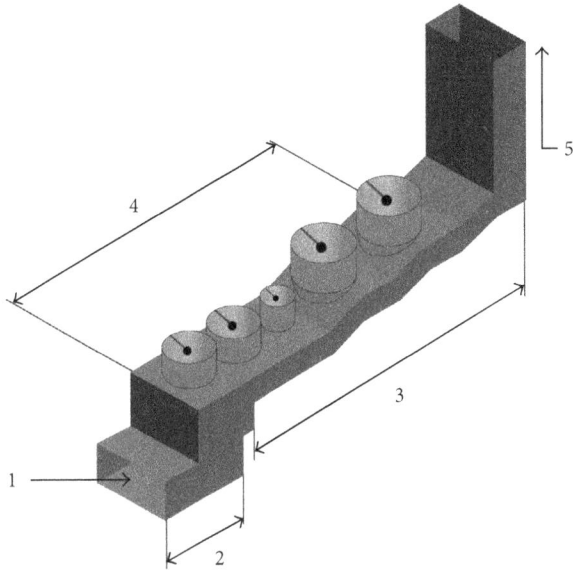

FIGURE 1: Traditional furnace for *panela* production: 1. Bagasse feed zone. 2. Combustion chamber. 3. Gas duct. 4. Heat exchangers (pans). 5. Chimney.

combustion chamber, gases circulation tunnel, heat exchangers, and chimney.

High energy loss in the traditional furnace is the principal disadvantage of this technology. Additionally, the overall productivity of the process is diminished by factors like exhaust gases leaving the chimney, poor combustion, and the low heat transfer through the system. This feature is also carrying environmental problems such as burning of tires, plastics, and wood [6]. In order to enhance the energy efficiency of the traditional furnace, some research projects have been performed: sugarcane concentration system using steam, multiple effect evaporators, and different CIMPA type multiefficient combustion chamber designs can be found in the scientific papers.

In order to evaluate the process used in Colombia for transforming sugarcane into panela, and propose a future intervention to strengthen of the most critical points in the technological system that directly affect the energy indices and productivity yield of the furnaces, the goal of this work was the energy and productivity yield assessment of a traditional furnace employed in the noncentrifugal brown sugar production, considering the following six indices: energy efficiency (%), exergy efficiency (%), energy loss through chimney (%), yield ($kg_{panela}/kg_{bagasse}$), productivity ($kg_{panela}/h$), and bagasse consumption ($kg_{bagasse\ consumed}/kg_{bagasse\ produced}$).

## 2. Materials and Methods

Experimental runs for *panela* productions were developed in a traditional furnace, placed in the municipality of Supía, Caldas, Colombia. The raw materials used were the most uniform possible, acquiring cane sugar juices, bagasse, and clarifying agent with a sole provider. Five noncentrifugal brown sugar productions were realized. In each one of them, the end product (*panela*) was obtained at 120°C of

temperature and a dissolved solids concentration equal to 90° Brix. Sugarcane juices used in the process had an average dissolved solids concentration between 17 and 18° Brix [7]. The clarification process of cane juices was performed using mucilage from *cadillo* plant (*Triumfetta láppulal*) like clarifying agent, added during the clarification process at 70°C of temperature [8]. In order to reach the stable state during the experimental runs, the traditional furnace was operated initially heating arbitrary volumes of water for one hour, or until the temperature of the chimney gases gotten constant values.

The following parameters were controlled and registered in each treatment: bagasse consumption rate, sugarcane juice quantity, clarifying extract dosage (flocculant), collected mud (suspended matter obtained during the clarification process), work time, dissolved solids concentration in juice and *panela* (ATAGO PAL-3 Digital Refractometer), temperature for the addition of mucilage from *cadillo*, temperature at the final point of *panela* (Type K Thermocouple, associated with a ThermoWorks Microtherma 2), room temperature, and relative humidity (EXTECH Thermo-hygrometer).

An evaluation of gases emissions form chimney was executed using the methodologies established by EPA [9], taking the following information: temperature and velocity of fluid, humidity, excess air, and concentration of carbon monoxide ($CO$), carbon dioxide ($CO_2$), oxygen ($O_2$), nitrogen ($N_2$), sulfur dioxide ($SO_2$), and nitrogen dioxide ($NO_2$). A combustion gas analyzer (E-Instruments 5500) and an isokinetic sampling console (ES—Environmental Supply Company C-5000) were used for measuring these variables. Moreover, samples of bagasse from each production were collected. They were analyzed in the Fuels and Combustion Laboratory of the *Universidad del Valle*, located in Cali, Colombia, in order to determine the moisture, carbon, hydrogen, nitrogen, sulfur, oxygen, and ash content by element analysis and the lower calorific value (MJ/kg) through proximate analysis.

*2.1. Mass Balance.* Assessment of the traditional furnace was performed using the outcomes obtained from five experimental run productions. Thermal, physical, and rheological properties of sugarcane and *panela* were taken of Arias et al. [10]. Mass, energy, and exergy balance were solved using Microsoft Excel 2013 and Matlab R2013a software. The mass balance of the traditional furnace for manufacturing *panela* was adapted from the methodology described by Velásquez et al. [6] and Shiralkar et al. [11], supposing mass balance without chemical reaction and steady state conditions. Figure 2 shows the material flows involved in the process. Global mass balance is described by (1). Equations (2)–(4) were used for the combustion chamber, gas duct, and the chimney (Figure 1).

$$m_{bs} + m_{ab} + m_{as} + m_{aa} + m_{jc} + m_{fl}$$
$$= m_{gs} + m_{at} + m_{mp} + m_r + m_p + m_{ae} + m_{ch}, \quad (1)$$

$$m_{bs} + m_{ab} + m_{as} + m_{aa} = m_{gs} + m_{at} + m_{mp} + m_r, \quad (2)$$

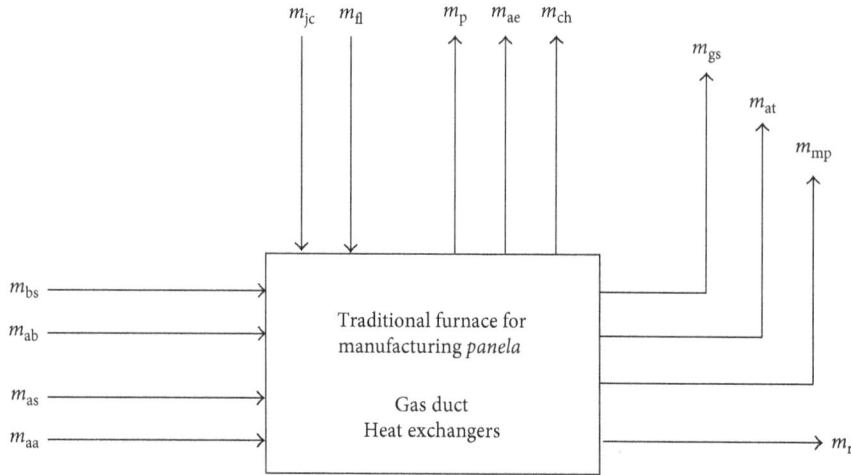

FIGURE 2: Traditional furnace mass balance, mass in kg. $m_{bs}$: dry bagasse (husk), $m_{ab}$: quantity of water in the bagasse (moisture), $m_{as}$: dry air for combustion, $m_{aa}$: water inlet with the air, $m_{jc}$: mass of sugarcane juice, $m_{fl}$: flocculant extract, $m_p$: noncentrifugal brown sugar (*panela*) obtained, $m_{ae}$: water evaporated during the concentration, $m_{ch}$: mud removed from clarification, $m_{gs}$: dry combustion gases leaving through chimney, $m_{at}$: steam that accompanies the gases through the chimney, $m_{mp}$: particulate material, and $m_r$: unburned cinder collected from the traditional furnace floor.

$$m_{bs} = m_{bh} * (1 - w), \tag{3}$$

$$m_{ab} = m_{bh} * (w). \tag{4}$$

Knowing the excess air of combustion for each trial, and considering the elemental composition of the dry bagasse, the air supply into the traditional furnace was estimated. The initial oxygen content in the bagasse and the stoichiometry of the reaction were taken into account to quantify the theoretical oxygen required for the complete combustion of C, H, and S. With the molar composition for the standard air, the oxygen and nitrogen mass inlet to combustion chamber were determined as dry air. The air absolute humidity, in $kg_{H_2O}/kg_{as}$, was calculated according to Geankoplis [12], and mass of water in the air intake was determined. Exhaust gases volumetric flow, given in $m^3/s$, was quantified as the product between the chimney gases velocity and the cross-sectional area of the duct at the sampling point. On the other hand, the total mass of chimney gases was calculated knowing their density in $kg/m^3$. The methodology proposed by Seader et al. [13] was used for density estimation. The quantity of dry exhaust gases and the humidity leaving the chimney were determined using the data provided by the emissions analysis, through (5) and (6). At the end of each production, the unburned residue that was accumulated into the gas duct was collected in order to quantify its mass.

$$m_{at} = m_{gh} * w_{gh}, \tag{5}$$

$$m_{gs} = m_{gh} - m_{at}. \tag{6}$$

For determining evaporated water mass in the concentration process, (7) was solved. The collected mud during the clarification of each production was decanted for one hour and later weighed. The solid material ($m_{sch}$) and the remnant juice in the mud ($m_{lch}$) were split, according to (8).

The *panela* obtained from each experimental runs was weighed, packaged, and stored.

$$m_{jc} + m_{fl} = m_p + m_{ch} + m_{ae}, \tag{7}$$

$$m_{ch} = m_{sch} + m_{lch}. \tag{8}$$

*2.2. Energy and Exergy Balances.* Energy balance for the traditional furnace was solved following partially the model presented by Velásquez et al. [6], keeping steady state conditions (according to Figure 3). The reference temperature and pressure were 0°C and 1 atm, respectively. Equation (9) develops global energy balance for the process:

$$E_1 + E_2 + E_3 + E_4 + E_5 = E_6 + E_7 + E_8 + E_9 + E_{10}. \tag{9}$$

Using the lower calorific value of the bagasse, expressed in MJ/kg, the energy quantity associated with this material was calculated. The enthalpy of the ambient humid air (kJ/$kg_{as}$) was found according to Geankoplis [12].

$$E_1 = (m_{bs} + m_{ab} - m_r - m_{mp}) * PCN,$$
$$E_2 = m_{as} * H_Y. \tag{10}$$

For the sugarcane juice, solid contaminants, flocculant extract, and *panela*, the following mathematical expressions were used:

$$\begin{aligned} E_3 &= (m_{jc} - m_{sch}) * Cp_{jc} * T_{jc}, \\ E_4 &= m_{sch} * Cp_{sch} * T_{sc}, \\ E_5 &= m_{fl} * Cp_{fl} * T_{fl}, \\ E_6 &= m_p * Cp_p * T_{point}. \end{aligned} \tag{11}$$

120°C was established like the final temperature at which noncentrifugal brown sugar was obtained ($T_{point} = 120°C$).

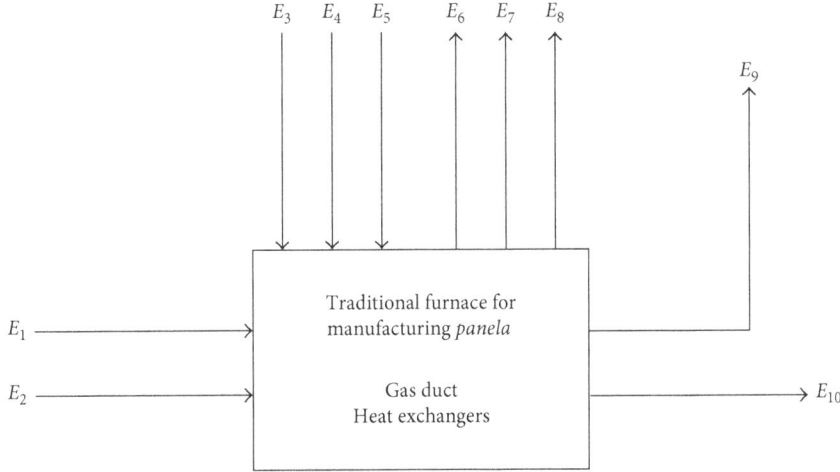

FIGURE 3: Energy balance in traditional furnace for making *panela*, energy in kJ. $E_1$: bagasse, $E_2$: air, $E_3$: sugarcane juice, $E_4$: solid contaminants in the sugarcane juice, $E_5$: flocculant extract, $E_6$: noncentrifugal brown sugar produced, $E_7$: mud, $E_8$: steam released during the evaporation, $E_9$: chimney gases, and $E_{10}$: other energy losses.

The solid contaminants specific heat (flocculant) was assumed as 2.2 kJ/kg·°C. This property was calculated as a mix of water and carbohydrates. According to Montoya and Giraldo [14], the mud-specific heat was taken as 2.8 kJ/kg·°C. The steam energy and energy of chimney gases were given by (13) and (14). For determining the thermodynamic properties of evaporated water and chimney gases, the Soave–Redlich–Kwong (SRK) model was used [13]. The roots of the equations were found with Matlab software version R2013a. For the case of the chimney gases, the rules of mixing proposed by Seader et al. [13] were used to estimate specific volume, molar enthalpy, and molar entropy. The $k_{ij}$ value was predicted according to Coutinho et al. [15].

$$E_7 = m_{ch} * 2.8 * T_{cl}, \tag{12}$$

$$E_8 = m_{ae} * \left(\frac{h_{mae}}{M_{ae}}\right), \tag{13}$$

$$E_9 = m_{gh} * \left(\frac{h_{mgh}}{M_{gh}}\right). \tag{14}$$

Exergy balance, according to Velásquez et al. [16], was developed for combustion chamber, gas duct, and chimney in steady state. In addition, exergy analysis involves the pans system where the cane juice, evaporated water, and panela streams were considered (see Figure 4 and (15)).

$$Ex_{bh} + Ex_{jc} = Ex_{gh} + Ex_{ae} + Ex_p + Ex_{dp}, \tag{15}$$

$$Ex_{ap} = Ex_{ae} + Ex_p. \tag{16}$$

The physical exergy of chimney gases was calculated in order to establish their available energy, as shown in (17). The kinetic and potential exergy were neglected [17].

$$\varphi = (h - h_o) - [T_o * (s - s_o)]. \tag{17}$$

Physical exergy for the air, the bagasse, and the raw juice were taken as zero, since both materials enter to the system at room temperature. Ashes exergy was neglected, because its low mass has a minimal level of energy. The air does not have chemical exergy, given that this substance forms part of the natural environment [16]. According to Kotas [18], bagasse chemical exergy ($Ex_{Qbh}$) was given by the expression

$$Ex_{bh} = m_{bh} * Ex_{Qbh},$$

$$\frac{Ex_{Qbh}}{PCN} = \frac{[1.0438 + 1.882(x_H/x_C) - 0.2509(1 + 0.7256(x_H/x_C)) + 0.0383(x_N/x_C)]}{1 - 0.3035(x_O/x_C)}. \tag{18}$$

Velásquez et al. [16] reported a chemical exergy for the sucrose of 7.06 kJ/kg and (20) to calculate the chemical exergy of sugarcane juice.

$$Ex_{jc} = m_{jc} * Ex_{Qjc}, \tag{19}$$

$$Ex_{Qjc} = (x_{H_2O} * Ex_{QH_2O}) + (x_{sucrose} * Ex_{Qsucrose}). \tag{20}$$

Chimney gases exergy have both physical and chemical components. The physical availability is known by (17), while chemical exergy was calculated through [16]:

$$Ex_{gh} = m_{gh} * (Ex_{Fgh} + Ex_{Qgh}),$$

$$Ex_{Qgh} = R * T_o * \sum_{j=1}^{n} \left[ \gamma_j * \ln\left(\frac{\gamma_j}{\gamma_j^{ambient}}\right) \right]. \tag{21}$$

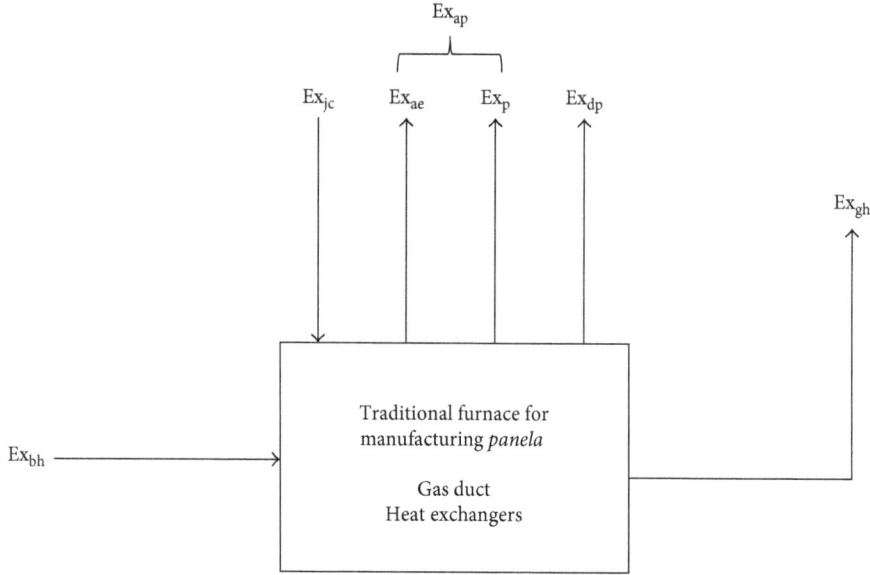

FIGURE 4: Exergy balance for the traditional furnace, exergy in kJ. Humid bagasse ($Ex_{bh}$), sugarcane juice ($Ex_{jc}$), evaporated water ($Ex_{ae}$), panela ($Ex_p$), chimney gases ($Ex_{gh}$), and exergy destruction ($Ex_{dp}$). $Ex_{ap}$ represents the exergy harnessed and consumed during the operation and corresponds to the sum of $Ex_{ae}$ and $Ex_p$.

TABLE 1: Performance indices used to assess the traditional furnace.

| Performance indices | Equation |
|---|---|
| Energy efficiency (%) | $\%e = [(E_6 + E_7 + E_8)/E_1] * 100$ |
| Exergy efficiency (%) | $\%e_x = [(E_{xae} + E_{xp})/(E_{xbh} + E_{xjc})] * 100$ |
| Energy loss through the chimney (%) | $\%n = (E_9/E_1) * 100$ |
| Yield ($kg_{panela}/kg_{bagasse}$) | $R = (m_p/m_{bh})$ |
| Productivity ($kg_{panela}/h$) | $P = (m_p/t_{production})$ |
| Bagasse consumption ($kg_{bagasse\ consumed}/kg_{produced\ bagasse}$) | $B = (m_{bh}/m_{bp})$ |

To determine total exergy efficiency of the production process, water and panela exergy values may be determined. The physical and chemical exergies were obtained from Kotas [18].

$$Ex_{ae} = m_{ae} * (Ex_{Fae} + Ex_{Qae}),$$

$$Ex_{Qae} = -R * T_o * \ln \left( \gamma_{H_2O}^{ambient} \right),$$

$$Ex_{Fae} = (h_{ae} - h_o) - [T_o * (s_{ae} - s_o)],$$

$$Ex_p = m_p * (Ex_{Fp} + Ex_{Qp}),$$

$$Ex_{Qp} = (x_{H_2O} * Ex_{QH_2O}) + (x_{sucrose} * Ex_{Qsucrose}),$$

$$Ex_{Fp} = Cp_p * (T_{point} - T_o) - \left[ T_o * Cp_p * \ln\left(\frac{T_{point}}{T_o}\right) \right].$$
$$(22)$$

2.3. Performance Indices. The traditional furnace for manufacturing panela was analyzed considering the indices presented in Table 1, proposed by Velásquez et al. [6] and Velásquez et al. [16].

## 3. Results and Discussion

3.1. Sugarcane Bagasse Characterization. Table 2 shows the results of the elemental and proximate analyses of sugarcane bagasse. The bagasse characterization outcomes are similar to those reported by Shiralkar et al. [11]. Nevertheless, it is worth noting that bagasse composition depends on cane variety, soil conditions, and crop nutrition [5]. For this reason, and considering that the cane harvest was done in a single cut, some differences with respect to available information in the literature can be found. For example, the lower calorific values reported by Shiralkar et al. [11] were found between 15.20 and 16.40 MJ/kg, determined for bagasse samples from different locations.

According to Sánchez et al. [19], in a fixed-bed combustion, gases composition and combustion rate can be optimized using bagasse with a moisture content between 10% and 30%, as used in this work. Low water content in solid fuel drives an appropriate carbon into $CO_2$ conversion, increasing the volatile compounds release rate and the material oxidation. In this way, the combustion efficiency can be increased in a range between 49% and 55%, compared to the use of bagasse with humidity greater than 40%. In addition, the combustion temperature is 16% higher [19].

TABLE 2: Elemental analysis of bagasse and determination of its lower calorific value (PCN).

| Property | Lot 01 | Lot 02 | Lot 03 | Lot 04 | Lot 05 | Mean* |
|---|---|---|---|---|---|---|
| Total moisture (%) | 16.96 | 12.72 | 16.58 | 14.24 | 14.39 | $14.98 \pm 1.77$ |
| Ash (%) | 4.51 | 7.20 | 3.60 | 3.08 | 3.36 | $4.35 \pm 1.68$ |
| Carbon (%) | 40.09 | 42.17 | 40.62 | 41.73 | 42.03 | $41.33 \pm 0.92$ |
| Hydrogen (%) | 5.61 | 5.80 | 5.52 | 5.57 | 5.53 | $5.61 \pm 0.11$ |
| Nitrogen (%) | 0.12 | 0.18 | 0.17 | 0.12 | 0.22 | $0.16 \pm 0.04$ |
| Sulfur (%) | 0.08 | 0.06 | 0.06 | 0.04 | 0.11 | $0.07 \pm 0.03$ |
| Oxygen (%) | 32.63 | 31.87 | 33.45 | 35.22 | 34.36 | $33.51 \pm 1.33$ |
| PCN (MJ/kg) | 13.85 | 14.75 | 13.90 | 14.47 | 14.59 | $14.31 \pm 0.41$ |

*Mean values for the five production lots ± standard deviation.

### 3.2. Isokinetic Sampling and Analysis.

Table 3 shows the results for the exhaust gases analysis. The excess air was very high, compared with the parameters indicated by Kuprianov et al. [20] and Sánchez et al. [19], who suggested percentages of excess air between 55% and 61%. For this reason, a decrease in the combustion flame temperature was presented [11]. Among the different treatments evaluated, a reduction of temperature up to 50°C was found in the worst cases. Moreover, this additional air can be placed on the pans forming an isolating layer, affecting the heat exchange efficiency [21].

High excess air, in the case of lots 03 and 04, promoted lower carbon monoxide formation and a complete combustion phenomenon [22–24]. When the flame front is generated in a uniform way over bagasse (not shallow), the CO concentrations decrease due to oxidation of both volatile and carbonized materials [20]. The chimney gases temperature in experiments 03 and 04 were greater in comparison with productions 01, 02, and 05, independent of the existence of an additional airflow that cooled the system. This fact enhances the heat transfer in the two mentioned cases, which is governed mainly by the mechanisms of radiation and forced convection.

According to Parra [25], the optimal gases velocity through the duct is equal to 4.5 m/s. In all experimental runs, the exhaust gases velocity was below of this value, which clearly indicates a chimney draft deficiency. Considering that the principal heat transfer mechanisms in the process are radiation and convection, the fluid slow circulation through the furnace affects its energy efficiency, which might increase with the redesign of combustion chamber and furnace [24].

### 3.3. Energy, Exergy, and Productivity.

Table 4 shows the furnace performance indices. For obtaining these indices the mass, energy, and exergy balances established in this work were solved.

All performance indices, except energy efficiency index (%$e$), were found within the ranges established by Velásquez et al. [6], Velásquez et al. [16], and Sardeshpande et al. [24]. In the traditional furnace, an average efficiency of 12.73% was obtained, while the minimal efficiency reported by the cited authors was 28%. Nevertheless, the references mentioned that this parameter can fall down to levels of 15%. The low energy efficiency of the traditional furnace used in the current study can be attributed mainly to a wrong design of

TABLE 3: Results of the isokinetic testing for chimney gases.

| Parameter | Lot 01 | Lot 02 | Lot 03 | Lot 04 | Lot 05 |
|---|---|---|---|---|---|
| Excess air (%) | 129.920 | 216.210 | 173.710 | 288.130 | 127.930 |
| Temperature (°C) | 434.650 | 473.470 | 451.410 | 465.850 | 417.390 |
| Velocity (m/s) | 2.700 | 2.900 | 2.700 | 3.270 | 2.800 |
| Humidity ($kg_{water}/kg_{gh}$) | 0.200 | 0.254 | 0.188 | 0.191 | 0.171 |
| % CO | 5.000 | 2.600 | 1.200 | 0.700 | 1.900 |
| % $CO_2$ | 10.69 | 11.2 | 8.91 | 7.38 | 10 |
| % $O_2$ | 9.92 | 9.4 | 11.75 | 13.31 | 10.7 |
| % $N_2$ | 74.129 | 76.614 | 78.039 | 78.529 | 77.234 |
| % $NO_2$ | 0.079 | 0.128 | 0.099 | 0.080 | 0.104 |
| % $SO_2$ | 0.183 | 0.057 | 0.001 | 0 | 0.062 |

the combustion chamber and the heat transfer section [6, 21]. Additionally, in the pilot traditional furnace, bagasse inlet was located on a lower level to that of the construction, limiting the uniform contact between the primary air and solid fuel (bagasse). Moreover, installed heat exchangers correspond to traditional semicylindrical pans, placed in parallel flow with respect to the combustion chamber, which makes that technology inefficient by its design. It is highlighted that this type of pan exhibits low overall heat transfer coefficients [11, 21].

Thermal loss through the furnace walls and chimney is other feature that aids to further decrease energy efficiency. Likewise, excess air in traditional furnace demands a complementary energy transfer to achieve its preheating. To guarantee drying, devolatilization, and oxidation during the combustion, a portion of bagasse energy available was used, denoting an irreversibility in the process, which reduced energy and exergy efficiencies in the traditional furnace [11, 19].

Despite having the lowest energy loss through the chimney, lots 01 and 05 also present a lower energy efficiency. The energy efficiency index, by definition, refers to the amount of heat lost with the gases leaving the line. In turn, this parameter is a function of the outgoing gases enthalpy and temperature [26]. In these two experimental cases, the lowest temperatures in the chimney were found; as a consequence, the energy efficiency in lots 01 and 05 were smaller. In addition, minor air excess during the combustion phenomena for these two experimental runs is the main cause of the few energy and exergy in the gases flow [20, 22]. Also, the lowest excess air drives an incomplete combustion generating a high carbon monoxide concentration. The CO has lower enthalpy and thermal conductivity than the

TABLE 4: Efficiency, productivity, and environmental indices for the traditional furnace.

| Indices | Lot 01 | Lot 02 | Lot 03 | Lot 04 | Lot 05 | Mean* |
|---|---|---|---|---|---|---|
| %$e$ | 11.450 | 13.846 | 14.008 | 12.794 | 11.533 | $12.726 \pm 1.091$ |
| %$e_x$ | 8.168 | 9.349 | 9.882 | 9.492 | 8.172 | $9.013 \pm 0.710$ |
| %$n$ | 61.924 | 75.898 | 73.757 | 91.152 | 58.736 | $72.293 \pm 11.507$ |
| $R$ ($kg_{panela}/kg_{bagasse}$) | 0.123 | 0.144 | 0.141 | 0.183 | 0.127 | $0.144 \pm 0.021$ |
| $P$ ($kg_{panela}/h$) | 7.098 | 8.223 | 6.753 | 7.356 | 7.821 | $7.450 \pm 0.520$ |
| $B$ ($kg_{used\ bagasse}/kg_{produced\ bagasse}$) | 1.443 | 1.270 | 1.149 | 1.063 | 1.366 | $1.258 \pm 0.139$ |

The indices are %$e$, energy efficiency, %$e_x$, exergy efficiency, %$n$, index of energy loss through chimney, $R$, process yield, $P$, productivity, $B$, bagasse consumption. *Mean values for the five production lots $\pm$ standard deviation.

TABLE 5: Technological alternatives for improving performance of traditional furnace.

| Alternative/Technological improvement | Description | Advantage |
|---|---|---|
| Furnace operation in combined flow | Fusion between the operation in counter-current and parallel flow: juice clarification near the chimney, evaporation above the combustion chamber, and final concentration of the product in the center of the furnace. | (i) Increase of energy efficiency: Use of high amounts of heat for the evaporation of the water present in the juice (phase change) (ii) Preservation of *panela* quality: The final product is protected from burning by the action of the maximum heat transfer in the concentration zone |
| Use of improved combustion chambers: WARD-type chamber (CWC), developed by CIMPA (Colombia) | Combustion chamber with a drying ramp for wet bagasse. It has and independent entrance section, both for primary and secondary combustion air. | (i) Reaction volume, three times higher than a traditional chamber (ii) Allows wet bagasse and works with better excess air (iii) Facilitates the air circulation and prevents the formation of high amounts of CO (iv) Range of temperatures up to 1200°C |
| Implementation of more efficient pans (heat exchangers) | Mainly, there are three significant improvements to traditional pans: adjusted semicylindrical, finned exchanger, and pyrotubular. | (i) Increase in the overall heat transfer coefficient and in the area/volume ratio (ii) Improves the heat exchange between the combustion gases and the juice, achieving greater energy and exergy efficiencies |
| Chimney draft control | Utilization of blowers and valves, to ensure the suction of the necessary air to achieve complete combustion. Speed control of the combustion gases, in order to favor convective and radiant heat transfer. | (i) Complete combustion. Reduction in the appearance of gaseous species such as CO and $NO_x$ (ii) Generation of desired temperatures (minimum of 500°C). (iii) Improvement of the energy, exergy and productive efficiency of the process |
| Energy integration | Use of the chimney gases exergy for some operation in the process. For example, the bagasse drying, the preheating of water or juice, among others, can be considered. | (i) Increase in the amount of energy used within the process (ii) Presence of better energy, exergy and productive indices. (iii) Operational costs reduction (iv) Possibility of achieving fuel self-sufficiency in the traditional furnace |
| Use of steam in industrial operations | Replacement of the combustion chamber by a boiler that generates steam. This fluid is used as energy source in the heat exchangers. | (i) Improvement in the heat transfer rate to the juice, as the steam is a cleaner fluid (ii) Possibility of using natural gas as fuel in the boiler, making the process more convenient from the environmental point of view (iii) Greater control and automation potential (iv) Increase in the scale of production, also allowing the reduction of associated costs |

species obtained from complete combustion. This fact causes the heat transfer velocity reduction inside the furnace [1, 17]. The highest CO concentration from incomplete combustion implies a physical exergy 30% lower in comparison with the physical exergy when the $CO_2$ formation predominates during the complete combustion. In this way, in an incomplete combustion, the exergy from the gases is lower, contributing to a less energy availability to be used for carrying out the heating and evaporation of sugarcane juice and *panela* [16, 19].

The average exergy efficiency for the experiments was $9.013 \pm 0.710$. Considering the values reported by Velásquez et al. [16], between 7.33% for an industrial process working with steam and 22.06% for an improved counter-current furnace (called GIPUN), it can be concluded that the exergetic performance of the traditional furnace was within standard values. For the same type of technology used in this research, cataloged as traditional and artisanal, the authors found an exergy efficiency of 10.94%.

Because of low energy efficiency index found in the process assessment, none of the experiments presented a self-sustaining fuel ($B < 1$). This fact indicates a low utilization of biomass energy resource. In all experimental runs, exergy flow in exhaust gases with available potential was proved. Thus, exergy available can be performing subsequent heating operations using the hot chimney gases as the energy main source.

High standard deviation presented by the results was due to the minimal control maintained over the excess air and the combustion process, which directly affects the composition, temperature, and velocity of the chimney gases [6]. In cases of minor oxidizing flow (Lot 01, 02 and 05), it causes a biomass incomplete burn, a low heat transfer via convection, and a great emission of particulate material [20].

As can clearly be seen, the combustion phenomenon directly affects the energy, exergy, and productivity indices for the traditional panela-making furnace. The operation efficiency also depends on the way in which the heat transfer is carried out between the energy resource and the evaporated juice. Therefore, its behavior depends directly on the area and the heat transfer coefficient, as well as on the temperature difference between the gases and the pans [11, 16, 21, 22, 24, 26]. According to Gutiérrez et al. [27], certain modifications can be made to the traditional process, in order to improve its performance from different the points of view. Table 5 presents some of these technological options.

## 4. Conclusions

The assessment of traditional furnace for manufacturing *panela* indicates that this technological configuration offers certain performance limitations and control over some operations, such as bagasse combustion, concentration of sugarcane juice, and noncentrifugal brown sugar obtaining. Among these problems, are highlighted the inappropriate location of the bagasse inlet, deficient furnace wall isolation, selection and use of inefficient pans (heat exchangers), high energy loss with exhaust gases, poor chimney draft, and the solid fuel uncontrolled burning. These last two aspects affect the traditional furnace performance, due to the fact that it gives way to the existence of an incomplete combustion phenomenon, generating low heat transfer rates through the juices, additionally producing carbon monoxide, particulate material, nitrous oxide and sulfur oxide. According to this, it can be concluded that the artisanal methods are inefficient from an energy and productivity point of view and generate a high environmental impact on the areas around the sugar mills.

The excess air is the most important factor that must be analyzed and controlled to enhance energy and productivity performances in the *panela* manufacturing process, since contact between the solid fuel and air allows using efficient advantage of the bagasse energy resource. Chimney gases composition depends on the factors temperature and velocity air through the furnace. The production process can be detained due to the loss through chimney draft and furnace duct clogging. In a direct manner, the excess air and chimney draft control the heat transfer rate by convection and radiation among the juices and fluids.

## Nomenclature

| | |
|---|---|
| $B$: | Index for bagasse use and consumption ($kg_{used\ bagasse}/kg_{produced\ bagasse}$) |
| $Cp_{fl}$: | Flocculant specific heat (kJ/kg·°C) |
| $Cp_{jc}$: | Cane juice specific heat (kJ/kg·°C) |
| $Cp_p$: | *Panela* specific heat (kJ/kg·°C) |
| $Cp_{sch}$: | Specific heat of mud (kJ/kg·°C) |
| $Ex_{ae}$: | Exergy of evaporated water during the juice concentration (kJ) |
| $Ex_{ap}$: | Harnessed exergy (kJ) |
| $Ex_{bh}$: | Humid bagasse exergy (kJ) |
| $Ex_{dp}$: | Exergy destruction in the process (kJ) |
| $Ex_{Fae}$: | Evaporated water physical exergy (kJ/kg) |
| $Ex_{Fgh}$: | Chimney gases physical exergy (kJ/kg) |
| $Ex_{Fp}$: | *Panela* physical exergy (kJ/kg) |
| $Ex_{gh}$: | Chimney gases exergy (kJ) |
| $Ex_{jc}$: | Sugarcane juice exergy (kJ) |
| $Ex_p$: | *Panela* exergy (kJ) |
| $Ex_{Qae}$: | Water evaporated chemical exergy (kJ/kg) |
| $Ex_{Qbh}$: | Humid bagasse chemical exergy (kJ/kg) |
| $Ex_{Qgh}$: | Chimney gases chemical exergy (kJ/kg) |
| $Ex_{Qjc}$: | Raw juice chemical exergy (kJ/kg) |
| $Ex_{QH2O}$: | Water chemical exergy (kJ/kg) |
| $Ex_{Qp}$: | *Panela* chemical exergy (kJ/kg) |
| $Ex_{Qsucrose}$: | Sucrose chemical exergy (kJ/kg) |
| $E_1$: | Cane bagasse energy (kJ) |
| $E_2$: | Air energy (kJ) |
| $E_3$: | Cane juice energy (kJ) |
| $E_4$: | Energy of solid contaminants presents in cane juice (kJ) |
| $E_5$: | Clarification extract energy (kJ) |
| $E_6$: | *Panela* energy (kJ) |
| $E_7$: | Mud energy (kJ) |
| $E_8$: | Energy of steam removed during concentration (kJ) |
| $E_9$: | Chimney gases energy (kJ) |
| $E_{10}$: | Other energy losses (kJ) |
| $h$: | Mass enthalpy (kJ/kg) |
| $h_{mae}$: | Molar enthalpy of water evaporated from juices (kJ/kmol) |
| $h_{mgh}$: | Molar enthalpy of humid chimney gases (kJ/kmol) |
| $H_Y$: | Enthalpy of humid ambient air (kJ/kg$_{as}$) |
| $h_o$: | Mass enthalpy evaluated at room temperature (kJ/kg) |
| $m_{aa}$: | Water mass, with the combustion air (kg) |

$m_{ab}$:        Water mass contained in the bagasse (kg)
$m_{ae}$:        Evaporated water mass (kg)
$M_{ae}$:        Molecular weight of evaporated water (kg/kmol)
$m_{as}$:        Dried air mass used in combustion (kg)
$m_{at}$:        Steam mass leaving the system with the chimney gases (kg)
$m_{bh}$:        Humid bagasse mass (kg)
$m_{bs}$:        Dried bagasse mass (kg)
$m_{ch}$:        Removed mud mass (kg)
$m_{fl}$:        Mass extract flocculant (kg)
$M_{gh}$:        Molecular weight of humid gases in chimney (kg/kmol)
$m_{gh}$:        Humid gases mass in chimney (kg)
$m_{gs}$:        Dry gases mass through chimney (kg)
$m_{jc}$:        Cane juice mass (kg)
$m_{lch}$:        Juice mass remaining from mud (kg)
$m_{mp}$:        Particulate material (kg)
$m_{p}$:        *Panela* obtained at the end of process (kg)
$m_{r}$:        Unburned residues (kg)
$m_{sch}$:        Mass of solids presents in mud (kg)
$P$:        Furnace productivity (kg$_{panela}$/h)
PCN:        Lower calorific value of the bagasse (MJ/kg)
$R$:        Universal constant of ideal gases (8.3140 kPa·m$^3$/kmol·K)
$R$:        Yield (kg$_{panela}$/kg$_{bagasse}$)
$s$:        Entropy (kJ/kg·K)
$s_{o}$:        Entropy evaluated at room temperature (kJ/kg·K)
$T_{cl}$:        Temperature at cane juice clarification (°C or K, according to equation)
$T_{fl}$:        Flocculant extract temperature (°C or K, according to equation)
$T_{jc}$:        Cane juice temperature (°C or K, according to equation)
$t_{production}$:        Total production time (s)
$T_{point}$:        Temperature of *Panela*-making point (°C or K, according to equation)
$T_{sc}$:        Temperature of contaminant solids in cane juice (°C or K, according to equation)
$T_0$:        Reference temperature (K)
$T_{o}$:        Room temperature (K)
$w$:        Raw bagasse mass fraction of humidity (kg$_{H_2O}$/kg)
$w_{gh}$:        Mass fraction of humidity in chimney gases (kg$_{H_2O}$/kg)
$x$:        Mass fraction
%$e$:        Energy efficiency
%$ex$:        Exergy efficiency
%$n$:        Energy loss via furnace chimney
$\gamma_{H_2O}$:        Molar fraction of water
$\gamma_i$:        Molar fraction of material $i$
$\gamma_j$:        Molar fraction of material $j$
$\varphi$:        Physical exergy for a gas flow (kJ/kg).

## Conflicts of Interest

The authors declare that there are no conflicts of interest regarding the publication of this paper.

## Acknowledgments

Thanks are due to Universidad de Caldas and the project "Implementation of the Research, Innovation and Technology Center for the Panela Sector of the Department of Caldas, BEKDAU Center," financed by the General System of Royalties (SGR).

## References

[1] G. N. Tiwari, S. Kumar, and O. Prakash, "Study of heat and mass transfer from sugarcane juice for evaporation," *Desalination*, vol. 159, no. 1, pp. 81–96, 2003.
[2] J. Singh, S. Solomon, and D. Kumar, "Manufacturing jaggery, a product of sugarcane, as health food," *Agrotechnology*, vol. 11, no. S11, pp. 1–3, 2013.
[3] Revista Dinero, *El negocio de la panela crece y se derrite a la vez [OL]*, 2014, http://www.dinero.com/empresas/articulo/balance-del-sector-panelero-colombia-2014/202561.
[4] Periódico El País, *Campaña para consumo de panela recibió premio internacional en Argentina [OL]*, 2015, http://www.elpais.com.co/elpais/economia/noticias/campana-para-consumo-panela-recibio-premio-internacional-argentina.
[5] H. R. García, L. C. Albarracín, A Toscano et al., *Guía Tecnológica para el Manejo Integral del Sistema Productivo de Caña Panelera*, Corpoica, Bogotá, Colombia, 2007.
[6] H. I. Velásquez, F. Chejne, and A. F. Agudelo, "Diagnóstico energético de los procesos productivos de la panela en Colombia," *Revista Facultad Nacional de Agronomía Medellín*, vol. 57, no. 2, pp. 1–15, 2004.
[7] P. V. K. Jagannadha Rao, M. Das, and S. K. Sas, "Changes in physical and thermo-physical properties of sugarcane, palmyra-palm and date-palm juices at different concentration of sugar," *Journal of Food Engineering*, vol. 90, no. 4, pp. 559–566, 2009.
[8] P. Laksameethanasan, N. Somla, S. Janprem et al., "Clarification of sugarcane juice for syrup production," *Procedia Engineering*, vol. 32, pp. 141–147, 2012.
[9] EPA United States Environmental Protection Agency, *Code of Federal Regulations, Title 40, Protection of Environment, Part 60 (Appendix)*, US EPA, Washington, DC, USA, 1991.
[10] S. Arias, A. M. Ceballos, and L. F. Gutiérrez, "Determinación experimental de propiedades térmicas y físicas para jugo de caña, miel y panela," *Vitae*, vol. 23, no. 1, pp. 145–148, 2016.
[11] J. Y. Shiralkar, S. K. Kancharla, N. G. Shah et al., "Energy improvements in jaggery making process," *Energy for Sustainable Development*, vol. 18, pp. 36–48, 2014.
[12] C. J. Geankoplis, *Transport Processes and Separation Process Principles (Includes Unit Operations)*, Prentice Hall, Upper Saddle River, NJ, USA, 4th edition, 2003.
[13] J. D. Seader, E. J. Henley, and D. K. Roper, *Separation Process Principles. Chemical and Biochemical Operations*, John Wiley and Sons Inc., New York, NY, USA, 3rd edition, 2010.
[14] C. F. Montoya and P. A. Giraldo, *Propuesta de Diseño de Planta de Procesamiento de Caña para la Elaboración de Panela en Yolombo—Antioquia*, Universidad Nacional de Colombia, Medellín, Colombia, 2009.
[15] J. Coutinho, G. Kontogeorgis, and E. Stenby, "Binary interaction parameters for nonpolar systems with cubic equations of state: a theoretical approach. CO$_2$/hydrocarbons using SRK equation of state," *Fluid Phase Equilibria*, vol. 102, no. 1, pp. 31–60, 1994.
[16] H. I. Velásquez, F. Chejne, and A. F. Agudelo, "Diagnóstico exergético de los procesos productivos de panela en Colombia," *Energética*, vol. 35, pp. 15–22, 2006.

equations of state: a theoretical approach. $CO_2$/hydrocarbons using SRK equation of state," *Fluid Phase Equilibria*, vol. 102, no. 1, pp. 31–60, 1994.

[16] H. I. Velásquez, F. Chejne, and A. F. Agudelo, "Diagnóstico exergético de los procesos productivos de panela en Colombia," *Energética*, vol. 35, pp. 15–22, 2006.

[17] Y. A. Cengel and M. A. Boles, *Thermodynamics: an Engineering Approach*, McGraw-Hill College, Boston, MA, USA, 5th edition, 2006.

[18] T. J. Kotas, *The Exergy Method of Thermal Plant Analysis*, Paragon Publishing, London, UK, 2012.

[19] Z. Sánchez, H. R. García, and O. A. Mendieta, "Efecto del precalentamiento del aire primario y la humedad del bagazo de caña de azúcar durante la combustión en lecho fijo," *Corpoica Ciencia y Tecnología Agropecuaria*, vol. 14, no. 1, pp. 5–16, 2013.

[20] V. I. Kuprianov, W. Permchart, and K. Janvijitsakula, "Fluidized bed combustion of pre-dried thai bagasse," *Fuel Processing Technology*, vol. 86, no. 8, pp. 849–860, 2005.

[21] S. I. Anwar, "Fuel and energy saving in open pan furnace used in jaggery making through modified juice boiling/concentrating pans," *Energy Conversion and Management*, vol. 51, no. 2, pp. 360–364, 2010.

[22] M. Baratieri, P. Baggio, L. Fiori et al., "Biomass as an energy source: thermodynamic constraints on the performance of the conversion process," *Bioresource Technology*, vol. 99, no. 15, pp. 7063–7073, 2008.

[23] L. Wang, C. L. Weller, D. D. Jones et al., "Contemporary issues in thermal gasification of biomass and its application to electricity and fuel production," *Biomass and Bioenergy*, vol. 32, no. 7, pp. 573–581, 2008.

[24] V. R. Sardeshpande, D. J. Shendage, and I. R. Pillai, "Thermal performance evaluation of a four pan jaggery processing furnace for improvement in energy optimization," *Energy*, vol. 35, no. 12, pp. 4740–4747, 2010.

[25] J. A. Parra, "Análisis térmico de una paila panelera," *Revista Ingenio Libre*, vol. 5, pp. 44–50, 2006.

[26] J. A. Osorio, H. J. Ciro, and A. Espinosa, "Evaluación térmica y validación de un modelo por métodos computacionales para la hornilla panelera GP150," *Dyna*, vol. 77, no. 162, pp. 237–247, 2010.

[27] L. F. Gutiérrez, S. Arias, and A. M. Ceballos, "Advances in traditional production of panela in Colombia: analysis of technological improvements and alternatives," *Ingeniería y competitividad*, vol. 20, no. 1, pp. 107–123, 2018.

# 10

# Impact of Additives on Heterogeneous Crystallization of Acetaminophen

**Hsinyun Hsu,**[1] **Lynne S. Taylor,**[2] **and Michael T. Harris** ⓘ[1]

[1]*Department of Chemical Engineering, Purdue University, 480 Stadium Mall Drive, West Lafayette, IN 47907, USA*
[2]*Department of Industrial and Physical Pharmacy, Purdue University, 575 Stadium Mall Drive, West Lafayette, IN 47907, USA*

Correspondence should be addressed to Michael T. Harris; mtharris@purdue.edu

Academic Editor: M. G. Ierapetritou

Introducing foreign substrates or additives is the common way to regulate polymorphism or kinetics of crystallization. Most present studies consider the substrate factor and additive factor separately. Here, the interplay between the additive, crystallizing molecules, and the substrate was investigated. Acetaminophen (APAP) was used as the model compound. 5 wt.% dioctyl sodium sulfosuccinate (AOT), poly(acrylic) acid (PAA), and hydroxypropyl methylcellulose (HPMC) were employed as additives. The interfacial crystal growth rate of APAP in the presence of additives was studied between slides coated with chitosan (CS) film. The crystallization kinetics of the additive/APAP mixture on CS substrate was also investigated. The additive/APAP was characterized by differential scanning calorimetry (DSC), and the interfacial molecular interaction was studied by Fourier transform infrared spectroscopy (FTIR). The results indicate that the additive-substrate interaction can change the interfacial growth behavior observed in the additive-APAP binary system. Nevertheless, crystallizing without confinement, the additive-APAP interaction is more effective at controlling the crystallization of APAP, and the substrates did not have much effect.

## 1. Introduction

Controlling crystallization is important in many areas of science and technology, such as manufacturing of pharmaceuticals, production of semiconductors and nonlinear optics, as well as the formation of biominerals. To obtain crystals with the desired structure and properties, foreign substrates or additives are often introduced.

In most practical circumstances, crystallization begins with heterogeneous nucleation on foreign substrates [1]. Therefore, foreign surfaces can be used to dictate crystallization of molecules and form crystals with the desired properties. Carters et al. [2] had shown that crystalline substrates which are lattice-matched with particular crystal faces have been used for direct nucleation of organic crystals and even control polymorph selectivity by epitaxy. Studies discussing the effects of porous substrates with pores that have various shapes and spherical aggregation on heterogeneous nucleation have been done by Diao et al. [3] and Quon et al. [4], and their results demonstrate the crucial impact of surface morphology on crystallization. Additionally, although the mechanism is not clear, different polymorphs of compounds have been successfully formed on amorphous polymeric films. Examples of small molecules [5], supramolecular complexes [5], and proteins [6] had been shown.

In addition to substrates, introducing additives is another popular approach to control crystallization. Tailor-made auxiliaries that can have molecular recognition with crystallizing molecules have been used to change crystallization behavior [7]. Generally, the tailor-made auxiliaries are classified into two categories: inhibitors and promoters. A variety of purposes including morphology control [8], assignment of the absolute structure of chiral molecules and polar crystals [9], elucidation of the impact of solvent on crystal growth [10], and crystallization of desired polymorph [11] can be achieved by using the inhibitor. As for the promoter, monolayers of amphiphilic molecules on water have been employed to induce the growth of a variety of three-dimensional crystals at the monolayer-solution interface by structural match [12], molecular complementarity [13], or electrostatic interaction [14]. In pharmaceutical sciences, compatible polymers usually serve as additives to stabilize the amorphous state of poorly water-soluble active pharmaceutical ingredients (APIs) and increase their bioavailability. Polymers are thought to inhibit crystallization of the amorphous solid by increasing the glass transition temperature ($T_g$) of the resultant molecular level blend (often termed an amorphous solid dispersion), leading to a decrease in mobility of the API molecules, and through the formation of API-polymer specific interactions which act to disrupt self-assembly [15].

Given the influence of substrates [16] and additives, exclusively, it would be interesting to investigate their combinational effects on crystallization. In this study, crystallization of acetaminophen (APAP) on chitosan (CS) substrates [16] in the presence of additives was examined. The impact of additives on crystallization of APAP on the citric CS film was compared. Dioctyl sodium sulfosuccinate (AOT), polyacrylic acid (PAA), and hydroxypropyl methylcellulose (HPMC), which have various extents of molecular interaction with APAP, were used as the additives. The interplay between APAP, additives, and substrates was studied.

## 2. Materials and Methods

*2.1. Materials.* Acetaminophen (APAP) and dioctyl sodium sulfosuccinate (AOT) were purchased from Sigma-Aldrich (St. Louis, MO, USA). Chitosan (CS) with a degree of deacetylation of approximately 85% and molecular weight 250 kDa was purchased from Koyo Chemical Co. Ltd. (Tokyo, Japan). Poly(acrylic acid) (PAA, Mv 450,000 g·mol$^{-1}$ and Mw 1800 g·mol$^{-1}$ was purchased from Sigma-Aldrich (St. Louis, MO, USA). Hydroxypropyl methylcellulose (HPMC) E3 was obtained from the Dow Chemical Company (Midland, MI, USA).

The chemical structure of the materials is given in Figure 1.

*2.2. Solution of CS Films Preparation.* CS was dissolved in a 1.5 M of citric acid solution at a concentration of 2% (wt/vol) concentration. Films made by the solution were uniform and equivalent according to the AFM characterization.

*2.3. Sample Preparation.* Additive/APAP mixtures were prepared at 5% additive by first melting the mixture above the melting point and subsequently quenching the molten mixture in liquid nitrogen. The resultant solid was ground by a mortar and pestle for five minutes and followed by 15 minutes milling in a cryogenic mill (6750 freezer mill, Spex Sampleprep, Metuchen, NJ) to ensure complete mixing.

*2.4. Crystal Growth Rate.* The chitosan solution prepared in Section 2.2 was spin-coated on cover slides using a KW-4A spin coater (Chemat Technology Inc., Northridge, CA) at a speed of 2,000 rpm and then dried in a desiccator to prepare the films.

The isothermal crystal growth rate of additive/APAP at 40°C on CS substrates was determined and performed by hot stage microscopy [16]. 3–5 mg of additive/APAP mixture was placed between two cover slips, melted at 185°C, and subsequently quenched in liquid nitrogen. The samples were then examined using a polarizing microscope (Nikon Eclipse E600 POL microscope, Nikon Corp, Tokyo, Japan). The temperature of the sample was controlled by a hot stage (Linkam THMS600, Surrey, UK). As shown in Figure 2, the crystal of APAP grew as a spherulite. The pictures of the growing spherulite were taken at constant intervals using time-lapse photography. When the radius of the spherulite was plotted against time, it resulted in a linear line ($R^2 > 0.99$), the slope of which was the growth rate.

*2.5. Time-Resolved Wide-Angle X-Ray Scattering.* The time-resolved WAXS experiments were conducted to study crystallization kinetics of additive/APAP on different substrates. The data were collected at the Advanced Photon Source Beam Station 12-ID-B, Argonne National Laboratory. The WAXS system was equipped with a Pilatus 300 detector. The energy of the X-ray source was 13.9984 keV ($\lambda = 0.88$ Å), and the distance between the sample to the detector was 455.26 mm. The $q$ (scattering vector) range was 0.9–2.12 Å$^{-1}$. The WAXS range was calibrated using silver behenate (AgBeh), and the absolute intensity was calibrated using glassy carbon.

400 $\mu l$ of chitosan solution prepared in Section 2.2 was poured into small aluminum pans (Tzero DSC sample pans, TA Instruments, New Castle, DE) to prepare the chitosan films. The additive/APAP mixture was melted in a syringe by a syringe heater (Braintree Scientific, Inc., Braintree, MA), and 0.1 mL of molten additive/APAP was deposited on the film by a syringe pump, and the sample was quenched in liquid nitrogen. Subsequently, the pan was mounted on a Linkam THMS600 stage to control the crystallization temperature at 40°C, and the WAXS patterns were collected at intervals of 5 seconds.

*2.6. Thermal Analysis.* At the crystallization temperature, 40°C, which is much lower than the melting point of APAP, crystal growth of APAP is under diffusion control [17]. Therefore, the molecular mobility of APAP is one of the important keys that could affect crystallization. Here, thermal analysis was conducted to determine the glass transition temperatures ($T_g$) of APAP, and the value is used to evaluate the mobility of APAP in each system.

The physical mixture of 5% wt/wt additive/APAP was analyzed by a differential scanning calorimeter (DSC). The thermal analysis was carried out using a TA Q100 DSC equipped with a refrigerated cooling accessory (TA Instruments, New Castle, DE). The instrument was calibrated for enthalpy and temperature using indium. 50 mL/min of nitrogen was used as the purge gas. 3–5 mg of each sample was sealed into hermetic aluminum Tzero sample pans (TA instruments, New Castle, DE). $T_g$ of APAP and additive/APAP mixtures were determined by cooling the melt at 20°C/min to −20°C and then reheating at 10°C/min.

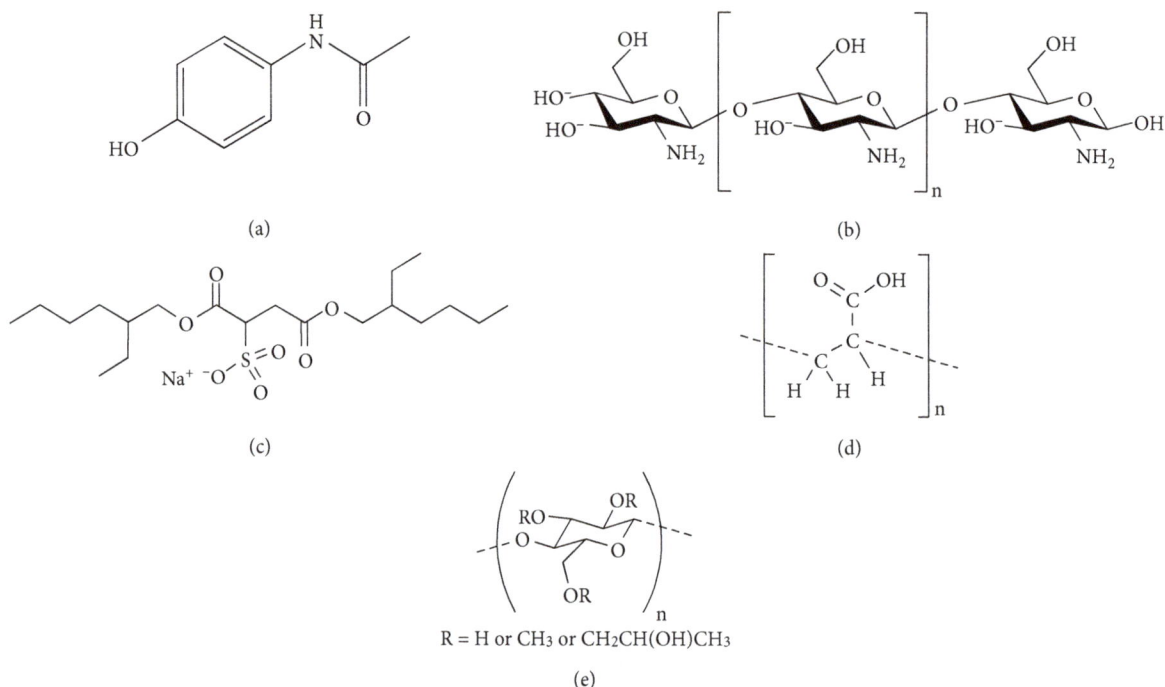

R = H or CH₃ or CH₂CH(OH)CH₃

(e)

FIGURE 1: Molecular structures of (a) APAP, (b) CS, (c) AOT, (d) PAA, and (e) HPMC.

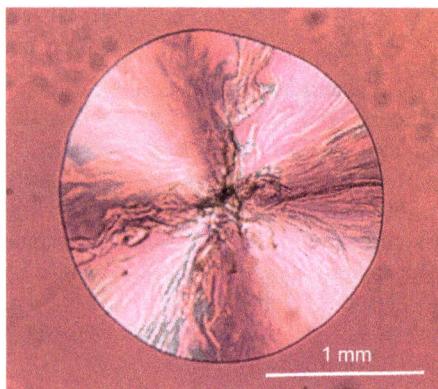

FIGURE 2: Photomicrographs of APAP crystal taken at 40°C. The scale bar represents 1 mm.

*2.7. Fourier Transform Infrared (FTIR) Spectroscopy.* FTIR was used to investigate interfacial interaction between the CS substrate and additive/APAP [16]. It was known that around 20–50% w/w additive is necessary to detect noticeable shifts in the infrared spectra arising from intermolecular interactions [18]. Therefore, solutions consisting of 30% wt/wt additive and 70% wt/wt APAP were prepared by dissolving the components in ethanol.

The samples were spin-coated onto ZnSe discs using a KW-4A spin coater (Chemat Technology Inc., Northridge, CA) set at a speed of 2,000 rpm. Spectra of pure APAP, CS substrate, pure additive, additive/APAP, and additive/APAP on CS were collected by a Nicolet Protégé 460 Fourier transform infrared spectrometer, equipped with a mercury-cadmium-telluride (MCT) detector cooled with liquid nitrogen, in transmission mode. The spectral contributions

from water vapor and carbon dioxide were minimized by continuously purging the instrument's sample chamber with dry air from a Balston purge gas generator. All spectra were taken at an interval of about $1\,\mathrm{cm}^{-1}$ with Happ-Genzel apodization and averaged over 256 scans.

## 3. Results and Discussion

*3.1. Effect of Additives on Crystal Growth Rate of APAP.* APAP is known to have three polymorphs [15]. Under physical constraint such as between two cover slips, melted APAP tends to yield form III [19]. This was confirmed by characterizing APAP that crystallized between two CS-coated quartz cover slips via Raman spectroscopy.

Compared with samples without additives, from Figure 3, AOT, PAA, and HPMC were effective at decreasing the growth rate. AOT and HPMC are the least and most effective additives, respectively.

The effect of surfactants on crystal growth of celecoxib has been reported by Mosquera-Giraldo et al. [20]. Surfactants decreased $T_g$ of celecoxib and increased its mobility and crystal growth rate. Nevertheless, in our study, AOT decreased the growth rate of APAP on CS substrates. In the experiments, the AOT/APAP mixture was sandwiched between the films; therefore, the impact of the films on crystal growth should be considered. The IR spectra of AOT-APAP-citric CS systems (Figure 4), show that the N–H and C=O stretch of APAP, and the combined C=O stretch of AOT and APAP shifted to lower wavenumbers, indicating the presence of the CS film interfered with the binary interaction between AOT and APAP. AOT is an anionic surfactant, and it can have strong interactions with CS, which is a cationic polymer. As deposited on the CS film, the electrostatic interaction between AOT and

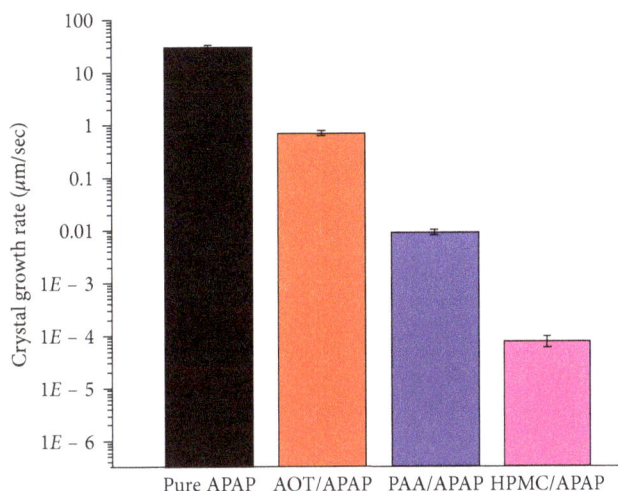

FIGURE 3: Interfacial crystal growth rate of APAP on citric CS films in the presence of different additives.

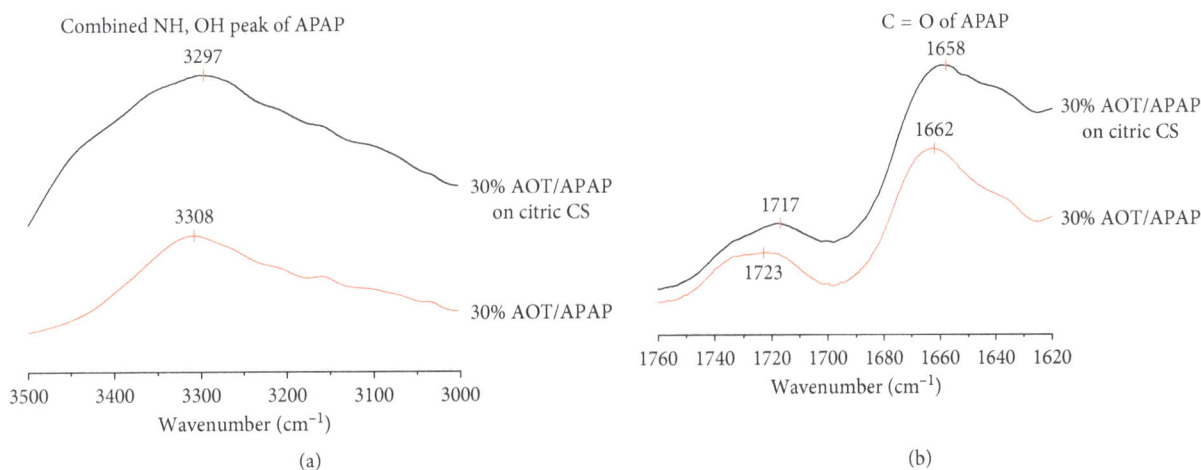

(a)

(b)

FIGURE 4: IR spectra of the AOT-APAP-citric CS system in the (a) combined N–H and O–H stretch region of APAP and (b) combined C=O stretch region of AOT and citric acid and C=O stretch region of APAP.

CS could promote the phase separation of the AOT/APAP mixture. So, the faster mobility resulted by the surfactant is not expected and the increased growth rate was not observed. Besides, the CS film was found to increase the growth rate of APAP (paper under review). When blending, AOT can block some sites on the surface of the CS films that APAP can interact with, so the growth rate was slower compared with pure APAP grown on the CS film.

Previously, among a series of hydrophilic polymers, in the polymer-APAP binary system in the absence of the substrates, PAA was found to have the greatest effect on reducing the growth rate of APAP, whereas the impact from HPMC was less [18]. The stronger PAA-APAP molecular interaction was claimed to be the cause of the higher effectiveness. However, in the ternary additive-APAP-substrate system, the additive can interact with both APAP and the substrate. Similar to the AOT system, when blended, PAA can interact with APAP through H-bonding. Besides, it can also form H-bonding between the carboxylic dimer and citric molecules doped on the CS film. Figure 5 shows the IR spectra of the N–H, O–H,

and C=O stretching regions of the samples containing PAA on citric the CS film. It can be seen that the functional groups all shifted to a lower wavenumber when the PAA/APAP mixture was coated on citric CS. Evidently, PAA in the blend was found to interact with the substrates, and the PAA-substrate interaction can weaken the binary PAA-APAP interaction and decreases the effectiveness of PAA at decreasing the growth rate. On the contrary, as Figure 6 shows, the peaks in the HPMC-APAP-CS systems did not have noticeable shifts, indicating HPMC does not interact with the substrate much. The previous report also pointed out that the interaction between HPMC and APAP was very weak [18]. Here is a possible explanation for the slow growth rate when blending APAP with HPMC: at 40°C, the additives are not miscible with APAP, and the additive-rich phase can also act like a substrate and affect crystallization of APAP. To confirm this hypothesis, the interfacial growth rate of APAP was characterized on PAA and HPMC films (Figure 7), and it was found when growing on HPMC film, the crystal growth rate was much lower compared with on glass or PAA film.

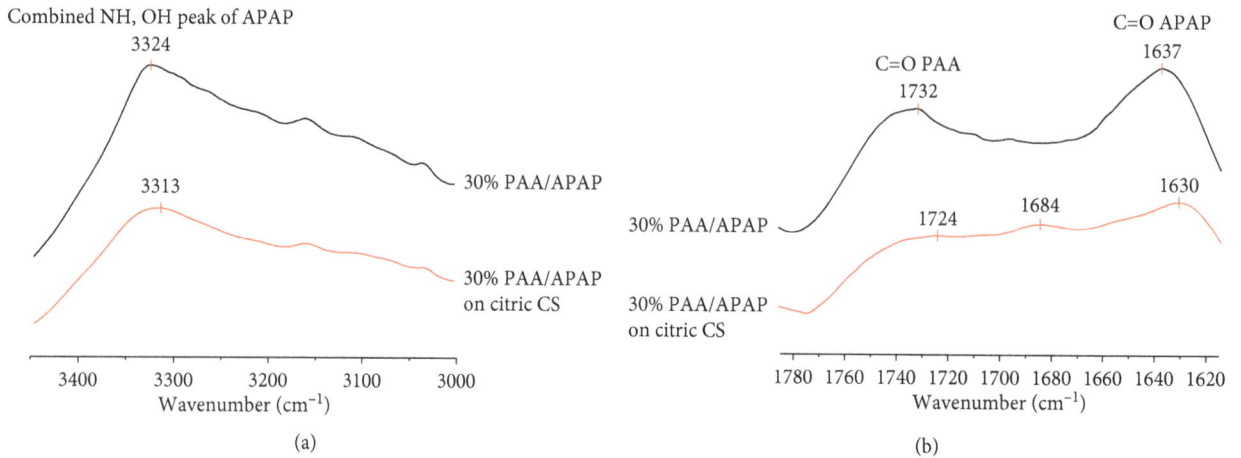

FIGURE 5: IR spectra of the PAA-APAP-citric CS system in the (a) combined N–H and O–H stretch region of APAP and (b) combined C=O stretch region of PAA and citric acid and C=O stretch region of APAP.

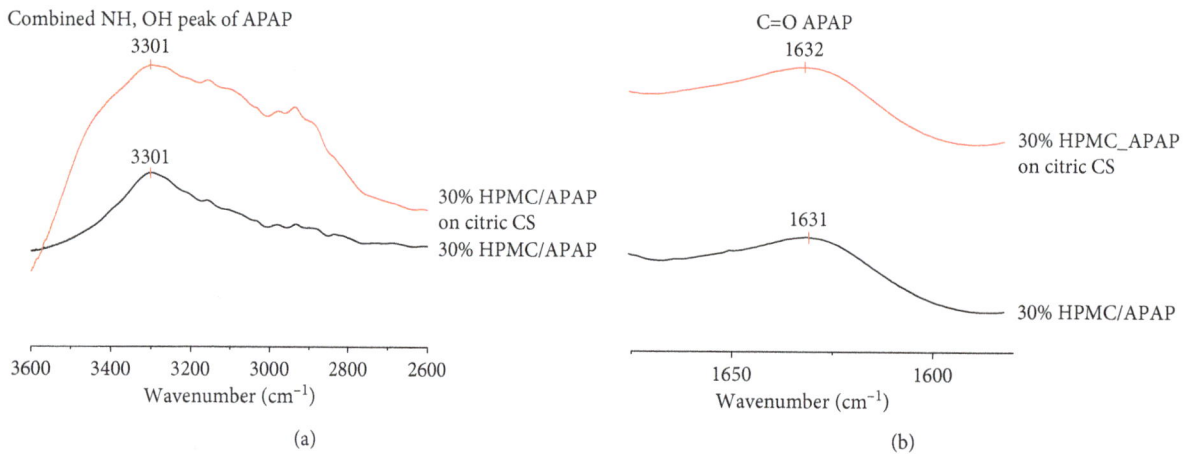

FIGURE 6: IR spectra of the HPMC-APAP-citric CS system in the (a) combined N–H and O–H stretch region of APAP and (b) C=O stretch region of citric acid and APAP.

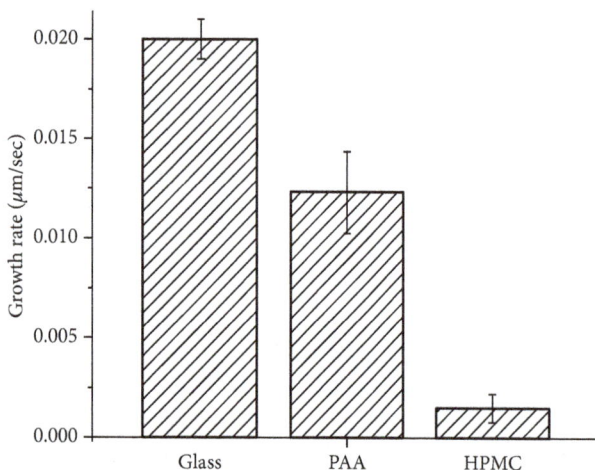

FIGURE 7: Interfacial crystal growth rate of APAP on glass, PAA, and HPMC films.

### 3.2. Effects of Additives on Crystallization Kinetics of APAP.

In addition to the interfacial study of crystal growth rate, the isothermal crystallization kinetics of additive/APAP on the CS film without confinement was also investigated using the time-resolved WAXS at the Argonne National Lab. The dynamic WAXS patterns of the samples containing additives and CS substrate were collected, and it was observed that APAP crystallized into Form I.

The WAXS patterns of AOT/APAP on CS at different time points are shown in Figure 8. Initially, the small peaks appeared at 0 second were contributed by the aluminum pan, and peaks corresponded to crystalline APAP shown up at about 1 minute. The percentage of crystalline APAP at different time periods, denoted as $X(t)$, was calculated by integrating the area under the APAP diffraction peaks at time $t$ ($\phi_{APAP}(t)$) and normalized by the area under the APAP diffraction peaks when crystallization was complete ($\phi_{APAP}(\infty)$) [4, 12, 21–24].

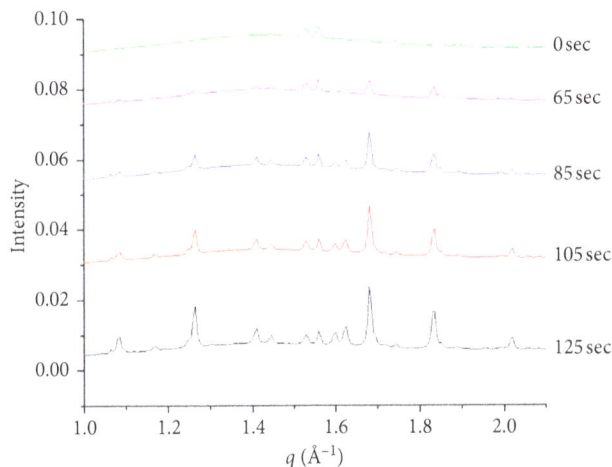

FIGURE 8: WAXS patterns of 5% AOT/APAP on the CS film at different time points.

TABLE 1: Summary of crystallization parameters of each system on CS films.

| Pure APAP | | AOT /APAP | | PAA /APAP | | HPMC/ APAP | |
|---|---|---|---|---|---|---|---|
| $t_{in}$ | $t_{50}^*$ | $t_{in}$ | $t_{50}^*$ | $t_{in}$ | $t_{50}^*$ | $t_{in}$ | $t_{50}^*$ |
| 1 | 1.5 | 1 | 0.5 | 5 | 9.57 | 30 | 7.76 |

$^*t_{in}$ is not included in $t_{50}$. $^{**}$The unit is minute.

TABLE 2: Glass transition temperature ($T_g$) of 5% additive/APAP.

| | APAP | AOT/APAP | PAA/APAP | HPMC/APAP |
|---|---|---|---|---|
| $T_g$ (°C) | $23.9 \pm 0.2$ | $19.3 \pm 0.4$ | $25.4 \pm 0.3$ | $24.6 \pm 0.4$ |

$$X(t) = \frac{\phi_{APAP}(t)}{\phi_{APAP}(\infty)}. \tag{1}$$

The induction time ($t_{in}$) and the time until 50% of the sample crystallized ($t_{50}$) for each sample are summarized in Table 1. HPMC was found to increase the induction time, $t_{in}$, the most compared with AOT and PAA, which is consistent with the previous study, which showed that HPMC was much more effective at inhibiting nucleation of APAP than PAA [18]. Samples containing PAA and HPMC had longer 50% crystallization times, $t_{50}$, compared with pure APAP on CS substrates.

In contrast to the interfacial growth rate study, the observed trend of $t_{50}$ is consistent with the observed binary additive-APAP systems which were published [18, 20]. AOT decreased $t_{50}$ of the samples. From the thermal analysis of additive/APAP mixture (Table 2), AOT decreased $T_g$ of APAP, and therefore, the increased mobility was suspected to be the reason for the shorter $t_{50}$ for AOT-added samples. PAA and HPMC both extended $t_{50}$, and the impact of PAA was stronger than HPMC.

Without confinement, only additives close to the interface can be affected by the substrate, and the influence was not large enough to affect the additive-APAP interaction in the bulk. It is likely that when allowed to crystallize in three dimensions, the additive-APAP interaction is more significant at controlling crystallization of APAP. Therefore, to confirm this hypothesis, crystal growth rate of APAP with additive and on different CS films was characterized without being covered by the top layer. In our previous research (paper under review), CS films prepared by different acids can have a profound impact on the crystallization of APAP. However, as Figure 9 demonstrates, when growing without confinement, the effect of substrates was negligible, and the impact of additives was consistent with the observed binary additive-APAP systems. AOT was found to increase the growth rate, and PAA had a stronger effect at reducing the growth rate than HPMC. This confirmed that additive-APAP interactions are more important at determining the

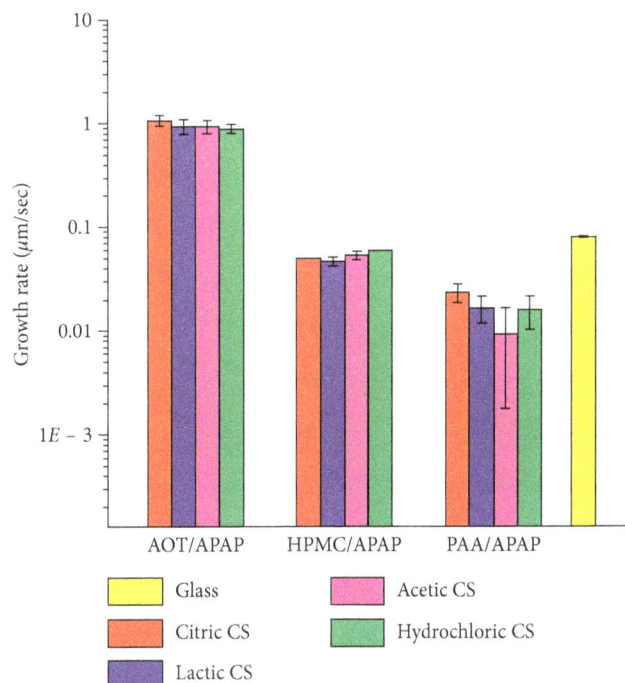

FIGURE 9: Growth rate of 5% additive/APAP on CS films without being sandwiched.

crystallization of APAP when it is allowed to crystallize in three dimensions.

## 4. Conclusion

With the presence of three different additives, the crystallization behavior of APAP on the CS film was characterized. In the study of interfacial crystal growth rate, the strong additive-substrate interaction can interfere with the interactions between the additive and APAP and change the crystallization behavior observed in the binary additive-APAP system. On the contrary, when allowed to crystallize in three dimensions, the additive-APAP interaction is more significant in controlling the crystallization of APAP. To conclude, when study crystallization at the interface, additive-substrate interaction can be important, and it can weaken the additive-APAP interaction; whereas, without confinement, the additive-APAP interaction is more significant at controlling crystallization.

## Conflicts of Interest

The authors declare that they have no conflicts of interest.

## References

[1] P. Debenedetti, *Metastable Liquids: Concepts and Principles*, Princeton University, Princeton, NJ, USA, 1996.

[2] P. W. Carter and M. D. Ward, "Topographically directed nucleation of organic crystals on molecular single-crystal substrates," *Journal of the American Chemical Society*, vol. 115, no. 24, pp. 11521–11535, 1993.

[3] Y. Diao, A. S. Myerson, T. A. Hatton, and B. L. Trout, "Surface design for controlled crystallization: the role of surface chemistry and nanoscale pores in heterogeneous nucleation," *Langmuir*, vol. 27, no. 9, pp. 5324–5334, 2011.

[4] J. L. Quon, K. Chadwick, G. P. F. Wood et al., "Templated nucleation of acetaminophen on spherical excipient agglomerates," *Langmuir*, vol. 29, no. 10, pp. 3292–3300, 2013.

[5] A. L. Grzesiak and A. J. Matzger, "New form discovery for the analgesics flurbiprofen and sulindac facilitated by polymer—induced heteronucleation," *Journal of Pharmaceutical Sciences*, vol. 96, no. 11, pp. 2978–2986, 2007.

[6] L. M. Foroughi, Y. N. Kang, and A. J. Matzger, "Polymer-induced heteronucleation for protein single crystal growth: structural elucidation of bovine liver catalase and concanavalin a forms," *Crystal Growth & Design*, vol. 11, no. 4, pp. 1294–1298, 2011.

[7] I. Weissbuch, R. Popovitz-Biro, M. Lahav, L. Leiserowitz, and Rehovot, "Understanding and control of nucleation, growth, habit, dissolution and structure of two-and three-dimensional crystals using tailor-made auxiliaries," *Acta Crystallographica Section B Structural Science*, vol. 51, no. 2, pp. 115–148, 1995.

[8] P. Hartman and W. G. Perdok, "On the relations between structure and morphology of crystals," *Acta Crystallographica*, vol. 8, no. 1, pp. 49–52, 1955.

[9] D. M. Walba, H. A. Razavi, N. A. Clark, and D. S. Parmar, "Design and synthesis of new ferroelectric liquid crystals. 5. Properties of some chiral fluorinated FLCs: a direct connection between macroscopic properties and absolute configuration in a fluid phase," *Journal of the American Chemical Society*, vol. 110, no. 26, pp. 8686–8691, 1988.

[10] L. J. W. Shimon, M. Vaida, L. Addadi, M. Lahav, and L. Leiserowitz, "Molecular recognition at the solid-solution interface: a relay mechanism for the effect of solvent on crystal growth and dissolution," *Journal of the American Chemical Society*, vol. 112, no. 17, pp. 6215–6220, 1990.

[11] I. Weissbuch, D. Zbaida, L. Addadi, L. Leiserowitz, and M. Lahav, "Design of polymeric inhibitors for the control of crystal polymorphism. Induced enantiomeric resolution at racemic histidine by crystallization at 25. Degree. C," *Journal of the American Chemical Society*, vol. 109, no. 6, pp. 1869–1871, 1987.

[12] E. M. Landau, S. G. Wolf, M. Levanon, L. Leiserowitz, M. Lahav, and J. Sagiv, "Stereochemical studies in crystal nucleation. Oriented crystal growth of glycine at interfaces covered with Langmuir and Langmuir-Blodgett films of resolved. alpha.-amino acids," *Journal of the American Chemical Society*, vol. 111, no. 4, pp. 1436–1445, 1989.

[13] I. Weissbuch, F. Frolow, L. Addadi, M. Lahav, and L. Leiserowitz, "Oriented crystallization as a tool for detecting ordered aggregates of water-soluble hydrophobic. alpha.-amino acids at the air-solution interface," *Journal of the American Chemical Society*, vol. 112, no. 21, pp. 7718–7724, 1990.

[14] S. Mann, B. Heywood, S. Rajam, and V. Wade, "Molecular Recognition in Biomineralization," in *Proceedings of Mechanisms and Phylogeny of Mineralization in Biological Systems*, pp. 47–55, Springer, Tokyo, Japan, 1991.

[15] M. L. Peterson, S. L. Morissette, C. McNulty et al., "Iterative high-throughput polymorphism studies on acetaminophen and an experimentally derived structure for form III," *Journal of the American Chemical Society*, vol. 124, no. 37, pp. 10958-10959, 2002.

[16] H. Hsu, O. O. Adigun, L. S. Taylor, S. Murad, and M. T. Harris, "Crystallization of acetaminophen on chitosan films blended with different acids," *Chemical Engineering Science*, vol. 126, pp. 1–9, 2015.

[17] V. Andronis and G. Zografi, "Crystal nucleation and growth of indomethacin poly-morphs from the amorphous state," *Journal of Non-Crystalline Solids Solids*, vol. 271, no. 3, pp. 236–248, 2000.

[18] N. S. Trasi and L. S. Taylor, "Effect of polymers on nucleation and crystal growth of amorphous acetaminophen," *CrystEngComm*, vol. 14, no. 16, pp. 5188–5197, 2012.

[19] A. L. Grzesiak and A. J. Matzger, "Selection and discovery of polymorphs of platinum complexes facilitated by polymer-induced heteronucleation," *Inorganic Chemistry*, vol. 46, no. 2, pp. 453–457, 2007.

[20] L. I. Mosquera-Giraldo, N. S. Trasi, and L. S. Taylor, "Impact of surfactants on the crystal growth of amorphous celecoxib," *International Journal of Pharmaceutics*, vol. 461, no. 1-2, pp. 251–257, 2014.

[21] M. S. Lisowski, Q. Liu, J. Cho, J. Runt, F. Yeh, and B. S. Hsiao, "Crystallization behavior of poly (ethylene oxide) and its blends using time-resolved wide-and small-angle X-ray scattering," *Macromolecules*, vol. 33, no. 13, pp. 4842–4849, 2000.

[22] F. J. Baltá-Calleja and C. G. Vonk, *X-Ray Scattering of Synthetic Polymers*, Elsevier Science, New York, NY, USA, 1989.

[23] J. C. Burley, M. J. Duer, R. S. Stein, and R. M. Vrcelj, "Enforcing Ostwald's rule of stages: isolation of paracetamol forms III and II," *European Journal of Pharmaceutical Sciences*, vol. 31, no. 5, pp. 271–276, 2007.

[24] Q. Zhu, S. J. Toth, G. J. Simpson, H. Y. Hsu, L. S. Taylor, and M. T. Harris, "Crystallization and dissolution behavior of naproxen/polyethylene glycol solid dispersions," *Journal of Physical Chemistry B*, vol. 117, no. 5, pp. 1494–1500, 2013.

# 11

# Determination of Cell Permeabilization and Beta-Galactosidase Extraction from *Aspergillus oryzae* CCT 0977 Grown in Cheese Whey

Caroline dos Santos Viana,[1] Denise Renata Pedrinho,[2] Luiz Rodrigo Ito Morioka (ID),[1] and Hélio Hiroshi Suguimoto[1]

[1]*Center for Research and Pos-Graduate, University of Pythagoras Unopar, Marseille Street 591, 86041-140 Londrina, PR, Brazil*
[2]*Anhanguera University, Uniderp, Alexandre Herculano Street 1400, 79037-280 Campo Grande, MS, Brazil*

Correspondence should be addressed to Luiz Rodrigo Ito Morioka; luiz.morioka@kroton.com.br

Academic Editor: Doraiswami Ramkrishna

*Aspergillus oryzae* grown in cheese whey has the ability to produce beta-galactosidase. The objective of this work was to define the parameters for the determination of cell permeabilization and extraction of the enzyme from *Aspergillus oryzae* CCT 0977 biomass, with high enzymatic activity. The Box–Behnken design was used to determine cell permeabilization and extraction of beta-galactosidase conditions. The fermentation was carried out for a period of 5 days at 28°C, having as substrate the deproteinized cheese whey. To determine the effect of the variables on beta-galactosidase activity, enzymatic activity was determined by the lactose hydrolysis reaction. The most efficient condition for cell permeabilization was 25% ethanol at 30°C for 90 min, obtaining an enzymatic activity of $0.44\,U\cdot mL^{-1}$. For beta-galactosidase extraction from the biomass, the most efficient condition was 5.3% chloroform at 48°C, with an enzymatic activity of $0.17\,U\cdot mL^{-1}$. The use of ethanol was most efficient to promote cell permeability of *Aspergillus oryzae* CCT 0977.

## 1. Introduction

Biotechnological processes are widely used to obtain products with high added value [1]. The use of agroindustrial wastes in bioconversion by microorganisms has been the subject of extensive research, especially with reference to the production of metabolites of interest such as proteins, enzymes, organics acids, and secondary metabolites [2]. Cheese whey has been widely used as a fermentation medium, as it is a low-cost cheese residue and is nutritionally rich [3, 4]. Beta-galactosidase can be obtained by fermentation process by various microorganisms, such as *Aspergillus oryzae*, a thermotolerant fungus, without many environmental requirements for cultivation [5, 6].

The enzyme beta-galactosidase is one among other enzymes with industrial potential used in the hydrolysis of lactose in milk and cheese whey, generating food with low levels of lactose, which results in a better solubility and digestibility of milk and dairy products, making them ideal for consumers intolerant to this sugar [7].

When a metabolite is intracellular, the rupture of the cell wall is the first step performed in the downstream process. This allows separating the substance for later purification. Various methods may be used, but the process will depend on the location and stability of the metabolite. There are mechanical methods of cell disruption such as high-pressure homogenizers, ultrasonic waves, and glass beads. Chemical disruption methods use alkalis, detergents, and organic solvents. Enzymatic methods, however, consist of enzymatic lysis or inhibition of cell wall synthesis [1, 8, 9].

The use of organic solvents for cellular permeabilization and enzymatic extraction is the most common methodology, which was applied after the cell fermentation process. The permeabilization is a simple and fast method that allows

TABLE 1: The Box–Behnken design used to determine the effects of ethanol, temperature, and time on the enzymatic activity of permeabilized cells of *Aspergillus oryzae* CCT 0977.

| Run | Variables | | | Beta-galactosidase activity (U·mL$^{-1}$) | |
|-----|-----------|---|---|---|---|
| | $X_1$ Ethanol (%) | $X_2$ Temperature (°C) | $X_3$ Time (min) | Experimental | Predicted |
| 1 | −1 (15) | −1 (20) | −1 (30) | 0.22 | 0.17 |
| 2 | −1 (15) | 0 (30) | 1 (90) | 0.36 | 0.40 |
| 3 | −1 (15) | 1 (40) | 0 (60) | 0.29 | 0.29 |
| 4 | 0 (25) | −1 (20) | 1 (90) | 0.38 | 0.38 |
| 5 | 0 (25) | 0 (30) | 0 (60) | 0.44 | 0.39 |
| 6 | 0 (25) | 1 (40) | −1 (30) | 0.17 | 0.22 |
| 7 | 1 (35) | −1 (20) | 0 (60) | 0.25 | 0.29 |
| 8 | 1 (35) | 0 (30) | −1 (30) | 0.23 | 0.23 |
| 9 | 1 (35) | 1 (40) | 1 (90) | 0.40 | 0.35 |

measuring the enzymatic activity [10]. For the extraction of enzymes, it is necessary to promote the chemical breakdown of the cell wall. It is a simple and efficient method that does not leave cellular fragments and does not require high investments with the mechanical methods, but it demands more time, due to the chemical reactions. These solvents modify the cell wall structures causing disorganizations in their pores allowing the passage of small molecules, such as substrates or other molecules present in the fermentative medium [11].

Thus, the aim of this work was to cultivate the filamentous fungus *Aspergillus oryzae* CCT 0977 in cheese whey and determine the conditions of cell permeabilization and extraction of the enzyme beta-galactosidase in the biomass.

## 2. Materials and Methods

### 2.1. Microorganism, Inoculum, and Medium.
The *Aspergillus oryzae* CCT 0977 strain was obtained from Tropical Culture Collection of Foundation Andre Tosello. The culture was maintained in tubes containing PDA (potato dextrose agar, Acumedia®) and stored at 4°C. For the inoculum, 0.85% saline solution with 1% Tween 80 was used, and the spore count was performed by using the Neubauer chamber at a count of $1 \times 10^6$ cells·mL$^{-1}$. An inoculum concentration of 1% v/v was used in relation to the culture medium. The cheese whey powder was obtained from a local dairy cooperative. The cheese whey powder was solubilized in distilled water at a concentration of 5% w/v. For deproteinization of the cheese whey, 85% lactic acid was added until pH 4.6 (isoelectric point of the milk caseins) and heated at 90°C for 30 min. After precipitation, the protein fraction was removed by filtration using Whatman no. 1 filter paper, and the medium was adjusted to pH 5.0. Pasteurization of the cheese whey was at 65°C for 30 min. The deproteinized and pasteurized cheese whey was used as the culture medium in the fermentation experiments.

### 2.2. Fermentation Conditions.
For culture, 1% v/v of the inoculum was added and the fermentation was performed on an orbital shaker (Tecnal®, TE-420) at 28°C and 120 rpm for 5 days. The biomass produced after five days of fermentation was ground with the homogenizer (IKA®,

T10 Basic), and 5 mL of the suspension was collected and transferred to falcon tubes, followed by centrifugation (Quimis®, Q222G) for 10 min at 1100 rpm, for separation of the biomass and supernatant. The Biomass was used for the two methods: cell permeabilization and extraction of beta-galactosidase.

### 2.3. Cell Permeabilization.
After fermentation of *Aspergillus oryzae* CCT 0977, the cells were collected by centrifugation (1100 rpm for 10 min) and washed once with distilled water. Cell permeabilization was performed under static mode and incubated at 30°C for 90 minutes in a water bath in Falcon tubes (15 mL) containing 5 mL of the reaction suspension consisting of ethanol, according to the experimental design (Table 1), approximately 50 mg wet biomass, and 0.1 M potassium phosphate buffer (pH 6.8). The flasks were incubated at temperature and time according to the statistical design (Table 1). The biomass was collected by centrifugation at 1100 rpm for 10 min for further analysis, and the cells were washed once with the same buffer. The biomass was suspended in 1 mL of the phosphate buffer, and the enzymatic activity was determined by the enzymatic hydrolysis of lactose.

### 2.4. Enzyme Extraction.
After fermentation of *Aspergillus oryzae* CCT 0977, the cells were collected by centrifugation (1100 rpm for 10 min) and washed once with distilled water. For the extraction of the beta-galactosidase, the biomass was transferred to Erlenmeyer flasks (50 mL) and resuspended in 0.1 M potassium phosphate buffer (pH 6.8) with the addition of chloroform at the concentrations according to the experimental design (Table 2), with a final volume of 10 mL. After addition of the reagents, the vials were incubated on the orbital shaker (Tecnal, TE-420) at 120 rpm, at the temperatures described in Table 2, overnight. The supernatant and the biomass were separated by centrifugation at 1100 rpm for 10 min. In the supernatant, the enzymatic activity was determined.

### 2.5. Determination of Beta-Galactosidase Activity.
The enzymatic activity was determined by the initial rates of the lactose hydrolysis reaction through the glucose dosage

TABLE 2: Analysis of variance (ANOVA) of the regression parameters for the Box–Behnken design used to determine the effect of ethanol, temperature, and time on the enzymatic activity of permeabilized cells of *Aspergillus oryzae* CCT 0977.

| | Seq. SS | df | MS | F | p value |
|---|---|---|---|---|---|
| Ethanol (%) (L + Q) | 0.00324 | 2 | 0.001619 | 0.251141 | 0.799271 |
| Temperature (°C) (L + Q) | 0.00617 | 2 | 0.003084 | 0.478510 | 0.676356 |
| Time (min) (L + Q) | 0.0451 | 2 | 0.022552 | 3.499239 | 0.222260 |
| Error | 0.01289 | 2 | 0.006445 | — | — |
| Total SS | 0.06739 | 8 | — | — | — |

Seq. SS: sequential sums of squares; df: the degrees of freedom; MS: adjusted mean square; F: F calculated; p: p value at 0.05%.

produced by the enzymatic-colorimetric glucose oxidase kit (Bioliquid®) method. The unit of activity used in the work was the glucose unit produced per minute, per milliliter of enzymatic suspension ($U \cdot mL^{-1}$), defined as $\mu mol$ of glucose produced per minute, per mL of enzymatic suspension at 47°C at a initial concentration of lactose solution equal to $10 \, g \cdot L^{-1}$, prepared in 0.1 M citrate-phosphate buffer (pH 6.5). For the hydrolysis, an enzymatic suspension with a concentration of 30% v/v was used, and the incubation in a water bath remained overnight. After the incubation time, the enzyme was inactivated at 90°C for 5 min followed by an ice bath. Analyses were performed in triplicate.

## 3. Results and Discussion

*3.1. Cell Permeabilization.* Taking into account that permeabilized cells are biocatalysts, that is, they function as a source of enzymes that naturally remains immobilized. The Box–Behnken design (BBD) was used to evaluate the effects among the significant variables and to determine their optimal values. BBD was developed to reduce the number of experimental runs and increase the efficiency. BBD has been applied and considered a very efficient statistical experimental design tool in several areas including chemical engineering optimization [12].

The determination of the enzymatic activity for permeabilized cells of *Aspergillus oryzae* CCT 0977 was carried out to find the optimal values of independent variables (ethanol, $X_1$; temperature, $X_2$; time, $X_3$), which would give maximum beta-galactosidase activity. Based on the Box–Behnken design, the experimental levels of beta-galactosidase activity under each set of conditions were determined and compared with the corresponding predicted levels suggested (Table 1). The maximum experimental value for beta-galactosidase activity was $0.44 \, U \cdot mL^{-1}$ while the value of predicted response was $0.39 \, U \cdot mL^{-1}$. Some authors report the supplementation of the culture medium during fermentation. Panesar et al. [13] used some salts, lactose, peptone and yeast extract and obtained a maximum activity of the *Aspergillus oryzae* NCIM 1212 beta-galactosidase of $0.50 \, U \cdot mL^{-1}$, at a temperature between 45 and 50°C, a value similar to that obtained in the present study. Senm et al. [14], under submerged fermentation at pH 4.5 with *Aspergillus alliaceus*, obtained an enzymatic activity of 0.0486 U/mL.

The coefficient of determination $R^2$ of the model was 0.81266, which indicated that the model adequately represented the real relationship between the variables under consideration. An $R^2$ value of 0.81266 means that 81.26% of the variability was explained by the model, which is acceptable for the biological system and only 18.74% was as a result of chance. Approximately 81.26% of validity was achieved, indicating that the model exerted an adequate prediction on the enzyme activity. The close correlation between the experimental and predicted data indicates the appropriateness of the experimental design. The maximum beta-galactosidase activity ($0.44 \, U \cdot mL^{-1}$) was achieved under the following conditions: ethanol (25%), temperature (30°C), and time (60 min) (Table 1). In Table 2, the analysis of variance (ANOVA) of the regression parameters for the Box–Behnken experimental design was evaluated.

Based on the results for high beta-galactosidase activity from permeabilized cells of *A. oryzae* CCT 0977, all linear (L) and quadratic (Q) effects of ethanol, temperature, and time were not significant ($p < 0.05$), indicating that the variables in the smaller range limit, 15%, 20°C, and 30 min, respectively, was enough for enzymatic activity.

In order to determine the ranges of variables that influence beta-galactosidase activity from permeabilized cells of *A. oryzae* CCT 0977, response surface plots were generated for this analysis. The response surfaces for beta-galactosidase activity evaluating the variables ethanol, temperature, and time were plotted as shown in Figure 1. Figure 1(c) shows the effects of temperature and ethanol on beta-galactosidase activity. Temperature and ethanol ranging from low to high values of process showed the high enzyme activity. The beta-galactosidase activity was higher in the temperature range of 26–33°C and ethanol between 20 and 30%. Figure 1(b) depicts the response surface plot as a function of time versus ethanol. Change of ethanol does not significantly affect the curvature of the surface. From a graphical representation, there is a dependence of beta-galactosidase activity on the time ranging from 60 to 90 min. Figure 1(c) shows high beta-galactosidase activity effectiveness within the temperature range of 20–40°C and time ranging from 60 to 90 min, while below and above these ranges, a significant decrease of activity can be observed. According to Kumari et al. [15], a lower temperature with the increase of the permeabilization time intensifies the activity of the enzyme, since an elevation of the temperature can generate the partial inactivation of the beta-galactosidase. This confirms that the range of these variables were chosen properly and sufficient for the process. A maximum beta-galactosidase activity of $0.44 \, U \cdot mL^{-1}$ for permeabilized cells was defined under the following conditions: ethanol 25%, at 30°C, and 60 min.

Based on the results obtained, the temperature and time are fundamental factors in the process of cell permeabilization and can be explained by the fact that the cell walls of the fungi are more rigid than others, requiring a lower concentration of ethanol with more reaction time and a higher temperature for cellular disorganization.

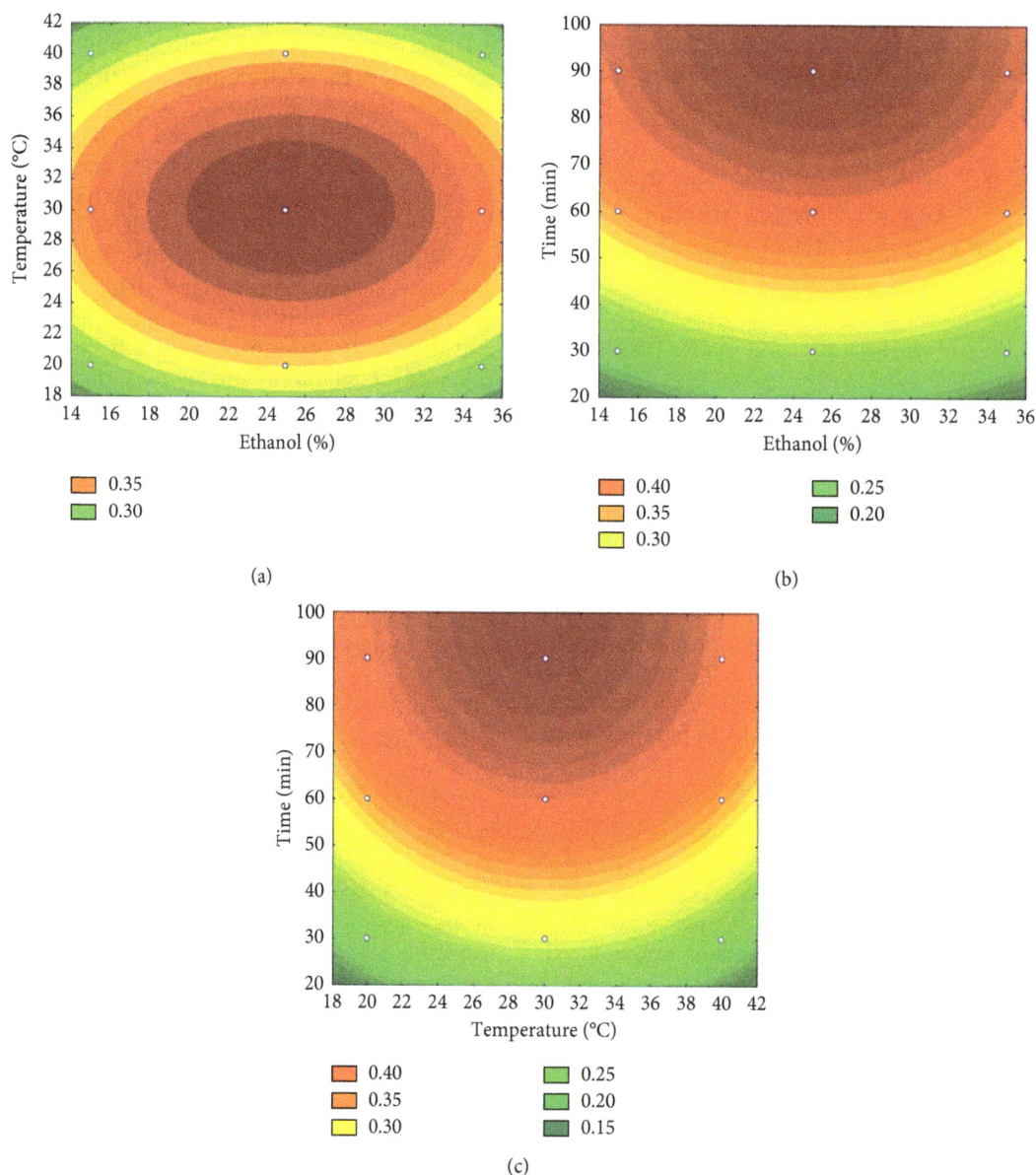

(a)

(b)

(c)

FIGURE 1: Response surface plot representing the effect of the variables on beta-galactosidase activity ($U \cdot mL^{-1}$) of permeabilized cells from *Aspergillus oryzae* CCT 0977.

### 3.2. Enzyme Extraction.

The determination of the enzymatic activity for extracted beta-galactosidase of *Aspergillus oryzae* CCT 0977 was carried out to find the optimal values of independent variables (chloroform ($X_1$) and temperature ($X_2$)), which would give maximum beta-galactosidase activity. Based on the Box–Behnken design, the experimental levels of beta-galactosidase activity under each set of conditions were determined and compared with the corresponding predicted levels suggested (Table 2). The maximum experimental value for beta-galactosidase activity was $0.17 \, U \cdot mL^{-1}$ while the value of predicted response was $0.15 \, U \cdot mL^{-1}$. The coefficient of determination $R^2$ of the model was 0.87248, which indicated that the model adequately represented the real relationship between the variables under consideration. An $R^2$ value of 0.87248

means that 87.24% of the variability was explained by the model, which is acceptable for biological system, and only 12.76% was as a result of chance. Approximately 87.24% of validity was achieved, indicating that the model exerted an adequate prediction on the enzyme activity. The close correlation between the experimental and predicted data indicates the appropriateness of the experimental design. The maximum beta-galactosidase activity ($0.17 \, U \cdot mL^{-1}$) was achieved under the following conditions: chloroform (5.0%) and temperature (45°C) (Table 3). In Table 4, the analysis of variance (ANOVA) of the regression parameters for the Box–Behnken experimental design was evaluated.

Based on the results for high beta-galactosidase activity from extracted beta-galactosidase of *Aspergillus oryzae* CCT

TABLE 3: The Box–Behnken design used to determine the effect of chloroform and temperature on the enzymatic activity of extracted beta-galactosidase of *Aspergillus orzyae* CCT 0977.

| Run | Variables | | Beta-galactosidase activity (U·mL$^{-1}$) | |
| | $X_1$ Chloroform (%) | $X_2$ Temperature (°C) | Experimental | Predicted |
|---|---|---|---|---|
| 1 | −1 (4.0) | −1 (40) | 0.09 | 0.08 |
| 2 | −1 (4.0) | 0 (45) | 0.12 | 0.13 |
| 3 | −1 (4.0) | 1 (50) | 0.15 | 0.14 |
| 4 | 0 (5.0) | −1 (40) | 0.10 | 0.11 |
| 5 | 0 (5.0) | 0 (45) | 0.17 | 0.15 |
| 6 | 0 (5.0) | 1 (50) | 0.16 | 0.16 |
| 7 | 1 (6.0) | −1 (40) | 0.12 | 0.12 |
| 8 | 1 (6.0) | 0 (45) | 0.16 | 0.16 |
| 9 | 1 (6.0) | 1 (50) | 0.15 | 0.15 |

TABLE 4: Analysis of variance (ANOVA) of the regression parameters for the Box–Behnken design used to determine the effects of chloroform and temperature on the enzymatic activity of extracted beta-galactosidase of *Aspergillus orzyae* CCT 0977.

| | SS | df | MS | F | p value |
|---|---|---|---|---|---|
| Chloroform (%) (L + Q) | 0.001089 | 2 | 0.000544 | 2.63677 | 0.218346 |
| Temperature (°C) (L + Q) | 0.004689 | 2 | 0.002344 | 11.35426 | 0.039863 |
| $X_1 * X_2$ | 0.0002225 | 1 | 0.000225 | 1.08969 | 0.373253 |
| Error | 0.000619 | 3 | 0.000206 | — | — |
| Total SS | 0.006622 | 8 | — | — | — |

Seq. SS: sequential sums of squares; df: the degrees of freedom; MS: adjusted mean square; F: F calculated; p: p value at 0.05%.

0977, the effect of temperature (L + Q) was significant ($p > 0.05$), indicating that higher temperature is more appropriate for high enzymatic activity. The linear (L) and quadratic (Q) effects of chloroform and the quadratic effect of temperature were not significant ($p > 0.05$), indicating that the variables remaining at the lowest levels, 4% and 40°C, respectively, were enough for enzymatic activity.

In order to determine the ranges of variables that influence enzymatic activity of beta-galactosidase extracted from *Aspergillus oryzae* CCT 0977, response surface plots were generated for this analysis. The response surfaces for beta-galactosidase activity evaluating the variables chloroform and temperature were plotted in Figure 2 that shows the effects of temperature and chloroform on beta-galactosidase activity. Chloroform ranging from lower to higher values of process showed the high enzyme activity. The beta-galactosidase activity was higher in the chloroform range of 5.0–6.0%. From a graphical representation, there is a dependence of beta-galactosidase activity on the temperature ranging from 45 to 50°C, while below these ranges, a significant decrease of activity can be noticed. A maximum beta-galactosidase activity of 0.17 U·mL$^{-1}$ was defined under the following conditions: chloroform 5.0% and 47°C.

Several authors have studied the effect of solvents on the extraction of enzymes [8, 16, 17]. Nagy et al. [18] determined the activity of the intracellular beta-galactosidase of

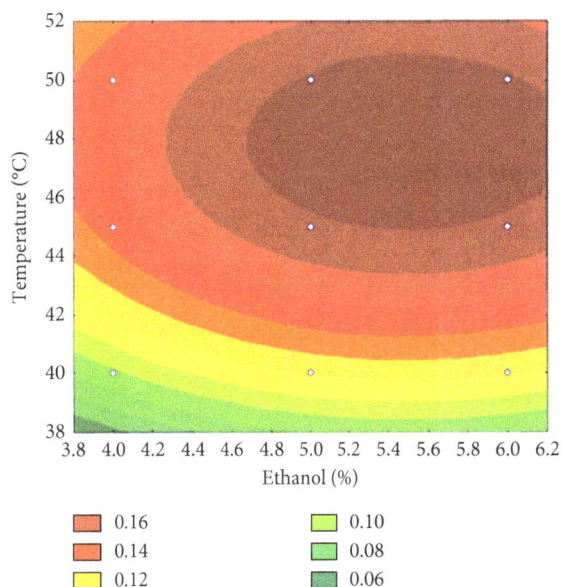

FIGURE 2: Response surface plot representing the effect of the variables on enzymatic activity (U mL$^{-1}$) of beta-galactosidase extracted from *Aspergillus oryzae* CCT 0977.

*Penicillium chrysogenum* NCAIM 00237 and verified an activity of approximately 0.16 U·mL$^{-1}$. Mirdamadi et al. [19] tested different compositions of liquid media for their efficacy in the production of beta-galactosidase by *Aspergillus orzyae* PTCC 5163 and verified that in cheese whey-based media, the maximum enzymatic activity obtained was approximately 0.13 U·mL$^{-1}$. Therefore, in our study, it was observed that, in the extraction process, there was an expressive reduction in the enzymatic activity. In the permeabilization, the activity was 0.44 U·mL$^{-1}$; in the extraction, this value decreases to 0.17 U·mL$^{-1}$, a reduction of 61.36% in beta-galactosidase activity. Probably, the process of extracting the enzyme from the inside of the cell by solvent causes damage to the enzyme that causes reduction in its activity.

## 4. Conclusion

*Aspergillus oryzae* CCT 0977 was successfully cultivated in deproteinized cheese whey for the production of the enzyme beta-galactosidase. Two chemical solvents were used: ethanol for cell permeabilization and chloroform for extraction of the enzyme. The ethanol as a permeabilizing agent obtained higher enzymatic activity when compared to the extraction by chloroform. The best condition in the permeabilization process, obtaining the highest activity of beta-galactosidase, was with ethanol 25%, temperature of 30°C, and for 60 minutes. Under these conditions, the maximum experimental value of beta-galactosidase activity was 0.44 U·mL$^{-1}$. For the extraction process, the condition was 5.3% chloroform at 48°C, with an enzymatic activity of 0.17 U·mL$^{-1}$.

## Conflicts of Interest

The authors declare that there are no conflicts of interest regarding the publication of this paper.

## Acknowledgments

The authors would like to thank Coordenacao de Aperfeicoamento de Pessoal de Nivel Superior–CAPES/Brazil for the financial support and KROTON/UNOPAR for school and masterships.

## References

[1] B. V. Kilikian and A. P. Junior, "Purificação de produtos biotecnológicos," in *Biotecnologia Industrial*, W. Schmidell, U. A. Lima, E. Aquarone, and W. Borzani, Eds., vol. 2, pp. 493–507, Edgard Blücher Ltd., São Paulo, Brazil, 2001.

[2] S. P. Saha and S. Ghosh, "Optimization of xylanase production by *Penicillium citrinum* xym2 and application in saccharification of agro-residues," *Biotechnology and Applied Biochemistry*, vol. 3, no. 4, pp. 188–196, 2014.

[3] G. Dragone, S. I. Mussatto, J. B. A. Silva, and J. A. Teixeira, "Optimal fermentation conditions for maximizing the ethanol production by *Kluyveromyces fragilis* from cheese whey powder," *Biomass and Bioenergy*, vol. 35, no. 5, pp. 1977–1982, 2011.

[4] S. R. Macwan, B. K. Dabhi, S. C. Parmar, and K. D. Aparnathi, "Whey and its utilization," *International Journal of Current Microbiology and Applied Sciences*, vol. 5, no. 8, pp. 134–155, 2016.

[5] K. Foley, G. Fazio, A. B. Jensen, and W. H. O. Hughes, "The distribution of *Aspergillus* spp. opportunistic parasites in hives and their pathogenicity to honey bees," *Veterinary Microbiology*, vol. 169, no. 3-4, pp. 203–210, 2014.

[6] P. Krijgsheld, R. Bleichrodt, G. J. Van Veluw et al., "Development in *Aspergillus*," *Studies in Mycology*, vol. 74, pp. 1–29, 2013.

[7] Q. Husain, "β–galactosidase and their potential applications: a review," *Critical Reviews in Biotechnology*, vol. 30, no. 1, pp. 41–62, 2010.

[8] M. Stred'Anský, M. Tomaska, E. Sturdík, and L. Kremnický, "Optimization of β-galactosidase extraction from *Kluyveromyces marxianus*," *Enzyme and Microbial Technology*, vol. 15, no. 12, pp. 1063–1065, 1993.

[9] F. O. Medeiros, F. G. Alves, C. R. Lisboa et al., "Ondas ultrassônicas e pérolas de vidro: um novo método de extração de β–galactosidase para o uso em laboratório," *Química Nova*, vol. 31, no. 2, pp. 336–339, 2008.

[10] P. S. Panesar, R. Panesar, R. S. Singh, and M. B. Bera, "Permeabilization of yeast cells with organic solvents for β-galactosidase activity," *Research Journal of Microbiology*, vol. 2, no. 1, pp. 34–41, 2007.

[11] G. A. Somkuti and D. H. Steinberg, "Permeabilization of *Streptococcus thermophilus* and the expression of beta-galactosidase," *Enzyme and Microbial Technology*, vol. 16, no. 7, pp. 573–576, 1994.

[12] L. Gámiz-Gracia, L. Cuadros-Rodríguez, E. J. Almansa-López, J. M. Soto-Chinchilla, and A. García-Campaña, "Use of highly efficient Draper–Lin small composite designs in the formal optimization of both operational and chemical crucial variables affecting a FIA-chemiluminescence detection system," *Talanta*, vol. 60, no. 2-3, pp. 523–534, 2003.

[13] P. S. Panesar, R. Kaur, and R. S. Singh, "Isolation and screening of fungal strains for β-galactosidase production," *International Journal of Biological, Biomolecular, Agricultural, Food and Biotechnological Engineering*, vol. 10, no. 7, pp. 390–394, 2016.

[14] S. Sen, L. Ray, and P. Chattopadhyay, "Production, purification, immobilization and characterization of a thermostable β-galactosidase from *Aspergillus alliaceus*," *Applied Biochemistry and Biotechnology*, vol. 167, no. 7, pp. 1938–1953, 2012.

[15] S. Kumari, P. S. Panesar, and M. B. Bera, "Statistical modeling for permeabilization of a novel yeast isolate for β-galactosidase activity using organic solvents," *International Journal of Biological, Biomolecular, Agricultural, Food and Biotechnological Engineering*, vol. 8, no. 6, pp. 567–572, 2014.

[16] S. Bansal, H. S. Oberoi, G. S. Dhillon, and R. T. Patil, "Production of β-galactosidase by *Kluyveromyces marxianus* MTCC 1388 using whey and effect of four different methods of enzyme extraction on β-galactosidase activity," *Indian Journal of Microbiology*, vol. 48, no. 3, pp. 337–341, 2008.

[17] P. S. Panesar, R. Panesar, R. S. Singh, J. F. Kennedy, and H. Kumar, "Microbial production, immobilization and applications of β-galactosidase," *Journal of Chemical Technology and Biotechnology*, vol. 81, no. 4, pp. 530–543, 2006.

[18] Z. Nagy, T. Kiss, A. Szentirmai, and S. Biró, "β-Galactosidase of *Penicillium chrysogenum*: production, purification, and characterization of the enzyme," *Protein Expression and Purification*, vol. 21, no. 1, pp. 24–29, 2001.

[19] S. Mirdamadi, N. Moazami, and M. N. Gorgani, "Production of beta-galactosidase in submerged media by *Aspergillus oryzae*, PTCC 5163," *Journal of Sciences Islamic Republic of Iran*, vol. 8, no. 1, pp. 23–27, 1997.

# 12

# *Tectona grandis* Capped Silver-Nanoparticle Material Effects on Microbial Strains Inducing Microbiologically Influenced Corrosion

**Joshua Olusegun Okeniyi**[ID],[1,2] **Abimbola Patricia Idowu Popoola**[ID],[2]
**Modupe Elizabeth Ojewumi**[ID],[3] **Elizabeth Toyin Okeniyi**[ID],[4]
**and Jacob Olumuyiwa Ikotun**[ID][5]

[1]*Mechanical Engineering Department, Covenant University, Ota, Nigeria*
[2]*Chemical and Metallurgical Engineering Department, Tshwane University of Technology, Pretoria, South Africa*
[3]*Chemical Engineering Department, Covenant University, Ota, Nigeria*
[4]*Petroleum Engineering Department, Covenant University, Ota, Nigeria*
[5]*Department of Civil Engineering and Building, Vaal University of Technology, Vanderbijlpark, South Africa*

Correspondence should be addressed to Joshua Olusegun Okeniyi; joshua.okeniyi@covenantuniversity.edu.ng

Academic Editor: Javier M. Ochando-Pulido

This paper investigates *Tectona grandis* capped silver nanoparticle material effects on the microbial strains inducing microbiologically influenced corrosion (MIC) of metals. Leaf-extract from *Tectona grandis* natural plant was used as a precursor for the synthesis of silver-nanoparticle material, which was characterised by a scanning electron microscopy having Energy Dispersion Spectroscopy (SEM + EDS) facility. Sensitivity and resistance studies by the synthesized *Tectona grandis* capped silver nanoparticle material on three Gram-positive and three Gram-negative, thus totalling six, MIC inducing microbial strains were then studied and compared with what was obtained from a control antibiotic chemical. Results showed that all the microbial strains studied were sensitive to the *Tectona grandis* capped silver nanoparticle materials whereas two strains of microbes, a Gram-positive and a Gram-negative strain, were resistant to the commercial antibiotic chemical. These results suggest positive prospects on *Tectona grandis* capped silver nanoparticle usage in corrosion control/protection applications on metallic materials for the microbial corrosion environment.

## 1. Introduction

Microbiologically influenced corrosion (MIC) can refer to changes in electrochemical reactions at the surface of a metallic material maintaining interface with system of microorganisms, conglomerated into biofilm, which induce corrosion process of the metallic material [1, 2]. Many metallic materials, including stainless steel, and many environments, for example, domestic water, wastewater, marine, food processing, and oil and gas and industrial chemical, are susceptible to the material degradation of MIC [1–4]. This has made MIC related crises a cost-gulping phenomenon, in billions of dollars of direct cost, in many countries, while in the natural gas industries alone, MIC has accounted for about a third of corrosion failures [3, 5]. Attachment of microbial strains unto metallic surfaces, leading to the formation of biofilm colony of microbes on the metal, has been identified as one of the major causative mechanism of microbiologically influenced corrosion (MIC) [1, 2, 4]. That the microbial attachment, the biofilm formation, and subsequent MIC induced corrosion damage of metallic material all occur on the material surface has been drawing research attention towards microbial and material surfaces interactions among the other conditions necessary for MIC attacks of metallic materials [1, 6, 7]. Microbes can refer to the entire evolutionary genus of microorganisms including bacteria, Archaea (methanogens), and Eukaryota (fungi), all of which are causative agents of MIC attacks on metallic materials [4]. However, motility of bacteria strains, especially the conveniently flagellated strains, enhances them with the special advantage of pioneering initial attachment to the metallic surface before attracting other secondary colonizers of microorganism, for forming MIC inducing biofilm [2, 7]. Once the mutual interactions of the colonies are established in the biofilm, resistance to MIC control approaches could be enhanced with consequence propagation of the corrosion degradation process. It is for this reason that attempts at making the initiation of MIC due to methods that could preclude or delay the initial attachment to the metallic surface are being opined as a corrosion-protection/control approach that will go a long way in MIC mitigation.

For instance, coating is a corrosion control method that can be applied to different types of metallic materials with suitable corrosion-protection additives [8] for which the use of nanotechnology based methods via nanoparticle usage is being proffered as a novel approach [9]. Notable research advances in this direction include the identification of the antimicrobial property of silver for its usage in the form of silver-nanoparticle dispersed in polymer or included in coating applications for metallic corrosion-protection and antibacterial resistance improvement [10, 11]. Of pertinent relevance to the current discourse is the use, in [12], of silver-nanoparticle material that had been synthesized from plants as reducing and capping agent for inhibition of both microbial growth and metallic corrosion in acidic chloride environment. Such synthesis of metallic nanoparticle via natural bioresource approach is known to offer benefits of environmental friendliness and ecocompatibility, especially due to the nonusage of toxic chemical for the synthesizing process.

*Tectona grandis* is a natural plant from which leaf-extract had been used successfully also for inhibiting mild steel and stainless steel corrosion in reported works [13, 14]. In addition to these, *Tectona grandis* leaf-extract has been employed recently for investigating the growth inhibition on MIC inducing microbes [15]. In spite of these, however, there is dearth of study in which the antimicrobial property from silver had been annexed with the green benefits that could be derivative from *Tectona grandis* usage as a precursor and capping agent for silver-nanoparticle synthesis. In contrast, extracts from many natural plants have been employed in studies for biologically mediated synthesis of silver-nanoparticle material [16–19]. Among these are leaves of *Prunus persica* by [16], *Aloe vera* by [17], and *Azadirachta indica* by [19]. In addition to these, the cited work in [18] detailed a review on the use of extracts from 40 plants, among which 27 are leaf-extracts, as a precursor for reducing silver from its compound to its nanoparticle. All these reports from the literature detailed antimicrobial effects of the biosynthesized silver nanoparticle, but none include use of *Tectona grandis* for the silver nanoparticle synthesis. Notable details of antimicrobial activity, from some of the cited works that also tested antimicrobial effect of the plant extract, for example, the report on *Azadirachta indica* in [19], include the indication that the plant extract usage alone resulted in "no zone" at inhibiting growth of the studied microbial strains. In sharp comparison, extracts from the fruit [20] and from the leaf [21] of *Tectona grandis* inhibited growth of different strains of microbes. Apart from these, *Tectona grandis* leaf-extract was also shown to be capable of improvement of

antimicrobial potency when it was used in conjunction with tetracycline in [22]. These garner supports for the choice of *Tectona grandis* leaf-extract for use as a precursor for plant extract capped silver nanoparticle in this study.

More especially, motivation for the present study was drawn from the reported work in [22] wherein *Tectona grandis* leaf-extract was actually used for silver-nanoparticle synthesis but with antibacterial study conducted on only one Gram-positive (*Staphylococcus aureus*) and only one Gram-negative (*Escherichia coli*) strains of microbes. The positive results from that study indicate that more works need to be carried out for assessing the effectiveness of this bionanoparticle material on more types of microbes, especially those known to induce MIC. Such more types of microbes need to necessarily include *Pseudomona aeruginosa* (a flagellated, motile Gram-negative microbial strain), for instance, for this strain is known to usually pioneer attachment to metals for biofilm formation and eventual MIC attack on metallic materials [2, 7, 23, 24]. Therefore the objective of this paper was to investigate effect of *Tectona grandis* capped silver-nanoparticle on the inhibition of the growth of microbial strains inducing microbiologically influenced corrosion of metallic materials.

## 2. Materials and Methods

*2.1. Tectona grandis Leaf-Extract Biosynthesis of Silver-Nanoparticle Material.* *Tectona grandis* (*T. grandis*), Verbenaceae, was subjected to standard procedure detailed in [15] for obtaining pasty form of leaf-extract. This procedure includes drying in a well aerated room maintained at $20°C$ (by a 5-ton air-conditioning unit) and grinding it into small bits before wrapping in Whatman® filter paper for placing into a condenser-equipped soxhlet extractor utilizing $CH_3OH$ (methanol) for solvent medium. Out of the *T. grandis* leaf-extract paste obtained from the procedure detailed, 25 mg was dissolved and made up to a volume of 1000 ml using 0.1 M $AgNO_3$, that is, silver nitrate (Sigma Aldrich®). As detailed in [5, 15], after 48 hrs, sample was obtained from the dissolution, which was then centrifuged at 3,500 rpm for 15 minutes in the Laboratory Centrifuge, Model SM-80-2 obtained from Surgifield® (England). The resultant supernatant was then poured out before the biosynthesized nanoparticle residue got transferred to a watch glass for air drying and afterwards collection in Eppendorf tubes. These were stored at room temperature for further use as the biosynthesized *T. grandis* capped silver nanoparticle in the study.

*2.2. Characterisation of the Biosynthesized Silver-Nanoparticle Material.* Sample from the biosynthesized *T. grandis* capped silver-nanoparticle was prepared and then characterised using a Pro X PHENOM™, Model 800-07334 (Phenomworld®, Netherlands) scanning electron microscopy and energy dispersive spectroscopy (SEM + EDS) instrument. For this characterisation, the sample was placed on a sample stub for loading onto the stage of the Quorum Sputter Coater, Model Q150R ES, obtained from Quorum Technologies Limited® (England), and for coating the sample with gold [5, 15]. This was followed by the removal of the sample stub, from the

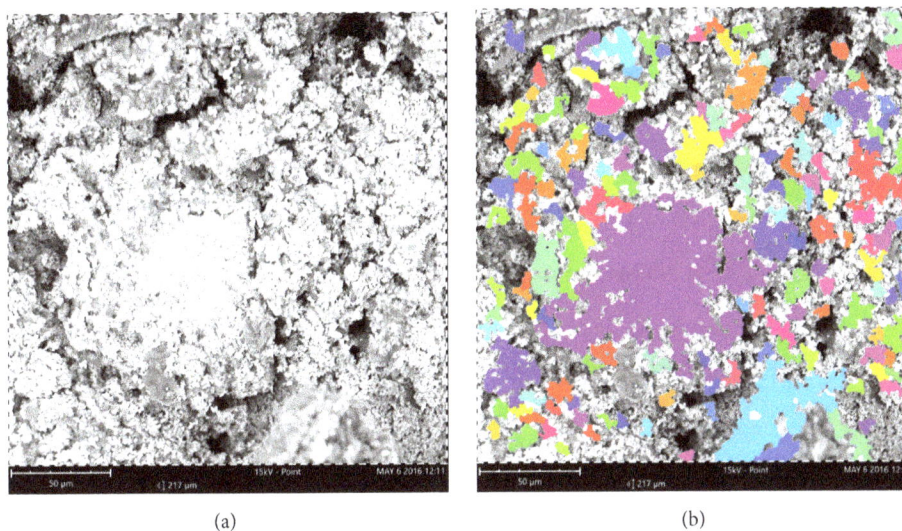

(a)                                                    (b)

FIGURE 1: SEM + EDS analyses of the *T. grandis* capped silver-nanoparticle material. (a) Image from scanned electron microscope. (b) Clusters of particles colouring by ParticleMetric rendering of the scanned microscopic image.

coating instrument, for placing on standard sample holder mounted on the Pro X PHENOM, Model 800-07334 obtained from Phenomworld (Netherlands). On this instrument, the biosynthesized *T. grandis* capped silver-nanoparticle was subjected to the scanning electron microscopy and energy dispersive spectroscopy (SEM + EDS) analyses.

*2.3. Inhibition Study of MIC Inducing Microbial Strains.* *Staphylococcus aureus (S. aureus)* and *Escherichia coli (E. coli)* were used, respectively, as Gram-positive and Gram-negative microbial strains in this study. Reasons for choosing these microbial strains include the motivation to facilitate comparison of test-results from this study with what obtained from the previously cited work by other researchers [22]. In addition, however, *Bacillus* spp. and *Micrococcus varians (M. varians)* were included among the Gram-positive while *Pseudomonas aeruginosa (P. aeruginosa)* and *Serratia* spp. were among the Gram-negative microbial strain in the present study.

These isolates were obtained from the culture collection centre in the Biotechnology Unit of Department of Applied Biological Sciences, Covenant University, Ota, Ogun State, Nigeria, and were maintained on nutrient broth and incubated overnight at 37°C between 18 and 20 hrs [25, 26]. From these cultures of microbial strains, 2 ml, of the microbes, was collected into sterile tube for making up with sterile distilled $H_2O$ until matching turbidity standard of 0.5% Mcfarland [27].

Each ensuing mixture of test-organisms was then used for seeding sterile nutrient agar plates, via the agar well diffusion method. Wells were bored into the seeded nutrient agar using 9 mm sterile cork borer. A gram by mass of the *T. grandis* capped silver nanoparticle was then dissolved and thoroughly mixed in 10 ml of $C_2H_6OS$ (Dimethyl sulfoxide; DMSO). From this dissolution, 0.2 ml was obtained using sterile pipette, for dispersing into the well that had been bored

onto the agar plate. This was followed by incubating the plates at 37°C for 24 hrs, before measuring, in mm unit, the zones of inhibition that resulted from this procedure [28].

For a positive control of antimicrobial effects, 10 $\mu$g Gentamicin, from Abtek Biologicals Limited™ (Liverpool, UK), was utilized. The antibiotic chemical was used also, that is, just as the synthesized bionanoparticle material, in the seeded agar plates for the microbial growth inhibition study. This approach was to facilitate the comparison of the inhibition results obtained from the biosynthesized nanoparticle with that obtained from the antibiotic chemical usage.

## 3. Results and Discussions

*3.1. Scanning-Electron-Microscopic (SEM) Analysis SEM.* The image of the *T. grandis* capped silver-nanoparticle by the SEM instrument is shown in Figure 1(a) and analyses employing ParticleMetric® rendering facility of the SEM instrument in Figure 1(b). Results from the ParticleMetric analysis indicated that for 157 particles the SEM facility could pick for the rendering in Figure 1(b), circle equivalent diameters, in $\mu$m unit, range from (median = 48.7: average = 57.5) to (median = 8.26: average = 9.95). Figure 1(b) however depicted regions of uncoloured clusters that represent regions of particles beyond the identifiable range of the SEM system and indicate that nanoscale range of particles was obtained.

*3.2. Energy-Dispersive-Spectroscopic (EDS) Characterisation.* Results from the EDS characterisation of elemental constituents from the *T. grandis* capped silver nanoparticle are presented in Figure 2.

From the EDS characterisation in Figure 2, up to three spikes of silver (Ag), among the prominent spikes in the figure, could be identified. By this, it was established that silver-nanoparticle material was obtained. The other spikes among which carbon, oxygen, and nitrogen are also prominent or

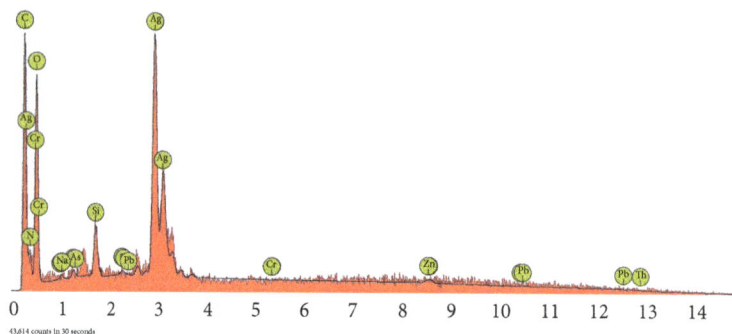

FIGURE 2: EDS characterisation of the bionanoparticle from the SEM facility.

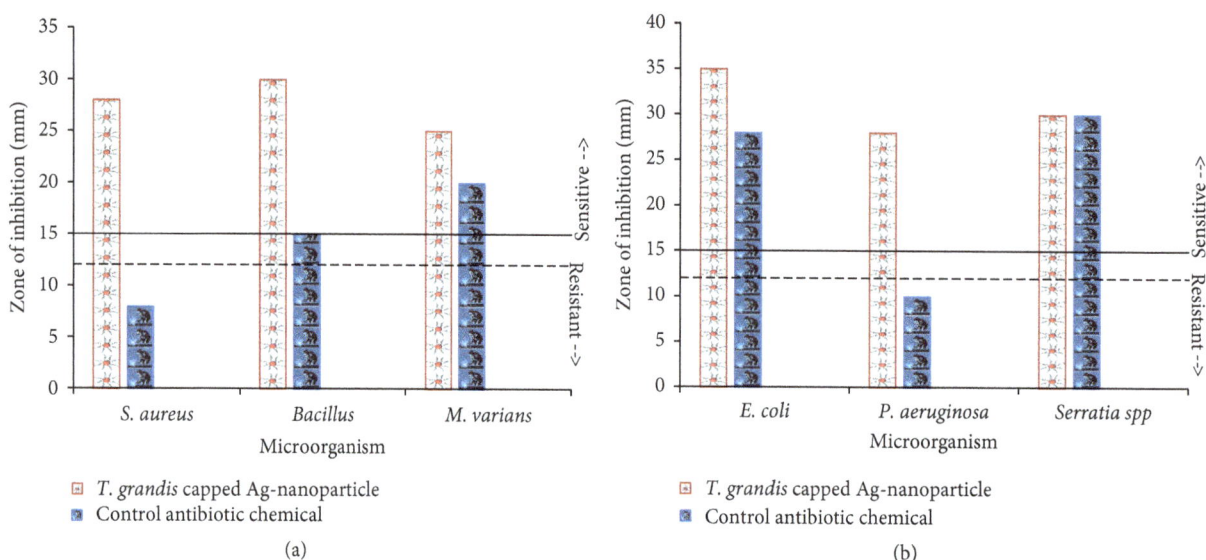

(a)

(b)

FIGURE 3: Growth inhibition effects by *T. grandis* capped silver-nanoparticle material on MIC inducing microbial strains. (a) Gram-positive microbial strains. (b) Gram-negative microbial strains.

prevalent corroborate elemental constituents of biocompatible compounds from the *T. grandis* leaf-extract as could be inferred from biochemical characterisation studies of other plant leaf-extracts in [29–33].

*3.3. T. grandis Capped Silver-Nanoparticle Effect on MIC Inducing Microbial Strains Studied.* Results of the growth inhibition effects by the *T. grandis* capped silver-nanoparticle on the studied MIC inducing microbial strains are presented in Figure 3, wherein Figure 3(a) depicts results from the Gram-positive and Figure 3(b) results from the Gram-negative microbial strains.

It could be noted that Figure 3 also includes linear plots of standard susceptibility criteria, as either sensitive or resistant, by the Gram-positive or Gram-negative microbes to the Gentamicin chemical, used as the antibiotic control for this study. These linear plots are for aiding direct interpretation of the zone of microbial growth inhibition obtained from the biosynthesized nanoparticle and the antibiotic control as per the susceptibility criteria from [34], according to microbial susceptibility interpretation for the standardized single-disc method [35]. From this, it could be deduced that all the

three Gram-positive microbial strains for the study exhibited were sensitive to the *T. grandis* capped silver-nanoparticle material, by the zone of inhibition that was greater than 15 mm obtained for these microbes. Compared to this, only two of the Gram-positive microbial strains studied were sensitive to the Gentamicin (control antibiotic chemical), among which the *Bacillus* spp. just reached the 15 mm bar of zone of inhibition. The third Gram-positive strain of microbe for the study, *S. aureus*, was resistant to the Gentamicin control. In an almost similar manner, all the three Gram-negative strains of microbes for the study were sensitive to the *T. grandis* capped silver-nanoparticle material compared to only two, *E. coli* and *Serratia* spp., that were sensitive to the Gentamicin control antibiotic. The *P. aeruginosa* was resistant to the Gentamicin chemical, as per the set susceptibility criteria interpretation for the single-disc method in [34, 35].

## 4. Conclusions

By the results in the study, it is established that the *T. grandis* capped silver nanoparticle material, to which all the microbial strains studied were sensitive, outperformed the Gentamicin chemical on the inhibition of the growth of MIC inducing microbial strains studied. By this, it is opined that the *T. grandis* capped silver nanoparticle material could be suitable for use in MIC control applications for the corrosion-protection of metallic materials designed to operate in the MIC environment. However, further studies are recommended for investigating specific applications to which the biosynthesized nanoparticle material would be applied for situating antimicrobial effects for that condition of application.

## Conflicts of Interest

The authors declare that there are no conflicts of interest regarding the publication of this paper.

## Acknowledgments

The authors appreciate part-funding of this research by the following institutions: The National Research Foundation-The World Academy of Sciences, NRF-TWAS [Grant no. 105552], Covenant University Centre for Research Innovation and Discovery, CUCRID, Covenant University, Ota, Nigeria, and Vaal University of Technology, Vanderbijlpark, South Africa.

## References

[1] M. A. Javed, W. C. Neil, P. R. Stoddart, and S. A. Wade, "Influence of carbon steel grade on the initial attachment of bacteria and microbiologically influenced corrosion," *Biofouling*, vol. 32, no. 1, pp. 109–122, 2016.

[2] J. Xia, C. Yang, D. Xu et al., "Laboratory investigation of the microbiologically influenced corrosion (MIC) resistance of a novel Cu-bearing 2205 duplex stainless steel in the presence of an aerobic marine Pseudomonas aeruginosa biofilm," *Biofouling*, vol. 31, no. 6, pp. 481–492, 2015.

[3] A. Y. Adesina, I. K. Aliyu, and F. M. Al-Abbas, "Microbiologically influenced corrosion (MIC) challenges in unconventional gas fields," in *CORROSION*, NACE International, Houston, TX, USA, 2015.

[4] B. J. Little and J. S. Lee, "Microbiologically influenced corrosion: An update," *International Materials Reviews*, vol. 59, no. 7, pp. 384–393, 2014.

[5] J. O. Okeniyi, G. S. John, T. F. Owoeye et al., "Effects of Dialium guineense Based Zinc Nanoparticle Material on the Inhibition of Microbes Inducing Microbiologically Influenced Corrosion," in *Proceedings of the 3rd Pan American Materials Congress*, pp. 21–31, Springer International Publishing, Cham, 2017.

[6] M. A. Javed, P. R. Stoddart, E. A. Palombo, S. L. McArthur, and S. A. Wade, "Inhibition or acceleration: Bacterial test media can determine the course of microbiologically influenced corrosion," *Corrosion Science*, vol. 86, pp. 149–158, 2014.

[7] H. Kanematsu, H. Ikigai, and M. Yoshitake, "Evaluation of various metallic coatings on steel to mitigate biofilm formation," *International Journal of Molecular Sciences*, vol. 10, no. 2, pp. 559–571, 2009.

[8] M. W. Kendig and R. G. Buchheit, "Corrosion inhibition of aluminum and aluminum alloys by soluble chromates, chromate coatings, and chromate-free coatings," *Corrosion*, vol. 59, no. 5, pp. 379–400, 2003.

[9] W. Fuerbeth, "New coatings for corrosion protection using nanoparticles or nanocapsules," in *CORROSION*, vol. 2015-, NACE International, Houston, Texas, USA, 2015.

[10] A. M. Atta, G. A. El-Mahdy, H. A. Al-Lohedan, and A. O. Ezzat, "Synthesis and application of hybrid polymer composites based on silver nanoparticles as corrosion protection for line pipe steel," *Molecules*, vol. 19, no. 5, pp. 6246–6262, 2014.

[11] R. Zeng, L. Liu, S. Li et al., "Self-assembled silane film and silver nanoparticles coating on magnesium alloys for corrosion resistance and antibacterial applications," *Acta Metallurgica Sinica (English Letters)*, vol. 26, no. 6, pp. 681–686, 2013.

[12] A. S. Johnson, I. B. Obot, and U. S. Ukpong, "Green synthesis of silver nanoparticles using Artemisia annua and Sida acuta leaves extract and their antimicrobial, antioxidant and corrosion inhibition potentials," *Journal of Materials and Environmental Science*, vol. 5, no. 3, pp. 899–906, 2014.

[13] S. Kadapparambil, K. Yadav, M. Ramachandran, and N. Victoria Selvam, "Electrochemical investigation of the corrosion inhibition mechanism of Tectona grandis leaf extract for SS304 stainless steel in hydrochloric acid," *Corrosion Reviews*, vol. 35, no. 2, pp. 111–121, 2017.

[14] N. Kasthuri and M. Priy, "Thermodynamic studies for the corrosion inhibition of mild steel by Tectona grandis leaves," *Journal of Applied Chemistry*, vol. 4, no. 4, pp. 1–11, 2016.

[15] J. O. Okeniyi, O. A. Omotosho, M. A. Inyang et al., "Investigating inhibition of microbes inducing microbiologically-influenced-corrosion by Tectona grandis based Fe-nanoparticle material," in *Proceedings of the AIP Conference Proceedings*, vol. 1814, AIP Publishing, 2017.

[16] R. Kumar, G. Ghoshal, A. Jain, and M. Goyal, "Rapid Green synthesis of silver nanoparticles (AgNPs) using (Prunus persica) plants extract: Exploring its antimicrobial and catalytic activities," *Journal of Nanomedicine and Nanotechnology*, vol. 8, no. 452, Article ID 1000452, 2017.

[17] P. Tippayawat, N. Phromviyo, P. Boueroy, and A. Chompoosor, "Green synthesis of silver nanoparticles in aloe vera plant extract prepared by a hydrothermal method and their synergistic antibacterial activity," *Peer*, vol. 2016, no. 10, Article ID e2589, 2016.

[18] S. Ahmed, M. Ahmad, B. L. Swami, and S. Ikram, "A review on plants extract mediated synthesis of silver nanoparticles for antimicrobial applications: a green expertise," *Journal of Advanced Research*, vol. 7, no. 1, pp. 17–28, 2016.

[19] S. Ahmed, M. Ahmad, B. L. Swami, and S. Ikram, "Green synthesis of silver nanoparticles using *Azadirachta indica* aqueous leaf extract," *Journal of Radiation Research and Applied Sciences*, vol. 9, no. 1, pp. 1–7, 2016.

[20] G. T. M. Bitchagno, L. S. Fonkeng, T. K. Kopa et al., "Antibacterial activity of ethanolic extract and compounds from fruits of *Tectona grandis* (Verbenaceae)," *BMC Complementary & Alternative Medicine*, vol. 15, no. 1, p. 265, 2015.

[21] S. Lanka and Parimala, "Antimicrobial activities of Tectona grandis leaf and bark extracts," *European Journal of Pharmaceutical and Medical Research*, vol. 4, no. 12, pp. 245–248, 2017.

[22] R. Nalvothula, V. B. Nagati, R. Koyyati, R. Merugu, and P. R. M. Padigya, "Biogenic synthesis of silver nanoparticles using *Tectona grandis* leaf extract and evaluation of their antibacterial potential," *International Journal of ChemTech Research*, vol. 6, no. 1, pp. 293–298, 2014.

[23] A. Abdolahi, E. Hamzah, Z. Ibrahim, and S. Hashim, "Micro-bially influenced corrosion of steels by Pseudomonas aeruginosa," *Corrosion Reviews*, vol. 32, no. 3-4, pp. 129–141, 2014.

[24] N. O. San, H. Nazir, and G. Dönmez, "Microbially influenced corrosion and inhibition of nickel-zinc and nickel-copper coatings by Pseudomonas aeruginosa," *Corrosion Science*, vol. 79, pp. 177–183, 2014.

[25] G. U. Akpan, G. Abah, and B. D. Akpan, "Correlation between microbial populations isolated from biofilms of oil pipelines and corrosion rates," *The International Journal of Engineering and Science*, vol. 2, no. 5, pp. 39–45, 2013.

[26] B. Mahesh and S. Satish, "Antimicrobial activity of some important medicinal plant against plant and human pathogens," *World Journal of Agricultural Sciences*, vol. 4, pp. 839–843, 2008.

[27] F. Olajubu, D. Ojo, I. Akpan, and S. Oluwalana, "Antimicrobial potential of Dialium guineense (Wild.) stem bark on some clinical isolates in Nigeria," *International Journal of Applied and Basic Medical Research*, vol. 2, no. 1, pp. 58–62, 2012.

[28] P. Logeswari, S. Silambarasan, and J. Abraham, "Synthesis of silver nanoparticles using plants extract and analysis of their antimicrobial property," *Journal of Saudi Chemical Society*, vol. 19, no. 3, pp. 311–317, 2015.

[29] J. O. Okeniyi, E. T. Okeniyi, O. O. Ogunlana, T. F. Owoeye, and O. E. Ogunlana, "Investigating biochemical constituents of *Cymbopogon citratus* leaf: Prospects on total corrosion of concrete steel-reinforcement in acidic-sulphate medium," in *TMS 146th Annual Meeting & Exhibition Supplemental Proceedings*, pp. 341–351, Springer International Publishing, 2017.

[30] J. Okeniyi, C. Loto, and A. Popoola, "Effects of Phyllanthus muellerianus Leaf-Extract on Steel-Reinforcement Corrosion in 3.5% NaCl-Immersed Concrete," *Metals*, vol. 6, no. 12, p. 255, 2016.

[31] J. O. Okeniyi, E. T. Okeniyi, and T. F. Owoeye, "Bio-characterisation of Solanum aethiopicum leaf: Prospect on steel-rebar total-corrosion in chloride-contaminated-environment," *Progress in Industrial Ecology , an International Journal*, vol. 10, no. 4, pp. 414–426, 2016.

[32] J. O. Okeniyi, O. O. Ogunlana, O. E. Ogunlana, T. F. Owoeye, and E. T. Okeniyi, "Biochemical characterisation of the leaf of Morinda lucida : Prospects for environmentally-friendly steel-rebar corrosion-protection in aggressive medium," in *TMS 2015 144th Annual Meeting & Exhibition*, pp. 637–644, Springer, Cham, 2015.

[33] J. O. Okeniyi, O. A. Omotosho, O. O. Ogunlana et al., "Investigating prospects of phyllanthus muellerianus as eco-friendly/sustainable material for reducing concrete steel-reinforcement corrosion in industrial/microbial environment," *Energy Procedia*, vol. 74, pp. 1274–1281, 2015.

[34] R. Franklin and M. D. Cockerill III, "Performance standards for antimicrobial susceptibility testing; twenty-first informational supplement M100-S21," *Clinical and Laboratory Standard Institute*, vol. 31, pp. 68–80, 2001.

[35] A. W. Bauer, W. M. Kirby, J. C. Sherris, and M. Turck, "Antibiotic susceptibility testing by a standardized single disk method," *American Journal of Clinical Pathology*, vol. 45, no. 4, pp. 493–496, 1966.

# 13

# A Comparative Study on Removal of Hazardous Anions from Water by Adsorption

**Yasinta John,[1] Victor Emery David Jr.,[1] and Daniel Mmereki[2]**

[1]*College of Urban Construction and Environmental Engineering, Chongqing University, Chongqing 400044, China*
[2]*National Centre for International Research of Low-Carbon and Green Buildings, Chongqing University, Chongqing 400045, China*

Correspondence should be addressed to Yasinta John; john_yasinta@yahoo.com

Academic Editor: Swapnil A. Dharaskar

This paper presents a comparative review of arsenite (As(III)), arsenate (As(V)), and fluoride (F$^-$) for a better understanding of the conditions and factors during their adsorption with focus on (i) the isotherm adsorption models, (ii) effects of pH, (iii) effects of ionic strength, and (iv) effects of coexisting substances such as anions, cations, and natural organics matter. It provides an in-depth analysis of various methods of arsenite (As(III)), arsenate (As(V)), and fluoride (F$^-$) removal by adsorption and the anions' characteristics during the adsorption process. The surface area of the adsorbents does not contribute to the adsorption capacity of these anions but rather a combination of other physical and chemical properties. The adsorption capacity for the anions depends on the combination of all the factors: pH, ionic strength, coexisting substances, pore volume and particles size, surface modification, pretreatment of the adsorbents, and so forth. Extreme higher adsorption capacity can be obtained by the modification of the adsorbents. In general, pH has a greater influence on adsorption capacity at large, since it affects the ionic strength, coexisting anions such as bicarbonate, sulfate, and silica, the surface charges of the adsorbents, and the ionic species which can be present in the solution.

## 1. Introduction

Hazardous anions are another group of pollutants in drinking water in addition to metal ions and organics which are known to be toxic and carcinogenic. The presence of these anions in ground and surface waters has resulted in severe contamination and has caused adverse health effects. Among the toxic anions, arsenite (As(III)), arsenate (As(V)), and fluoride (F$^-$) have shown concern in the treatment of wastewaters and drinking water. Adsorption process among several treatment technologies is applied for removal of fluoride (F$^-$), arsenite (As(III)), and arsenate (As(V)) ions due to the availability of local material and being easy to operate and maintain and in general it is a low-cost technology and is suitable for use in developing countries. The adsorption process is influenced by several environmental conditions, such as pH, salt effect, or ionic strength, the presence of other anions, cations, and organic matter, and adsorbate physical and chemical properties which have been reported in previous studies [1–5]. However, no previous study was done to compare the two

anions based on the influence factors such pH, salt effect, or ionic strength and the presence of other anions and their relationship on the adsorption isotherm. This review aims to present an evaluation of these parameters and their effects on arsenite (As(III)), arsenate (As(V)), and fluoride (F$^-$) by adsorption process on different adsorbents. It also seeks to find the relationship between the anions and their influencing factors. Therefore, this review focuses on the isothermal adsorption models, the effects of the pH and ionic strength, and the impact of coexisting substances including anions and cations to show the characteristics of fluoride (F$^-$), arsenite (As(III)), and arsenate (As(V)) anionic adsorption processes.

*1.1. Arsenic.* Arsenic is discharged into the environment by natural activities and anthropogenic activities [6, 7]. Arsenic occurs in both organic and inorganic forms. Arsenate (As(V)) and arsenite (As(III)) are the inorganic forms of arsenic which are considered to be more toxic and are more prevalent in water. The existence of arsenic species depends on the pH solution and redox conditions [8]. The ingestion of inorganic arsenic can result in both cancerous and noncancerous health effects. An arsenic concentration less than 0.05 mg/L in chronic exposure has been linked to skin diseases, neurological and cardiovascular system disorder, and skin, kidney, and lung cancer [6, 9, 10]. Arsenic affects physiological activities, such as the activities of essential cations, enzymes, and transcriptional events in cells [6, 9]. Elevated concentrations of arsenic in drinking water have also been reported to cause an increase in abortions and stillbirth [11].

The methods of arsenic removal from water include oxidation and filtration, adsorption, biological oxidation, coprecipitation, and membrane technologies [12]. Numerous arsenic treatment technologies require pH adjustment and are useful in removing arsenic in the pentavalent state. Thus oxidation is included as a pretreatment to convert As(III) to As(V) [13]. However, the process of coagulation and flocculation produces sludge, which makes it inefficient for As(III) removal [13]. Adsorption process with a wide range of adsorbents has been proven to be effective for the adsorption of arsenate (As(V)) and arsenite (As(III)) ions from water and has been reported by various researchers [14–16].

*1.2. Fluoride ($F^-$).* The contamination of fluoride in groundwater has been known worldwide to cause severe human health problems [17, 18]. Fluorides exist with iron, aluminum, and beryllium as fluoride ion ($F^-$) in natural waters [19]. It is caused by the discharge of mineral sediments and industrial (production of phosphate fertilizers (3.8% fluorine), bricks, tiles, and ceramics) effluent which contains fluoride in receiving water bodies [19]. The inorganic fluorine compounds are used in industry for aluminum production and as a flux in the steel and glass fiber, fluorosilicic acid, sodium hexafluorosilicate, and sodium fluoride are used in public water fluoridation treatment [18]. Fluoride is also essential in human health as the component for normal mineralization of bones and formation of dental enamel [20].

Fluoride is beneficial in drinking water at levels up to 0.7 mg/L but is harmful above 1.5 mg/L, according to the World Health Organization limit [21]. The excessive intake of fluorides may result in dental and skeletal fluorosis [20]. Fluorosis is prevalent in more than 20 developed and developing nations [20, 21]. Fluorides can also cause arthritis, infertility osteoporosis, brain damage cancer, Alzheimer syndrome, thyroid disorder, and brittle bones [17, 21]. High fluoride concentrations are known to interfere with carbohydrates, lipids, proteins, vitamins and mineral metabolism, and gastrointestinal irritation when by initially, acting on the intestinal mucosa, and then form hydrofluoric acid in the stomach at a later stage [17, 20]. Fluoride can also cause kidney disease in both animals and humans and interfere with the functions of the brain and pineal gland [17].

Due to the toxic effects of fluoride in human health, various methods have been developed for the removal of excess fluoride ions from drinking water. Currently, the common methods include adsorption into activated alumina (AA), bone char, and clay [22], precipitation with lime, dolomite, and aluminum sulfate, Nalgonda technique [23], ion exchange and reverse osmosis, electrodialysis, and nanofiltration [22]. The coagulation, adsorption, and ion-exchange techniques are widely opted as defluoridation techniques practiced in fluoride endemic areas [23]. The Nalgonda technique and adsorption by bone char have been used in many developing countries like Tanzania and India [23]. However, the disadvantages of this technique were reported by some researchers, for example, higher residual aluminum concentration (2–7 mg/L) in treated water than WHO standard of 0.2 mg/L [17]. The membrane processes provide good-quality water but have a higher cost of operation [23].

## 2. Adsorption Method for Arsenic and Fluoride Removal

Adsorption methods have been proven to remove fluoride and arsenic up to 90% from water [16, 22]. However, in adsorption process, the constituents of adsorbents are at large accountable for the elimination of pollutants [24, 25]. Many adsorbents are used in adsorption process, but the most common ones in the treatment of both water and wastewater are activated carbon, activated alumina, ion-exchange resins, metal oxides, hydroxides, carbonates, and clays [16, 22]. The adsorption process is described through equilibrium isotherm which occurs when an adsorbate comes in contact with the adsorbent for a period of time. Thus, the concentration of adsorbate in the solution is balanced with the interface concentration [25]. Equilibrium isotherm is essential for the effective design of sorption systems [26].

The adsorption capacity of the various adsorbents is influenced by pH, the existence of other adsorbing anions, ionic strength, temperature, properties of the adsorbents, initial concentration of the adsorbates, and so forth [16, 17]. Experimental results showed that the pH solution is the main controlling parameter in the adsorption processes [27]. pH affects adsorption by affecting the surface charge on the adsorbent [4, 24, 28, 29]. However, the evaluation of pH effects is governed by the specific interactions between the ions and the adsorption sites [30]. The adsorption of anions decreases by increasing pH due to the higher concentration of competitive anions, such as $OH^-$ and increases due to protonated surfaces. In aqueous solution, the adsorption process is primarily governed by the zero point charge (ZPC) of an adsorbent [31, 32]. At zero point charge, the pHzpc is the characteristic that determines the pH at which the adsorbent surface has net electrical neutrality. It is also where the acidic or basic functional groups no longer contribute to the pH of a solution [33]. At pH values above the ZPC, the surface has a net negative or anionic charge, and the surface would take part in cation attraction, as well as cation exchange reactions. The surface has a net positive charge at pH values below the ZPC. Therefore, it will attract anions USEPA [34]. Metal adsorption's dependence on pH is related to the nature and ionic state of the functional group existing in the adsorbent and the metal chemistry in the solution [35–37].

Many studies have been conducted on fluoride removal by using adsorption [22] and arsenic [16]. Moreover, pH has been reported to be a key factor affecting fluoride adsorption at the water adsorbent interface [38]. It has been reported that the removal of fluorine ions occurs between pH of 2.0

and 8.0 [27, 39, 40]. Also, the adsorption of arsenic species is highly dependent on pH due to its ability to exist as As(III) or As(IV) at different pH values. As(III) is highly removed at higher pH (basic) value, while As(IV) is removed in acidic pH and rapidly decreases in basic medium [41–43].

The adsorption capacity of the adsorbents is also influenced by the physical and chemical characteristics of the adsorbent [16]. The surface area and total pore volume determine the adsorption capacity based on the results of the kinetics and equilibrium on adsorption experiments [44–47]. The control of chemical and physical oxygen functional groups considerably affects their performance on the adsorption process [47]. However, the use of Extended X-ray Absorption Fine Structure (EXAFS), Fourier transform infrared spectroscopy (FTIR), and X-ray absorption near edge structure (XANES) has been employed for the removal of fluoride [48] and arsenic [49] to further study the adsorption mechanism and involvement of functional groups in different adsorbents. These have revealed the involvement of functional groups like hydroxyl and carbonates in the surface, which participate in the adsorption.

Different materials exist in water and wastewater, which have different absorption properties and may compete with each other for the limited amount of sites adsorption, and therefore the adsorption in the specific material is reduced [4, 50, 51]. These materials include anions and cations and organic matters [51–55]. Anions such as sulfate, nitrate, carbonate, chloride, bicarbonate, and phosphate influence adsorption by adjustment of the electrostatic charge at the solid surface because of the same negative ions [11, 56–61]. The effects of anions on fluoride removal were reported by [40, 62–66]. Both direct and indirect effects of these anions are influenced by pH, anions' concentrations, and intrinsic binding affinities [42, 59]. One of the common interfering anions is phosphates. Phosphates exist in four different forms depending on the pH value, phosphate ion ($PO_4^{3-}$), dihydrogen phosphate ion ($H_2PO_4^-$), and hydrogen phosphate ion ($HPO_4^{2-}$) in dilute aqueous solution and in aqueous phosphoric acid ($H_3PO_4(aq)$) [60]. $PO_4^{3-}$ ions are more predominant in strongly basic conditions, $HPO_4^{2-}$ ions in weakly basic conditions, $H_2PO_4^-$ ions in weak acidic condition, and aqueous $H_3PO_4$ [60]. In general, $SO_4^{2-}$, $NO_3^-$, and $Cl^-$ interfering ions have shown an insignificant effect on adsorption of these anions compared with $HPO_4^{2-}$ and $HCO_3^-$, which have a great impact in all anions adsorption [16, 67].

The effects of cations like $Al^{3+}$, $Fe^{3+}$, $Mg^{2+}$, $Ca^{2+}$, and $Mn^{2+}$ in adsorption and their interference with the adsorption capacity of these anions have been reported by [56, 62, 68–71]. The presence of cations like $Ca^{2+}$ and $Mg^{2+}$ increases the fluoride with increasing $Ca^{2+}$ and $Mg^{2+}$ concentration [72]. Ionic strength has been reported to increase with an increase in adsorption for arsenate [67] and it has no significant effect on the adsorption of fluoride [73].

Natural organic matter in water may delay sorption equilibrium and decrease the extent of arsenite and arsenate adsorption [16]. Natural organic matter contains a mixture of weak organic acids and organic compounds which do not have a clear chemical structure [50]. Fulvic acids (FA) and humic acids (HA) are hydrophobic in nature, and they represent almost 60% of the dissolved organic in aquatic systems [56]. Humic substances interfere with anionic adsorption through stable metal complex formations [23, 54, 56]. Due to the importance of these parameters in adsorption process, we evaluated the adsorptions isotherms, coexisting substances, the effects of pH, and ionic strength and their effects on arsenite (As(III)), arsenate (As(V)), and Fluoride ($F^-$) removal on different adsorbents and aimed to find the relationship between them. The insight of anion adsorption in water basically confined to the adsorption of a single ion and on the specific adsorbent. Therefore this review summarizes the common characteristics of anions adsorption and explains the roles for the relevant forces and the adsorption control conditions and features of the adsorbents suitable for the arsenite (As(III)), arsenate (As(V)), and fluorine ($F^-$) adsorption in water.

## 3. Adsorption Isotherm Studies on As(III) and As(V)

Freundlich isotherm, Langmuir isotherm, and BET isotherm are examples of the most frequently used isotherm models in adsorption process [25, 26, 37, 52]. In this review, only Freundlich and Langmuir were taken into consideration due to their wide applicability to gain an insight into the degree of the favorable adsorption [74].

Adsorption technology can reduce arsenic concentrations to less than 10 mg/L [75], and the common sorbents for arsenic removal are commercial and synthetic activated carbons, agricultural products and by-products, and industrial waste soils and constituents, for example, clay minerals, manganese dioxide, zeolite, activated alumina, ferrihydrite/iron and hydroxide/iron oxides, hydroxides, hydrotalcites, phosphates, and metal-based methods, for example, zerovalent iron [13, 16]. Iron-based sorbents (IBS) have been reported as promising adsorbents for arsenic removal, since they have higher affinity for arsenic under neutral conditions [76]. These include activated ferric oxide or ferric hydroxide and iron-coated sand [13]. The removal of arsenic by using iron-based adsorbents occurs through adsorption to a surface hydroxyl group, coprecipitation, and ion exchange [77]. However, with the iron oxides adsorbents, the removal of As(V) at neutral pH occurs through ligand exchange by the formation of monodentate complexes at low surface coverage, while at high surface coverage, As(V) species bind to the oxides through the formation of bidentate complexes while occupying two adsorption sites at the same time [78]. Few activated carbons are selective for the adsorption of As(III) and As(V) in concentrations <0.5 mg/L from water [51].

As(V) can be easily removed by a wide range of adsorbents compared to arsenic (III) (Table 1) and because arsenic (III) is easily adsorbed to arsenic (V) and is oxidized before or during the adsorption process. As(III) oxidation occurs naturally or by using iron, hydrogen peroxide, chlorine, hypochlorite, and manganese, as shown in a study conducted by Giles et al. [79]. Also, another reason is the fact that adsorption of arsenic significantly depends on arsenic speciation, which

TABLE 1: Adsorption isotherm for arsenic.

| Adsorbent | Adsorption capacity | | Adsorption isotherm | Initial concentration | References |
| --- | --- | --- | --- | --- | --- |
| | Arsenic III | Arsenic V | | | |
| Zirconium Polyacrylamide hybrid (ZrPACM-43) | 0.20 mg/g 0.80 mg/g | | Freundlich | 10 mg/L | [14] |
| Manganese oxide-coated alumina (MOCA) | 42.48 mg/g | — | Sips | — | [2] |
| Nano zerovalent | 18.2 mg/g | 12 mg/g | Freundlich and Langmuir | — | [3] |
| MnFe$_2$O$_4$ | 94 mg/g | 90 mg/g | Langmuir | 10 mg/L | [42] |
| CoFe$_2$O$_4$ | 100 mg/g | 74 mg/g | Langmuir | 10 mg/L | [42] |
| Manganese(II, III) oxide (Mn$_3$O$_4$) | | 101 $\mu$g/m$^2$ | Langmuir Freundlich | 1 mg/L | [52] |
| Activated alumina | 0.0545 mg/g | — | Langmuir, Freundlich | 0.5 mg/L | [82] |
| Copper(II) oxide nanoparticles | — | 1.0862 mg/g | Langmuir | 100 $\mu$g/L | [57] |
| Chitosan-coated biosorbent | 56.5 mg/g | 96.46 mg/g | Langmuir | 100 ppm | [83] |
| Cellulose-carbonated hydroxyapatite nanocomposites | | 12.72 mg/g | Langmuir | 10 mg/L | [84] |
| Magnetiteemaghemite nanoparticles | 3.69 mg/g | 3.71 mg/g | Freundlich | 1.5 mg/L | [85] |
| Polymeric Al/Fe modified montmorillonite | 19.11 mg/g | 21.23 mg/g | Freundlich | 10 mg/L | [86] |
| Activated alumina grains | — | 15.9 mg/g | Langmuir | 2.85 and 11.5 mg/L | [87] |
| Activated alumina grains | 3.48 mg/g | | Langmuir | 0.79 and 4.90 mg/L | [87] |
| Iron (Fe$_2$O$_3$) | — | 0.66 mg/g | Langmuir | 200 $\mu$g/L | [60] |
| Aluminum oxide (Al$_2$O$_3$) | — | 0.17 mg/g | Langmuir | — | [60] |
| Oxide-coated sand | 0.0411 mg/g | 0.0426 mg/g | Langmuir | 100 $\mu$g/L | [88] |
| Shirasu-zeolite P1 (-SZP1) | — | 65.93 mg/g | Freundlich | — | [89] |
| Aluminum-loaded Shirasu-zeolite P1 (Al-SZP1) | — | 10.47 mg/g | Freundlich | 0.13 mM | [89] |
| Alum-impregnated activated alumina (AIAA) | — | 0.0314 mmol/g | Langmuir | 1–25 mg/L | [90] |
| Copper oxide incorporated | 2.161 mg/g | 2.017 mg/g | Langmuir | — | [91] |
| Unmodified alumina | 0.925 mg/g | 0.637 mg/g | Langmuir | — | [91] |
| Goethite | — | 15 mg/g | Langmuir | — | [92] |
| Jarosite | — | 21 mg/g | Langmuir | — | [92] |
| Iron-coated zeolite (ICZ) | — | 0.68 mg/g | Langmuir | — | [93] |
| Iron-impregnated granular activated carbon | — | 0.6 mg/g and 1.95 mg/g | Langmuir | — | [93] |
| Reclaimed iron oxide-coated sands | 6.7–8.7 g/g | — | Langmuir | 301 g/L | [94] |
| Maghemite nanoparticles | — | 16.7 mg/g | Langmuir | — | [95] |
| Acid modified carbon black | — | 46.3 mg/g | Langmuir | 100 mg/L | [96] |
| Modified red mud | — | 68.5 mg/g | Langmuir | — | [69] |
| MnO$_2$-modified natural clinoptilolite zeolite | — | 1 $\mu$g/g | Freundlich | 20 $\mu$g/L | [97] |
| Unmodified natural clinoptilolite | — | 0.38 $\mu$g/g | Freundlich | 20 $\mu$g/L | [97] |
| Novel hybrid material | 0.25 mg/g | | Langmuir | 10 mg/L, 50 mg/L, and 100 mg/L | [98] |
| Synthetic siderite | — | 31 mg/g | Langmuir | | [99] |
| Iron oxide-coated fungal biomass | 880 $\mu$g/g | 1080 $\mu$g/g | Langmuir | 100 g/L | [100] |
| Cupric oxide nanoparticles | 26.9 mg/g | 22.6 mg/g | Langmuir isotherm | | [101] |

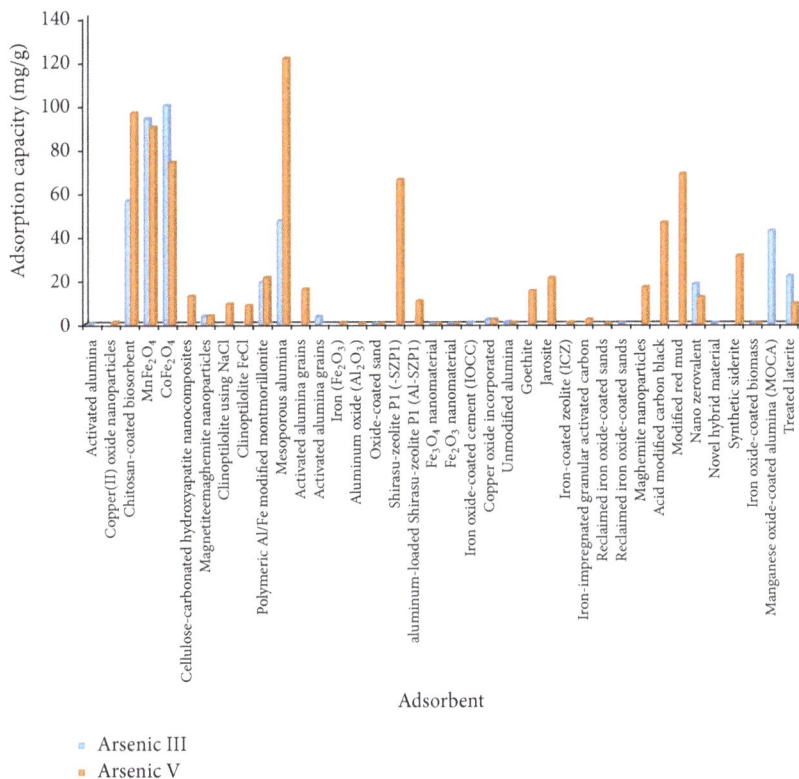

FIGURE 1: Comparison of the adsorption capacity for arsenic removal.

also depends on pH and potential. Modified adsorbents have been shown to have a higher capacity for arsenic adsorption (Figure 1) compared to unmodified adsorbents. The modified adsorbents include, but are not limited to, acid modified carbon black, modified red mud, chitosan-coated biosorbent, bimetal oxide magnetic nanomaterials ($MnFe_2O_4$ and $CoFe_2O_4$), polymeric Al/Fe modified montmorillonite, manganese oxide-coated alumina and synthetic siderite. During arsenic adsorption, surface area is insignificant (Figure 2) and does not contribute to the adsorption capacity of the adsorbents [30, 76, 80]. The equilibrium adsorption capacity depends on the concentration of ionic species in solution and the sorbent properties [30]. The choice of the adsorbents to agree with Langmuir or Freundlich is still unclear. However, a study by [81] suggested that if As(V) followed Freundlich isotherm on arsenic adsorption onto Iron-Zirconium Binary Oxide-Coated Sand (IZBOCS), it showed that the adsorption of As(V) followed multilayer and heterogeneous adsorption process. Langmuir model was the best fit for As(III) and As(V) adsorption on iron–aluminum hydroxide coated onto macroporous and, according to the author, the adsorptions occur through chemisorption.

Studies by using X-ray photoelectron spectroscopy (XPS) have revealed that As(III) was oxidized and absorbed in the form of As(V) on the surface of cupric oxide nanoparticles (CuO) [101]. FTIR spectroscopy has also been used to elucidate the arsenic adsorption mechanism on iron mineral oxide which involves ligand exchange reactions where the anions displace $OH^-$ and $H_2O$ from the surface [102]. Fourier transform infrared (FTIR) spectra demonstrated that Ca–OH functional group was involved in As(V) removal of bone char and coprecipitation and ion exchange was involved before and after As(V) adsorption [103]. Another study suggested that the adsorption of As(III) occurs through electrostatic attraction on bismuth-impregnated biochar [104]. Similar results were obtained by [105] on arsenic adsorption on feldspars, but the electrostatic attraction was between the luminol function groups and predominant form of arsenate in low pH. Another EXAFS analysis concluded that the adsorption of As(III) and As(V) on iron–aluminium hydroxide coated onto macroporous was involved through formation of inner-sphere complexes to the iron hydroxide. However, for As(III) adsorption, the bidentate complex was formed, which corresponded with the pseudo-second-order kinetic model [106].

The adsorption of As(V) anions on activated carbon was associated with the concentration of hard acid functional groups and not the carbon surface area [30]. Similar results were reported by [80] on arsenic adsorption on hydrous ferric oxide. Figure 2 shows that adsorption capacity of various adsorbents does not depend on the surface area for arsenic removal. However, it has been reported that changes in the type of arsenic complex binding with the surface area depend on pH and the number of available sites for adsorption for specific adsorbents [80]. In arsenic ions adsorption, pore diffusion is important, since the particle size, solubility, pH, and metal ions concentration affect adsorption capacity and the adsorption capacity increases with a decrease in particle size [76, 107].

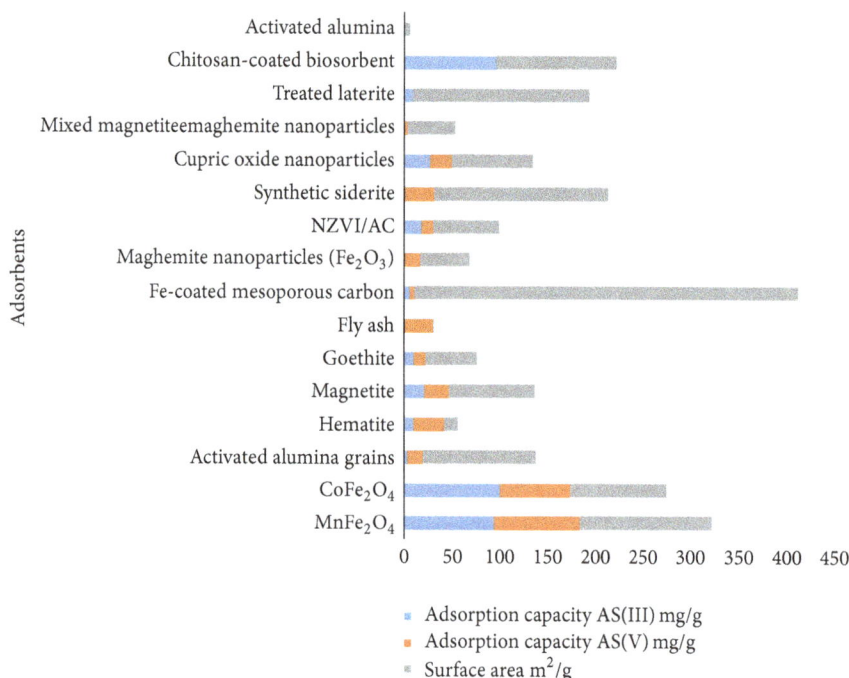

FIGURE 2: Adsorbents sorption capacity and surface area for arsenic removal.

Hassan et al. [108] investigated arsenic removal on potassium hydroxide activated carbon-based apricot stone, calcium alginate beads, and calcium alginate/activated carbon composite beads. The adsorbents materials have surface areas of $1621 \, m^2/g$, $32.9 \, m^2/g$, and $733.6 \, m^2/g$ and the adsorption capacity at $20°C$ was $26.3 \, mg/g$, $39.4 \, mg/g$, and $54.8 \, mg/g$, respectively. The authors concluded that the As(V) adsorption depends largely on pH, surface chemistry of solid adsorbent, and textural properties of the adsorbate besides its surface area. Another study by using magnetic graphene oxide as an adsorbent for arsenic removal showed that initial concentration of arsenic and the adsorbent dosage were the major factors affecting the adsorption capacity [109].

A maximum adsorption capacity of 3.1 mg/g was reported for activated carbon prepared from oat hulls which has a specific area of $522 \, m^2/g$ at initial arsenic concentrations of 25 to 200 $\mu g/L$ [110]. Similarly, a low adsorption capacity for adsorbents with a high surface area was observed by Roy et al. [111] on thioglycolated sugarcane carbon (TSCC) as an adsorbent for arsenic removal. TSCC has a surface area of $5690 \, m^2/g$. The maximum reported As(III) and As(V) removal by TSCC was 85.01 and 83.82 $\mu g/g$, respectively, with an initial arsenic concentration of 1,500 $\mu g/L$ [111].

Low surface areas with high adsorption capacity have also been observed in numerous research works. For example, the removal of As(V) and As(III) from water by a Zr(IV)-loaded orange waste gel had an initial concentration of 20 mg/L and surface area of $7.25 \, m^2/g$. The maximum adsorption capacity of the Zr(IV)-loaded sol-gel was 88 mg/g and 130 mg/g for As(V) and As(III), respectively [112]. Similar results were reported by [83] on chitosan-coated biosorbent and by [14] on zirconium polyacrylamide hybrid material (ZrPACM-43). A high adsorption capacity has also been reported

for an adsorbent with a high surface area. For example, Liu et al. [113] investigated arsenic adsorption on $Fe_3O_4$-loaded activated carbon prepared from waste biomass with surface area of $349 \, m^2/g$; a very high adsorption capacity of 204.2 mg/g was observed with arsenic initial concentration of 40 mg/L [113]. With the reported observation, selection of adsorbent for arsenic adsorption should be based not only on surface area of the adsorbent but also on the combination of chemical properties of the adsorbent and adsorbate.

## 4. Effects of pH on Arsenic Removal

Removal of arsenic from water depends on pH as the adsorption capacity changes with changes in pH. The reduced trivalent form of arsenic (arsenite (As(III))) primarily exists in natural waters and anaerobic environment, while arsenate (As(V)), an oxidized pentavalent form, is found in aerobic environment conditions such as surface waters [79]. Arsenate species exist as $AsO_4^{3-}$, $HAsO_4^{2-}$, and $H_2AsO_4^-$ and arsenate species exist as $As(OH)_3$, $As(OH)_4^-$, and $AsO_3^{3-}$ depending on the pH solution [114]. Chemical speciation affects arsenic removal in aqueous solution (USEPA) [115]. As(V) exists as a charged species in water across a wide range of pH while As(III) exists as a charged species across a much narrower range of pH [16, 42, 87, 89, 116, 117], and it has been reported that many adsorbents prefer to adsorb charged species [16]. Arsenite (As(III)) is uncharged (i.e., $H_3AsO_3$) at natural pH levels (6–9) (Figure 3); thus it it difficult to remove compared to charged arsenate (As(V)) (i.e., $H_2AsO_4^-$ or $HAsO_4^{2-}$) (USEPA) [115]. As(III) is oxidized to As(V) by an oxidizing agent such as chlorine or permanganate in most arsenic treatment processes [79, 89]. Adsorbents like iron-based and

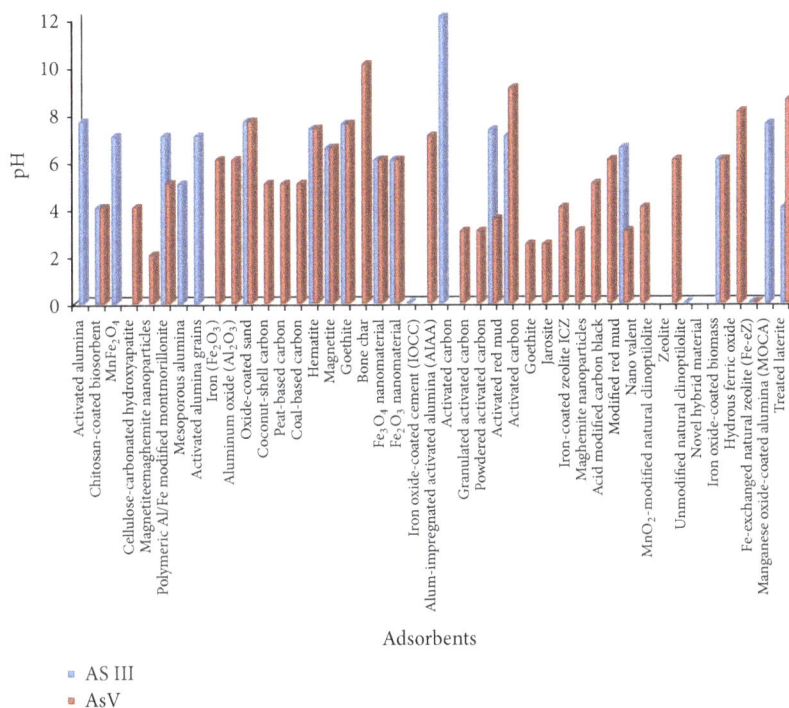

FIGURE 3: Effects of pH on As(V) and As(III).

iron minerals are characterized by their point of zero charges at a certain pH, where the mineral surface charge is equal to zero [42, 50, 84]. However, in other adsorbents, the strong adsorption of arsenic at pH > pHpzc indicates that the adsorption process is influenced by surface complexation and not electrostatic interaction [86].

Arsenate and arsenite adsorption onto activated alumina was governed by both the surface charge of activated alumina and the form of arsenic species in the water. Since alumina is sensitive to pH, adjustment of pH is done prior to treatments. Therefore, the adsorption of arsenite (As(III)) is much less than that of arsenate (As(V)) for activated alumina in most pH conditions, because arsenate is present in the negatively ionic state and arsenite is in nonionic state. For example, Singh and Pant [82] studied the removal of As(III) by activated alumina (AA), and the result showed that 94.4% maximum adsorption was reached in the pH range of 6.0–8.0 and it decreased at higher pH values [82]. Similar results were observed by [80, 87].

A study by Mamindy-Pajany et al. [118] on adsorption of As(V) on commercial hematite and goethite as a function of pH (2 to 12) and ionic strength showed that As(V) adsorption was higher at low pH value and insignificant at higher pH values. Similar results were reported by [119] on As(V) adsorption on goethite from aqueous solution and [120] on As(III) and As(V) adsorption on hematite, magnetite, and goethite. However, hematite was reported to be the appropriate adsorbent for As(V) removal in natural medium because it can exist at a wide pH range [99]. Mamindy-Pajany et al. [121] observed a similar trend on As(V) adsorption by hematite but, for As(III), the adsorption was highly dependent on

initial pH, and a decrease in adsorption was observed at the pH value of 3.

The adsorption of As(III) and As(V) has also been reported to be affected by pH higher than 10. For example, a study of As(III) and As(V) ions to the $Fe_3O_4$ nanomaterial [70, 122] observed similar trend on an iron oxide-coated cement (IOCC) adsorption capacity of As(III) removal and [123] on activated carbon. Also, a high adsorption percentage was observed at higher pH for As(V) adsorption over activated charcoal and bone char adsorbents. Adsorption of As(V) has been reported to be maximum at low pH values. Ansari and Sadegh [123] reported a maximum adsorption of As(V) at pH of 3 and decreased with the decrease of pH. That is, adsorption by activated carbon is thought to be carried out through anion exchange process and physicochemical adsorption due to the highly porous structure of activated carbon. The surface of the activated carbon has a positive charge under acidic conditions, where these positive charges are balanced with their associated anions. Therefore, these anions can be exchanged with the other anionic species present in the solution. Arsenic (III) adsorption on synthetic siderite has also been related to electrostatic attraction as well as physicochemical reactions [124]. In general, As(III) is adsorbed at neutral pH (Figure 3) and As(V) at a wide range of pH (2–10) depending on the adsorbent material used. A similar trend on arsenic adsorption by various adsorbents was reported by [50, 57, 60, 103].

Another different result on effects of pH on As(V) has been reported by [83] on removal of arsenic (III) and arsenic (V) from aqueous medium using chitosan-coated biosorbent (CCB), whereby a decrease in pH increases

As(V) adsorption and the mechanisms involved in As(V) removal were ionic attraction, nodule formation, and absorption.

## 5. Effects of Coexisting Anion, Cation, and Organic Matter on Arsenic

*5.1. Effects of Coexisting Anions.* Coexisting anions such as $Cl^-$, $NO_3^-$, $SO_4^{2-}$, $CO_3^{2-}$, and $PO_4^{3-}$ ions, in water sources, might compete with arsenic for active adsorption sites and significantly reduce the arsenic removal performance by the adsorbents. These coexisting anions have been reported to form inner-sphere complexes with arsenic and reduce the interaction between the two anions. The effects tend to be significant with a change in pH and with the increase of the anions concentration [3, 78].

*5.1.1. Sulfate.* Arsenic removal using $KMnO_4$–Fe(II) showed that, in the presence of 50–100 mg/l of sulfate, adsorption of arsenic was reduced by 6.5–36.0% over pH 6–9. The reasons were the competition for surface sites, the weak affinity of arsenic to the adsorbent, and the mechanisms for the adsorption of arsenic and sulfate being not the same. Guan et al. [56] studied the adsorption of arsenic into hydrous ferric oxide, and the results showed a decreasing trend on adsorption of As(III) after the pH was less than 4 [58]. A similar trend was observed by [60] on As(V) removal by iron oxide and aluminum oxide and by [92] on arsenic removal by goethite and jarosite in acidic conditions. However, Maiti et al. [50] showed that sulfate ion did not have any effect on As(V) and As(III) ions adsorption due to higher adsorption affinity of treated laterite surface on As(V). A similar result was observed by [42, 125].

*5.1.2. Phosphate.* The effect of phosphate is different for both As(III) and As(V). The decrease in adsorption capacity of the adsorbents is more for As(III) at lower concentration and As(V) at higher concentration. However, the influence of phosphate is also affected by pH. Arsenic removal using $KMnO_4$–Fe(II) showed that arsenic removal was reduced by 29.8%, 34.2%, and 47.3% at pH of 4, 5, and 6, respectively, in the presence of phosphate. The reasons were the competition of arsenate and phosphate for binding sites and inhibition of forming Fe precipitates and reduction of surface sites [56]. And, by increasing phosphate concentration to 2.5 mg/L, the arsenic removal was reduced by 59.6% [56]. As(III) and As(V) removal from water by copper oxide-incorporated mesoporous alumina showed that the presence of phosphate significantly reduces the As(V) adsorption due to the competition between phosphate and As(V) [45, 91]. Jeong et al. [60] reported that the increase of phosphate concentration decreases As(V) adsorption capacity on the adsorption of As(V) by $Al_2O_3$ compared to $Fe_2O_3$. A similar observation was reported by [89] on aluminum-loaded Shirasu-zeolite as an adsorbent. Arsenic adsorption was reduced by more than 20% when phosphate concentration increased by more than 0.2 mM on copper (II) oxide nanoparticles as adsorbents [57]. Maiti et al. [50] reported that arsenic was reduced to 72.9% removal when phosphate concentration was increased

to 5 mg/L on synthetic siderite as an adsorbent for As(V) removal. However, at low concentration of 0.2 mg/L phosphate on Fe(VI)/Al(III) chloride salts, arsenic removal was reduced to 73% [61].

*5.1.3. Silicate.* In natural water, silicate originates from the weathering of minerals and is always dissolving and precipitating at the earth's surface. Silicate is usually oxyanions in natural water, with their concentration between 0.45 and 14 mg Si/L [56, 60]. A 50% arsenite removal reduction was observed on Fe(VI), Fe(VI)/Fe(III), and Fe(VI)/Al(III) salt as adsorbent for arsenite removal in presence of 10 mg Si/L [61], whereas with 10 mg Si/L silicate concentration showed no effect with $KMnO_4$–Fe(II) as adsorbent at pH 5, but adsorption was observed to be decreased by 42% at pH 9 [56]. Therefore effects of silicate on arsenic removal are more obvious with an increase in pH and can be associated with the weak silicic acid species distribution with pH [56]. Similar results were reported by [60] on adsorption of As(V) on $Fe_2O_3$ and $Al_2O_3$. Silicate in water has been reported to have no significant effects on adsorption of As(V) by cupric oxide nanoparticles but only slightly inhibited adsorption of As(III) [101]. 1 mg Si/L silicate concentration was reported to increase Arsenic adsorption due to electrostatic effects of $Al_2O_3$ compared to $Fe_2O_3$ adsorbent [60].

*5.1.4. Bicarbonate and Carbonate.* Dissolved carbonate exists as $HCO_3^-$ (bicarbonate) and carbonate ($CO_3^{2-}$) in groundwater and is likely to interfere with arsenic adsorption. Bicarbonate ions are known to form inner-sphere monodentate complexes with surface functional groups of Fe and Al hydroxides. With the existence of 400 mg/L of carbonate concentration in solutions, the percentage of As(III) adsorption decreases to 77%, as it was observed on treated laterite as adsorbents. This was due to the competition between $HCO_3^-$ and $HASO_4^{2-}$ for positively charged adsorption sites [50]. According to Zhang et al. [42], the addition of $CO_3^{2-}$ decreased arsenic adsorption moderately on arsenite and arsenate adsorption by $MnFe_2O_4$ and $CoFe_2O_4$. The decrease of adsorption was due to the basic condition when $Na_2CO_3$ was added and formation of arsenic–carbonate complexes $As(CO_3)^{2-}$, $As(CO_3)(OH)^{2-}$, and $AsCO^{3+}$ in the presence of high concentration of $CO_3^-$ [42].

*5.1.5. Chloride and Nitrate.* Jeong et al. [60] showed that chloride and nitrate anions do not have visible effects on the adsorption of As(V) on $Fe_2O_3$ and $Al_2O_3$ [60]. This was due to the fact that complexes of chloride and nitrate with $Fe_2O_3$ and $Al_2O_3$ are much weaker than those between arsenate and $Al_2O_3$ or $Fe_2O_3$. Similar result was obtained by [61], where nitrate ions were increased from 1 to 10 $mgL^{-1}$ concentrations on the arsenite removal by Fe(VI), Fe(VI)/Fe(III), and Fe(VI)/Al(III) salts, due to lack of competition between $NO_3^-$ and the adsorbent for adsorption sites [61].

*5.2. Effect of Cations.* $Ca^{2+}$ was reported to significantly enhance the adsorption capacity of modified red mud at pH 7.3 [69]. Increasing $Ca^{2+}$ concentration from 0 mg/L to

40 mg/L resulted in an increase of 1 mg/g of arsenate adsorbed. $Ca^{2+}$ was claimed to bond the modified red mud particle with arsenate and form a metal–arsenate complex or metal–$H_2O$–arsenate complex [69]. Effects of $Mg^{2+}$, $Ca^{2+}$, and $Fe^{2+}$ on arsenate As(V) and arsenite AS(III) were studied on nano zerovalent iron on activated carbon, and it was reported that the cations considerably increase the adsorption of arsenate As(V) by increasing the pH [3]. The existence of metal cations in the solution shifted the surface of the adsorbent to more positively charged nature, which on the other hand allowed the adsorbent to show higher affinity for arsenate anions [3]. Meanwhile, $Fe^{2+}$ suppressed arsenite AS(III) adsorption by 4.1% at pH 3.5 and by 22% at pH 6.5 within the presence of $Fe^{2+}$ [3]. The deprotonation of $H_3AsO_3$ at pH < 9 controlled the arsenite adsorption, whereas surface charge was not important. $Fe^{2+}$ ions form complexes with arsenite in aqueous solution, and, as a result, the degree of deprotonation/dissociation is repressed, and the adsorption was reduced [3]. As(III) removal was increased with an increase of $Mg^{2+}$, $Ca^{2+}$, and $Fe^{2+}$ concentration on iron oxide-coated cement (IOCC): the enhancement of As(III) adsorption onto the adsorbent surface was attributed to the formation of a surface species, whereby $Mg^{2+}$, $Ca^{2+}$, and $Fe^{2+}$ ions cause the negative charges to be weak and act as a link between the adsorbent surface and the As(III) ions, as it was reported in the study conducted by Kundu and Gupta [70].

*5.3. Effect of Organic Matter.* Natural organic matter (NOM), which is a carbon source, comprises the combination of acidic organic molecules, which does not originate from a diversity of natural sources including sediments, water, and soil [126]. NOM concentration in natural waters ranges between 1 and 50 mg/l and may compete with arsenic for sorption sites [127]. The theory was proven by Guan et al. [56] on As(III) removal by $KMnO_4$–Fe(II), whereby, due to the presence of the NOM, the adsorption of other solute decreased by competing for adsorption sites; hence the adsorption sites were reduced [56]. However, it was reported that the NOM adsorption for magnetite nanoparticles is through the interaction of organic functional groups OH and COOH and ligand exchange of surface hydroxyl groups, electrostatic attraction, and hydrogen bonding [127]. NOM has been found to assist in the As(III) oxidation to As (V) under alkaline conditions in the absence of both $O_2$ and light [128]. It was observed that 8 mg/L of NOM reduces the adsorption of arsenic, whereas the adsorption increases with increasing pH in the pH range of 4.0–9.4 [128]. Another study on the adsorption of As(V) on alumina in the presence of fulvic acid (FA) showed that, at a pH between 3 and 7.5, FA reduces As(V) adsorption due to attraction forces and negatively charged surfaces were formed because of the deprotonation of functional groups at pH above pKa. Arsenic removal in the $KMnO_4$–Fe(II) process was reported not to be affected at pH 4, but at pH 5 in 1 mg/L or 4 mg/L humic acid (HA), the adsorption dropped by 9% [56]. Arsenic removal was reported to be decreased by 24.8% at pH of 6 and by 58.4% at pH 7 in the presence of 1 mg/L HA [56].

*5.4. Effects of Ionic Strength on Arsenic.* The ionic strength has the ability to affect the binding of the adsorbed species [1, 54] and therefore compete for adsorption sites. By determining the effects of ionic strength, the inner-sphere and outer-sphere ion-surface complexes can be distinguished [129]. Outer-sphere complexes are predictable to be more vulnerable to ionic strength variations than inner-sphere complexes because the background electrolyte ions are positioned in the same plane for outer-sphere complexes [1]; therefore a decrease in the adsorption is observed when conversely the electrolyte concentration is increased due to competition for adsorption sites [130]. Ions that form inner-sphere complexes are straight synchronized to surface groups and may not compete or compete at lower percentage with electrolyte ions [130]. Thus, the adsorption is less affected by changing the ionic. In many cases of inner-sphere complex formation, the adsorption increases with increasing electrolyte concentration [130]. This effect is usually attributed to changes in the electric potential in the interface, whereby the electrostatic repulsion between the charged surface and the anion is decreased, and the adsorption is favored [130].

The effect of ionic strength on the As(V) sorption was studied at 0.02 and 0.15 mol dm$^{-3}$ NaCl solutions goethite and jarosite and the result shows that As(V) sorption is independent of ionic strength [92]. The researchers suggested that As(V) adsorption on goethite could proceed via the formation of inner-sphere surface. References [92, 131] studied the effects of ionic strength in magnetite nanoparticles on arsenate and arsenite adsorption, and it was reported that the adsorption decreased by 4% by increasing the ionic strength from 0.01 to 0.1. Guo et al. [99] investigated the effect of background electrolyte (NaCl) concentration on synthetic siderite with 0.001–0.1 mol/L NaCl as background electrolyte. The result showed that As(V) removal was not affected by (NaCl). The increase of ionic strength was reported to increase As(V) adsorption onto $TiO_2$ under alkaline conditions (pH 7.0–11.0). However, under acidic conditions, the adsorption of As(V) onto $TiO_2$ decreased with increasing ionic strength in NaCl electrolyte [126]. Therefore ionic strength is also affected by pH of the solution.

# 6. Fluoride ($F^-$) Adsorption Isotherm and Adsorption Capacity

The adsorption behavior of fluoride by various adsorbents varies depending on the bonding between fluoride species and active sites on the surface of the adsorbent [132]. From the various adsorbents compared (Table 2), the Langmuir and Freundlich equations were the best fit for many adsorbents. The highest adsorption capacity for fluoride removal (400 mg/g) was observed on hydroxyl aluminum oxalate with a surface area of 68.34 m$^2$/g [133] and (600 mg/g) for aluminum fumarate with a surface area of 1156 m$^2$/g [134]. There is no general conclusion on how the type of adsorbent fits the chosen isotherm, since the extent of adsorption depends on many factors including the nature of the adsorbate and adsorbent, surface area, activation of the adsorbent, and experimental conditions. The isotherm may differ on the

TABLE 2: Adsorption isotherm for fluoride ($F^-$) removal.

| Adsorbent | Adsorption capacity mg/g | Surface area m²/g | Applicable isotherm models | Initial fluoride conc. mg/L | References |
|---|---|---|---|---|---|
| Acidic alumina | 8.4 | 144.27 | Langmuir | 15 | [32] |
| Alkoxide origin alumina | 2 | 100 | Langmuir | 5 | [4] |
| Al(III) modified calcium hydroxyapatite | 32.57 | 258.6 | Langmuir | 10 | [146] |
| Bauxite | 5.16 | 38 | Langmuir | 4–24 | [147] |
| Ceramic | 2.16 | 80.94 | Freundlich and Langmuir | 10 | [40] |
| Chitosan-based mesoporous alumina | 8.264 | 413.65 | Langmuir | 5 | [66] |
| Granular red mud | 0.851 | 10.2 | Redlich–Peterson, Freundlich | 15 | [148] |
| Calcite | 0.39 | 0.057 | Freundlich | $2.5 \times 10^{-5}$ | [149] |
| Quartz | 0.19 | 0.06 | Freundlich | $t0$ | [149] |
| Fluorspar | 1.79 | 0.048 | Freundlich | $6.34 \times 10^{-2}$ | [149] |
| Hydroxyapatite | 4.54 | 0.052 | Freundlich | — | |
| Activated quartz | 1.16 | 0.06 | Freundlich | — | [149] |
| Nano alumina | 14 | 151.7 | Langmuir | — | [29] |
| Magnesia-amended activated alumina granules | 10.12 | 193.5 | Sips (S) | — | [150] |
| Manganese oxide-coated alumina | 2.85 | 170.39 | Langmuir | 2–30 | [151] |
| Mesoporous alumina (meso-Al-400) | 39 | 361 | Langmuir | 10 | [31] |
| Cerium-impregnated fibrous protein | 17.5 | 3.65 | Langmuir | — | [152] |
| Aluminium titanate (AT) | 0.85 | — | Freundlich, Langmuir | 4 | [153] |
| Bismuth aluminate (BA) | 1.55 | — | Freundlich, Langmuir | 4 | [153] |
| Alum-impregnated activated alumina (AIAA) | 40.3 | 176 | Bradley equation | — | [154] |
| Alumina/chitosan (AlCs) composite | 3.809 | 55.23 | Freundlich and Langmuir | — | [155] |
| Alumina cement granules | 4.75 and 3.91 | 4.385 | Langmuir | 20 8.65 | [23] |
| Aluminum hydroxide-coated rice husk ash (AH-coated RHA) | 15 | 50.4 | Freundlich | 10 | [156] |
| Aluminum hydroxide, $Al(OH)_3$ | 25 | 50.4 | Freundlich | 10 | [156] |
| Activated alumina | 2.41 | — | Langmuir | — | [39] |
| Hydrated cement | 2.6788 | — | Freundlich, Langmuir | — | [157] |
| Waste mud | 27.2 | — | Langmuir | 5.0 and 950 | [64] |
| Sol-gel-derived activated alumina, CaO-AA | 96.23 | 255.42 | Langmuir | 0.99 | [135] |
| Sol-gel-derived activated alumina, $MnO_2$-AA | 0.99 | 218.0 | Langmuir | 432 | [135] |
| Aluminum impregnation of activated carbon | 2.549 | — | Langmuir | — | [158] |
| Brushite | 6.59 | — | Langmuir and Freundlich | 25 | [159] |
| Ceramic | 2.16 | 80.94 | Freundlich and Langmuir isotherms | 10 | [40] |
| Granular ceramic | 12.12 | 73.67 | Freundlich | — | [40] |
| Ca650/C charcoal that contains calcium compounds | 19.05 | – | Langmuir | 10 | [160] |

TABLE 2: Continued.

| Adsorbent | Adsorption capacity mg/g | Surface area m²/g | Applicable isotherm models | Initial fluoride conc. mg/L | References |
|---|---|---|---|---|---|
| Titanium dioxide (TiO₂) | 0.27 | 56. | Temkin, Dubinin–Radushkevich (DR) | — | [68] |
| Magnetic-chitosan particle | 20.96–23.98 | — | Langmuir, Bradley's | — | [161] |
| Mg–Al–Fe hydrotalcite-like compound | 14 | — | Langmuir | — | [65] |
| Mixed rare earth oxide | 12.5 | 6.75 | Langmuir | 50 | [162] |
| Plaster of Paris | 0.366 | — | Freundlich and Langmuir | 2–10 | [163] |
| Polypyrrole/Fe₃O₄ magnetic nanocomposite | 17.6–22.3 | — | Freundlich and Langmuir | — | [164] |
| Cross-linked chitosan particles | 8.1 | 4.37 | Freundlich and Langmuir | 11.8 to 59.0 | [165] |
| Carbon derived from *Sargassum* sp. by lanthanum | 94.34 | | Langmuir | | [166] |
| Hydroxyl aluminum oxalate | 400 | | Langmuir | | [133] |
| Hierarchical Ce–Fe bimetal oxides | 60.97 | 164.9 m²/g | Langmuir | | [167] |
| Hydrous ferric oxide doped alginate beads | 8.90 | 25.80 | Langmuir | | [168] |
| Al(III)–Zr(IV) binary oxide | 114.54 mg/g | | Langmuir | | [169] |
| Hierarchical MgO microspheres | 115.5 mg/g | 33.7 | Freundlich model | | [143] |
| Micron-sized magnetic adsorbent (MMA) | 41.8 mg/g | | Langmuir | 200 mg/L | [170] |
| Hydroxyapatite nanowires | 40.65 mg/gat | | | 200 mg/L | [171] |
| Aluminium fumarate | 600 | 1156 m²/g | Freundlich isotherm | 1000 | [134] |

same fluoride concentration according to the type of the adsorbents. For example, [135] reported that Freundlich fitted the sol-gel-derived activated alumina modified with calcium oxide and Langmuir sol-gel-derived activated alumina modified with magnesium oxide.

One of the most widely used adsorbents on fluoride removal is activated alumina and it has widely been applied for fluoride adsorption [22]. However, the adsorption capacity of fluoride on activated alumina depends on pH at large and not on the surface areas of the adsorbent. This can be proven in Figure 4, whereby different adsorbents capacities were compared with their surface area. However, modified activated alumina has shown a considerate higher adsorption capacity compared to unmodified alumina even at high fluoride concentration. For example, [135] reported the removal of fluoride by activated alumina up to 96.23 mg/g for water with a fluoride concentration of 432 mg/L in a sol-gel-derived activated alumina modified with calcium oxide. The reason for the higher adsorption capacity was the alkalinity of the adsorbent and the technique used for modification of the activated alumina. Another study by using sol-gel alumina adsorbents was done on modified immobilized activated alumina (MIAA), and adsorption capacity of 0.76 mg/g was reported compared to 0.47 mg/g for activated charcoal on fluoride concentration of 12 mg/L [136]. Another comparison was made between alkoxide origin alumina and activated alumina on fluoride removal, and higher fluoride uptake

was seen with alkoxide alumina because of $Fe_2O_3$ and $SiO_2$, activated carbon pores, and the increased electropositivity of the material [4]. The effectiveness of modified material was also reported by [137] on magnetite modified with aluminum and lanthanum ions for the adsorption of fluoride. The adsorption of fluoride was reported to increase by 60% and 66% for the aluminum and lanthanum modified materials, respectively [137].

Fourier transform infrared (FTIR), scanning electron microscopy (SEM), and Extended X-ray Absorption Fine Structure (EXAFS) spectroscopy are used to clarify the mechanism of fluoride adsorption [133]. Most of the FTIR studies on fluoride removal by different adsorbents have shown the involvement and the participation of the surface sites and potential functional groups in the adsorption, especially the hydroxyl group [138, 139]. It was observed that fluoride interacts with the OH and NH groups on surface aluminum-impregnated coconut fiber surface [140]. However, the adsorption of fluoride on Fe-impregnated chitosan was reported to occur due to ion exchange between fluoride and chloride [141]. Other studies using FTIR have revealed the formation of inner-spherically bonded complexes on $\gamma$−Fe2O3 nanoparticles with fluoride [142]. By using FTIR and XPS analyses, it was concluded that hydroxyl and the surface carbonates participate in the coexchange with fluoride ions on hierarchical magnesium oxide microspheres [143] and on magnesium oxide nanoplates adsorbents [144].

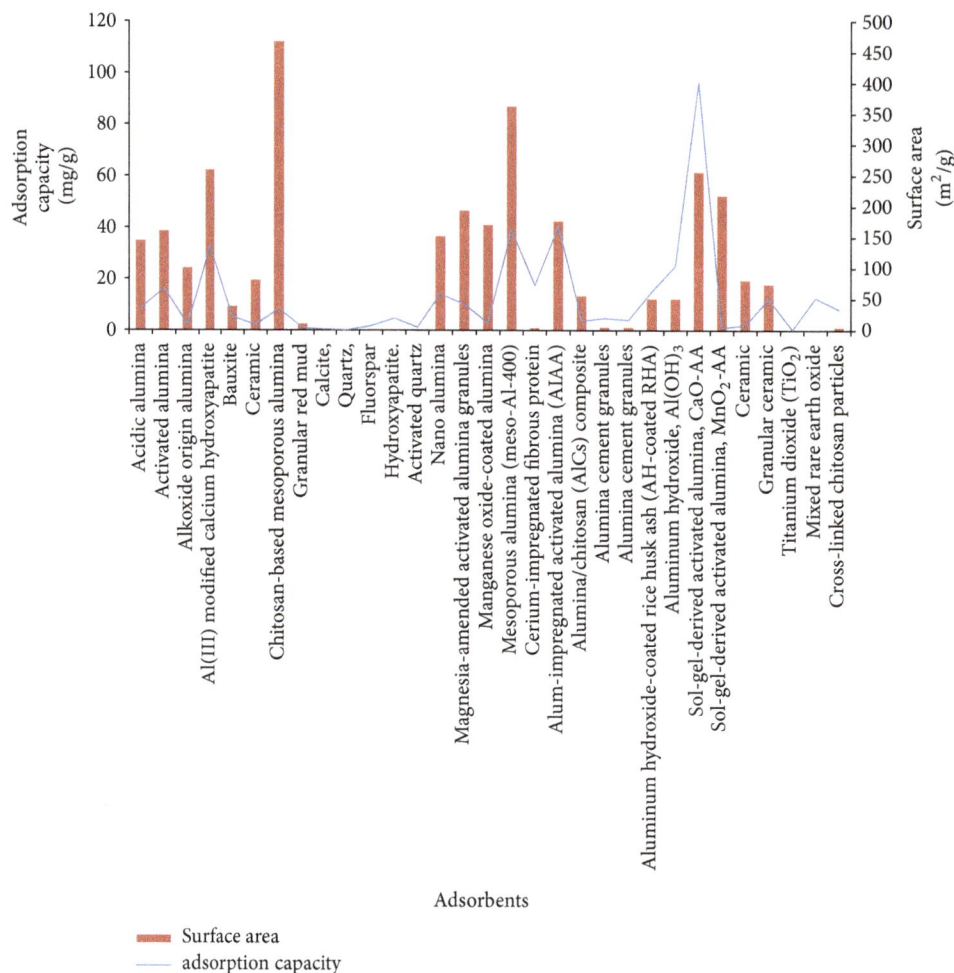

FIGURE 4: Relationship between the adsorption capacity and surface area of the adsorbents for fluoride removal.

Another study revealed that the coexistence of C–O and C=O functional groups on lanthanum-loaded magnetic cationic hydrogel composite contributed to the fluoride adsorption by using FTIR [145].

Both Langmuir and Freundlich isotherm seem to fit well for adsorptions of these anions. However, the choice of the adsorbents to agree with Langmuir or Freundlich is still unclear. Higher adsorption capacity has been observed for fluoride ions compared to arsenic ions. Both modified adsorbents have been shown to have higher adsorption capacity compared to unmodified adsorbents, and, up to date, iron-based adsorbents (IBS) are still suitable material for arsenic adsorption and aluminum for fluoride.

This is the first report to show the relationship between the adsorption capacity and surface area of the adsorbents for arsenic and fluoride removal. It can be concluded that high surface area of the adsorbents does not often cause higher adsorption capacity on these anions' removal but rather the chemical properties of the adsorbents. Modified adsorbents have been proven to be effective for the anions removal because modifications of adsorbents alter the physical and chemical properties of the adsorbents. For example, Gong et al. [132] studied different types of alumina and concluded

that the acidity and basicity properties of the alumina significantly affect the fluoride adsorption and not the surface area.

Chen et al. [172] reported that adsorption of fluoride on $PPy/TiO_2$ was not affected by surface area. $PPy/TiO_2$ has a surface area of $95.71 \, m^2/g$, and its maximum adsorption capacity was $33.178 \, mg/g$ at $25°C$ with an initial fluoride concentration of $11.678 \, mg/L$. The author concluded that physical adsorption was not the primary adsorption mechanism for the adsorption of fluoride on $PPy/TiO_2$. Another study on adsorption of fluoride by a synthetic iron(III)–aluminum(III) mixed oxide with a specific surface area of $195.6 \, m^2/g$ reached a maximum fluoride removal of $17.73 \, mg/g$ [173]. Zhang et al. [170] investigate the fluoride adsorption on micron-sized magnetic $Fe_3O_4@Fe$-Ti composite adsorbent (MMA) with a specific surface area of $99.2 \, m^2/g$ and pore size of $0.38 \, cm^3/g$. The model followed Langmuir in isotherm study, and a high maximum adsorption value of $41.8 \, mg/g$ was observed [170]. Adsorption of fluoride on aluminum-impregnated coconut fiber ash (AICFA) was investigated by Mondal et al. [140]. AICFA has a surface area of $26.3 \, m^2/g$ and $3.192 \, mg/g$ adsorption capacity. Another study reported $40.3 \, mg/g$ fluoride removal from drinking water by adsorption onto alum-impregnated activated alumina with a surface area of $176 \, m^2/g$ [154]. Low adsorption capacity for adsorbents with the higher surface area was also reported. For example, a study on fluoride adsorption into nanoparticle goethite anchored onto graphene oxide (FeOOH + Ac/GO) with $202.60 \, m^2/g$ reached a maximum adsorption capacity of 17.65 mg/g. Meanwhile, a rice spike like akaganeite anchored onto graphene oxide (FeOOH/GO) which has $255.24 \, m^2/g$ reached 19.82 mg/g maximum adsorption capacity [174]. Another study reported a higher adsorption capacity of 93.84 mg/g when the fluoride concentration was 200 mg/L on $Al(OH)_3$ nanoparticles modified hydroxyapatite (Al-HAP) nanowires with a surface area of $104.05 \, m^2/g$ on fluoride removal [175].

## 7. Effects of pH on Fluoride Removal

The extent of fluoride adsorption is influenced by pH at large [4, 29]. Since the protonated surface is accountable for anions adsorption, the maximum fluoride adsorption for many adsorbents occurs at acidic pH [32, 63] and decreases at higher pH values. For adsorbents like activated alumina, fluoride adsorption is controlled by pH at point zero charges (pHzpc) [32, 73]. At a certain selected pH, the adsorption showed an increasing trend due to the positively charged alumina complexes $AlF^{2+}$ and $AlF_2^+$ on fluoride removal by acidic alumina [32]. The fluoride adsorption decreased after pHzpc because the concentration of protonated surface sites decreases with increasing pH [4, 73], which causes strong competition of hydroxide ions [57]. A similar observation was reported by [66] on defluoridation of drinking water using chitosan-based mesoporous alumina. Nie et al. [146] reported maximum fluoride removal of 75% at pH of 5 and the adsorption decreased further as the pH increases on Al (III) modified calcium hydroxyapatite.

The adsorption of fluoride decreases below or above a certain pH value, as it has been reported by various researchers. For example, Li et al. [31] studied the effect of pH on amorphous alumina supported on carbon nanotubes. The result showed that the maximum fluoride adsorption occurred at pH 5.0–9.0 but decreased at pH less than 3.0 and more than 11.0. However, the distribution of $F^-$ and HF on the surface was reported to be the reason for a decrease of fluoride at pH less than 5 on fluoride removal by aluminum-impregnated coconut fiber [140]. Similar results were reported by Tang et al. [63] on the adsorption of fluoride by granular ferric hydroxide, where at pH below 3, the HF were predominant.

Fluoride adsorption capacity was steady within a pH range of 2–11 on porous MgO nanoplates; however, it decreases abruptly at higher pH value above 12 [31, 144]. Tor et al. [148] also reported similar results in the removal of fluoride from water by using granular red mud (GRM). However, other studies reported maximum fluoride removal at neutral pH [29]. Huang et al. [165] observed that the maximum adsorption capacity by using protonated crosslinked chitosan particles was at pH 7. Moreover, a decrease below pH 7 and above was also observed on iron oxyhydroxide nanoparticles [176]. In acidic conditions, the decrease in

adsorption capacity was caused by weak hydrofluoric acid. A similar observation was made by [177] on magnesium substituted hydroxyapatite absorbent at a pH below 3 by using different adsorbents as it was reported by [151, 154, 159, 161].

pH is the main parameter that affects the adsorption of these anions for both arsenic and fluoride. However, the adsorption of fluoride into various adsorbents has been reported to vary significantly at both high and low pH values due to the pH at point zero charges (pHzpc) which controls the adsorptions of adsorbents like activated alumina in fluoride adsorption and iron-based adsorbents for arsenic adsorption. The adsorption capacity of arsenic has been changed with changes in pH due to the chemical speciation of arsenic. Many adsorbents tend to remove charged species, which is As(V), which can exist at a wide range of pH compared to As(III).

## 8. Effects of Anion, Cation, and Organic Matter on Fluoride Removal

*8.1. Effects of Anions.* Drinking water contaminated with fluoride always exists with other coions like phosphate, bicarbonate, chloride, carbonate, sulfate, and nitrate [40, 62, 63, 65, 66]. In fluoride adsorption, it is expected that the presence of anions in solution would enhance coulombic repulsion forces between the anions and fluoride or would compete with fluoride for the active adsorption sites and therefore fluoride adsorption is reduced or increased [66, 135, 136, 178, 179]. Eskandarpour et al. [180] reported that, in fluoride adsorption, chloride and nitrate ions form outer-sphere complexes and sulfate and phosphate ions form inner-sphere complexes with binding surfaces. However, sulfate ions partially form outer-sphere complexes or inner-sphere complexes [180]. The adsorption of fluoride will increase due to an increase in ionic strength of the solution and weakening of lateral repulsion between adsorbed fluoride ions. For example, Chen et al. [40] investigated the presence of 20–200 mg/L salt solutions of chloride, nitrate, sulfate, carbonate, and phosphate in fluoride adsorption. Fluoride removal slightly increased in the presence of chloride and nitrate ions. The authors concluded that it was due to an increase in the ionic strength of the solution or weakening of lateral repulsion between adsorbed fluoride ions. Fluoride sorption was slightly decreased by sulfate ion due to the high coulombic repulsive forces, which reduce the probability of fluoride interactions with the active sites [40, 179].

Carbonate and phosphate ions showed a most significant effect on fluoride sorption due to the competition for the same active sites with fluoride and the high affinity and capacity for carbonate and phosphate ions [40]. The divalent nature of sulfate ion in solution may influence strong coulombic repulsive forces which lead to lessening fluoride interaction with the active sites. Similar results were reported by [38, 178, 179]. The effects of these anions on fluoride adsorption have also been reported to be different depending on experiment conditions which include pH, the anions concentrations, and the characteristics of the adsorbent. Recently, Zhang et al. [141] reported that the increasing concentrations of sulfate and carbonate to 200 mg/L cause a decrease in the

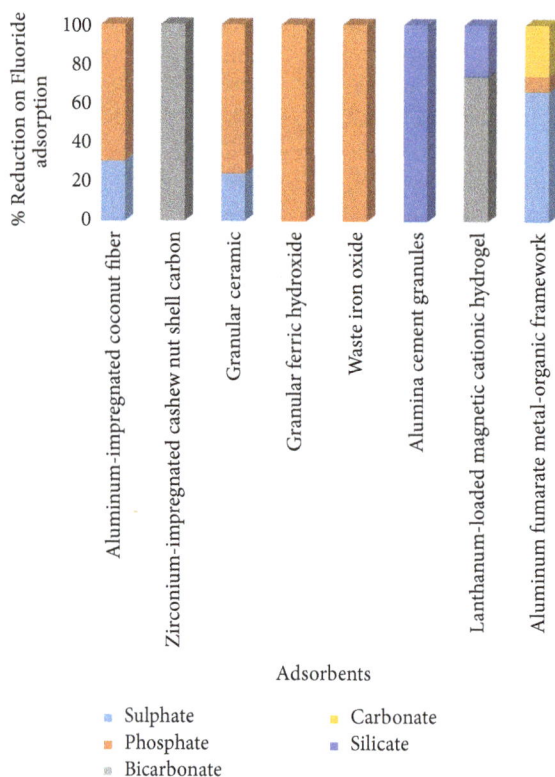

FIGURE 5: Effects of coexisting anions on fluoride removal on different adsorbents.

fluoride adsorption by 43.24% and 18.77%, respectively, while 100 mg $Cl^-$/L increases the fluoride adsorption capacity of 0.58 mg/g by 2.53 mg/g with 5 mg of $HCO_3$. For the former, the decrease was because of the electrostatic interactions of the anions and the adsorbent, and for the latter, the reaction of bicarbonate with ammonium acetate leads to the decrease of the ammonium acetate concentration.

Other researchers have reported a decrease depending on pH. At pH between 4.4 and 12, 500 mg/L of carbonate and bicarbonate ions concentration was reported to decrease the fluoride adsorption on acidic alumina [32]. Fluoride adsorption was reduced significantly as bicarbonate was acting as a pH buffering agent, and its existence in solution increases and buffered the pH; therefore, the adsorption of fluoride was decresed [4, 155, 166] Kamble et al. [4] reported that the addition of $CO_3^{2-}$, $SO_4^{2-}$, and $HCO_3^-$ ions increases pH of fluoride solution in adsorption of fluoride on alumina of alkoxide nature. Similar results were observed by [29, 147, 165, 181]. Adsorption of fluoride is more affected by the presence of phosphate (Figure 5) compared to other coexisting anions.

Changing of pH to alkaline condition (11.5) was reported to decrease the adsorption capacity of fluoride by more than 25% in the presence of carbonate [134]. The decrease in fluoride adsorption observed by the presence of $HCO_3^-$, $CO_3^{2-}$, and $SO_4^{2-}$ was due to the competition for active sorption sites or due to the change in pH or combination of them [4, 29, 147, 160, 161, 163–167, 181], where the decrease of fluoride was related to anions competitions or effects of pH;

however, in the same study, it was reported that sulfate did not have significant effect on fluoride removal [157]. A similar study was reported on MgO nanoplates [144].

The coexisting anions silicate and phosphate have been reported to have a negligible effect on the fluoride removal on a micron-sized magnetic $Fe_3O_4$@Fe-Ti composite adsorbent compared to 160 mg/L $SiO_3^{2-}$ and 20 mg/L $PO_4^3$ which decreased the adsorption efficiency [170]. Effects of phosphate on fluoride removal were also reported by [146]. The tendency of the anions to form inner-sphere complexes was reported by [28, 29, 32, 63, 170, 174, 175, 177, 178, 180–183]. Chloride ions formed outer-sphere surface complexes and had a minor effect on fluoride adsorption; thus they are less absorbed on the absorbent surface [65]. Anion affinity for a surface site is associated with the physicochemical characteristics of the adsorption mechanism [40, 147, 179]. Another study demonstrated effects of coexisting ions on the removal of fluoride due to changing pH after salt addition and not the competitive adsorption of coexisting ions [184].

8.2. Effects of Cations. Only a few studies to date have reported the effects of cations on fluoride removal. At higher cations concentrations, effects become significant. Contradictory results have been presented for cations removal. Some researchers reported increased fluoride adsorption, while others reported a decrease in fluoride adsorption. An increase in adsorption capacity was observed with a higher concentration of magnesium and calcium due to an increase in surface positive charges and attraction of negatively charged ions onto various metal oxide surfaces on fluoride removal by nano magnesia (NM), as shown in the study conducted by Maliyekkal et al. [62]. Similar effects were observed by [157] at 400 mg/L of $Ca^{2+}$ by using hydrated cement and on alumina cement granules [23]. Also, the formation of insoluble $CaF_2$ and $MgF_2$ was the reason for the increase of adsorption by granular ceramic at 200 mg/L and $Ca^{2+}$ and $Mg^{2+}$ concentration [40]; however, $Mn^{2+}$ and $Fe^{3+}$ at 200 mg/L were reported to decrease fluoride removal capacity up to 10% on nano magnesia [62]. Babaeivelni and Khodadoust [68] reported that the effects of calcium and magnesium ions on fluoride removal onto crystalline titanium dioxide were insignificant.

*8.3. Effects of Organic Matter on Fluoride Adsorption.* Natural organic matter (NOM) molecules consist of combinations of functional groups, including ester, phenolic, amino, nitroso, carboxylic, sulfhydryl, quinone, and hydroxyl, and most of them are negatively charged at neutral pH [23]. Thus NOM can compete with fluoride due to the principal anionic character together with high reactivity on both metals and surfaces [23]. The presence of NOM influences the fluoride adsorption either by decreasing the adsorption or increasing it. Only a few studies have presented the influence of NOM on fluoride adsorption, and the results presented vary at large.

The presence of humic acid was reported to cause a 50% reduction in fluoride removal efficiency by manganese oxide-coated alumina as an adsorbent; this was due to the blocking of active $MnO_2$ sites by the larger humic acid molecule [151]. According to Ayoob and Gupta [23], the adsorption by using alumina cement granules was reduced with increasing NOM concentration because the adsorption capacity was dependent on the availability of the organic molecules and the inner surface of the adsorbent; thus the NOM can access mesopores of the adsorbent and the small molecules can access micropores. However, a study on carbon derived from *Sargassum* sp. by lanthanum indicated that there were insignificant effects on the adsorption of fluoride because the humic acid can be adsorbed on the surface of the activated carbon [166]. Samarghandi et al. [185] reported that, with 20 mg/L of NOM, fluoride adsorption by activated alumina was enhanced at a pH value of 5.5 to 6.

Adsorption of fluoride is more affected by the presence of phosphate compared to other coexisting anions. However, for arsenic adsorption, the influence of phosphate and other existing anions is affected by pH which controls the effects of arsenic to be adsorbed by the adsorbents if they are present in the water. Ionic strength can also interfere with the adsorption of these anions in the presence of other existing anions, and physicochemical characteristics of the adsorption mechanism affect the anion affinity for a surface site. Cations like $Mg^{2+}$, $Ca^{2+}$, and $Fe^{2+}$ have been reported to increase the adsorption of the fluoride and arsenic by using different adsorbents. The presence of NOM influences the fluoride adsorption either by decreasing the adsorption or increasing it. However, pH also affects arsenic adsorption if NOM are also present by decreasing or increasing the extent of arsenic adsorption into the adsorbents.

*8.4. Effects of Ionic Strength Fluoride.* So far, very few studies have been done on the effects of ionic strength on fluoride removal. However, the importance of ionic strength lies in the fact that it can be used to distinguish whether inner-sphere or outer-sphere surface complexes are formed during the adsorption. It has been related to pH of the solution and characteristics of the adsorbent. Ionic strength has been reported to have insignificant effects on the fluoride removal by micron-sized magnetic $Fe_3O_4$@Fe-Ti composite and was confirmed to form an inner-sphere complex on the surface of the adsorbent [170]. However, the fluoride adsorption was inhibited by the increase of ionic strength, and therefore the adsorption process occurred through the outer-sphere complex adsorption mechanism on fluoride removal by carbon derived from *Sargassum* sp. by lanthanum [166]. Similar results were reported by [73] on fluoride adsorption onto granular ferric hydroxide and [73] on the adsorption of fluoride on activated alumina.

In general, presence of ionic strength can affect the adsorption of both arsenic and fluoride through the formation of inner-sphere or outer-sphere surface complexes. When the adsorption occurs through the inner sphere, there is less competition for adsorption sites; therefore, increasing the salt concentrations increases the adsorption. However, the ionic strength is also affected by pH of the solution when present with arsenic and fluoride anion in water.

# 9. Conclusion

Adsorption of both arsenic and fluoride is largely affected by environmental factors such as pH solution, ionic strength, and coexisting substances such as anions, cations, and organic matter. The adsorption capacity of the adsorbents depends not only on the surface area, pore volume, and particles size but also on a combination of all factors, surface chemistry, and pore structure. Specific area of the adsorbents does not contribute to the adsorption capacity on the removal of these anions from water. Therefore selection of the adsorbents for these anions removal should be based on a combination of all factors for the adsorbents and adsorbate. However, surface modification of the adsorbents increases the adsorption capacity of the adsorbents for removal of these anions due to the presence of more protonated surface sites which favor the removal of these anions. In general, pH has a greater influence on adsorption capacity at large, since it affects the ionic strength and the presence of coexisting anions such as bicarbonate, sulfate, and silica and it affects the surface charges of the adsorbents and the ionic species which can be present in the solution. Modified aluminum adsorbents have shown higher adsorption capacity for fluoride and modified iron oxide and aluminum minerals for arsenic. The adsorption of fluoride and arsenic is inhibited by the presence of phosphate followed by sulfate and silicate. Fourier transform infrared (FTIR) and Extended X-ray Absorption Fine Structure (EXAFS) spectroscopy have revealed new functional groups on adsorbents surface, which participate in the adsorption of arsenic and fluoride.

## Disclosure

This research did not receive any specific grant from funding agencies in the public or commercial or nonprofit organizations.

## Conflicts of Interest

The authors declare that they have no conflicts of interest.

## References

[1] O. Ajouyed, C. Hurel, M. Ammari, L. B. Allal, and N. Marmier, "Sorption of Cr(VI) onto natural iron and aluminum (oxy)hydroxides: Effects of pH, ionic strength and initial concentration," *Journal of Hazardous Materials*, vol. 174, no. 1-3, pp. 616–622, 2010.

[2] S. M. Maliyekkal, L. Philip, and T. Pradeep, "As(III) removal from drinking water using manganese oxide-coated-alumina: Performance evaluation and mechanistic details of surface binding," *Chemical Engineering Journal*, vol. 153, no. 1-3, pp. 101–107, 2009.

[3] H. Zhu, Y. Jia, X. Wu, and H. Wang, "Removal of arsenic from water by supported nano zero-valent iron on activated carbon," *Journal of Hazardous Materials*, vol. 172, no. 2-3, pp. 1591–1596, 2009.

[4] S. P. Kamble, G. Deshpande, P. P. Barve et al., "Adsorption of fluoride from aqueous solution by alumina of alkoxide nature: Batch and continuous operation," *Desalination*, vol. 264, no. 1-2, pp. 15–23, 2010.

[5] A. Maiti, H. Sharma, J. K. Basu, and S. De, "Modeling of arsenic adsorption kinetics of synthetic and contaminated groundwater on natural laterite," *Journal of Hazardous Materials*, vol. 172, no. 2-3, pp. 928–934, 2009.

[6] B. K. Mandal and K. T. Suzuki, "Arsenic round the world: a review," *Talanta*, vol. 58, no. 1, pp. 201–235, 2002.

[7] USEPA, *Arsenic Occurrence in Public Water Supplies*, epa-815-r-00-023, USEPA, Washington, DC, USA, 2000.

[8] C. Oh, S. Rhee, M. Oh, and J. Park, "Removal characteristics of As(III) and As(V) from acidic aqueous solution by steel making slag," *Journal of Hazardous Materials*, vol. 213-214, pp. 147–155, 2012.

[9] Y. Lee, I.-H. Um, and J. Yoon, "Arsenic (III) oxidation by iron(VI) (ferrate) and subsequent removal of arsenic (V) by iron (III) coagulation," *Environmental Science & Technology*, vol. 37, no. 24, pp. 5750–5756, 2003.

[10] Y. Wu, X. Ma, M. Feng, and M. Liu, "Behavior of chromium and arsenic on activated carbon," *Journal of Hazardous Materials*, vol. 159, no. 2-3, pp. 380–384, 2008.

[11] J. Youngran, M. FAN, J. Van Leeuwen, and J. F. Belczyk, "Effect of competing solutes on arsenic (V) adsorption using iron and aluminum oxides," *Journal of Environmental Sciences*, vol. 19, no. 8, pp. 910–919, 2007.

[12] M. Chiban, M. Zerbet, G. Carja, and F. Sinan, "Application of low-cost adsorbents for arsenic removal: A review," *Journal of Environmental Chemistry and Ecotoxicology*, vol. 4, no. 5, pp. 91–102, 2012.

[13] C. K. Jain and R. D. Singh, "Technological options for the removal of arsenic with special reference to South East Asia," *Journal of Environmental Management*, vol. 107, no. 1, pp. 1–18, 2012.

[14] S. Mandal, M. K. Sahu, and R. K. Patel, "Adsorption studies of arsenic(III) removal from water by zirconium polyacrylamide hybrid material (ZrPACM-43)," *Water Resources and Industry*, vol. 4, pp. 51–67, 2013.

[15] Y. Glocheux, M. M. Pasarín, A. B. Albadarin, S. J. Allen, and G. M. Walker, "Removal of arsenic from groundwater by adsorption onto an acidified laterite by-product," *Chemical Engineering Journal*, vol. 228, pp. 565–574, 2013.

[16] D. Mohan and C. U. Pittman Jr., "Arsenic removal from water/wastewater using adsorbents—a critical review," *Journal of Hazardous Materials*, vol. 142, no. 1-2, pp. 1–53, 2007.

[17] A. Bhatnagar, E. Kumar, and M. Sillanpää, "Fluoride removal from water by adsorption—a review," *Chemical Engineering Journal*, vol. 171, no. 3, pp. 811–840, 2011.

[18] WHO, Fluoride in drinking-water. WHO/SDE/WSH/03.04/96, 2004.

[19] S. J. Mahmood, N. Taj, F. Parveen, T. H. Usmani, R. Azmat, and F. Uddin, "Arsenic, Fluoride and Nitrate in Drinking Water: The Problem and its Possible Solution," *Research Journal of Environmental Sciences*, vol. 1, no. 4, pp. 179–184, 2007.

[20] Meenakshi and R. C. Maheshwari, "Fluoride in drinking water and its removal," *Journal of Hazardous Materials*, vol. 137, no. 1, pp. 456–463, 2006.

[21] WHO, Fluoride in drinking water. ISBN, 2006, 1900222965.

[22] M. Mohapatra, S. Anand, B. K. Mishra, D. E. Giles, and P. Singh, "Review of fluoride removal from drinking water," *Journal of Environmental Management*, vol. 91, no. 1, pp. 67–77, 2009.

[23] S. Ayoob and A. K. Gupta, "Performance evaluation of alumina cement granules in removing fluoride from natural and synthetic waters," *Chemical Engineering Journal*, vol. 150, no. 2-3, pp. 485–491, 2009.

[24] M. Grassi, G. Kaykioglu, V. Belgiorno, and G. Lofrano, "Removal of Emerging Contaminants from Water and Wastewater by Adsorption Process," in *Emerging Compounds Removal from Wastewater*, SpringerBriefs in Molecular Science, pp. 15–37, Springer Netherlands, Dordrecht, 2012.

[25] A. Dąbrowski, "Adsorption—from theory to practice," *Advances in Colloid and Interface Science*, vol. 93, no. 1–3, pp. 135–224, 2001.

[26] Y. S. Ho, J. F. Porter, and G. McKay, "Equilibrium isotherm studies for the sorption of divalent metal ions onto peat: copper, nickel and lead single component systems," *Water, Air, & Soil Pollution*, vol. 141, no. 1–4, pp. 1–33, 2002.

[27] V. M. C. a. K. J. Vardhan, "Removal of fluoride from water using low cost materials," in *Proceedings of the Fifteenth International Water Technology Conference*, vol. IWTC-15, 2011.

[28] W. Li, C.-Y. Cao, L.-Y. Wu, M.-F. Ge, and W.-G. Song, "Superb fluoride and arsenic removal performance of highly ordered mesoporous aluminas," *Journal of Hazardous Materials*, vol. 198, pp. 143–150, 2011.

[29] E. Kumar, A. Bhatnagar, U. Kumar, and M. Sillanpää, "Defluoridation from aqueous solutions by nano-alumina: Characterization and sorption studies," *Journal of Hazardous Materials*, vol. 186, no. 2-3, pp. 1042–1049, 2011.

[30] F. Di Natale, A. Erto, A. Lancia, and D. Musmarra, "Experimental and modelling analysis of As(V) ions adsorption on granular activated carbon," *Water Research*, vol. 42, no. 8-9, pp. 2007–2016, 2008.

[31] Y.-H. Li, S. Wang, A. Cao et al., "Adsorption of fluoride from water by amorphous alumina supported on carbon nanotubes," *Chemical Physics Letters*, vol. 350, no. 5-6, pp. 412–416, 2001.

[32] A. Goswami and M. K. Purkait, "The defluoridation of water by acidic alumina," *Chemical Engineering Research and Design*, vol. 90, no. 12, pp. 2316–2324, 2012.

[33] S. Chen, Q. Yue, B. Gao, and X. Xu, "Equilibrium and kinetic adsorption study of the adsorptive removal of Cr(VI) using modified wheat residue," *Journal of Colloid and Interface Science*, vol. 349, no. 1, pp. 256–264, 2010.

[34] USEPA, *Technologies and Costs for Removal of Arsenic from Drinking Water*, vol. EPA, 815-R-00-028., USEPA, Washington, DC, USA, 2000.

[35] S. Mor, K. Ravindra, and N. R. Bishnoi, "Adsorption of chromium from aqueous solution by activated alumina and activated charcoal," *Bioresource Technology*, vol. 98, no. 4, pp. 954–957, 2007.

[36] H. Demiral, I. Demiral, F. Tümsek, and B. Karabacakoğlu, "Adsorption of chromium(VI) from aqueous solution by activated carbon derived from olive bagasse and applicability of different adsorption models," *Chemical Engineering Journal*, vol. 144, no. 2, pp. 188–196, 2008.

[37] Y. Li, B. Gao, T. Wu et al., "Hexavalent chromium removal from aqueous solution by adsorption on aluminum magnesium mixed hydroxide," *Water Research*, vol. 43, no. 12, pp. 3067–3075, 2009.

[38] X. Xu, Q. Li, H. Cui et al., "Adsorption of fluoride from aqueous solution on magnesia-loaded fly ash cenospheres," *Desalination*, vol. 272, no. 1–3, pp. 233–239, 2011.

[39] S. Ghorai and K. K. Pant, "Equilibrium, kinetics and breakthrough studies for adsorption of fluoride on activated alumina," *Separation and Purification Technology*, vol. 42, no. 3, pp. 265–271, 2005.

[40] N. Chen, Z. Zhang, C. Feng et al., "An excellent fluoride sorption behavior of ceramic adsorbent," *Journal of Hazardous Materials*, vol. 183, no. 1-3, pp. 460–465, 2010.

[41] K. B. Payne and T. M. Abdel-Fattah, "Adsorption of arsenate and arsenite by iron-treated activated carbon and zeolites: effects of pH, temperature, and ionic strength," *Journal of Environmental Science and Health—Part A: Toxic/Hazardous Substances and Environmental Engineering*, vol. 40, no. 4, pp. 723–749, 2005.

[42] S. Zhang, H. Niu, Y. Cai, X. Zhao, and Y. Shi, "Arsenite and arsenate adsorption on coprecipitated bimetal oxide magnetic nanomaterials: MnFe$_2$O$_4$ and CoFe$_2$O$_4$," *Chemical Engineering Journal*, vol. 158, no. 3, pp. 599–607, 2010.

[43] K. B. Vu, M. D. Kaminski, and L. Nunez, "Review of arsenic removal technologies for contaminated groundwaters.," Tech. Rep. ANL-CMT-03/2, 2003.

[44] W. T. Tsai, C. W. Lai, and K. J. Hsien, "Effect of particle size of activated clay on the adsorption of paraquat from aqueous solution," *Journal of Colloid and Interface Science*, vol. 263, no. 1, pp. 29–34, 2003.

[45] H. Liu, J. Zhang, H. H. Ngo et al., "Effect on physical and chemical characteristics of activated carbon on adsorption of trimethoprim: Mechanisms study," *RSC Advances*, vol. 5, no. 104, pp. 85187–85195, 2015.

[46] S.-J. Park, Y.-S. Jang, J.-W. Shim, and S.-K. Ryu, "Studies on pore structures and surface functional groups of pitch-based activated carbon fibers," *Journal of Colloid and Interface Science*, vol. 260, no. 2, pp. 259–264, 2003.

[47] W. Shen, Z. Li, and Y. Liu, "Surface Chemical Functional Groups Modification of Porous Carbon," *Recent Patents on Chemical Engineering*, vol. 1, no. 1, pp. 27–40, 2010.

[48] V. Tomar and D. Kumar, "A critical study on efficiency of different materials for fluoride removal from aqueous media," *Chemistry Central Journal*, vol. 7, no. 1, article no. 51, 2013.

[49] M. Gallegos-Garcia, K. Ramírez-Muñiz, and S. Song, "Arsenic removal from water by adsorption using iron oxide minerals as adsorbents: A review," *Mineral Processing and Extractive Metallurgy*, vol. 33, no. 5, pp. 301–315, 2012.

[50] A. Maiti, B. K. Thakur, J. K. Basu, and S. De, "Comparison of treated laterite as arsenic adsorbent from different locations and performance of best filter under field conditions," *Journal of Hazardous Materials*, vol. 262, pp. 1176–1186, 2013.

[51] X.-J. Gong, W.-G. Li, D.-Y. Zhang, W.-B. Fan, and X.-R. Zhang, "Adsorption of arsenic from micro-polluted water by an innovative coal-based mesoporous activated carbon in the presence of co-existing ions," *International Biodeterioration & Biodegradation*, vol. 102, pp. 256–264, 2015.

[52] K. Babaeivelni, A. P. Khodadoust, and D. Bogdan, "Adsorption and removal of arsenic (V) using crystalline manganese (II,III) oxide: Kinetics, equilibrium, effect of pH and ionic strength," *Journal of Environmental Science and Health, Part A: Toxic/Hazardous Substances and Environmental Engineering*, vol. 49, no. 13, pp. 1462–1473, 2014.

[53] G. Yang, Y. Liu, and S. Song, "Competitive adsorption of As(V) with co-existing ions on porous hematite in aqueous solutions," *Journal of Environmental Chemical Engineering (JECE)*, vol. 3, no. 3, pp. 1497–1503, 2015.

[54] X. Lv, Y. Hu, J. Tang, T. Sheng, G. Jiang, and X. Xu, "Effects of co-existing ions and natural organic matter on removal of chromium (VI) from aqueous solution by nanoscale zero valent iron (nZVI)-Fe$_3$O$_4$ nanocomposites," *Chemical Engineering Journal*, vol. 218, pp. 55–64, 2013.

[55] A. Afkhami, T. Madrakian, and Z. Karimi, "The effect of acid treatment of carbon cloth on the adsorption of nitrite and nitrate ions," *Journal of Hazardous Materials*, vol. 144, no. 1-2, pp. 427–431, 2007.

[56] X. Guan, H. Dong, J. Ma, and L. Jiang, "Removal of arsenic from water: Effects of competing anions on As(III) removal in KMnO4-Fe(II) process," *Water Research*, vol. 43, no. 15, pp. 3891–3899, 2009.

[57] A. Goswami, P. K. Raul, and M. K. Purkait, "Arsenic adsorption using copper (II) oxide nanoparticles," *Chemical Engineering Research and Design*, vol. 90, no. 9, pp. 1387–1396, 2012.

[58] J. A. Wilkie and J. G. Hering, "Adsorption of arsenic onto hydrous ferric oxide: effects of adsorbate/adsorbent ratios and co-occurring solutes," *Colloids and Surfaces A: Physicochemical and Engineering Aspects*, vol. 107, pp. 97–110, 1996.

[59] A. Maiti, J. K. Basu, and S. De, "Experimental and kinetic modeling of As(V) and As(III) adsorption on treated laterite using synthetic and contaminated groundwater: effects of phosphate, silicate and carbonate ions," *Chemical Engineering Journal*, vol. 191, pp. 1–12, 2012.

[60] Y. Jeong, M. Fan, S. Singh, C.-L. Chuang, B. Saha, and J. Hans van Leeuwen, "Evaluation of iron oxide and aluminum oxide as potential arsenic(V) adsorbents," *Chemical Engineering and Processing: Process Intensification*, vol. 46, no. 10, pp. 1030–1039, 2007.

[61] A. Jain, V. K. Sharma, and O. S. Mbuya, "Removal of arsenite by Fe(VI), Fe(VI)/Fe(III), and Fe(VI)/Al(III) salts: Effect of pH

and anions," *Journal of Hazardous Materials*, vol. 169, no. 1-3, pp. 339–344, 2009.

[62] S. M. Maliyekkal, Anshup, K. R. Antony, and T. Pradeep, "High yield combustion synthesis of nanomagnesia and its application for fluoride removal," *Science of the Total Environment*, vol. 408, no. 10, pp. 2273–2282, 2010.

[63] Y. Tang, X. Guan, J. Wang, N. Gao, M. R. McPhail, and C. C. Chusuei, "Fluoride adsorption onto granular ferric hydroxide: effects of ionic strength, pH, surface loading, and major co-existing anions," *Journal of Hazardous Materials*, vol. 171, no. 1-3, pp. 774–779, 2009.

[64] B. Kemer, D. Ozdes, A. Gundogdu, V. N. Bulut, C. Duran, and M. Soylak, "Removal of fluoride ions from aqueous solution by waste mud," *Journal of Hazardous Materials*, vol. 168, no. 2-3, pp. 888–894, 2009.

[65] W. Ma, N. Zhao, G. Yang, L. Tian, and R. Wang, "Removal of fluoride ions from aqueous solution by the calcination product of Mg-Al-Fe hydrotalcite-like compound," *Desalination*, vol. 268, no. 1-3, pp. 20–26, 2011.

[66] S. Jagtap, M. K. N. Yenkie, N. Labhsetwar, and S. Rayalu, "Defluoridation of drinking water using chitosan based mesoporous alumina," *Microporous and Mesoporous Materials*, vol. 142, no. 2-3, pp. 454–463, 2011.

[67] T. B. Mlilo, L. R. Brunson, and D. A. Sabatini, "Arsenic and fluoride removal using simple materials," *Journal of Environmental Engineering*, vol. 136, no. 4, pp. 391–398, 2010.

[68] K. Babaeivelni and A. P. Khodadoust, "Adsorption of fluoride onto crystalline titanium dioxide: Effect of pH, ionic strength, and co-existing ions," *Journal of Colloid and Interface Science*, vol. 394, no. 1, pp. 419–427, 2013.

[69] S. Zhang, C. Liu, Z. Luan, X. Peng, H. Ren, and J. Wang, "Arsenate removal from aqueous solutions using modified red mud," *Journal of Hazardous Materials*, vol. 152, no. 2, pp. 486–492, 2008.

[70] S. Kundu and A. K. Gupta, "Adsorptive removal of As(III) from aqueous solution using iron oxide coated cement (IOCC): Evaluation of kinetic, equilibrium and thermodynamic models," *Separation and Purification Technology*, vol. 51, no. 2, pp. 165–172, 2006.

[71] S. Gupta and B. V. Babu, "Removal of toxic metal Cr(VI) from aqueous solutions using sawdust as adsorbent: equilibrium, kinetics and regeneration studies," *Chemical Engineering Journal*, vol. 150, no. 2-3, pp. 352–365, 2009.

[72] W. Ding, H. Turson, and X. Huang, "Primary study of fluoride removal from aqueous solution by activated ferric hydroxide," in *Proceedings of the 2nd International Conference on Bioinformatics and Biomedical Engineering, iCBBE 2008*, pp. 2911–2913, China, May 2008.

[73] Y. Tang, X. Guan, T. Su, N. Gao, and J. Wang, "Fluoride adsorption onto activated alumina: modeling the effects of pH and some competing ions," *Colloids and Surfaces A: Physicochemical and Engineering Aspects*, vol. 337, no. 1-3, pp. 33–38, 2009.

[74] N. Öztürk and T. E. Bektaş, "Nitrate removal from aqueous solution by adsorption onto various materials," *Journal of Hazardous Materials*, vol. 112, no. 1-2, pp. 155–162, 2004.

[75] J. C. Saha, K. Dikshit, and M. Bandyopadhyay, "Omparative studies for selection of technologies for arsenic removal from drinking water. UNU Inter, Workshop," in *Proceedings of the UNU Inter, Workshop*, pp. 76–84, Dhaka, Bangladesh, 2001.

[76] Y. Kim, C. Kim, I. Choi, S. Rengaraj, and J. Yi, "Arsenic Removal Using Mesoporous Alumina Prepared via a Templating Method," *Environmental Science & Technology*, vol. 38, no. 3, pp. 924–931, 2004.

[77] R. Singh, S. Singh, P. Parihar, V. P. Singh, and S. M. Prasad, "Arsenic contamination, consequences and remediation techniques: a review," *Ecotoxicology and Environmental Safety*, vol. 112, pp. 247–270, 2015.

[78] A. Sigdel, J. Park, H. Kwak, and P.-K. Park, "Arsenic removal from aqueous solutions by adsorption onto hydrous iron oxide-impregnated alginate beads," *Journal of Industrial and Engineering Chemistry*, 2015.

[79] D. E. Giles, M. Mohapatra, T. B. Issa, S. Anand, and P. Singh, "Iron and aluminium based adsorption strategies for removing arsenic from water," *Journal of Environmental Management*, vol. 92, no. 12, pp. 3011–3022, 2011.

[80] M. Streat, K. Hellgardt, and N. L. R. Newton, "Hydrous ferric oxide as an adsorbent in water treatment: part 2. Adsorption studies," *Process Safety and Environmental Protection*, vol. 86, no. 1, pp. 11–20, 2008.

[81] S. A. Chaudhry, Z. Zaidi, and S. I. Siddiqui, "Isotherm, kinetic and thermodynamics of arsenic adsorption onto Iron-Zirconium Binary Oxide-Coated Sand (IZBOCS): Modelling and process optimization," *Journal of Molecular Liquids*, vol. 229, pp. 230–240, 2017.

[82] T. S. Singh and K. K. Pant, "Equilibrium, kinetics and thermodynamic studies for adsorption of As(III) on activated alumina," *Separation and Purification Technology*, vol. 36, no. 2, pp. 139–147, 2004.

[83] V. M. Boddu, K. Abburi, J. L. Talbott, E. D. Smith, and R. Haasch, "Removal of arsenic (III) and arsenic (V) from aqueous medium using chitosan-coated biosorbent," *Water Research*, vol. 42, no. 3, pp. 633–642, 2008.

[84] M. Islam, P. C. Mishra, and R. Patel, "Arsenate removal from aqueous solution by cellulose-carbonated hydroxyapatite nanocomposites," *Journal of Hazardous Materials*, vol. 189, no. 3, pp. 755–763, 2011.

[85] S. R. Chowdhury and E. K. Yanful, "Arsenic and chromium removal by mixed magnetite-maghemite nanoparticles and the effect of phosphate on removal," *Journal of Environmental Management*, vol. 91, no. 11, pp. 2238–2247, 2010.

[86] A. Ramesh, H. Hasegawa, T. Maki, and K. Ueda, "Adsorption of inorganic and organic arsenic from aqueous solutions by polymeric Al/Fe modified montmorillonite," *Separation and Purification Technology*, vol. 56, no. 1, pp. 90–100, 2007.

[87] T.-F. Lin and J.-K. Wu, "Adsorption of arsenite and arsenate within activated alumina grains: equilibrium and kinetics," *Water Research*, vol. 35, no. 8, pp. 2049–2057, 2001.

[88] O. S. Thirunavukkarasu, T. Viraraghavan, and K. S. Subramanian, "Arsenic removal from drinking water using iron oxide-coated sand," *Water, Air, & Soil Pollution*, vol. 142, no. 1-4, pp. 95–111, 2003.

[89] Y.-H. Xu, T. Nakajima, and A. Ohki, "Adsorption and removal of arsenic(V) from drinking water by aluminum-loaded Shirasu-zeolite," *Journal of Hazardous Materials*, vol. 92, no. 3, pp. 275–287, 2002.

[90] S. S. Tripathy and A. M. Raichur, "Enhanced adsorption capacity of activated alumina by impregnation with alum for removal of As(V) from water," *Chemical Engineering Journal*, vol. 138, no. 1-3, pp. 179–186, 2008.

[91] P. Pillewan, S. Mukherjee, T. Roychowdhury, S. Das, A. Bansiwal, and S. Rayalu, "Removal of As(III) and As(V) from water

by copper oxide incorporated mesoporous alumina," *Journal of Hazardous Materials*, vol. 186, no. 1, pp. 367–375, 2011.

[92] M. P. Asta, J. Cama, M. Martínez, and J. Giménez, "Arsenic removal by goethite and jarosite in acidic conditions and its environmental implications," *Journal of Hazardous Materials*, vol. 171, no. 1-3, pp. 965–972, 2009.

[93] C.-S. Jeon, K. Baek, J.-K. Park, Y.-K. Oh, and S.-D. Lee, "Adsorption characteristics of As(V) on iron-coated zeolite," *Journal of Hazardous Materials*, vol. 163, no. 2-3, pp. 804–808, 2009.

[94] J.-C. Hsu, C.-J. Lin, C.-H. Liao, and S.-T. Chen, "Removal of As(V) and As(III) by reclaimed iron-oxide coated sands," *Journal of Hazardous Materials*, vol. 153, no. 1-2, pp. 817–826, 2008.

[95] T. Tuutijärvi, J. Lu, M. Sillanpää, and G. Chen, "As(V) adsorption on maghemite nanoparticles," *Journal of Hazardous Materials*, vol. 166, no. 2-3, pp. 1415–1420, 2009.

[96] D. Borah, S. Satokawa, S. Kato, and T. Kojima, "Sorption of As(V) from aqueous solution using acid modified carbon black," *Journal of Hazardous Materials*, vol. 162, no. 2-3, pp. 1269–1277, 2009.

[97] L. M. Camacho, R. R. Parra, and S. Deng, "Arsenic removal from groundwater by MnO2-modified natural clinoptilolite zeolite: Effects of pH and initial feed concentration," *Journal of Hazardous Materials*, vol. 189, no. 1-2, pp. 286–293, 2011.

[98] S. Mandal, T. Padhi, and R. K. Patel, "Studies on the removal of arsenic (III) from water by a novel hybrid material," *Journal of Hazardous Materials*, vol. 192, no. 2, pp. 899–908, 2011.

[99] H. Guo, Y. Li, and K. Zhao, "Arsenate removal from aqueous solution using synthetic siderite," *Journal of Hazardous Materials*, vol. 176, no. 1-3, pp. 174–180, 2010.

[100] D. Pokhrel and T. Viraraghavan, "Arsenic removal from aqueous solution by iron oxide-coated fungal biomass: A factorial design analysis," *Water, Air, & Soil Pollution*, vol. 173, no. 1-4, pp. 195–208, 2006.

[101] C. A. Martinson and K. J. Reddy, "Adsorption of arsenic(III) and arsenic(V) by cupric oxide nanoparticles," *Journal of Colloid and Interface Science*, vol. 336, no. 2, pp. 406–411, 2009.

[102] M. Gallegos-Garciaa, K. Ramírez-Muñiza, and S. Songa, "Arsenic removal from water by adsorption using iron oxide minerals as adsorbents: A review," *Mineral Processing & Extractive Metall Rev*, vol. 33, 2012.

[103] Y.-N. Chen, L.-Y. Chai, and Y.-D. Shu, "Study of arsenic(V) adsorption on bone char from aqueous solution," *Journal of Hazardous Materials*, vol. 160, no. 1, pp. 168–172, 2008.

[104] N. Zhu, T. Yan, J. Qiao, and H. Cao, "Adsorption of arsenic, phosphorus and chromium by bismuth impregnated biochar: Adsorption mechanism and depleted adsorbent utilization," *Chemosphere*, vol. 164, pp. 32–40, 2016.

[105] M. R. Yazdani, T. Tuutijärvi, A. Bhatnagar, and R. Vahala, "Adsorptive removal of arsenic(V) from aqueous phase by feldspars: kinetics, mechanism, and thermodynamic aspects of adsorption," *Journal of Molecular Liquids*, vol. 214, pp. 149–156, 2016.

[106] P. Suresh Kumar, R. Q. Flores, C. Sjöstedt, and L. Önnby, "Arsenic adsorption by iron-aluminium hydroxide coated onto macroporous supports: insights from X-ray absorption spectroscopy and comparison with granular ferric hydroxides," *Journal of Hazardous Materials*, vol. 302, pp. 166–174, 2016.

[107] Y. Chen, Y. Zhu, Z. Wang et al., "Application studies of activated carbon derived from rice husks produced by chemical-thermal process - A review," *Advances in Colloid and Interface Science*, vol. 163, no. 1, pp. 39–52, 2011.

[108] A. F. Hassan, A. M. Abdel-Mohsen, and H. Elhadidy, "Adsorption of arsenic by activated carbon, calcium alginate and their composite beads," *International Journal of Biological Macromolecules*, vol. 68, pp. 125–130, 2014.

[109] A. Sherlala, A. Raman, and M. Bello, "Adsorption of arsenic from aqueous solution using magnetic graphene oxide," *IOP Conference Series: Materials Science and Engineering*, vol. 210, p. 012007, 2017.

[110] C. L. Chuang, M. Fan, M. Xu et al., "Adsorption of arsenic(V) by activated carbon prepared from oat hulls," *Chemosphere*, vol. 61, no. 4, pp. 478–483, 2005.

[111] P. Roy, N. K. Mondal, S. Bhattacharya, B. Das, and K. Das, "Removal of arsenic(III) and arsenic(V) on chemically modified low-cost adsorbent: batch and column operations," *Applied Water Science*, vol. 3, no. 1, pp. 293–309, 2013.

[112] B. K. Biswas, J.-I. Inoue, K. Inoue et al., "Adsorptive removal of As(V) and As(III) from water by a Zr(IV)-loaded orange waste gel," *Journal of Hazardous Materials*, vol. 154, no. 1-3, pp. 1066–1074, 2008.

[113] Z. Liu, F.-S. Zhang, and R. Sasai, "Arsenate removal from water using Fe3O4-loaded activated carbon prepared from waste biomass," *Chemical Engineering Journal*, vol. 160, no. 1, pp. 57–62, 2010.

[114] V. K. Sharma and M. Sohn, "Aquatic arsenic: toxicity, speciation, transformations, and remediation," *Environment International*, vol. 35, no. 4, pp. 743–759, 2009.

[115] USEPA, *Arsenic Treatment Technology Evaluation Handbook for Small Systems*, vol. EPA 816-R-03-014, USEPA, Washington,DC, USA, 2003.

[116] S. Kundu and A. K. Gupta, "Sorption kinetics of As(V) with iron-oxide-coated cement - A new adsorbent and its application in the removal of arsenic from real-life groundwater samples," *Journal of Environmental Science and Health, Part A: Toxic/Hazardous Substances and Environmental Engineering*, vol. 40, no. 12, pp. 2227–2246, 2005.

[117] H. Fakour, Y.-F. Pan, and T.-F. Lin, "Effect of humic acid on arsenic adsorption and pore blockage on iron-based adsorbent," *Water, Air, & Soil Pollution*, vol. 226, no. 2, article no. 2224, 2015.

[118] Y. Mamindy-Pajany, C. Hurel, N. Marmier, and M. Roméo, "Arsenic adsorption onto hematite and goethite," *Comptes Rendus Chimie*, vol. 12, no. 8, pp. 876–881, 2009.

[119] P. Lakshmipathiraj, B. R. V. Narasimhan, S. Prabhakar, and G. Bhaskar Raju, "Adsorption of arsenate on synthetic goethite from aqueous solutions," *Journal of Hazardous Materials*, vol. 136, no. 2, pp. 281–287, 2006.

[120] J. Giménez, M. Martínez, J. de Pablo, M. Rovira, and L. Duro, "Arsenic sorption onto natural hematite, magnetite, and goethite," *Journal of Hazardous Materials*, vol. 141, no. 3, pp. 575–580, 2007.

[121] Y. Mamindy-Pajany, C. Hurel, N. Marmier, and M. Roméo, "Arsenic (V) adsorption from aqueous solution onto goethite, hematite, magnetite and zero-valent iron: Effects of pH, concentration and reversibility," *Desalination*, vol. 281, no. 1, pp. 93–99, 2011.

[122] S. Luther, N. Borgfeld, J. Kim, and J. G. Parsons, "Removal of arsenic from aqueous solution: A study of the effects of pH and interfering ions using iron oxide nanomaterials," *Microchemical Journal*, vol. 101, pp. 30–36, 2012.

[123] R. Ansari and M. Sadegh, "Application of activated carbon for removal of arsenic ions from aqueous solutions," *E-Journal of Chemistry*, vol. 4, no. 1, pp. 103–108, 2007.

[124] H. Guo, Y. Li, K. Zhao, Y. Ren, and C. Wei, "Removal of arsenite from water by synthetic siderite: Behaviors and mechanisms," *Journal of Hazardous Materials*, vol. 186, no. 2-3, pp. 1847–1854, 2011.

[125] S. Yao, Z. Liu, and Z. Shi, "Arsenic removal from aqueous solutions by adsorption onto iron oxide/activated carbon magnetic composite," *Journal of Environmental Health Science and Engineering*, vol. 12, no. 1, article no. 58, 2014.

[126] G. J. Liu, X. R. Zhang, L. Mcwilliams, J. W. Talley, and C. R. Neal, "Influence of ionic strength, electrolyte type, and NOM on As(V) adsorption onto TiO2," *Journal of Environmental Science and Health, Part A: Toxic/Hazardous Substances and Environmental Engineering*, vol. 43, no. 4, pp. 430–436, 2008.

[127] S. Yean, L. Cong, C. T. Yavuz et al., "Effect of magnetite particle size on adsorption and desorption of arsenite and arsenate," *Journal of Materials Research*, vol. 20, no. 12, pp. 3255–3264, 2005.

[128] X.-H. Guan, J. Wang, and C. C. Chusuei, "Removal of arsenic from water using granular ferric hydroxide: macroscopic and microscopic studies," *Journal of Hazardous Materials*, vol. 156, no. 1-3, pp. 178–185, 2008.

[129] R. Weerasooriya and H. J. Tobschall, "Mechanistic modeling of chromate adsorption onto goethite," *Colloids and Surfaces A: Physicochemical and Engineering Aspects*, vol. 162, no. 1-3, pp. 167–175, 2000.

[130] J. Antelo, M. Avena, S. Fiol, R. López, and F. Arce, "Effects of pH and ionic strength on the adsorption of phosphate and arsenate at the goethite-water interface," *Journal of Colloid and Interface Science*, vol. 285, no. 2, pp. 476–486, 2005.

[131] H. J. Shipley, S. Yean, A. T. Kan, and M. B. Tomson, "Adsorption of arsenic to magnetite nanoparticles: Effect of particle concentration, pH, ionic strength, and temperature," *Environmental Toxicology and Chemistry*, vol. 28, no. 3, pp. 509–515, 2009.

[132] W.-X. Gong, J.-H. Qu, R.-P. Liu, and H.-C. Lan, "Adsorption of fluoride onto different types of aluminas," *Chemical Engineering Journal*, vol. 189-190, pp. 126–133, 2012.

[133] S. Wu, K. Zhang, J. He et al., "High efficient removal of fluoride from aqueous solution by a novel hydroxyl aluminum oxalate adsorbent," *Journal of Colloid and Interface Science*, vol. 464, pp. 238–245, 2016.

[134] S. karmakar, J. Dechnik, C. Janiak, and S. De, "Aluminium fumarate metal-organic framework: A super adsorbent for fluoride from water," *Journal of Hazardous Materials*, vol. 303, pp. 10–20, 2016.

[135] L. M. Camacho, A. Torres, D. Saha, and S. Deng, "Adsorption equilibrium and kinetics of fluoride on sol-gel-derived activated alumina adsorbents," *Journal of Colloid and Interface Science*, vol. 349, no. 1, pp. 307–313, 2010.

[136] A. Rafique, M. A. Awan, A. Wasti, I. A. Qazi, and M. Arshad, "Removal of fluoride from drinking water using modified immobilized activated alumina," *Journal of Chemistry*, Article ID 386476, 2013.

[137] J. J. García-Sánchez, M. Solache-Ríos, J. M. Martínez-Gutiérrez, N. V. Arteaga-Larios, M. C. Ojeda-Escamilla, and I. Rodríguez-Torres, "Modified natural magnetite with Al and la ions for the adsorption of fluoride ions from aqueous solutions," *Journal of Fluorine Chemistry*, vol. 186, pp. 115–124, 2016.

[138] L. Xu, G. Chen, C. Peng et al., "Adsorptive removal of fluoride from drinking water using porous starch loaded with common metal ions," *Carbohydrate Polymers*, vol. 160, pp. 82–89, 2017.

[139] M. Mohapatra, K. Rout, P. Singh et al., "Fluoride adsorption studies on mixed-phase nano iron oxides prepared by surfactant mediation-precipitation technique," *Journal of Hazardous Materials*, vol. 186, no. 2-3, pp. 1751–1757, 2011.

[140] N. K. Mondal, R. Bhaumik, and J. K. Datta, "Removal of fluoride by aluminum impregnated coconut fiber from synthetic fluoride solution and natural water," *Alexandria Engineering Journal*, vol. 54, no. 4, pp. 1273–1284, 2015.

[141] J. Zhang, N. Chen, Z. Tang, Y. Yu, Q. Hu, and C. Feng, "A study of the mechanism of fluoride adsorption from aqueous solutions onto Fe-impregnated chitosan," *Physical Chemistry Chemical Physics*, vol. 17, no. 18, pp. 12041–12050, 2015.

[142] L. Jayarathna, A. Bandara, W. J. Ng, and R. Weerasooriya, "Fluoride adsorption on $\gamma - fe2o3$ nanoparticles," *Journal of Environmental Health Science and Engineering*, vol. 13, no. 54, 2015.

[143] Z. Jin, Y. Jia, T. Luo et al., "Efficient removal of fluoride by hierarchical MgO microspheres: Performance and mechanism study," *Applied Surface Science*, vol. 357, pp. 1080–1088, 2015.

[144] Z. Jin, Y. Jia, K.-S. Zhang et al., "Effective removal of fluoride by porous MgO nanoplates and its adsorption mechanism," *Journal of Alloys and Compounds*, vol. 675, pp. 292–300, 2016.

[145] S. Dong and Y. Wang, "Characterization and adsorption properties of a lanthanum-loaded magnetic cationic hydrogel composite for fluoride removal," *Water Research*, vol. 88, pp. 852–860, 2016.

[146] Y. Nie, C. Hu, and C. Kong, "Enhanced fluoride adsorption using Al (III) modified calcium hydroxyapatite," *Journal of Hazardous Materials*, vol. 233-234, pp. 194–199, 2012.

[147] M. G. Sujana and S. Anand, "Fluoride removal studies from contaminated ground water by using bauxite," *Desalination*, vol. 267, no. 2-3, pp. 222–227, 2011.

[148] A. Tor, N. Danaoglu, G. Arslan, and Y. Cengeloglu, "Removal of fluoride from water by using granular red mud: batch and column studies," *Journal of Hazardous Materials*, vol. 164, no. 1, pp. 271–278, 2009.

[149] X. Fan, D. J. Parker, and M. D. Smith, "Adsorption kinetics of fluoride on low cost materials," *Water Research*, vol. 37, no. 20, pp. 4929–4937, 2003.

[150] S. M. Maliyekkal, S. Shukla, L. Philip, and I. M. Nambi, "Enhanced fluoride removal from drinking water by magnesia-amended activated alumina granules," *Chemical Engineering Journal*, vol. 140, no. 1-3, pp. 183–192, 2008.

[151] S. M. Maliyekkal, A. K. Sharma, and L. Philip, "Manganese-oxide-coated alumina: A promising sorbent for defluoridation of water," *Water Research*, vol. 40, no. 19, pp. 3497–3506, 2006.

[152] H. Deng and X. Yu, "Adsorption of fluoride, arsenate and phosphate in aqueous solution by cerium impregnated fibrous protein," *Chemical Engineering Journal*, vol. 184, pp. 205–212, 2012.

[153] M. Karthikeyan and K. P. Elango, "Removal of fluoride from water using aluminium containing compounds," *Journal of Environmental Sciences*, vol. 21, no. 11, pp. 1513–1518, 2009.

[154] S. S. Tripathy, J.-L. Bersillon, and K. Gopal, "Removal of fluoride from drinking water by adsorption onto alum-impregnated activated alumina," *Separation and Purification Technology*, vol. 50, no. 3, pp. 310–317, 2006.

[155] N. Viswanathan and S. Meenakshi, "Enriched fluoride sorption using alumina/chitosan composite," *Journal of Hazardous Materials*, vol. 178, no. 1-3, pp. 226–232, 2010.

[156] V. Ganvir and K. Das, "Removal of fluoride from drinking water using aluminum hydroxide coated rice husk ash," *Journal of Hazardous Materials*, vol. 185, no. 2-3, pp. 1287–1294, 2011.

[157] S. Kagne, S. Jagtap, P. Dhawade, S. P. Kamble, S. Devotta, and S. S. Rayalu, "Hydrated cement: A promising adsorbent for the removal of fluoride from aqueous solution," *Journal of Hazardous Materials*, vol. 154, no. 1-3, pp. 88–95, 2008.

[158] R. Leyva Ramos, J. Ovalle-Turrubiartes, and M. A. Sanchez-Castillo, "Adsorption of fluoride from aqueous solution on aluminum-impregnated carbon," *Carbon*, vol. 37, no. 4, pp. 609–617, 1999.

[159] M. Mourabet, H. El Boujaady, A. El Rhilassi et al., "Defluoridation of water using Brushite: Equilibrium, kinetic and thermodynamic studies," *Desalination*, vol. 278, no. 1-3, pp. 1–9, 2011.

[160] E. Tchomgui-Kamga, E. Ngameni, and A. Darchen, "Evaluation of removal efficiency of fluoride from aqueous solution using new charcoals that contain calcium compounds," *Journal of Colloid and Interface Science*, vol. 346, no. 2, pp. 494–499, 2010.

[161] W. Ma, F.-Q. Ya, M. Han, and R. Wang, "Characteristics of equilibrium, kinetics studies for adsorption of fluoride on magnetic-chitosan particle," *Journal of Hazardous Materials*, vol. 143, no. 1-2, pp. 296–302, 2007.

[162] A. M. Raichur and M. J. Basu, "Adsorption of fluoride onto mixed rare earth oxides," *Separation and Purification Technology*, vol. 24, no. 1-2, pp. 121–127, 2001.

[163] V. Gopal and K. P. Elango, "Equilibrium, kinetic and thermodynamic studies of adsorption of fluoride onto plaster of Paris," *Journal of Hazardous Materials*, vol. 141, no. 1, pp. 98–105, 2007.

[164] M. Bhaumik, T. Y. Leswifi, A. Maity, V. V. Srinivasu, and M. S. Onyango, "Removal of fluoride from aqueous solution by Polypyrrole/Fe3O4 magnetic nanocomposite," *Journal of Hazardous Materials*, vol. 186, no. 1, pp. 150–159, 2011.

[165] R. Huang, B. Yang, Q. Liu, and K. Ding, "Removal of fluoride ions from aqueous solutions using protonated cross-linked chitosan particles," *Journal of Fluorine Chemistry*, vol. 141, pp. 29–34, 2012.

[166] Y. Yu, C. Wang, X. Guo, and J. P. Chen, "Modification of carbon derived from *Sargassum* sp. by lanthanum for enhanced adsorption of fluoride," *Journal of Colloid and Interface Science*, vol. 441, pp. 113–120, 2015.

[167] D. Tang and G. Zhang, "Efficient removal of fluoride by hierarchical Ce-Fe bimetal oxides adsorbent: thermodynamics, kinetics and mechanism," *Chemical Engineering Journal*, vol. 283, pp. 721–729, 2016.

[168] M. G. Sujana, A. Mishra, and B. C. Acharya, "Hydrous ferric oxide doped alginate beads for fluoride removal: adsorption kinetics and equilibrium studies," *Applied Surface Science*, vol. 270, pp. 767–776, 2013.

[169] J. Zhu, X. Lin, P. Wu, Q. Zhou, and X. Luo, "Fluoride removal from aqueous solution by Al(III)-Zr(IV) binary oxide adsorbent," *Applied Surface Science*, vol. 357, pp. 91–100, 2015.

[170] C. Zhang, Y. Li, T.-J. Wang, Y. Jiang, and H. Wang, "Adsorption of drinking water fluoride on a micron-sized magnetic Fe3O4@Fe-Ti composite adsorbent," *Applied Surface Science*, vol. 363, pp. 507–515, 2016.

[171] J. He, K. Zhang, S. Wu et al., "Performance of novel hydroxyapatite nanowires in treatment of fluoride contaminated water," *Journal of Hazardous Materials*, vol. 303, pp. 119–130, 2016.

[172] J. Chen, C. Shu, N. Wang, J. Feng, H. Ma, and W. Yan, "Adsorbent synthesis of polypyrrole/TiO2 for effective fluoride removal from aqueous solution for drinking water purification: Adsorbent characterization and adsorption mechanism," *Journal of Colloid and Interface Science*, vol. 495, pp. 44–52, 2017.

[173] K. Biswas, S. K. Saha, and U. C. Ghosh, "Adsorption of fluoride from aqueous solution by a synthetic iron(III)-aluminum(III) mixed oxide," *Industrial & Engineering Chemistry Research*, vol. 46, no. 16, pp. 5346–5356, 2007.

[174] L. Kuang, Y. Liu, D. Fu, and Y. Zhao, "FeOOH-graphene oxide nanocomposites for fluoride removal from water: Acetate mediated nano FeOOH growth and adsorption mechanism," *Journal of Colloid and Interface Science*, vol. 490, pp. 259–269, 2017.

[175] J. He, K. Chen, X. Cai et al., "A biocompatible and novelly-defined Al-HAP adsorption membrane for highly effective removal of fluoride from drinking water," *Journal of Colloid and Interface Science*, vol. 490, pp. 97–107, 2017.

[176] M. Chaudhary, P. Bhattacharya, and A. Maiti, "Synthesis of iron oxyhydroxide nanoparticles and its application for fluoride removal from water," *Journal of Environmental Chemical Engineering (JECE)*, vol. 4, no. 4, pp. 4897–4903, 2016.

[177] P. Garg and C. Chaudhari, "Adsorption of fluoride from drinking water on magnesium substituted hydroxyapatite," in *Proceedings of the 2012 International Conference on Future Environment and Energy IPCBEE*, vol. 28, 2012.

[178] M. S. Onyango, Y. Kojima, O. Aoyi, E. C. Bernardo, and H. Matsuda, "Adsorption equilibrium modeling and solution chemistry dependence of fluoride removal from water by trivalent-cation-exchanged zeolite F-9," *Journal of Colloid and Interface Science*, vol. 279, no. 2, pp. 341–350, 2004.

[179] N. Chen, Z. Zhang, C. Feng, N. Sugiura, M. Li, and R. Chen, "Fluoride removal from water by granular ceramic adsorption," *Journal of Colloid and Interface Science*, vol. 348, no. 2, pp. 579–584, 2010.

[180] A. Eskandarpour, M. S. Onyango, A. Ochieng, and S. Asai, "Removal of fluoride ions from aqueous solution at low pH using schwertmannite," *Journal of Hazardous Materials*, vol. 152, no. 2, pp. 571–579, 2008.

[181] S. S. Tripathy and A. M. Raichur, "Abatement of fluoride from water using manganese dioxide-coated activated alumina," *Journal of Hazardous Materials*, vol. 153, no. 3, pp. 1043–1051, 2008.

[182] A. H. Mahvi, B. Heibati, A. Mesdaghinia, and A. R. Yari, "Fluoride adsorption by pumice from aqueous solutions," *E-Journal of Chemistry*, vol. 9, no. 4, pp. 1843–1853, 2012.

[183] S. Rajkumar, S. Murugesh, V. Sivasankar, A. Darchen, T. Msagati, and T. Chaabane, "Low-cost fluoride adsorbents prepared from a renewable biowaste: Syntheses, characterization and modeling studies," *Arabian Journal of Chemistry*, 2015.

[184] S. Peng, Q. Zeng, Y. Guo, B. Niu, X. Zhang, and S. Hong, "Defluoridation from aqueous solution by chitosan modified natural zeolite," *Journal of Chemical Technology and Biotechnology*, vol. 88, no. 9, pp. 1707–1714, 2013.

[185] M. R. Samarghandi, M. Khiadani, M. Foroughi, and H. Zolghadr Nasab, "Defluoridation of water using activated alumina in presence of natural organic matter via response surface methodology," *Environmental Science and Pollution Research*, vol. 23, no. 1, pp. 887–897, 2016.

# 14

# $CO_2$ Adsorption from Biogas using Amine-Functionalized MgO

**Preecha Kasikamphaiboon** ⓘ **and Uraiwan Khunjan**

*Department of Science, Faculty of Science and Technology, Prince of Songkla University, Pattani Campus, Pattani 94000, Thailand*

Correspondence should be addressed to Preecha Kasikamphaiboon; preecha.kas@psu.ac.th

Academic Editor: Bhaskar Kulkarni

Biogas is a renewable fuel source of methane ($CH_4$), and its utilization as a natural gas substitute or transport fuel has received much interest. However, apart from $CH_4$, biogas also contains carbon dioxide ($CO_2$) which is noncombustible, thus reducing the biogas heating value. Therefore, upgrading biogas by removing $CO_2$ is needed for most biogas applications. In this study, an amine-functionalized adsorbent for $CO_2$ capture from biogas was developed. Mesoporous MgO was synthesized and functionalized with different tetraethylenepentamine (TEPA) loadings by wet impregnation technique. The prepared adsorbents (MgO-TEPA) were characterized by X-ray diffraction (XRD) and $N_2$ adsorption-desorption. The $CO_2$ adsorption performance of the prepared MgO-TEPA was tested using simulated biogas as feed gas stream. The results show that the $CO_2$ adsorption capacities of the adsorbents increase with increasing TEPA loading. The optimum TEPA loading is 40 wt.%, which gives the highest $CO_2$ adsorption capacity of 4.98 mmol/g. A further increase in TEPA loading to 50 wt.% significantly reduces the $CO_2$ adsorption capacity. Furthermore, the stability and regenerability of the adsorbent with 40% TEPA loading (MgO-TEPA-40) were studied by performing ten adsorption-desorption cycles under simulated biogas and real biogas conditions. After ten adsorption-desorption cycles, MgO-TEPA-40 shows slight decreases of only 5.42 and 5.75% of $CO_2$ adsorption capacity for the simulated biogas and biogas, respectively. The results demonstrate that MgO-TEPA-40 possesses good stability and regenerability which are important for the potential application of this amine-based adsorbent.

## 1. Introduction

Due to rising fossil fuel prices, greenhouse gas emissions from fossil fuel combustion, and high energy demands, sustainable and renewable energy sources are needed [1, 2]. Biogas, considered to be one of the alternative sources of renewable energy, has the potential to supplement the current energy requirements. Biogas produced from anaerobic digestion processes is mainly composed of 50–70% of methane ($CH_4$) and 30–50% of carbon dioxide ($CO_2$) [3]. Apart from these two gases, biogas contains smaller amounts of hydrogen sulfide ($H_2S$) and trace amounts of ammonia ($NH_3$), nitrogen ($N_2$), hydrogen ($H_2$), carbon monoxide (CO), and oxygen ($O_2$) [4, 5]. The energy content in terms of calorific value of pure methane is $36 \, MJ/m^3$ (at STP conditions), whereas that of the biogas containing 60–65% $CH_4$ is approximately 20–25 $MJ/m^3$ [3, 6]. Therefore, being noncombustible, $CO_2$ also causes a reduction in the heating value and energy density of biogas on volume basis. In addition, high $CO_2$ content

increases energy demand for compression and transportation of biogas. Biogas can be upgraded to a higher fuel standard by increasing the energy density by reducing major noncombustible gas ($CO_2$) and other impurities. The final product consisting of 95–99% $CH_4$ and 1–5% $CO_2$ [7], which is called biomethane, can be used as a substitute for natural gas and a vehicle fuel. Therefore, upgrading biogas to biomethane is one of the technologies that receive much attention in the bioenergy industry.

To remove acid gases such as $CO_2$ and $H_2S$ from gas streams, several technologies have been developed, including physical absorption, chemical absorption, adsorption, membrane separation, and cryogenic separation [8]. Among these technologies, the chemical absorption using liquid amine absorbents is one of the most widely used technologies for separating $CO_2$ from gas streams [9, 10]. However, amine scrubbing processes have some disadvantages such as high energy consumption for solvent regeneration, high equipment corrosion rate, solvent degradation, and fouling of the process equipment [11–13]. As a result, adsorption-based technologies, usually involving solid $CO_2$ adsorbents, are suggested and studied to overcome those problems [14]. Adsorbents for $CO_2$ capture are mainly porous materials, such as activated carbon [15, 16], mesoporous silica [17], zeolites [18, 19], and metal-organic frameworks (MOFs) [20, 21], which usually have a large surface area in mesopores and micropores. Compared to liquid amine absorption, the solid adsorbents possess significant advantages for energy efficiency. However, there are some existing problems such as low $CO_2$ adsorption capacities and selectivity. Therefore, modification or functionalization of the solid adsorbents by introducing amines into their porous structure for $CO_2$ adsorption has also attracted great interest [22]. The amine-functionalized adsorbents combine the high $CO_2$ affinity of amines and porous materials (large pore volume and large surface area) which exhibit advantageous properties such as high $CO_2$ adsorption rate and high thermal stability. These adsorbents not only have the potential to reduce the energy consumption compared to the large amount of energy required to heat bulk water for the regeneration of absorbent in the aqueous amine absorbent process but also have high $CO_2$ adsorption capacity and the ability to reduce the corrosion of equipment caused by high concentrated aqueous amine absorbent [10, 23]. Several research groups have studied and reported the performances of mesoporous materials functionalized with amines for $CO_2$ removal. These materials can be grouped together depending on the preparation methods which are grafting and impregnation with diverse amine species such as aminosilanes, polyethylenimine (PEI), tetraethylenepentamine (TEPA), and diethanolamine (DEA) [24–31]. The adsorbents obtained by chemical grafting of amine onto the support are relatively more stable than those obtained by physical impregnation. However, the amount of grafted amine is limited, resulting in a relatively lower amine loading which leads to lower $CO_2$ adsorption capacity. Unlike the grafting method, the impregnation method possesses several advantages such as easier preparation, higher amine loading, and lower cost [32].

Apart from the solid adsorbents mentioned above, magnesium oxide (MgO) has been widely studied for $CO_2$ capture and has been reported as a potential adsorbent for $CO_2$ adsorption [33–35]. However, pure MgO exhibits fairly small $CO_2$ adsorption capacity [36]. To the best of our knowledge, introducing amines into the porous structure of MgO to form a composite with high $CO_2$ adsorption capacity has not been reported. Therefore, in the present study, a new adsorbent for $CO_2$ removal from biogas was developed. MgO with mesoporous structure was synthesized and functionalized with tetraethylenepentamine (TEPA) via an impregnation process. The prepared adsorbents with different amine loadings were investigated for $CO_2$ adsorption capacity. Multiple adsorption-desorption cyclic stability of the adsorbents was also tested by using simulated biogas and biogas as feed gas streams.

## 2. Experimental

### 2.1. Synthesis of Mesoporous MgO.
Mesoporous MgO was synthesized using the sol-gel method as reported in the literature [37]. Magnesium nitrate hexahydrate ($Mg(NO_3)_2 \cdot 6H_2O$) and oxalic acid (($COOH)_2 \cdot 2H_2O$), purchased from Sigma-Aldrich, in the molar ratio of 1 : 1, were first dissolved separately in ethanol and used for the synthesis of MgO. These solutions were then mixed together with vigorous magnetic stirring to yield a thick white gel and kept stirred for 12 h. Subsequently, the gel product was dried at 100°C for 24 h. The dried mixture was ground and calcined under atmospheric pressure at 600°C for 2 h. The chemical reactions in the sol-gel and calcination step can be expressed by equations (1) and (2), respectively [37]:

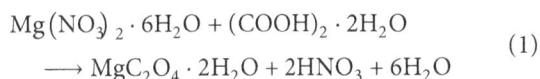

$$Mg(NO_3)_2 \cdot 6H_2O + (COOH)_2 \cdot 2H_2O$$
$$\longrightarrow MgC_2O_4 \cdot 2H_2O + 2HNO_3 + 6H_2O \quad (1)$$

$$MgC_2O_4 \cdot 2H_2O + 0.5O_2 \longrightarrow MgO + 2CO_2 + 2H_2O \quad (2)$$

### 2.2. Preparation of Amine-Functionalized Adsorbents.
In this study, the amine-functionalized adsorbents were prepared by the wet impregnation method described by Xu et al. [38] with some modification. The desired amount of tetraethylenepentamine (TEPA), obtained from Sigma-Aldrich, was dissolved in 25 ml of methanol under stirring for 15 min. Then, 5 g of the synthesized MgO was added to the TEPA-methanol solution, and the resultant slurry was stirred continuously for approximately 30 min. The slurry was then dried at 70°C for 2 h. The adsorbents impregnated with TEPA were designated as MgO-TEPA-$x$, where $x$ represents the weight percentage of TEPA in the adsorbents.

### 2.3. Characterization of the Adsorbents.
The crystal structures of the prepared adsorbents were analyzed by powder X-ray diffraction (PXRD) by using a Philips X'Pert MPD diffractometer with Cu-K$\alpha$ radiation. The $N_2$ adsorption-desorption isotherms were measured at −196°C using a Micromeritics ASAP2060 volumetric analyzer. The isotherms were used for calculating the surface area, pore volume, and average pore diameter of the adsorbents via the Brunauer–Emmett–Teller (BET) method and the Barrett–Joyner–Halenda (BJH) method.

### 2.4. CO2 Adsorption.
The $CO_2$ adsorption-desorption tests were performed in a fixed-bed reactor of which a simplified flow diagram is given in Figure 1. The reactor was a stainless-steel adsorption column with an inner diameter of 15 mm and a length of 160 mm. Typically, 5 g of the adsorbent was packed into the adsorption column and supported by quartz wool from both sides. Before each experimental run, the sample was heated to 100°C in a $N_2$ (99.99% purity) stream at the flow rate of 40 mL/min for 60 min. After cooling to the adsorption temperature (30°C), the adsorption tests began by introducing simulated biogas into the adsorption column at a flow rate of 40 mL/min. The simulated biogas was produced by mixing $CO_2$ (99.99% purity) and $N_2$ (99.99% purity) from gas cylinders at the desired flow rates to obtain a concentration of 40% $CO_2$ and 60% $N_2$. The adsorbent-bed

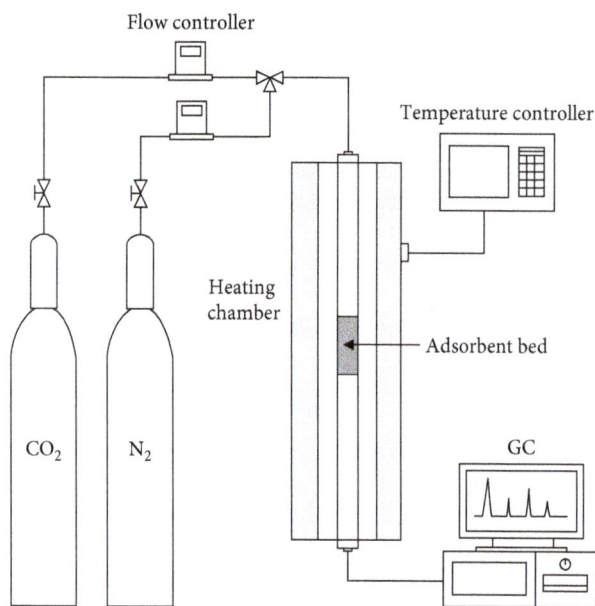

FIGURE 1: Schematic diagram of the $CO_2$ adsorption system.

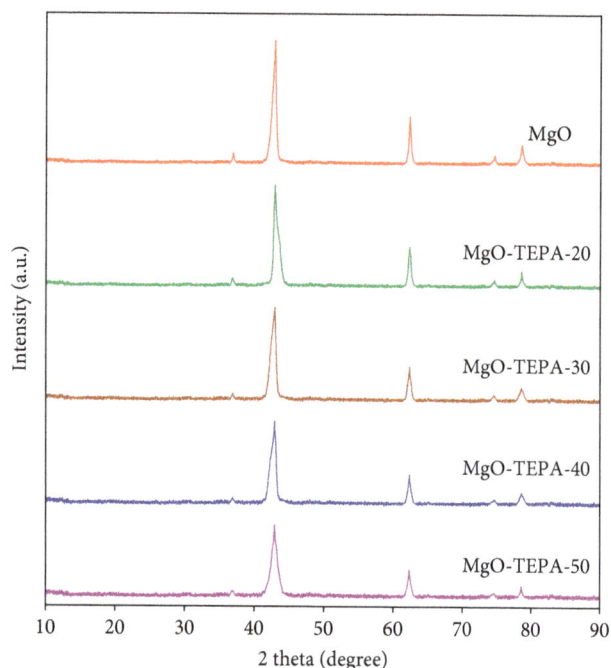

FIGURE 2: XRD patterns of MgO with different TEPA loadings.

temperature was measured by thermocouples and controlled by a temperature controller. After completing the adsorption process, desorption of the adsorbed $CO_2$ was performed by heating the sample to 150°C in a $N_2$ gas stream for 60 min. In order to test the regenerability of the adsorbent, ten cycles of adsorption-desorption were performed on the same sample. In this study, biogas produced from a local swine farm was also used as feed gas to test the $CO_2$ adsorption performance of the prepared adsorbent using the same experimental procedure. The biogas consists of 38.4% $CO_2$ and 61.5% $CH_4$, which are in the range of the typical biogas compositions found in the literature [39]. The concentrations of $CO_2$ in the feed and treated gas streams were determined by a gas chromatograph equipped with a TCD detector.

The breakthrough curve data obtained from the $CO_2$ adsorption tests were used for calculating the equilibrium capacity of the adsorbents according to equations (3) and (4) [40]:

$$q = \frac{Qt_sC_0}{22.4m}, \tag{3}$$

$$t_s = \int_0^t \left(1 - \frac{C_t}{C_0}\right)dt, \tag{4}$$

where $t_s$ is the mean residence time (min), $C_0$ and $C_t$ are the inlet and outlet concentrations of $CO_2$ (vol.%), respectively, $q$ is the equilibrium adsorption capacity of $CO_2$ (mmol/g), $t$ is the adsorption time (min), $Q$ is the feed volumetric flow rate (ml/min) at standard temperature and pressure (STP), and $m$ is the mass of the adsorbent (g).

## 3. Results and Discussion

*3.1. Characterization.* Figure 2 shows the XRD patterns of the prepared MgO with different TEPA loadings. The pristine MgO exhibits five well-resolved diffraction peaks at

36.9°, 42.9°, 62.3°, 74.7°, and 78.6°. These are characteristic peaks of MgO, which are consistent with the results of previous reports [35, 41, 42]. This confirms a high degree of purity and suggests that the obtained MgO samples are highly crystalline. From comparing the diffraction pattern of MgO with those of MgO-TEPA-$x$ with different TEPA loadings; the diffraction angles are nearly identical, indicating that the structure of MgO is preserved after loading the TEPA. However, the diffraction intensity of MgO decreases with increasing TEPA loadings. It has been reported that diffraction intensities are relative to the degrees of pore filling [43]. Therefore, the incorporation of amine into the pore channels of the support possibly causes the loss of intensity [28]. This indicates that TEPA is loaded into the pore of the MgO support, which is similar to that observed in MCM-41 loaded with polyethylenimine (PEI) [38].

To investigate the porous structure of the prepared samples, $N_2$ adsorption-desorption isotherms were measured. Figure 3 shows the $N_2$ adsorption-desorption isotherms of the prepared MgO with different TEPA loadings. According to the International Union of Pure and Applied Chemistry (IUPAC) classification [44], all of the samples display type IV isotherms with the hysteresis loop. The hysteresis loop present in the isotherm is due to the capillary condensation taking place in mesopores. The pristine MgO exhibits a large hysteresis loop compared with the other four TEPA-functionalized samples, indicating the presence of highly porous structure. In addition, it can be seen that the $N_2$ uptake decreases with increasing TEPA loading. As the TEPA loading increases from 20 to 50 wt.%, the mesopores of MgO are almost completely filled with TEPA molecules at 50 wt.% which results in restricting the access of $N_2$ into the pores, thus causing MgO-TEPA-50 to become a nonporous material [45].

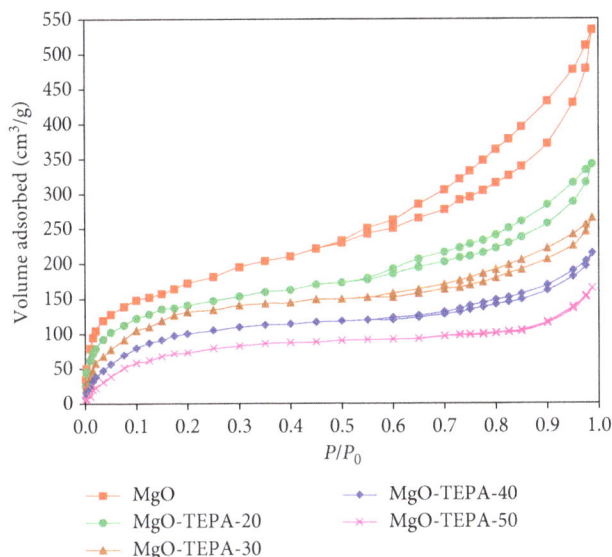

FIGURE 3: Adsorption-desorption isotherms of MgO with different TEPA loadings.

TABLE 1: Surface area, pore volume, and average pore diameter of MgO-TEPA-$x$.

| Adsorbent | BET surface area ($m^2$/g) | Pore volume ($cm^3$/g) | Pore diameter (nm) |
|---|---|---|---|
| MgO | 207 | 0.81 | 7.3 |
| MgO-TEPA-20 | 118 | 0.53 | 4.6 |
| MgO-TEPA-30 | 54 | 0.39 | 3.2 |
| MgO-TEPA-40 | 19 | 0.22 | 1.9 |
| MgO-TEPA-50 | 4 | 0.08 | 1.1 |

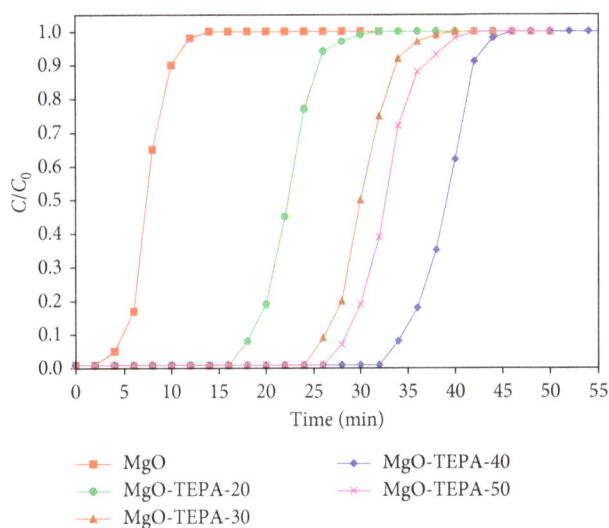

FIGURE 4: Breakthrough curves of MgO with different TEPA loadings.

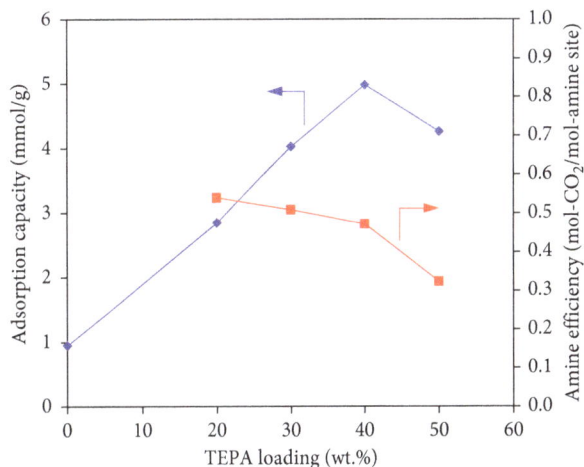

FIGURE 5: $CO_2$ adsorption capacity and amine efficiency of MgO with different TEPA loadings.

From the $N_2$ adsorption-desorption isotherms, the surface area, pore volume, and average pore diameter of the MgO samples were calculated according to the BET and BJH method and are listed in Table 1. Loading TEPA onto MgO significantly reduces the surface area and pore volume. For example, when the TEPA loading is 20 wt.%, the surface area and the pore volume decrease from 207 to 118 $m^2$/g and from 0.81 to 0.53 $cm^3$/g, respectively. A further increase in TEPA loading to 50 wt.% causes a great decrease in the surface area and the pore volume. This is because, when the TEPA loading reaches 50 wt.% or higher, excessive amount of TEPA blocks most of the pores and covers the surface of the adsorbent [46, 47]. The same trend as the surface area and pore volume is observed for the pore size of the samples.

3.2. CO_2 Adsorption Behaviors. The effect of TEPA loading on the $CO_2$ adsorption performance of MgO-TEPA was investigated using simulated biogas as the feed gas. Figures 4 and 5 show the breakthrough curves and the adsorption capacities (mmol of $CO_2$/g of adsorbent), respectively. As can be seen, the $CO_2$ adsorption of the pristine MgO reaches the saturation point within only 12 min, resulting in a low adsorption capacity of 0.95 mmol/g. Increasing TEPA loading within a suitable range increases the breakthrough time and adsorption capacity of the adsorbents, which is consistent with the results of previous reports [48, 49]. This is because, as the loading increases, more amine active sites for $CO_2$ adsorption are provided. When the TEPA loading increases to 40 wt.% (MgO-TEPA-40), the adsorbent shows the maximum breakthrough time and adsorption capacity of 34 min and 4.98 mmol/g, respectively. However, as the TEPA loading increases from 40 to 50 wt.%, the breakthrough time and the adsorption capacity significantly drop from 34 to 28 min and from 4.98 to 4.26 mmol/g, respectively. Too high loadings may cause the TEPA molecules to aggregate in the pore or coat the surface of the adsorbent,

which hinder the diffusion of $CO_2$ to react with the active sites in the pores [48, 50]. The above results suggest that the optimum TEPA loading for MgO modification is 40 wt.%.

The most common mechanism for $CO_2$ capture in supported amines under dry conditions involves the formation of carbamates between two amine groups [51, 52]. Two types of these amine groups exist within TEPA including primary ($R_1$-$NH_2$) and secondary ($R_1$-NH-$R_2$)

TABLE 2: Comparison of $CO_2$ adsorption capacity of amine-impregnated adsorbents.

| Support | Amine | | Gas composition | Adsorption temperature (°C) | $CO_2$ adsorption capacity (mmol/g) | Reference |
|---------|-------|------|------|------|------|------|
| | Type | wt.% | | | | |
| MCM-41 | PEI | 75 | 100% $CO_2$ | 75 | 3.02 | [28] |
| SBA-15 | PEI | 50 | 15% $CO_2$ | 75 | 3.18 | [53] |
| Carbon black | PEI | 50 | 100% $CO_2$ | 75 | 3.07 | [54] |
| SBA-15 | PEI | 50 | 100% $CO_2$ | 75 | 2.89 | [55] |
| KIT-6 | PEI | 50 | 100% $CO_2$ | 75 | 3.07 | [55] |
| KIT-6 | PEI | 50 | 100% $CO_2$ | 105 | 3.10 | [56] |
| SBA-15 | TEPA | 70 | 100% $CO_2$ | 75 | 3.93 | [57] |
| HMS | PEI | 60 | 100% $CO_2$ | 75 | 4.18 | [58] |
| AC | DETA | 39 | 100% $CO_2$ | 25 | 0.91 | [59] |
| $Al_2O_3$ | DETA | 40 | 100% $CO_2$ | 25 | 1.41 | [60] |
| $SiO_2$ | APTES | 70 | 10% $CO_2$ | 100 | 2.03 | [61] |
| PMMA | PEI | 40 | 15% $CO_2$ | 45 | 3.65 | [62] |
| SBA-15 | PEI | 60 | 15% $CO_2$ | 75 | 3.14 | [63] |
| MMSV | PEI | 60 | 100% $CO_2$ | 90 | 4.73 | [64] |
| MSU-J | TEPA | 50 | 100% $CO_2$ | 25 | 3.73 | [50] |
| MgO | — | — | 40% $CO_2$ | 30 | 0.95 | This study |
| MgO | TEPA | 40 | 40% $CO_2$ | 30 | 4.98 | This study |

amines. The reactions of $CO_2$ with the primary and secondary amines in TEPA can be represented as in equations (5)–(7) [52]:

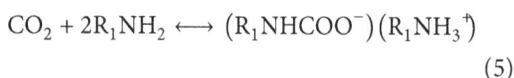

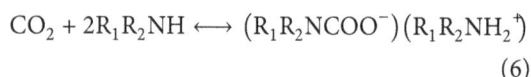

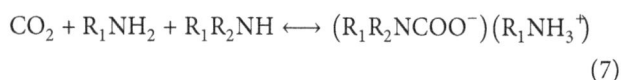

$$CO_2 + 2R_1NH_2 \longleftrightarrow (R_1NHCOO^-)(R_1NH_3^+)$$
(5)

$$CO_2 + 2R_1R_2NH \longleftrightarrow (R_1R_2NCOO^-)(R_1R_2NH_2^+)$$
(6)

$$CO_2 + R_1NH_2 + R_1R_2NH \longleftrightarrow (R_1R_2NCOO^-)(R_1NH_3^+)$$
(7)

In order to explore the degree of utilization of amine groups in TEPA by $CO_2$, the amine efficiency was defined as the molar ratio of the adsorbed $CO_2$ to all amine groups present in the adsorbents [48]. The number of amine groups was calculated from the molar amount of nitrogen atoms in TEPA. The molecular weight of TEPA is 189.3 g/mol, and each TEPA molecule contains 5 amine groups. Therefore, the amine efficiency was calculated and is presented in Figure 5. It can be seen that the amine efficiency decreases with an increase in TEPA loading. The highest amine efficiency of 0.54 is obtained at 20 wt.% TEPA loading, suggesting that two moles of amine groups reacts with one mole of $CO_2$. At low TEPA loading, the amine chains are more dispersible, making the adsorption sites easily available. However, as the amine loading increases, amines would begin to conglomerate within the pores. This leads to poor distribution of amine sites, resulting in a reduction in the number of the accessible amine sites for $CO_2$ sorption, although the potential amine sites are more for higher TEPA loading. With a further increase in TEPA loading to 50 wt.%, the amine efficiency decreases significantly to 0.32, indicating that some of amine sites are unoccupied.

The $CO_2$ adsorption capacities of MgO and MgO-TEPA-40 in this study and those of amine-impregnated adsorbents reported in the literature are compared in Table 2. As can be seen, various supports and amines are used for adsorbent preparation. The comparison shows that the $CO_2$ adsorption capacity of MgO-TEPA-40 is higher than those previously reported in the literature.

### 3.3. Regenerability of the Adsorbents.
In industrial applications, the stability and regenerability of $CO_2$ adsorbents are critical parameters for long-term operation. In this study, the cyclic $CO_2$ adsorption capacity of MgO-TEPA-40 was investigated by performing ten $CO_2$ adsorption-desorption cycles using simulated biogas and real biogas as feed gas streams. Figure 6 shows the cyclic $CO_2$ adsorption capacities of the adsorbent in ten consecutive runs. For the simulated biogas, the $CO_2$ adsorption capacity decreases slightly with increasing number of adsorption-desorption cycle. After ten cycles, the $CO_2$ adsorption capacity decreases from 4.98 to 4.71 mmol/g, which is a decrease of only 5.42%. The $CO_2$ adsorption capacity for the biogas exhibits the same trend, decreasing from 4.87 to 4.59 mmol/g after ten cycles, meaning the capacity loss of only 5.75%. These capacity drops are lower than those of some amine-based adsorbents reported in the literature [49, 65]. The loss of adsorption capacity during the adsorption-desorption cycles could be due to the volatilization of the impregnated TEPA [51, 66].

## 4. Conclusions

Mesoporous MgO was synthesized and functionalized by impregnation with different tetraethylenepentamine (TEPA) loadings. The prepared MgO-TEPA was used as an adsorbent for $CO_2$ separation from simulated biogas. The $CO_2$ adsorption capacity of the adsorbents was found to increase with increasing TEPA loading. MgO-TEPA-40 with 40 wt.% TEPA exhibits the best $CO_2$ adsorption performance, with the $CO_2$ adsorption capacity of 4.98 mmol/g. However, a further increase in TEPA loading to 50 wt.% causes a significant reduction of $CO_2$ adsorption capacity. The stability

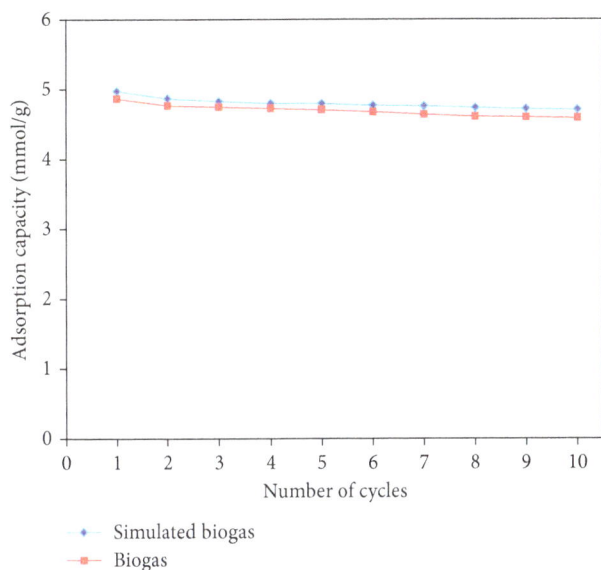

FIGURE 6: Cyclic $CO_2$ adsorption capacities of MgO-TEPA-40.

and regenerability of MgO-TEPA-40 were examined by performing ten consecutive $CO_2$ adsorption-desorption runs under simulated biogas and real biogas conditions. MgO-TEPA-40 shows slight decreases of only 5.42 and 5.75% of $CO_2$ adsorption capacity for the simulated biogas and biogas, respectively, after ten cycles. These results indicate that MgO-TEPA-40 has a stable $CO_2$ adsorption capacity and good regenerability both in simulated biogas and biogas, which are beneficial properties for practical applications as $CO_2$ adsorbent.

## Conflicts of Interest

The authors declare that there are no conflicts of interest regarding the publication of this paper.

## Acknowledgments

The authors gratefully acknowledge the Faculty of Science and Technology, Prince of Songkla University, for the instruments and facilities used in this study.

## References

[1] M. Poeschl, S. Ward, and P. Owende, "Environmental impacts of biogas deployment-part II: life cycle assessment of multiple production and utilization pathways," *Journal of Cleaner Production*, vol. 24, pp. 184–201, 2012.

[2] W. Fan, Y. Liu, and K. Wang, "Detailed experimental study on the performance of monoethanolamine, diethanolamine, and diethylenetriamine at absorption/regeneration conditions," *Journal of Cleaner Production*, vol. 125, pp. 296–308, 2016.

[3] I. Angelidaki, L. Treu, P. Tsapekos et al., "Biogas upgrading and utilization: current status and perspectives," *Bio-technology Advances*, vol. 36, no. 2, pp. 452–466, 2018.

[4] L. Lombardi and E. Carnevale, "Economic evaluations of an innovative biogas upgrading method with $CO_2$ storage," *Energy*, vol. 62, pp. 88–94, 2013.

[5] S. Rasi, A. Veijanen, and J. Rintala, "Trace compounds of biogas from different biogas production plants," *Energy*, vol. 32, no. 8, pp. 1375–1380, 2007.

[6] J. Arroyo, F. Moreno, M. Muñoz, C. Monné, and N. Bernal, "Combustion behavior of a spark ignition engine fueled with synthetic gases derived from biogas," *Fuel*, vol. 117, pp. 50–58, 2014.

[7] O. W. Awe, Y. Zhao, A. Nzihou, D. P. Minh, and N. Lyczko, "A Review of biogas utilisation, purification and upgrading technologies," *Waste and Biomass Valorization*, vol. 8, no. 2, pp. 267–283, 2017.

[8] E. Ryckebosch, M. Drouillon, and H. Vervaeren, "Techniques for transformation of biogas to biomethane," *Biomass and Bioenergy*, vol. 35, no. 5, pp. 1633–1645, 2011.

[9] G. T. Rochelle, "Amine scrubbing for $CO_2$ capture," *Science*, vol. 325, no. 5948, pp. 1652–1654, 2009.

[10] B. Dutcher, M. Fan, and A. G. Russell, "Amine-based $CO_2$ capture Technology development from the beginning of 2013-A review," *ACS Applied Materials and Interfaces*, vol. 7, no. 4, pp. 2137–2148, 2015.

[11] C.-H. Yu, C.-H. Huang, and C.-S. Tan, "A review of $CO_2$ capture by absorption and adsorption," *Aerosol and Air Quality Research*, vol. 12, no. 5, pp. 745–769, 2012.

[12] S. Ma'mun, R. Nilsen, and H. F. Svendsen, "Solubility of carbon dioxide in 30 mass % monoethanolamine and 50 mass % methyldiethanolamine solutions," *Journal of Chemical and Engineering Data*, vol. 50, no. 2, pp. 630–634, 2005.

[13] B. P. Mandal and S. S. Bandyopadhyay, "Absorption of carbon dioxide into aqueous blends of 2-amino-2-methyl-1-propanol and monoethanolamine," *Chemical Engineering Science*, vol. 61, no. 16, pp. 5440–5447, 2006.

[14] G.-P. Hao, W.-C. Li, and A.-H. Lu, "Novel porous solids for carbon dioxide capture," *Journal of Materials Chemistry*, vol. 21, no. 18, pp. 6447–6451, 2011.

[15] R. V. Siriwardane, M.-S. Shen, E. P. Fisher, and J. A. Poston, "Adsorption of $CO_2$ on molecular sieves and activated carbon," *Energy and Fuels*, vol. 15, no. 2, pp. 279–284, 2001.

[16] A. E. Ogungbenro, D. V. Quang, K. Al-Ali, and M. R. M. Abu-Zahra, "Activated carbon from date seeds for CO2 capture applications," *Energy Procedia*, vol. 114, pp. 2313–2321, 2017.

[17] S. Loganathan, M. Tikmani, and A. K. Ghoshal, "Novel pore-expanded MCM-41 for $CO_2$ capture: synthesis and characterization," *Langmuir*, vol. 29, no. 10, pp. 3491–3499, 2013.

[18] R. V. Siriwardane, M.-S. Shen, E. P. Fisher, and J. Losch, "Adsorption of $CO_2$ on zeolites at moderate temperatures," *Energy and Fuels*, vol. 19, no. 3, pp. 1153–1159, 2005.

[19] M. Mofarahi and F. Gholipour, "Gas adsorption separation of $CO_2/CH_4$ system using zeolite 5A," *Microporous and Mesoporous Materials*, vol. 200, pp. 1–10, 2014.

[20] D. Britt, H. Furukawa, B. Wang, T. G. Glover, and O. M. Yaghi, "Highly efficient separation of carbon dioxide by a metal-organic framework replete with open metal sites," *Proceedings of National Academy of Sciences*, vol. 106, no. 49, pp. 20637–20640, 2009.

[21] A. R. Millward and O. M. Yaghi, "Metal–Organic frameworks with exceptionally high capacity for storage of carbon dioxide at room temperature," *Journal of American Chemical Society*, vol. 127, no. 51, pp. 17998-17999, 2005.

[22] N. Gargiulo, F. Pepe, and D. Caputo, "$CO_2$ adsorption by functionalized nanoporous materials: a review," *Journal of Nanoscience and Nanotechnology*, vol. 14, no. 2, pp. 1811–1822, 2014.

[23] G. Qi, Y. Wang, L. Estevez et al., "High efficiency nanocomposite sorbents for $CO_2$ capture based on amine-functionalized mesoporous capsules," *Energy and Environmental Science*, vol. 4, no. 2, pp. 444–452, 2011.

[24] H. Y. Huang, R. T. Yang, D. Chinn, and C. L. Munson, "Amine-grafted MCM-48 and silica xerogel as superior sorbents for acidic gas removal from natural gas," *Industrial and Engineering Chemistry Research*, vol. 42, no. 12, pp. 2427–2433, 2003.

[25] P. J. E. Harlick and A. Sayari, "Applications of pore-expanded mesoporous silica. 5. Triamine grafted material with exceptional $CO_2$ Dynamic and equilibrium adsorption performance," *Industrial and Engineering Chemistry Research*, vol. 46, no. 2, pp. 446–458, 2007.

[26] N. Hiyoshi, K. Yogo, and T. Yashima, "Adsorption characteristics of carbon dioxide on organically functionalized SBA-15," *Microporous and Mesoporous Materials*, vol. 84, no. 1–3, pp. 357–365, 2005.

[27] A. C. C. Chang, S. S. C. Chuang, M. Gray, and Y. Soong, "In-situ infrared study of $CO_2$ Adsorption on SBA-15 grafted with $\gamma$-(aminopropyl)triethoxysilane," *Energy and Fuels*, vol. 17, no. 2, pp. 468–473, 2003.

[28] X. Xu, C. Song, J. M. Andresen, B. G. Miller, and A. W. Scaroni, "Novel polyethylenimine-modified mesoporous molecular sieve of MCM-41 type as high-capacity adsorbent for $CO_2$ Capture," *Energy and Fuels*, vol. 16, no. 6, pp. 1463–1469, 2002.

[29] M. B. Yue, Y. Chun, Y. Cao, X. Dong, and J. H. Zhu, "$CO_2$ capture by as-prepared SBA-15 with an occluded organic template," *Advanced Functional Materials*, vol. 16, no. 13, pp. 1717–1722, 2006.

[30] M. B. Yue, L. B. Sun, Y. Cao et al., "Promoting the $CO_2$ adsorption in the amine-containing SBA-15 by hydroxyl group," *Microporous and Mesoporous Materials*, vol. 114, no. 1–3, pp. 74–81, 2008.

[31] R. Sanz, G. Calleja, A. Arencibia, and E. S. Sanz-Pérez, "$CO_2$ adsorption on branched polyethyleneimine-impregnated mesoporous silica SBA-15," *Applied Surface Science*, vol. 256, no. 17, pp. 5323–5328, 2010.

[32] K. Zhou, S. Chaemchuen, and F. Verpoort, "Alternative materials in technologies for Biogas upgrading via $CO_2$ capture," *Renewable and Sustainable Energy Reviews*, vol. 79, pp. 1414–1441, 2017.

[33] M. Bhagiyalakshmi, P. Hemalatha, M. Ganesh, P. M. Mei, and H. T. Jang, "A direct synthesis of mesoporous carbon supported MgO sorbent for $CO_2$ capture," *Fuel*, vol. 90, no. 4, pp. 1662–1667, 2011.

[34] A. Hanif, S. Dasgupta, and A. Nanoti, "Facile synthesis of high-surface-area mesoporous MgO with excellent high-temperature $CO_2$ adsorption potential," *Industrial and Engineering Chemistry Research*, vol. 55, no. 29, pp. 8070–8078, 2016.

[35] Y.-D. Ding, G. Song, X. Zhu, R. Chen, and Q. Liao, "Synthesizing MgO with a high specific surface for carbon dioxide adsorption," *RSC Advances*, vol. 5, no. 39, pp. 30929–30935, 2015.

[36] G. Song, Y.-D. Ding, X. Zhu, and Q. Liao, "Carbon dioxide adsorption characteristics of synthesized MgO with various porous structures achieved by varying calcination temperature," *Colloids and Surfaces A: Physicochemical and Engineering Aspects*, vol. 470, pp. 39–45, 2015.

[37] A. Kumar and J. Kumar, "On the synthesis and optical absorption studies of nano-size magnesium oxide powder," *Journal of Physics and Chemistry of Solids*, vol. 69, no. 11, pp. 2764–2772, 2008.

[38] X. Xu, C. Song, J. M. Andrésen, B. G. Miller, and A. W. Scaroni, "Preparation and characterization of novel $CO_2$ "molecular basket" adsorbents based on polymer-modified mesoporous molecular sieve MCM-41," *Microporous and Mesoporous Materials*, vol. 62, no. 1-2, pp. 29–45, 2003.

[39] L. Appels, J. Baeyens, J. Degrève, and R. Dewil, "Principles and potential of the anaerobic digestion of waste-activated sludge," *Progress in Energy and Combustion Science*, vol. 34, no. 6, pp. 755–781, 2008.

[40] H. Chang and Z.-X. Wu, "Experimental study on adsorption of carbon dioxide by 5A molecular sieve for helium purification of high-temperature gas-cooled reactor," *Industrial and Engineering Chemistry Research*, vol. 48, no. 9, pp. 4466–4473, 2009.

[41] W. Zhou, S. Upreti, and M. S. Whittingham, "High performance Si/MgO/graphite composite as the anode for lithium-ion batteries," *Electrochemistry Communications*, vol. 13, no. 10, pp. 1102–1104, 2011.

[42] K. K. Han, Y. Zhou, W. G. Lin, and J. H. Zhu, "One-pot synthesis of foam-like magnesia and its performance in $CO_2$ adsorption," *Microporous and Mesoporous Materials*, vol. 169, pp. 112–119, 2013.

[43] X. Wang, Q. Guo, J. Zhao, and L. Chen, "Mixed amine-modified MCM-41 sorbents for $CO_2$ capture," *International Journal of Greenhouse Gas Control*, vol. 37, pp. 90–98, 2015.

[44] K. S. W. Sing, D. H. Everett, R. A. W. Haul et al., "Reporting physisorption data for gas/solid systems with special reference to the determination of surface area and porosity (recommendations 1984)," *Pure and Applied Chemistry*, vol. 57, no. 4, pp. 603–619, 1985.

[45] S. Ahmed, A. Ramli, and S. Yusup, "$CO_2$ adsorption study on primary, secondary and tertiary amine functionalized Si-MCM-41," *International Journal of Greenhouse Gas Control*, vol. 51, pp. 230–238, 2016.

[46] X. Wang, X. Ma, C. Song et al., "Molecular basket sorbents polyethylenimine-SBA-15 for $CO_2$ capture from flue gas: characterization and sorption properties," *Microporous and Mesoporous Materials*, vol. 169, pp. 103–111, 2013.

[47] Q. Ye, J. Jiang, C. Wang, Y. Liu, H. Pan, and Y. Shi, "Adsorption of low-concentration carbon dioxide on amine-modified carbon nanotubes at ambient temperature," *Energy and Fuels*, vol. 26, no. 4, pp. 2497–2504, 2012.

[48] M. Yao, Y. Dong, X. Hu et al., "Tetraethylenepentamine-Modified silica nanotubes for low-temperature $CO_2$ capture," *Energy and Fuels*, vol. 27, no. 12, pp. 7673–7680, 2013.

[49] M. Irani, K. A. M. Gasem, B. Dutcher, and M. Fan, "$CO_2$ capture using nanoporous TiO(OH) 2/tetraethylenepentamine," *Fuel*, vol. 183, pp. 601–608, 2016.

[50] J. Jiao, J. Cao, Y. Xia, and L. Zhao, "Improvement of adsorbent materials for $CO_2$ capture by amine functionalized mesoporous silica with worm-hole framework structure," *Chemical Engineering Journal*, vol. 306, pp. 9–16, 2016.

[51] W. Wang, J. Xiao, X. Wei, J. Ding, X. Wang, and C. Song, "Development of a new clay supported polyethylenimine composite for $CO_2$ capture," *Applied Energy*, vol. 113, pp. 334–341, 2014.

[52] L. He, M. Fan, B. Dutcher et al., "Dynamic separation of ultradilute $CO_2$ with a nanoporous amine-based sorbent," *Chemical Engineering Journal*, vol. 189-190, pp. 13–23, 2012.

[53] X. Ma, X. Wang, and C. Song, ""Molecular basket" sorbents for separation of $CO_2$ and $H_2S$ from various gas streams," *Journal of the American Chemical Society*, vol. 131, no. 16, pp. 5777–5783, 2009.

[54] D. Wang, C. Sentorun-Shalaby, X. Ma, and C. Song, "High-capacity and low-cost carbon-based "molecular basket" sorbent for CO2Capture from flue gas," *Energy and Fuels*, vol. 25, no. 1, pp. 456–458, 2011.

[55] W.-J. Son, J.-S. Choi, and W.-S. Ahn, "Adsorptive removal of carbon dioxide using polyethyleneimine-loaded mesoporous silica materials," *Microporous and Mesoporous Materials*, vol. 113, no. 1–3, pp. 31–40, 2008.

[56] R. Kishor and A. K. Ghoshal, "High molecular weight polyethyleneimine functionalized three dimensional mesoporous silica for regenerable CO2 separation," *Chemical Engineering Journal*, vol. 300, pp. 236–244, 2016.

[57] M. B. Yue, L. B. Sun, Y. Cao, Y. Wang, Z. J. Wang, and J. H. Zhu, "Efficient $CO_2$ capturer derived from as-synthesized MCM-41 modified with amine," *Chemistry-A European Journal*, vol. 14, no. 11, pp. 3442–3451, 2008.

[58] C. Chen, W.-J. Son, K.-S. You, J.-W. Ahn, and W.-S. Ahn, "Carbon dioxide capture using amine-impregnated HMS having textural Mesoporosity," *Chemical Engineering Journal*, vol. 161, no. 1-2, pp. 46–52, 2010.

[59] M. G. Plaza, C. Pevida, A. Arenillas, F. Rubiera, and J. J. Pis, "$CO_2$ capture by adsorption with nitrogen enriched carbons," *Fuel*, vol. 86, no. 14, pp. 2204–2212, 2007.

[60] M. G. Plaza, C. Pevida, B. Arias et al., "Application of thermogravimetric analysis to the evaluation of aminated solid sorbents for $CO_2$ capture," *Journal of Thermal Analysis and Calorimetry*, vol. 92, no. 2, pp. 601–606, 2008.

[61] D. V. Quang, T. A. Hatton, and M. R. M. Abu-Zahra, "Thermally stable Amine-grafted adsorbent prepared by impregnating 3-aminopropyltriethoxysilane on mesoporous silica for CO2 capture," *Industrial and Engineering Chemistry Research*, vol. 55, no. 29, pp. 7842–7852, 2016.

[62] M. L. Gray, J. S. Hoffman, D. C. Hreha et al., "Parametric study of solid amine sorbents for the capture of carbon dioxide†," *Energy and Fuels*, vol. 23, no. 10, pp. 4840–4844, 2009.

[63] J. Wang, M. Wang, B. Zhao, W. Qiao, D. Long, and L. Ling, "Mesoporous carbon-supported solid amine sorbents for low-temperature carbon dioxide capture," *Industrial and Engineering Chemistry Research*, vol. 52, no. 15, pp. 5437–5444, 2013.

[64] L. Zhang, N. Zhan, Q. Jin, H. Liu, and J. Hu, "Impregnation of polyethylenimine in mesoporous multilamellar silica vesicles for $CO_2$ capture: a kinetic study," *Industrial and Engineering Chemistry Research*, vol. 55, no. 20, pp. 5885–5891, 2016.

[65] L. Guo, J. Yang, G. Hu, X. Hu, H. DaCosta, and M. Fan, "$CO_2$ removal from flue gas with amine-impregnated titanate nanotubes," *Nano Energy*, vol. 25, pp. 1–8, 2016.

[66] Y. Wang, T. Du, Y. Song, S. Che, X. Fang, and L. Zhou, "Amine-functionalized mesoporous ZSM-5 zeolite adsorbents for carbon dioxide capture," *Solid State Sciences*, vol. 73, pp. 27–35, 2017.

# 15

# Synthesis of Wollastonite Powders by Combustion Method: Role of Amount of Fuel

Imarally V. de S. R. Nascimento [1,2] Willams T. Barbosa,[1,2] Raúl G. Carrodeguas,[1] Marcus V. L. Fook,[1] and Miguel A. Rodríguez [2]

[1]CERTBIO, Universidade Federal de Campina Grande, Campina Grande, Brazil
[2]Instituto de Cerámica y Vidrio (CSIC), Madrid, Spain

Correspondence should be addressed to Imarally V. de S. R. Nascimento; imarally.souza@hotmail.com

Academic Editor: Donald L. Feke

The objective of this work has been the synthesis of wollastonite by solution combustion method. The novelty of this work has been obtaining the crystalline phase without the need of thermal treatments after the synthesis. For this purpose, urea was used as fuel. Calcium nitrate was selected as a source of calcium and colloidal silica served as a source of silicon. The effect of the amount of fuel on the combustion process was investigated. Temperature of the combustion reaction was followed by digital pyrometry. The obtained products were characterized by scanning electron microscopy (SEM), X-ray diffraction (XRD), and specific surface area. The results showed that the combustion synthesis provides nanostructured powders characterized by a high surface area. When excess of urea was used, wollastonite-2M was obtained with a submicronic structure.

## 1. Introduction

The materials which are used to replace or supplement the functions of living tissues are known as biomaterials. To meet the requirement of an ideal biomaterial, it must be biocompatible, biodegradable, and bioactive, among others. One of the areas where biomaterials are used is in bone substitute or regeneration, which most often uses scaffolds. In this case, in addition to the intrinsic properties of the biomaterials, these must be osteoconductors and osteoinducers. Also, they must present an interconnected network of open porosity that is essential for cell nutrition, proliferation, and migration for tissue vascularization and providing mechanical support until formation of new tissues [1].

A variety of materials such as metals [2], ceramics [3], natural and synthetic polymers [4], and their combinations [5] have been explored for replacement and repair of damaged bone tissues. An extensive attention has been paid on the development of bioactive materials, including calcium phosphates, silicates, glasses, etc. which were used in

the tissue regeneration applications [5, 6]. More attention has been paid on the calcium silicate ceramic materials, particularly wollastonite ($CaSiO_3$), used as a biomaterial for bone regeneration because of its excellent *in vitro* bioactivity. A lot of authors have studied their ability to induce formation of the hydroxyapatite layer on their surface *in vitro* conditions when immersed in simulated body fluid (SBF) with ion concentrations, pH, and temperature almost equal to those of human blood plasma [7–13].

It was demonstrated by De Aza and Luklinska [14], De Aza, Guitian, and De Aza [15, 16], and Wu et al. [17] that materials which contain Ca and Si atoms in their composition and which are free of P atoms can also produce HA layer on their surface. Fiocco et al. [9] proposes that the Mg atoms can further enhance the bioactivity of calcium-silicate materials.

Generally, calcium silicates are synthesized at high temperatures, between 900 and 1100°C. Wollastonite can be synthesized by the solid state route [18], sol gel [19], or coprecipitation method [20] but these methods require high energy and a long time for the formation of products.

Solution combustion synthesis (SCS) is widely used to synthesize submicronic oxide ceramic powders. This synthesis consists in an exothermic redox reaction, associated with the decomposition of nitrates (oxidants) and the oxidation of the fuel. The heat evolved is larger than that needed to maintain the combustion process; thus, the system becomes self-sustained. The chemical energy released from the exothermic reaction provides spontaneous synthesis [20–22]. The SCS has become an excellent technique for synthesis because of their inherent superior qualities, which include the better homogeneity, composition control, lower processing temperature, and single step operation, which result in high reactive powders [20, 23–25]. The final product is usually a crystalline material with nanometric size clusters and has a large specific surface area as a consequence of the large amount of gases produced during the synthesis process. This high volume of gases produces high porous agglomerates. The ignition temperature ($T_{ig}$) is significantly lower than the combustion temperature ($T_c$) which results in the final solid phase formation [23].

Chakradhar et al. [26] prepared macroporous nanocrystalline wollastonite ($CaSiO_3$) ceramic powders by a simple, low-temperature initiated, self-propagating, and solution combustion process. The phases of $\beta$-$CaSiO_3$ and $\alpha$-$CaSiO_3$ were obtained only after heating at 950°C and 1200°C, respectively. Huang and Chang [22] synthesized wollastonite ($\alpha$-$CaSiO_3$) by a citrate-nitrate gel combustion method using citric acid as a fuel and nitrate as an oxidant. The formation of $\alpha$-$CaSiO_3$ was observed only after calcining powders at 650°C for 2 h.

The objective of this work was obtaining $\beta$-$CaSiO_3$ by combustion synthesis without the need of using further heating treatment. For this, $Ca(NO_3)_2$ and colloidal $SiO_2$ were employed as precursors and urea was used as fuel. The effect of the amount of fuel on the combustion process and characteristics of the products were studied as well.

## 2. Materials and Methods

Calcium nitrate tetrahydrate ($Ca(NO_3)_2.4H_2O$) (Neon Co., A.R. grade, 97%), colloidal $SiO_2$ (Sigma-Aldrich, Co., 50 wt% in water), urea ($NH_2CONH_2$) (Sigma-Aldrich, Co., A.R. grade, 99.5%), and nitric acid ($HNO_3$) (Sigma Aldrich, Co., A.R. grade 70%) were used as raw materials. The amounts of the reagents required to obtain 5 g of product were dissolved in 50 mL of deionized water.

The solution was dried and homogenized by stirring at 80°C to remove excess water. The beaker containing the solution was introduced into a muffle furnace preheated to 600°C. When placed in the furnace, the mixed solution soon started to boil, underwent dehydration, and the mass swelled to yield foam. The entire process required less than 15 min with flame duration of nearly one minute. The as-synthesized products were typically voluminous, fluffy foam-like mass that occupied a large volume. The resulting soft agglomerate was readily ground manually in an agate mortar/pestle into fine powder and was characterized.

*2.1. Characterization.* The phase composition of the prepared powders was determined by X-ray diffraction (XRD; Bruker D8) using nickel-filtered CuK$_\alpha$ radiation. Intensities were collected by step-scanning from 10° to 60° ($2\theta$) with a step size of 0.02° and counting time of 1 s for each step. Fourier-transform infrared spectroscopy was obtained using a SPECTRUM 400 FTIR spectrometer. The samples were prepared at potassium bromide (KBr) pellets (ca. 2% by mass in KBr). The infrared spectra of prepared samples between 400 and 4000 cm$^{-1}$ were recorded. The specific surface area ($S_{BET}$) of the powders was determined by nitrogen absorption (ASAP 2020M Analyzer, Micromeritics). The equivalent diameter ($d_{BET}$) of particle was calculated from the measured surface area ($S_{BET}$) values by using the following relationship:

$$d_{BET} = \frac{6}{S_{BET}\,\rho},\tag{1}$$

where $\rho$ is the theoretical density of the powder (2.8 g/cm$^3$ for the wollastonite). Field emission scanning electron microscopy (FESEM S-4700, Hitachi, Japan) was used to characterize the microstructure and morphology of the combustion product, which beforehand was coated with gold, by sputtering. Temperature profiles of the powder mixture were measured by a two-color pyrometer (IMPAC model IGAR 12-LO) that can measure temperature in the range 500°C–2200°C. The pyrometer was placed near the open door of the reactor furnace. The optics (with a glass fiber guide) was focused on the sample. Maximum flame temperatures were theoretically calculated for the combustion reactions with urea as fuel and for different ratios of fuel/oxidant. This was calculated from enthalpy of combustion reaction ($\Delta H°$) and the heat capacity of products ($C_p$) at constant pressure.

## 3. Results and Discussion

The synthesis of wollastonite by combustion reaction can be represented by the (2):

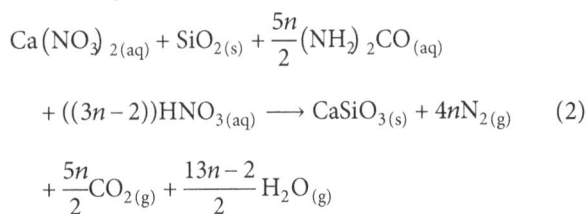

$$Ca(NO_3)_{2(aq)} + SiO_{2(s)} + \frac{5n}{2}(NH_2)_2CO_{(aq)}$$

$$+ ((3n-2))HNO_{3(aq)} \longrightarrow CaSiO_{3(s)} + 4nN_{2(g)}\tag{2}$$

$$+ \frac{5n}{2}CO_{2(g)} + \frac{13n-2}{2}H_2O_{(g)}$$

where $n$ is a coefficient that defines the total amount of fuel. This is related to the total released energy during the synthesis. The values of the coefficient $n$ were 1, 2, and 3 for samples: $n = 1$ implies that the amount of urea required is the stoichiometric ratio of fuel to oxidant, considering aluminium nitrate as the only oxidant. $n = 2$ or 3 implies that the amount of fuel is twice and thrice, respectively, the stoichiometric one; in this case of fuel excess, corresponding amount of $HNO_3$ was added to balance the rest of fuel.

Available thermodynamic data in literature [27] of various reactants and products are listed in Table 1. Under adiabatic condition, the flame temperature ($T_f$) of

TABLE 1: Thermodynamic data employed in calculations [27].

| Compound | $\Delta H_f$ (kJ/mol) | $C_p$ (J/mol.K) |
|---|---|---|
| $Ca(NO_3)_2.4H_2O$ | −2132.33 | 149.4 |
| $SiO_2$ | −903.49 | 43.975 |
| $(NH_2)_2CO$ | −333.51 | 92.79 |
| $HNO_3$ | −174.1 | — |
| $CaSiO_3$ | −1628.4 | 86.448 |
| $CO_2$ | −393.51 | 37.1 |
| $H_2O$ | −241.82 | 35.59 |
| $N_2$ | 0 | 29.124 |

TABLE 2: Influence of $n$ ratio on the maximum combustion temperature.

| $n$ ratio | Measured temperature ($T_c$) (°C) | Calculated temperature ($T_f$) (°C) |
|---|---|---|
| 1 | 803 | 1664 |
| 2 | 819 | 2220 |
| 3 | 825 | 2453 |

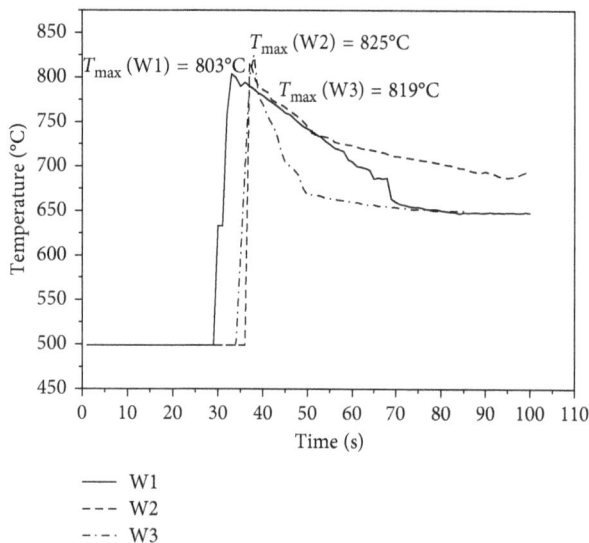

FIGURE 1: Time-temperature profile of aqueous solution combustion synthesis of wollastonite (initial furnace temperature of 600°C).

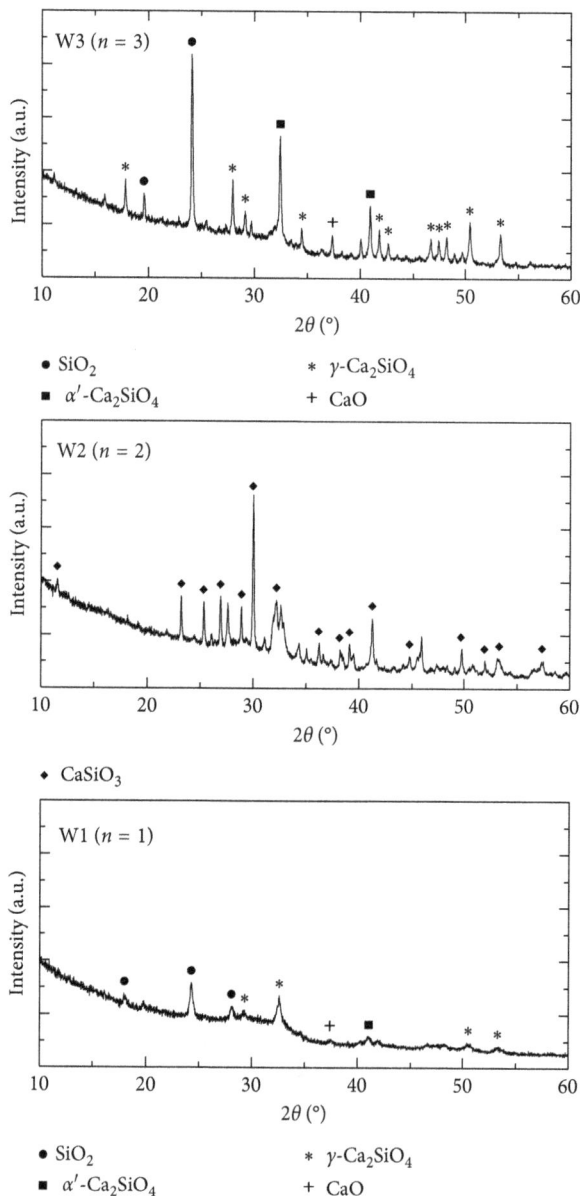

FIGURE 2: XRD patterns of combustion synthesis powders obtained from systems containing different amounts of urea.

combustion systems for different contents of fuel was calculated by Equation (3):

$$Q = -\Delta H_{298} = \int_{T_0}^{T_{ad}} \left( \sum nC_p \right) dT, \qquad (3)$$

where $\Delta H$ is given by $[\sum n\Delta H_f^\circ]_{products} - [\sum n\Delta H_f^\circ]_{reactants}$, $\Delta H_{298}^\circ$ is the standard enthalpy of formation, $C_p$ is the heat capacity of reaction products at constant pressure, and $T_{ad}$ is the adiabatic flame temperature, in Kelvin scale (K).

The typical temperature-time profile for the solution combustion synthesis of $CaSiO_3$ is shown in Figure 1. Initially, it is observed that a short constant temperature region is the referred region where the temperature of the sample is lower than the minimum temperature range of the pyrometer. During this time, water is partially evaporated. Later, a sudden increase in temperature takes place up to a maximum value ($T_{max}$), and the synthesis is terminated with slow cooling.

The measured ($T_c$) and calculated flame temperatures ($T_f$) of this reaction and those with different fuel to oxidizer ratios are shown in Table 2. In all the profiles, it is observed that in the initial instants, the temperature verified by the pyrometer is of 500°C, although the furnace was at 600°C, and this difference of temperature is related to the fact that,

initially, the sample was in a temperature lower than the minimum temperature measured by the pyrometer, which is 500°C. This period is also characterized by the evaporation of still remaining free water in the sample. When the material has not yet reached the furnace temperature, combustion begins. This increases rapidly from temperatures below 500°C to the maximum of combustion.

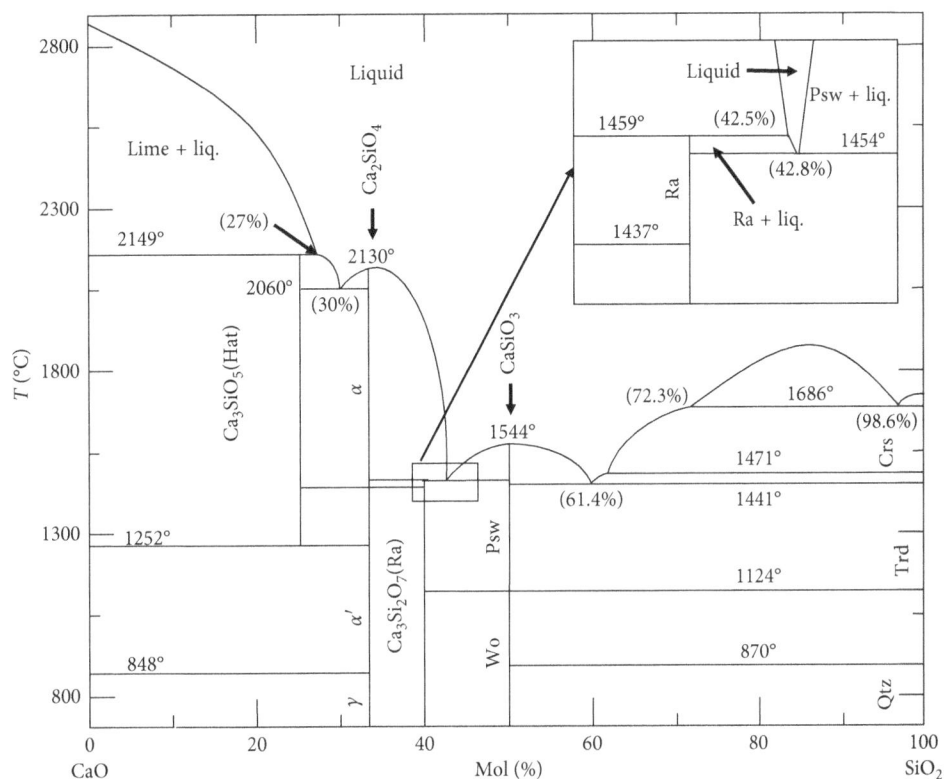

FIGURE 3: The CaO-SiO$_2$ phase diagram by Huang (1995) [28].

The measured temperatures are much lower than the corresponding calculated temperature. The main reason is that, in addition, the pyrometer used for measurement has area of measure approximately 2 mm in diameter, which means that it measures not only the combustion front temperature but also areas of preheating and cooling. In conclusion, the temperature measured corresponds to the average of the area of measurement.

The maximum combustion temperature was observed when an excess ($n = 3$) of fuel was used. The sharp increase in temperature indicates the ignition of the combustion reaction. The maximum temperature is referred to as combustion temperature, and after reaching the maximum temperature, the sample cools down rapidly. It can be observed that a dependence of the measured combustion temperature on the varying ratio, increasing the amount of fuel provides an increase in combustion temperature of the sample.

X-ray diffraction analysis of the combustion products obtained with different amounts of urea is shown in Figure 2. It can be observed that when $n = 1$, the powder showed poor crystallinity and the formed phases were $\alpha'$-Ca$_2$SiO$_4$ (JCPDS file no. 00-036-0642), $\gamma$-Ca$_2$SiO$_4$ (JCPDS file no. 00-049-1672), CaO (JCPDS file no. 01-075-0264), and SiO$_2$ (JCPDS file no. 01-082-1564); probably, the starting materials used did not react as expected and therefore did not form a desired phase, thus causing the formation of the SiO$_2$ phase.

When the amount of urea was increased ($n = 2$), fully crystalline phase of wollastonite-2M was observed (JCPDS

file no. 00-027-0088), indicating that the temperature achieved was sufficient to obtain the desired phase without subsequent calcination.

Further increase of amount urea ($n = 3$) provides the disappearance of wollastonite-2M phase and the appearance phases $\alpha'$-Ca$_2$SiO$_4$ (JCPDS file no. 00-036-0642), $\gamma$-Ca$_2$SiO$_4$ (JCPDS file no. 00-049-1672), CaO (JCPDS file no. 01-075-0264), and SiO$_2$ (JCPDS file no. 01-082-1564). These phases have high crystallinity due to the high temperature of the synthesis, confirmed by the temperature measurements using the pyrometer (Figure 1). The higher combustion temperature provided by the large amount of fuel probably exceeded the melting temperature of wollastonite (as the calculated temperature predicts) and arrives directly to the liquid phase. As can be seen in the phase diagram (Figure 3), at a given temperature ($T = 1544°$C), this phase has a congruent melting point. The direct passage to the liquid phase may have led to precipitation of the coexistent crystalline phases at the reached temperature during cooling.

Figure 4 shows infrared spectrum of all samples. The IR spectroscopic analysis data confirm that the IR peaks in the range of 480–1110 cm$^{-1}$ are due to CaSiO$_3$ [9, 28]. The peaks in ~480 cm$^{-1}$ can be attributed to the bending vibration Si–O–Si bond. Those bands within the range 900–1110 cm$^{-1}$ were assigned to Si–O–Si asymmetric stretching vibration and the O–Si–O stretching vibration absorption band [29, 30]. The peak around 1370 cm$^{-1}$ is attributed to the vibrations of ionic (NO$_3$), indicating the presence of trace amounts of nitrate from the starting material calcium nitrate in the sample; this peak decreases with the increase of the

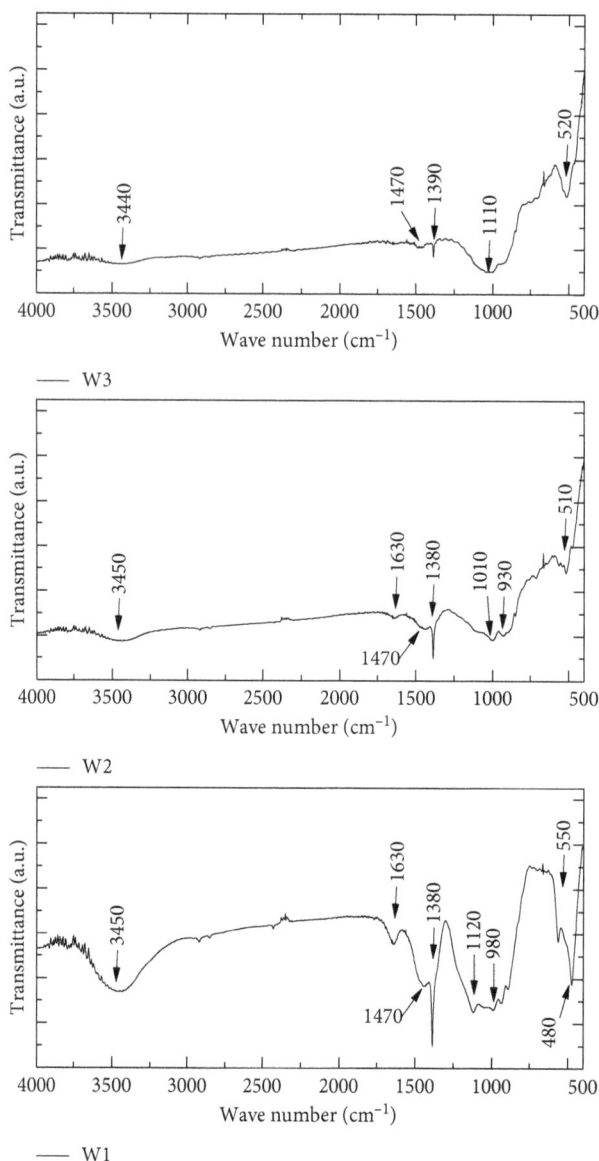

FIGURE 4: FTIR spectra of powders obtained from systems containing different amounts of urea.

TABLE 3: Influence of $n$ ratio on the specific surface area and particle size of as-synthesized powders.

| $n$ ratio | SSA ($m^2/g$) | $d_{BET}$ (nm) |
| --- | --- | --- |
| 1 | 15.7 | 130 |
| 2 | 7.2 | 280 |
| 3 | 15.6 | 130 |

amount of fuel [31]. The bands in the range $1470 \, cm^{-1}$, when a sample was formed, show the presence of CaO in the sample, with the increase amount of fuel, and these bands progressively decreased [32, 33]. The band around $1644 \, cm^{-1}$ can be ascribed to be the bending vibration of the H–O–H bond in molecule water. A broad band around $3400–3600 \, cm^{-1}$ is caused by the stretching vibration of different kinds of hydroxyls and the remaining adsorbed water. With the increase in the amount of urea and

consequently at the combustion temperature, these peaks disappear. The decrease could be attributed to the release of water molecules trapped inside the solid matrix [26, 34, 35].

Table 3 shows the specific surface area (SSA) and particle sizes (estimated from specific surface area ($d_{BET}$) of the powders obtained by combustion as a function of $n$. It is a well-established fact that, in combustion synthesis, the higher the combustion temperature, the longer the flame duration, leading to coarsening of the particles. Within the framework of this study, when $n = 1$, low flame temperature and high gas volume inhibited the particles growth and generated powder with higher surface area. As the amount of fuel increases, $n = 2$, the higher temperature during combustion reaction decreases the surface area due to grain growth. On the contrary, when $n = 3$, a possible melting and precipitation of new phases, caused by the high combustion temperature, provided a small particle size and high specific surface.

FIGURE 5: FESEM micrograph of the synthesized powders: (a, b) $n = 1$, (c, d) $n = 2$, and (e, f) $n = 3$.

FESEM observation of the product (Figure 4) shows that all samples exhibit particle nanostructures with nodular morphology which are quite agglomerated. At higher magnification (Figures 4(b), 4(d), and 4(f)), it is possible to observe that powders are structured submicronically and presented smaller particles were sintered together and form larger particles, which can be due to the high combustion temperature.

Figure 5(d) shows a greater presence of pores, when compared to samples shown in Figures 5(b) and 5(f), which is a microstructural characteristic of the products of combustion due to the escape of a large number of gases during the combustion; the increase of the amount of fuel (Figures 5(e) and 5(f)) provides a decrease in the presence of these pores, indicating a greater sintering of the particles.

In Figures 6(a) and 6(b), it is possible to observe another type of morphology of the samples obtained with $n = 1$ and

$n = 3$, respectively, and a heterogeneous morphology is formed by a crystalline phase (clear part) surrounded by an amorphous phase (dark part). The heterogeneity can be confirmed from the analysis of the X-ray results (Figure 2) of these samples that evidence the formation of distinct phases. In the case of samples obtained with $n = 3$, where the phases found are crystalline, this can be explained due to the fact that the high combustion temperature was obtained that favoured the dissociation and decomposition of the wollastonite obtained when $n = 2$.

## 4. Conclusions

Crystalline $CaSiO_3$ has been synthesized by the solution combustion method without any thermal treatment after the synthesis.

(a)

(b)

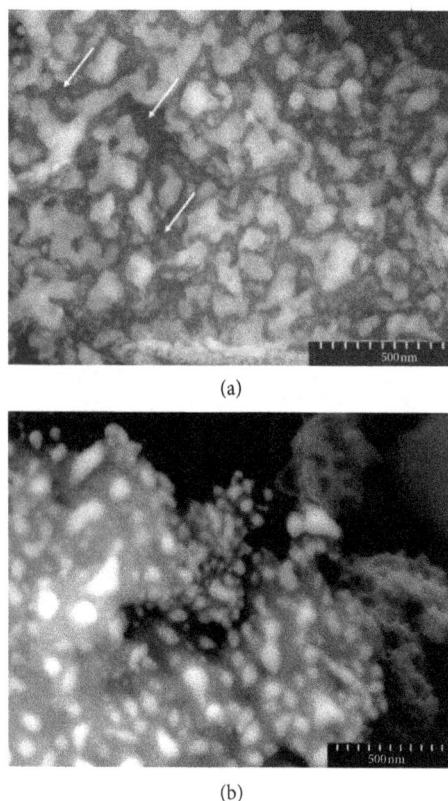

FIGURE 6: FESEM micrograph of the synthesized powders: (a) $n = 1$ and (b) $n = 3$.

It was observed that the amount of fuel significantly influences the composition of the phase.

When an excess of urea ($n = 2$) was used the wollastonite–2M was the only phase obtained.

The use of a stoichiometric amount of fuel ($n = 1$) did not favour the synthesis by combustion and consequently it was not possible to obtain crystalline phases of the material. While a higher amount of fuel ($n = 3$) provided the formation of calcium silicate phases, but it was not possible to obtain the wollastonite.

The combustion synthesis method for obtaining submicronic structured wollastonite, using urea as fuel, is an efficient method for obtaining powders without needing additional heat treatment.

## Conflicts of Interest

The authors declare that there are no conflicts of interest regarding the publication of this paper.

## Acknowledgments

This work was funded by the Ministry of Economy and Competitiveness of Spain under Project MAT2013-48426-C2-1R and Program Science without Borders MEC/MCTI/CAPES/CNPq/FAPs (Call no. 03/2014), Process no. 401220/2014-1.

## References

[1] Q. L. Loh and C. Choong, "Three-dimensional scaffolds for tissue engineering applications: role of porosity and pore size," *Tissue Engineering Part B: Reviews*, vol. 19, no. 6, pp. 485–502, 2013.

[2] N. K. Awad, S. L. Edwards, and Y. S. Morsi, "A review of TiO2 NTs on Ti metal: electrochemical synthesis, functionalization and potential use as bone implants," *Materials Science and Engineering: C*, vol. 76, pp. 1401–1412, 2017.

[3] S. Kunjalukkal Padmanabhan, F. Gervaso, M. Carrozzo, F. Scalera, A. Sannino, and A. Licciulli, "Wollastonite/hydroxyapatite scaffolds with improved mechanical, bioactive and biodegradable properties for bone tissue engineering," *Ceramics International*, vol. 39, no. 1, pp. 619–627, 2013.

[4] X. Liu and P. Ma, "Polymeric scaffolds for bone tissue engineering," *Annals of Biomedical Engineering*, vol. 32, no. 3, pp. 477–486, 2004.

[5] P. Khoshakhlagh, S. M. Rabiee, G. Kiaee et al., "Development and characterization of a bioglass/chitosan composite as an injectable bone substitute," *Carbohydrate Polymers*, vol. 157, pp. 1261–1271, 2017.

[6] B. Li, Z. Liu, J. Yang et al., "Preparation of bioactive β-tricalcium phosphate microspheres as bone graft substitute materials," *Materials Science and Engineering: C*, vol. 70, pp. 1200–1205, 2017.

[7] X. Wan, C. Chang, D. Mao, L. Jiang, and M. Li, "Preparation and in vitro bioactivities of calcium silicate nanophase materials," *Materials Science and Engineering: C*, vol. 25, no. 4, pp. 455–461, 2005.

[8] I. H. M. Aly, L. Abed Alrahim Mohammed, S. Al-Meer, K. Elsaid, and N. A. M. Barakat, "Preparation and characterization of wollastonite/titanium oxide nanofiber bioceramic composite as a future implant material," *Ceramics International*, vol. 42, no. 10, pp. 11525–11534, 2016.

[9] L. Fiocco, S. Li, M. M. Stevens, E. Bernardo, and J. R. Jones, "Biocompatibility and bioactivity of porous polymer-derived Ca-Mg silicate ceramics," *Acta Biomaterialia*, vol. 50, pp. 56–67, 2017.

[10] M. Magallanes-Perdomo, A. H. De Aza, I. Sobrados, J. Sanz, Z. B. Luklinska, and P. Pena, "Structural changes during crystallization of apatite and wollastonite in the eutectic glass of $Ca_3 (PO_4)_2$-$CaSiO_3$ system," *Journal of the American Ceramic Society*, vol. 100, no. 9, pp. 4288–4304, 2017.

[11] P. N. De Aza, Z. B. Luklinska, M. Anseau, F. Guitian, and S. De Aza, "Morphological studies of pseudowollastonite for biomedical application," *Journal of Microscopy*, vol. 182, no. 1, pp. 24–31, 1996.

[12] P. N. De Aza, F. Guitian, A. Merlos, E. Lora-Tamayo, and S. De Aza, "Bioceramics-simulated body fluid interfaces: PH and its influence of hydroxyapatite formation," *Journal of Materials Science: Materials in Medicine*, vol. 7, no. 7, pp. 399–402, 1996.

[13] P. N. De Aza, Z. B. Luklinska, M. R. Anseau, F. Guitian, and S. De Aza, "Bioactivity of pseudowollastonite in human saliva," *Journal of Dentistry*, vol. 27, no. 2, pp. 107–113, 1999.

[14] P. N. De Aza and Z. B. Luklinska, "Effect of glass-ceramic microstructure on its in vitro bioactivity," *Journal of Materials Science: Materials in Medicine*, vol. 14, no. 10, pp. 891–898, 2003.

[15] P. N. de Aza, F. Guitian, and S. de Aza, "Bioactivity of wollastonite ceramics: in vitro evaluation," *Scripta Metallurgica et Materialia*, vol. 31, no. 8, pp. 1001–1005, 1994.

[16] P. N. De Aza, F. Guitian, and S. De Aza, "Polycrystalline wollastonite ceramics. Biomaterials free of $P_2O_5$," in *Advances in Science and Technology: Materials in Clinical Applications*, P. Vicenzini, Ed., pp. 19–27, 1995.

[17] C. Wu, J. Chang, W. Zhai, and S. Ni, "A novel bioactive porous bredigite ($Ca_7MgSi_4O_{16}$) scaffold with biomimetic apatite layer for bone tissue engineering," *Journal of Materials Science: Materials in Medicine*, vol. 18, no. 5, pp. 857–864, 2007.

[18] R. Abd Rashid, R. Shamsudin, M. A. Abdul Hamid, and A. Jalar, "Low temperature production of wollastonite from limestone and silica sand through solid-state reaction," *Journal of Asian Ceramic Societies*, vol. 2, no. 1, pp. 77–81, 2014.

[19] A. Udduttula, S. Koppala, and S. Swamiappan, "Sol-gel combustion synthesis of nanocrystalline wollastonite by using glycine as a fuel and its in vitro bioactivity studies," *Transactions of the Indian Ceramic Society*, vol. 72, no. 4, pp. 257–260, 2013.

[20] R. Morsy, R. Abuelkhair, and T. Elnimr, "Synthesis and in vitro bioactivity mechanism of synthetic α-wollastonite and β-wollastonite bioceramics," *Journal of Ceramic Science and Technology*, vol. 7, no. 1, pp. 65–70, 2016.

[21] M. A. Aghayan and M. A. Rodríguez, "Influence of fuels and combustion aids on solution combustion synthesis of bi-phasic calcium phosphates (BCP)," *Materials Science and Engineering: C*, vol. 32, no. 8, pp. 2464–2468, 2012.

[22] X.-H. Huang and J. Chang, "Synthesis of nanocrystalline wollastonite powders by citrate–nitrate gel combustion method," *Materials Chemistry and Physics*, vol. 115, no. 1, pp. 1–4, 2009.

[23] S. L. González-Cortés and F. E. Imbert, "Fundamentals, properties and applications of solid catalysts prepared by solution combustion synthesis (SCS)," *Applied Catalysis A: General*, vol. 452, pp. 117–131, 2013.

[24] A. S. Mukasyan, P. Epstein, and P. Dinka, "Solution combustion synthesis of nanomaterials," *Proceedings of the Combustion Institute*, vol. 31, no. 2, pp. 1789–1795, 2007.

[25] K. C. Patil, S. T. Aruna, and T. Mimani, "Combustion synthesis: an update," *Current Opinion in Solid State and Materials Science*, vol. 6, no. 6, pp. 507–512, 2002.

[26] R. P. Sreekanth Chakradhar, B. M. Nagabhushana, G. T. Chandrappa, K. P. Ramesh, and J. L. Rao, "Solution combustion derived nanocrystalline macroporous wollastonite ceramics," *Materials Chemistry and Physics*, vol. 95, no. 1, pp. 169–175, 2006.

[27] J. A. Dean, *Lange's Handbook of Chemistry*, 15th ed., 1999.

[28] S. Atalay, H. I. Adiguzel, and F. Atalay, "Infrared absorption study of $Fe_2O_3$-CaO-$SiO_2$ glass ceramics," *Materials Science and Engineering: A*, vol. 304–306, no. 1–2, pp. 796–799, 2001.

[29] X. Chen, J. Jiang, F. Yan, S. Tian, and K. Li, "A novel low temperature vapor phase hydrolysis method for the production of nano-structured silica materials using silicon tetrachloride," *RSC Advances*, vol. 4, no. 17, p. 8703, 2014.

[30] R. K. Nariyal, P. Kothari, and B. Bisht, "FTIR Measurements of $SiO_2$ Glass Prepared by Sol-Gel Technique," *Chemical Science Transactions*, vol. 3, no. 3, pp. 1064–1066, 2014.

[31] P. Saravanapavan and L. L. Hench, "Mesoporous calcium silicate glasses. I. Synthesis," *Journal of Non-Crystalline Solids*, vol. 318, no. 1–2, pp. 1–13, 2003.

[32] K. A. Almasri, H. A. A. Sidek, K. A. Matori, and M. H. M. Zaid, "Effect of sintering temperature on physical, structural and optical properties of wollastonite based glass-ceramic derived from waste soda lime silica glasses," *Results in Physics*, vol. 7, pp. 2242–2247, 2017.

[33] R. Lakshmi, V. Velmurugan, and S. Sasikumar, "Preparation and phase evolution of Wollastonite by sol-gel combustion method using sucrose as the fuel," *Combustion Science and Technology*, vol. 185, no. 12, pp. 1777–1785, 2013.

[34] C. Wang, D. Wang, and S. Zheng, "Characterization, organic modification of wollastonite coated with nano-$Mg(OH)_2$ and its application in filling $PA_6$," *Materials Research Bulletin*, vol. 50, pp. 273–278, 2014.

[35] N. Tangboriboon, T. Khongnakhon, S. Kittikul, R. Kunanuruksapong, and A. Sirivat, "An innovative $CaSiO_3$ dielectric material from eggshells by sol–gel process," *Journal of Sol-Gel Science and Technology*, vol. 58, no. 1, pp. 33–41, 2011.

# 16

# Dynamic Study of VSA and TSA Processes for VOCs Removal from Air

**Rawnak Talmoudi** ⓘ,[1] **Amna AbdelJaoued,**[1] **and Mohamed Hachemi Chahbani**[1,2]

[1]*Laboratoire de Recherche Génie des procédés et systèmes industriels, Ecole Nationale d'Ingénieurs de Gabès, Université de Gabès, Gabès, Tunisia*
[2]*Institut Supérieur des Sciences Appliquées et de Technologie de Gabès, Rue Omar Ibn Elkhattab, Zrig, Gabès 6029, Tunisia*

Correspondence should be addressed to Rawnak Talmoudi; talmoudirawnak@hotmail.com

Academic Editor: Doraiswami Ramkrishna

Volatile organic compounds are air pollutants that necessitate to be eliminated for health and environment concerns. In the present paper, two VOCs, that is, dichloromethane and acetone are recovered by adsorption on activated carbon from a nitrogen gas stream. Experimental adsorption isotherms of the two VOCs are determined at three different temperatures 298, 313, and 323 K by the dynamic column breakthrough method. The dynamic mathematical model succeeds to predict satisfactorily the experimental breakthrough curves for pure VOCs and different binary mixtures for various conditions. Thus, the validated dynamic mathematical model has been used as a simulation tool for optimization purposes of VSA and TSA processes in order to achieve the highest performances under the given constraints. The effects of the adsorption step duration, the vacuum pressure, and the desorption temperature on the recovery of dichloromethane and acetone have been studied. A recovery of 100% of the two VOCs could be attained. However, the adsorption step duration should be determined precisely so as not to affect the recovery and alter the quality of air being purified due to the breakthrough of VOCs. The vacuum pressure and the desorption temperature should be carefully chosen in order to both reduce the energy consumption and shorten the purge step duration. Regeneration by hot nitrogen stream seems to be more efficient than regeneration by reducing pressure.

## 1. Introduction

Volatile organic compounds (VOCs) constitute a heterogeneous group of substances with different physical and chemical behaviours, including alkanes, ketones, aromatics, paraffins, olefins, alcohols, ethers, esters, and halogenated and sulfur hydrocarbons [1, 2]. VOCs are emitted from chemical industry gaseous effluents, oil sands process, food industries, paint industries, and petrochemical industries [2].

Dichloromethane, acetone, and ethyl formate are widely used in the oil sands process for the extraction of the bitumen from the oil sands [3]. After the separation of bitumen, these VOCs should be recovered to avoid serious effects on human health and environment.

Tropospheric ozone is not emitted into the air, but actually formed in the atmosphere through a photochemical process for which VOCs play a significant role. VOCs in the air react with oxides of nitrogen $NO_x$ in the presence of sunlight to form ozone. Tropospheric ozone has to be distinguished from ozone in the stratosphere which is part of what is commonly referred to as the "ozone layer" helping to block the sun's ultraviolet radiation. Unfortunately, most VOCs produce ozone, which inhabits the troposphere. Troposphere or ground level ozone is a harmful, photochemical oxidant that significantly contributes to the formation of smog. It is regularly measured as an indicator of smog levels in the atmosphere. Preventive measures should be taken to avoid serious problems for those with asthma and respiratory conditions for days when ozone level is high. Ozone concentrations of even 0.08 ppm contributed to lung disease and pulmonary function deterioration in children. There is evidence that prolonged exposure to ozone causes permanent damage to lung tissue and interferes with the functioning of the immune system. Other consequences of excessive ozone levels can include damage to the ecosystem including the retardation of plant growth and crop yields [4].

Being a precursor of photochemical oxidants, eye irritant and a suspected toxicant of respiratory, gastrointestinal and nervous system, acetone may have detrimental effects on human health and environment [5]. Human studies have observed associations between occupational exposure to dichloromethane and increased risk for several specific cancers, including brain cancer, liver and biliary tract cancer, non-Hodgkin lymphoma, and multiple myeloma [6].

Results from experimental studies in humans indicate that acute neurobehavioral deficits, measured, for example, by psychomotor tasks, tests of hand-eye coordination, visual evoked response changes, and auditory vigilance, may occur at concentrations >200 ppm with 4–8 hr of exposure [7].

As a result, the control of dichloromethane, acetone, and any VOC exhausts in ambient air is considered imperative.

There are several methods used to reduce the VOCs such as absorption, adsorption, incineration, catalytic oxidation, biological treatment, ionization, photocatalysis. Adsorption onto activated carbon is a highly efficient and relatively inexpensive technique employed for the removal and recovery of VOCs from air [8–11].

The most known cyclic adsorption operation is pressure swing adsorption (PSA) [12, 13], vacuum swing adsorption (VSA), and temperature swing adsorption (TSA) [14, 15]. The difference between these processes is their adsorption and regeneration mode. In TSA processes, the temperature changes in each step with the low temperature upon adsorption and the high temperature in regeneration. The PSA technique depends on the pressure where the adsorption step is carried out at the high pressure and the regeneration operation is at atmospheric pressure.

Many studies of PSA, TSA, and VSA have been widely investigated for recovery of VOCs from polluted air streams [16–18]. There are more papers dealing with dichloromethane [19, 20] than acetone [21]. Each VOC is treated separately and works dealing with mixtures of the two VOCs are lacking in the literature. In general, these works are just limited to the experimental determination of breakthrough curves and experiments of regeneration; the cyclic steady state is not considered. Furthermore, the models used for simulation often use simplifying assumptions such as negligible pressure drop, isothermal conditions.

It has to be noted that the adsorption step of VOCs has been widely studied in the literature for different objectives (selection of the appropriate adsorbent, effect of operating conditions). However, the regeneration step has been rarely investigated despite its importance for the performance of the whole process. Regeneration could be done by reducing the pressure or increasing the temperature. Regeneration is crucial so as to ensure that the adsorption process remains attractive in comparison with other processes. It has a notable effect on the duration of the total cycle and hence affects the size and the number of beds of the purification installation.

The principal aim of this work is to design an adsorption based separation process (Vacuum Swing Adsorption and temperature swing adsorption) for recovering two VOCs,

that is, dichloromethane and acetone, used extensively in oil sands process. The design of the process involves experimental determination of equilibrium isotherms of the two VOCs on a commercial activated carbon using the dynamic column breakthrough method and the realization of a simulation tool. The validated simulation tool, considering minimum simplifying assumptions, is to be used for testing the technical feasibility and optimizing the VSA and TSA process through varying the operating parameters. Results of simulations should give elements of responses concerning the most appropriate process to be used for the present case, that is, the capture of acetone and dichloromethane.

## 2. Experiments

The adsorbent used is a commercial activated carbon supplied by CECA (Communauté Européenne pour le Charbon et l'Acier); it is of plant origin and precisely of maritime pine.

The apparent density and the BET Surface are $250 \, kg \cdot m^3$ and $1706 \, m^2 \cdot g^{-1}$, respectively. The specific surface area of the adsorbent is determined by measuring a $N_2$ isotherm at 77 K using an ASAP 2020 apparatus from Micromeritics. The specific surface area is calculated according to the BET method [22].

The measurements of pure species adsorption equilibrium isotherms and breakthrough curves of pure components and binary mixtures of VOCs vapors are performed with a dynamic adsorption unit.

Figure 1 shows the experimental setup used. VOCs at the liquid state (a pure component or a binary mixture) are injected by one or two syringe pumps in the vaporization column. At the exit of the column, the liquid is completely vaporized.

A continuous nitrogen gas stream (inert gas) fixed by a mass flow controller (Brooks Instruments) is introduced in the unit. Nitrogen is then mixed with the VOCs vapor exiting the vaporization column before entering the adsorption column at a superficial velocity of $0.17 \, m \cdot s^{-1}$. The amount of activated carbon packed in the column is 2 g. Experiments are carried out at atmospheric pressure and a temperature of 25°C. The column is made of stainless steel and has a height of 10 cm and an internal diameter of 1 cm. The temperature of experiments is controlled by a thermostatic enclosure. Pure components isotherms are determined for concentrations ranging from 500 to 12000 ppm on activated carbon at atmospheric pressure and at three temperatures, 298 K, 313 K, and 323 K. Breakthrough curves for single solvent vapor of dichloromethane and acetone are carried out at 298 K, 313 K, and 323 K under atmospheric pressure. Concentrations of solvent vapors at the inlet of the bed are set from 500 to 12000 ppm.

Four different breakthrough curves of dichloromethane/ acetone vapor mixtures at various inlet concentrations, 3000 ppm/1500 ppm, 2000 ppm/500 ppm, 500 ppm/2000 ppm, and 750 ppm/400 ppm, are performed on activated carbon at 298 K and at atmospheric pressure.

The variation of VOCs concentrations with time at the exit of the adsorption bed are measured by a Gas

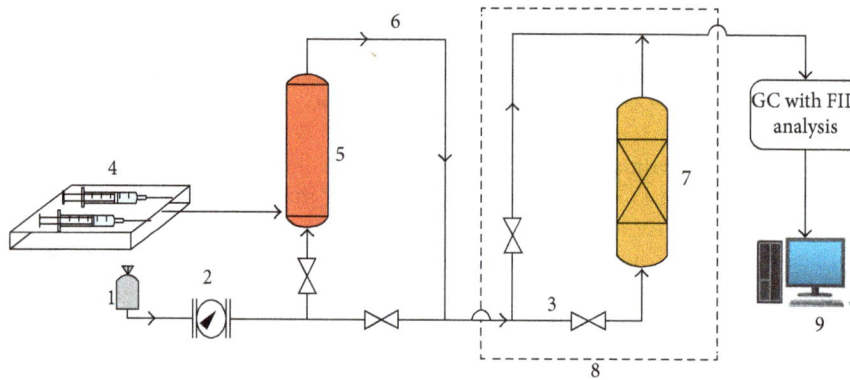

FIGURE 1: Schematic diagram of experimental unit for adsorption of VOCs vapors over activated carbon. (1) Feed nitrogen; (2) mass flow controller; (3) pure N2 stream; (4) syringe pump; (5) evaporator column; (6) main stream; (7) adsorption column; (8) thermostatic enclosure; (9) gas chromatography; (10) acquisition software.

Chromatograph (CompactGC$^{4.0}$ Interscience, G.A.S)) combined with a flame ionization detector (FID, sampling loop $20\,\mu L$, sampling from the main stream by vacuum pump). The sample is transported into chromatographic column which has an inner diameter of 0.32 mm and a length of 10 mm, and then, the sample is separated by the GC column coated with Poraplot Q stationary phase. The time for analysis of the single isotherm adsorption is about 3 min for both dichloromethane and acetone. Breakthrough curves were obtained by plotting the concentrations of each VOC at the bed exit over time. After each adsorption step, the feed stream is changed to pure nitrogen so as to regenerate the fixed adsorption bed.

## 3. Mathematical modeling

The mathematical model for a packed adsorption bed, assuming nonisothermal operation, involves mass balances for the bulk fluid phase, mass transfer kinetics, gas-phase and solid phase heat balances, and a momentum balance.

The following assumptions are considered:

(i) Non isothermal operation

(ii) Ideal gas behavior for the gas phase

(iii) Constant void fraction and velocity of gas

(iv) Radial concentration and temperature gradients negligible

(v) Diffusion and mass transfer within a particle was described by linear driving force (LDF)

(vi) Absence of temperature gradients within the solid particle

*3.1. Mass Balance.* A differential fluid phase mass balance for the component $i$ is given by the following axially dispersed plug flow equation [23]:

$$-D_z\frac{\partial^2 c_i}{\partial Z^2}+\frac{\partial c_i}{\partial t}+\frac{\partial\left(v_g c_i\right)}{\partial Z}+\frac{(1-\varepsilon)}{\varepsilon}\rho_p\frac{\partial q_i}{\partial t}=0 \quad i=1,2. \quad (1)$$

The overall mass balance for the bulk gas is given by

$$\frac{\partial C}{\partial t}+\frac{\partial\left(v_g C\right)}{\partial Z}+\frac{(1-\varepsilon)}{\varepsilon}\rho_p\sum_{i=1}^{n}\frac{\partial q_i}{\partial t}=0, \quad (2)$$

where $v_g$ is the interstitial gas velocity, $\varepsilon$ is the bed void fraction, $C$ is total bulk concentration, $\rho_p$ is the particle density, $q_i$ is the adsorbed concentration of component $i$, $Z$ is the axial co-ordinate, and $D_z$ is the axial dispersion coefficient, which can be estimated by [24]

$$D_z=0.73D_m+0.5\frac{d_p\mu_i}{1+9.7(D_m/d_p\mu_i)}, \quad (3)$$

where $d_p$ is the particle diameter, $\mu_i$ is the interstitial velocity, and $D_m$ is the molecular diffusivity that can be estimated from the Chapman–Enskog equation [25].

*3.2. Momentum Balance.* The pressure drop along the bed was determined from the Ergun equation [23]:

$$\frac{\partial P}{\partial Z}=150\frac{(1-\varepsilon)^2}{\varepsilon^2 d_p^2}\mu v_g+1.75\frac{(1-\varepsilon)}{\varepsilon d_p}\rho_g v_g^2, \quad (4)$$

where $\mu$ is the gas mixture viscosity, $\rho_g$ is the gas density, and $d_p$ is the particle diameter.

*3.3. Mass Transfer Kinetics.* The mass transfer rate between the gas and solid phases is given the linear driving force (LDF) model represented by the following equations [26]:

$$\frac{\partial q_i}{\partial t}=k_{i,\mathrm{LDF}}\left(q_i^*-q_i\right) \quad i=1,2, \quad (5)$$

where

$$k_{i,\mathrm{LDF}}=\frac{15D_{e,i}}{r_p^2}, \quad (6)$$

where $k_{i,\mathrm{LDF}}$ is the mass transfer coefficient, $q_i^*$ is the loading of component $i$, $D_{e,i}$ is the effective diffusivity of component $i$, and $r_p$ is the particle radius.

It is common practice to predict mixture isotherms from pure component isotherms. The multicomponent extension of Langmuir–Freundlich model of dichloromethane and acetone is used based on parameters obtained from pure components. The model is represented by the following equations [27]:

$$q_i^* = q_{m,i} \frac{(b_i C_i)^{k_{3i}}}{1 + \sum_j (b_j C_j)^{k_{3i}}} \quad i, j = 1, 2, \quad (7)$$

$$b_i = k_{1i} \exp \frac{k_{2i}}{T}, \quad (8)$$

where $q_{m,i}$ is the saturation capacity, $b_i$, $k_{1i}$, $k_{2i}$, and $k_{3i}$ are the Langmuir–Freundlich equation parameters for component $i$.

The fit between experimental data and isotherms equation is calculated according to Equation (9) which indicated the minimum sum of squares SS at three temperatures of pure 2 VOCs:

$$SS(\%) = \sum_{T_i}^{T_3} \sum_{j=1}^{N} (q_{\exp,i} - q_{\mathrm{mod},i})^2, \quad (9)$$

where $q_{\exp,i}$ and $q_{\mathrm{mod},i}$ are the experimental and predicted amounts adsorbed, respectively, $T_1$ to $T_3$ are the three tested temperatures, $j$ is the number of points per isotherm and gas component, $i$ represents the component (dichloromethane and acetone), and N is the total number of experimental data points.

*3.4. Heat Balance for the Gas Phase.* The gas-phase energy balance includes convection of energy, axial thermal conduction, accumulation of heat, gas-solid heat transfer, and gas-wall heat transfer. The governing partial differential equation is as follows [28]:

$$C_{pg} \rho_g \frac{\partial T_g}{\partial t} - \lambda_L \frac{\partial^2 T_g}{\partial Z^2} + \rho_g C_{pg} \frac{\partial (v_g T_g)}{\partial Z} + \left(\frac{1-\varepsilon}{\varepsilon}\right) a_s h_f (T_g - T_s)$$

$$+ \frac{4h_w}{\varepsilon_b d_{\mathrm{int}}} (T_g - T_w) = 0, \quad (10)$$

where $C_{pg}$ and $C_{ps}$ are the specific heats of the gas and solid phases, $\lambda_L$ is the effective axial thermal conductivity, $h_f$, $h_w$ are the heat transfer coefficient between the gas and the adsorbent and the heat transfer coefficient between the gas and the wall, and $T_g$, $T_s$, and $T_w$ are the gas, solid, and wall temperatures.

The film heat transfer $h_f$ is calculated from the following equation [29]:

$$Nu_g = 2 + 1,1 Re^{0,6} Pr^{1/3}, \quad (11)$$

$$Nu_g = \frac{h_f d_p}{k_g}, \quad (12)$$

$$Re = \frac{\rho_g v_g d_p}{\mu}, \quad (13)$$

$$Pr = \frac{C_{pg} v_g}{k_g}. \quad (14)$$

The axial heat dispersion $\lambda_L$ for each VOC can be estimated from the following equation [30]:

$$\lambda_L = D_Z C_{pg} \rho_g. \quad (15)$$

*3.5. Heat Balance for the Solid Phase.* The solid phase energy balance takes into account the transfer of heat over the film, axial thermal conduction, accumulation in solid phase, heat generated by the adsorption phenomenon. The energy balance equation for the bed can be expressed as in the following equation [28]:

$$-k_s \frac{\partial^2 T_s}{\partial z^2} + \rho_p C_{ps} \frac{\partial T_s}{\partial t} = a_s h_f (T_g - T_s) + \sum_{i=1}^{n} (-\Delta H_i) \frac{\partial q_i}{\partial t}, \quad (16)$$

where $k_s$ is the effective axial solid phase thermal conductivity, $a_s$ expresses the ratio of the particle external surface area to volume, and $\Delta H_i$ is the isosteric heat of adsorption.

*3.6. Numerical Solution.* Aspen Adsorption enables process simulation and optimization for a wide range of industrial gas adsorption processes. It permits to choose from various geometries to complete process simulation tasks such as vacuum swing adsorption modeling. Modeling includes a wide range of kinetic models including lumped resistance, micro/macropore, and general rate. It uses a wide range of standard equilibrium models for pure and multicomponent gases. It also uses a highly configurable energy balance to account for nonisothermal behavior, conduction, heat loss, and wall effects.

Thus, the set of aforementioned partial differential equations relative to mass, heat, and momentum balances, together with the appropriate initial and boundary conditions to fully describe the adsorption process, are numerically solved by Aspen Adsorption software. The set of partial differential equations change according to assumptions considered.

*3.7. VSA and TSA Simulation.* After the determination of needed data to be used in simulations (adsorption isotherms and mass transfer coefficients for dichloromethane and acetone), VSA and TSA processes were designed to evaluate the feasibility of acetone and dichloromethane recovery from polluted air with activated carbon.

*3.7.1. Four-Step VSA Cycle.* The 4-step VSA cycle considered consists of four steps: adsorption (I), countercurrent evacuation (II), countercurrent purge (III), and cocurrent pressurization with feed (IV). A schematic diagram of a one-bed VSA cycle is indicated in Figure 2(a).

In step I, a continuous feed gas (VOCs + N2) is introduced at atmospheric pressure. The desorption step of

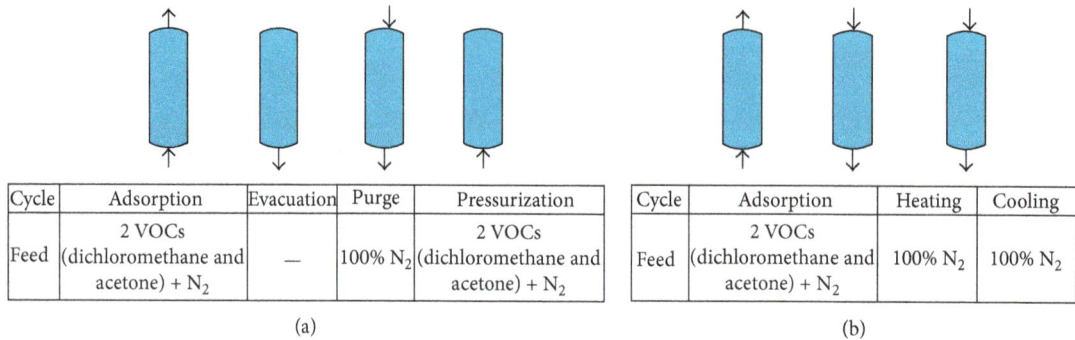

| Cycle | Adsorption | Evacuation | Purge | Pressurization |
|-------|------------|------------|-------|----------------|
| Feed | 2 VOCs (dichloromethane and acetone) + N$_2$ | — | 100% N$_2$ | 2 VOCs (dichloromethane and acetone) + N$_2$ |

(a)

| Cycle | Adsorption | Heating | Cooling |
|-------|------------|---------|---------|
| Feed | 2 VOCs (dichloromethane and acetone) + N$_2$ | 100% N$_2$ | 100% N$_2$ |

(b)

FIGURE 2: Schematic diagram and sequence of cycle steps in the operation of (a) 1-bed/4-step VSA cycle; (b) 1-bed/3-step TSA cycle for 2 VOCs removal.

VSA cycle consists of steps II and III. In the evacuation step, moles of VOCs adsorbed are removed from the bed by decreasing the bed pressure from 1 atm to 0.5 atm. In the purge step, pure nitrogen is used to regenerate the bed at a pressure of 0.5 atm. During step IV, the pressure in the column increases from 0.5 atm to the atmospheric pressure, the bed being pressurized by a mixture of VOCs and N2. The bed is initially filled with pure nitrogen. The duration of the adsorption step is chosen so as not to allow VOCs to breakthrough. The purge step is stopped when the entire adsorbed amount is cleared from the adsorption bed.

*3.7.2. Three-Step TSA Cycle.* The cycle step sequence for TSA includes adsorption (I), heating (II), and cooling (III) steps. The schematic diagram of TSA process is illustrated in Figure 2(b).

During the adsorption step, a gas mixture with 3000 ppm dichloromethane/1500 ppm acetone is fed to the bed at 298 K and atmospheric pressure. During the heating step, dichloromethane and acetone are desorbed from the bed by increasing the column temperature up to 443 K using a hot nitrogen stream. The direct heating with hot inert gas facilitates the thermal and concentration swing for removal of the two VOCs. During the cooling step, the temperature of the bed is decreased with the aid of inert nitrogen stream at 298 K which comes in direct contact with the activated carbon.

*3.7.3. VSA and TSA Performance.* The platform of Aspen Adsorption is widely used in the industry by engineers to optimize and simulate a wide range of PSA, TSA, VSA, and other variants of these with a wide range of solid adsorbents. Aspen Adsorption can be used for adsorption process design and can reduce significantly time and cost of laboratory and pilot plant trials. This is the major benefit of simulation works. The software gives the user the freedom to configure the process to improve the plant operations in order to determine the effect of various variables on plant performance.

Using Aspen Adsorption software, VSA and TSA cycle runs are performed for a binary VOCs mixture of dichloromethane/acetone (gas inlet concentrations 3000 ppm/ 1500 ppm) at 298 K and atmospheric pressure using activated carbon.

The mathematical modeling of each step of the VSA and TSA cycles are performed by the same equations described in Section 3.

The performance of the VSA and TSA processes is analyzed through the obtained values for the recovery of dichloromethane and acetone, and this parameter is defined as the amount of the VOCs (dichloromethane and acetone) recovered during the regeneration step divided by the total amount of VOCs used during the feed step. Equations (17) and (18) are used to calculate the recovery for each VOC component at cyclic steady state conditions [31].

For the VSA process, the regeneration step includes evacuation and purge steps and the feed step comprises adsorption and pressurization steps.

$$\text{Recovery}\,(\%) = \frac{\int_0^{II} c_i v|_{z=L}\,dt + \int_0^{III} c_i v|_{z=L}\,dt}{\int_0^I c_i v|_{z=0}\,dt + \int_0^{IV} c_i v|_{z=0}\,dt}. \tag{17}$$

For the TSA process, the regeneration step is the heating step and the feed step is the adsorption step.

$$\text{Recovery}\,(\%) = \frac{\int_0^{II} c_i v|_{z=L}\,dt}{\int_0^I c_i v|_{z=0}\,dt}. \tag{18}$$

## 4. Results and Discussion

*4.1. Adsorption Equilibrium Isotherms.* Reliable adsorption equilibrium isotherms are key elements in the design of an adsorption separation process. The dynamic column breakthrough method could be used safely for obtaining single-component equilibrium isotherm data. It is used herein to derive the equilibrium isotherms of dichloromethane and acetone on activated carbon. The experimental setup and analysis required for this method were previously described. The equilibrium results for these gases at atmospheric pressure and at three different temperatures are presented.

The dynamic column method of measuring single-component isotherm data involves monitoring a series of breakthrough curves in a column packed with the adsorbent.

Various feed concentrations of the adsorbable component in an inert carrier are normally used in order to obtain the isotherm over a large range of concentrations. From the breakthrough curves, the adsorbed amount of VOCs for that particular operating conditions in equilibrium with gas phase can be deduced according to the following equation:

$$q_i^* = \frac{Q_{\text{feed}}}{m_{\text{ad}}} \int_0^{t_f} (C_o - C(t))\, dt, \qquad (19)$$

where $C_o$ and $C(t)$ are the gas-phase concentration of VOC at the inlet and outlet of the column, respectively, $m_{\text{ad}}$ is the mass of the activated carbon sample, $Q_{\text{feed}}$ is the gas flow rate, and $t_f$ is the adsorption end time.

In the latter equation, the amounts of VOC in the interparticle and in the intraparticle voids per adsorbent mass are not considered because they are negligible in comparison with the adsorbed amount $q_i^*$.

Figures 3 and 4 show the experimental results relative to the adsorption equilibrium of pure dichloromethane and acetone on activated carbon at three temperatures (298, 313, and 323 K).

For each experimental isotherm, fitted curves obtained with the Langmuir–Freundlich model are represented. The figures show a good agreement between the values given by the two-parameter ($b_i$ and $q_{\text{mi}}$) model and the experimental data for the two VOCs at each temperature. The expressions of parameters $b_i$ and $q_{\text{mi}}$ in function of temperature according to Equations (6) and (7) are obtained by fitting. The $k_i$ parameters for dichloromethane and acetone are given in Table 1.

### 4.2. Breakthrough Curves.

The experimental breakthrough curves for pure solvents vapors of dichloromethane and acetone were determined at different temperatures and inlet concentrations.

Figures 5(a), 5(b), and 5(c) show a comparison between breakthrough curves obtained experimentally and by simulation using Aspen Adsorption software for dichloromethane at atmospheric pressure and for the temperatures 298, 313, and 323 K, respectively. Inlet concentrations chosen were 500, 1000, 2000, 4000, 8000, and 12000 ppm. It should be mentioned that these breakthrough curves were previously used to determine the equilibrium isotherms of dichloromethane at the different temperatures.

Figures 6(a), 6(b), and 6(c) give experimental and theoretical breakthrough curves of acetone for the same conditions.

It has to be noted that the dynamic mathematical model succeeds to predict satisfactorily the experimental data for the different conditions. This confirms the validity of the mathematical model along with the assumptions considered. The mass transfer coefficients $k_{\text{LDF}}$ used in the linear driving model for dichloromethane and acetone are assessed by adjusting simulation results to experimental breakthrough curves. The kinetic parameters ($k_{\text{LDF}}$) obtained for the two VOCs for adsorption on activated carbon are given in Table 2. The $k_{\text{LDF}}$ value retained corresponds to the minimum sum of squares (SS) considering the different breakthrough curves obtained for the different concentrations (similar to Equation (9)).

Figures 7(a), 7(b), 7(c), and 7(d) give breakthrough curves of binary mixtures of dichloromethane and acetone

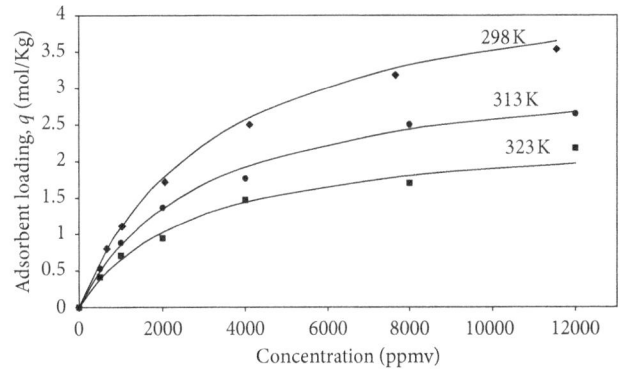

FIGURE 3: Adsorption isotherms of dichloromethane on activated carbon at 298, 313, and 323 K and at atmospheric pressure; symbols: experimental adsorption data; solid lines: curves fittings using extended Langmuir–Freundlich model.

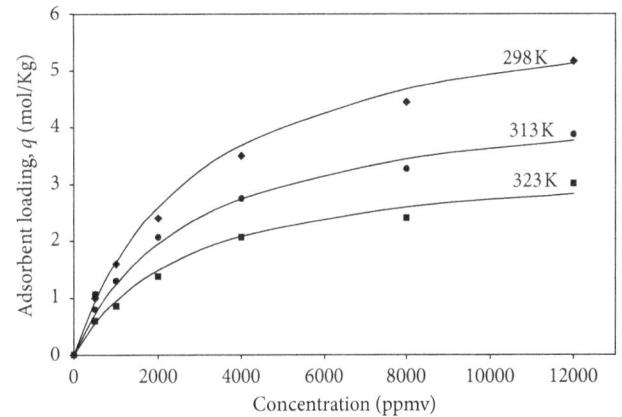

FIGURE 4: Adsorption isotherms of acetone on activated carbon at 298, 313, and 323 K and at atmospheric pressure; symbols: experimental adsorption data; solid lines: curves fittings using extended Langmuir–Freundlich model.

TABLE 1: Langmuir–Freundlich extended isotherm parameters for the adsorption of VOCs vapors on activated carbon.

|  | Dichloromethane | Acetone |
|---|---|---|
| $q_{\text{m,i}}$ (kmol/kg) | 0.0051 | 0.0044 |
| $K_1$ (bar) | 0.017 | 0.014 |
| $K_2$ (K) | 3235.5 | 4165.74 |
| $K_3$ (−) | 0.76 | 0.82 |

on activated carbon for various feed compositions (3000 ppm/1500 ppm, 2000 ppm/500 ppm, 500 ppm/2000 ppm, and 750 ppm/400 ppm) at atmospheric pressure and 298 K. Whatever the composition of the binary mixture, dichloromethane is the first compound to breakthrough. As expected, the results corroborate the ability of activated carbon to separate the dichloromethane and acetone from a nitrogen stream. For simulations, the same values of mass transfer coefficients determined previously for pure VOCs were used. Although experimental measurements can be carried out, they are laborious and time-consuming, so it is

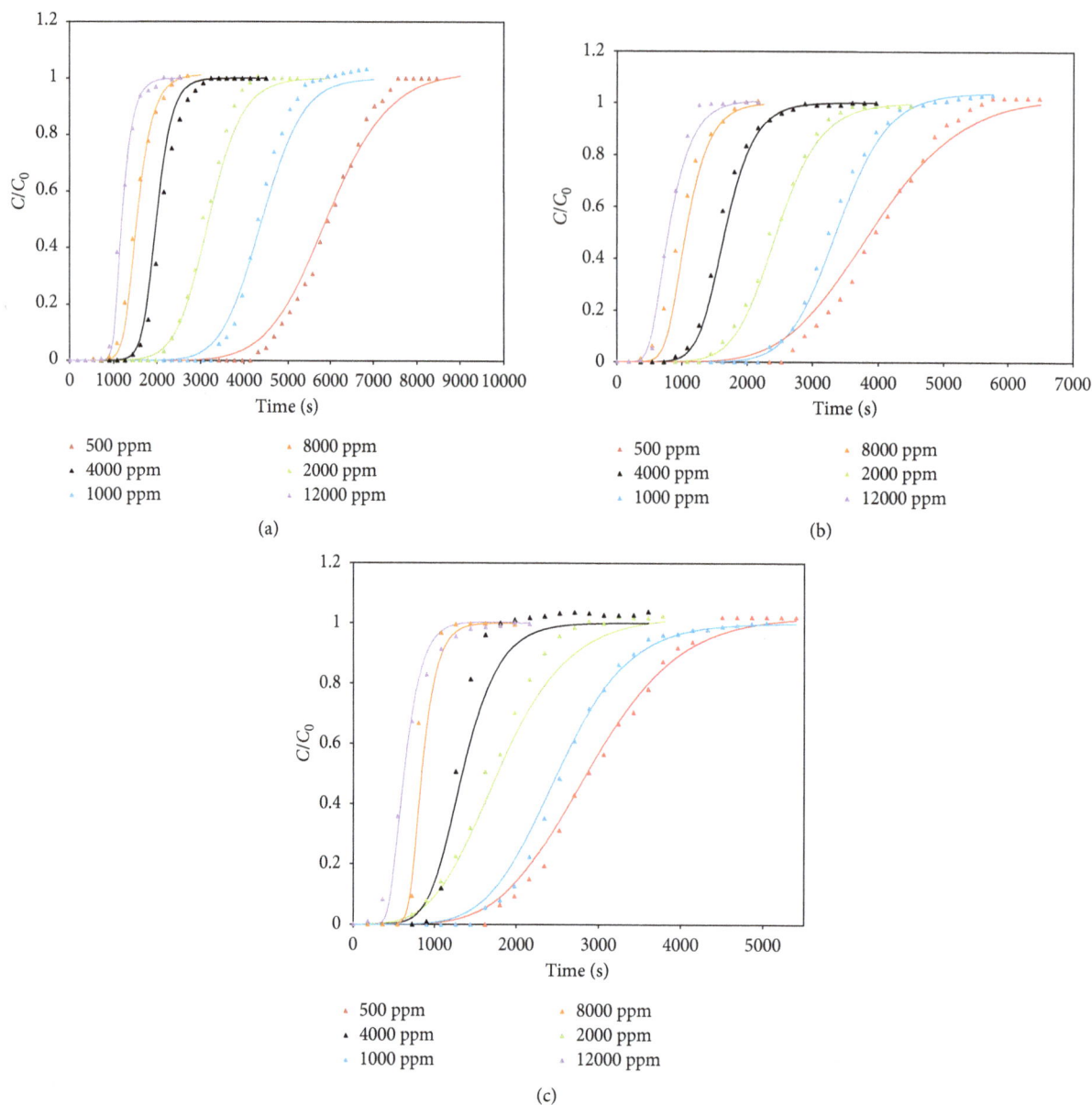

FIGURE 5: Breakthrough curves for dichloromethane adsorption on activated carbon at (a) 298 K, (b) 313 K, (c) 323 K. Symbols: experimental data. Lines: LDF model.

common practice to predict mixture isotherms from pure component isotherms. Several models for predicting mixture isotherms from pure component data have been used; the multicomponent Langmuir–Freundlich model has been used to predict mixture isotherms herein. From Figure 7, one can note that there is a good agreement between experimental and simulation results for the different binary mixtures. Thus, the dynamic mathematical model has been validated and could be used safely as a useful tool for optimization purposes of PSA or VSA processes in order to obtain highest achievable performances under the given constraints. It is worthwhile to note that the dynamic mathematical model could simulate any multicomponent gaseous mixture (more than two) provided that the adsorption equilibrium isotherms of the different components are available.

### 4.3. VSA Performance.

The VSA performance will be assessed via process simulation which is a useful tool allowing time and money savings. The accuracy of simulations will depend, of course, on the mathematical modeling and, in particular, on the validity of assumptions considered. For the present case, the model considered permits to account for all the main mechanisms involved in the process of adsorption; it involves the mass, momentum, and energy balances that govern the process, taking into account the axial mixing and mass transfer resistances. The resulting model, which is a system of coupled algebraic and partial differential equations over time and space domains, is numerically solved by Aspen Adsorption software. As indicated previously, the performance of the VSA process will be evaluated according to product recovery.

(a)

(b)

(c)

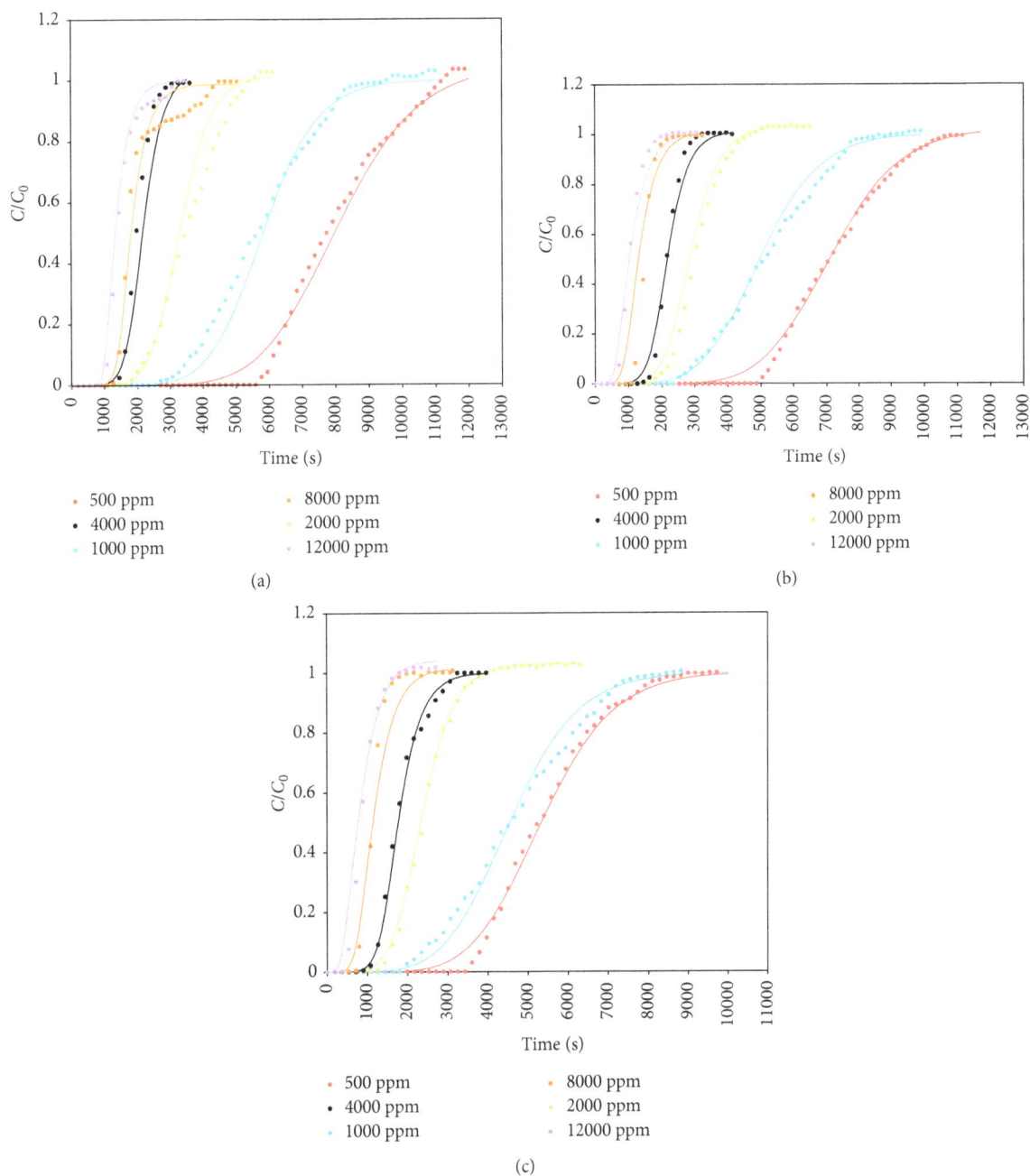

FIGURE 6: Breakthrough curves for acetone adsorption on activated carbon at (a) 298 K, (b) 313 K, (c) 323 K. Symbols: experimental data. Lines: LDF model.

TABLE 2: Mass transfer coefficients $k_{LDF}$ for VOCs vapors for dichloromethane and acetone.

| VOC | $k_{LDF}$ (s$^{-1}$) |
| --- | --- |
| Dichloromethane | 0.008 |
| Acetone | 0.01 |

The aim of simulations is to test the feasibility of the VSA cycle in removing and recovering dichloromethane and acetone from polluted air with activated carbon and find the optimal operational parameters in order to provide useful information for future industrial design and application. The

simulation results obtained for a one-bed VSA process are represented in this section. The conditions of simulations of the VSA cycle are summarized in Table 3.

The following simulation results are relative to a feed whose composition is 3000 and 1500 ppm for dichloromethane and acetone, respectively. The corresponding simulation results are obtained for the cyclic steady state which is attained after nearly 10 cycles. Figure 8 shows the pressure history at one end of the adsorption column (closed end for pressurization and evacuation steps) during one cycle at steady state.

The simulated axial profiles of the adsorbed amount of dichloromethane and acetone at the end of different steps of

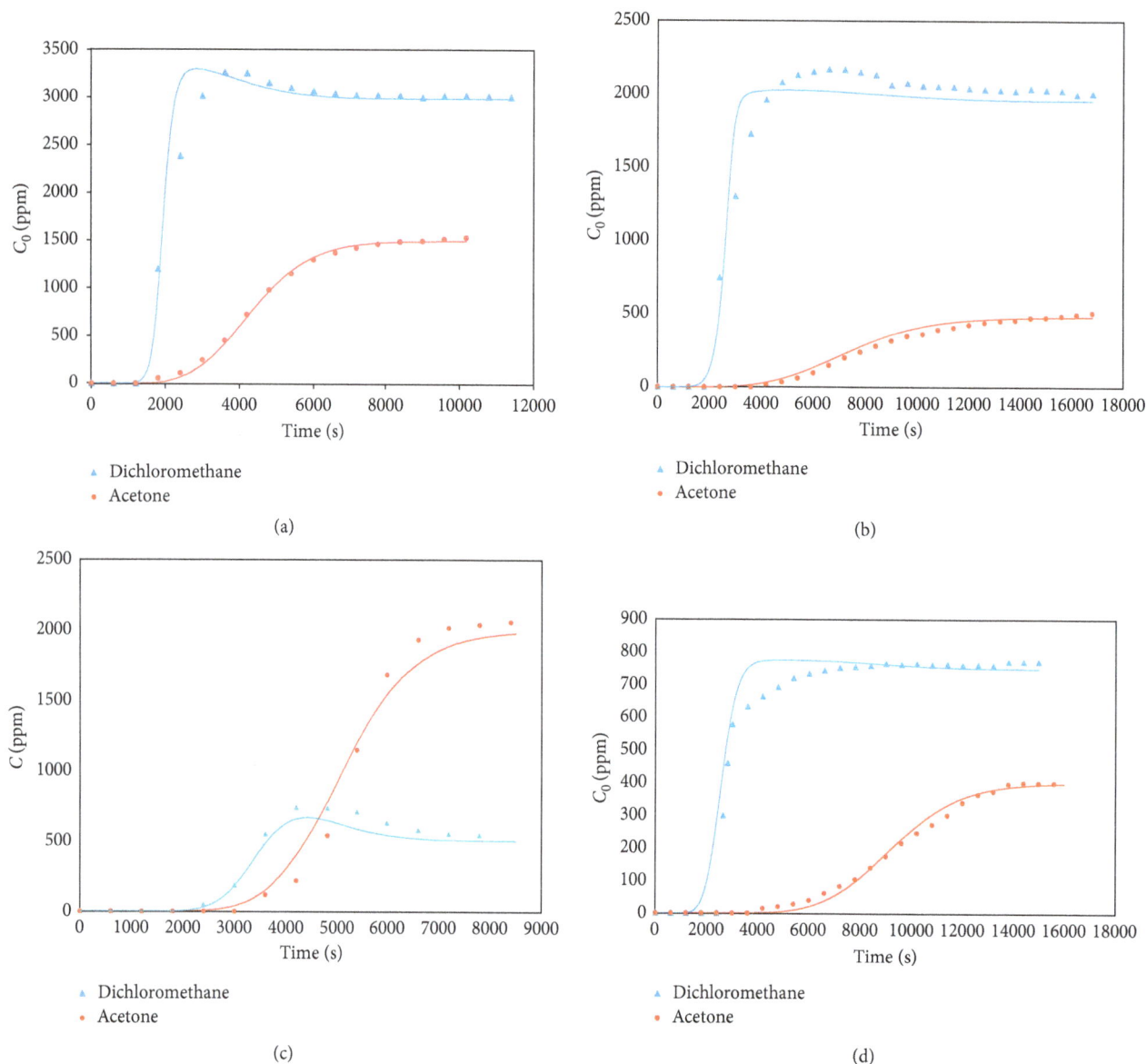

FIGURE 7: Breakthrough curves of mixtures of dichloromethane and acetone for different compositions: (a) 3000 ppm/1500 ppm, (b) 2000 ppm/500 ppm, (c) 500 ppm/2000 ppm, and (d) 750 ppm/400 ppm on activated carbon, at 298 K and atmospheric pressure. Symbols: experimental data. Lines: LDF model.

the cycle at steady state are given in Figure 9. One can note that despite the decrease of pressure from 1.0 (pressure prevailing at the end of the adsorption step) to 0.5 atm (value of pressure at the end of the evacuation step), the axial profiles of the adsorbed amount of dichloromethane and acetone have decreased slightly. In fact, mass transfer between solid and gas phases which is controlled by intraparticle diffusion, being very slow, does not permit the two phases to be in equilibrium given the rapid change in pressure.

On the contrary, the small amount of dichloromethane and acetone desorbed during the evacuation step due to the pressure decrease results in a notable increase in the concentration in the gas phase at the end of this step as can be seen in Figure 10. The concentration of dichloromethane

and acetone has nearly doubled in the saturated zone of the bed, passing from 3000 to 6000 ppm for dichloromethane and from 1500 to 3000 for acetone. If the equilibrium model was used instead of the linear driving force model to account for intraparticle mass transfer resistances, the adsorbed amount of dichloromethane obtained at the end of the evacuation step would have been much lower.

The corresponding axial temperature profiles at the end of adsorption, evacuation, and purge steps are shown in Figure 11. Once again, because of the small amount of VOCs desorbed during the evacuation step, the decrease in temperature due to desorption during this step is very slight. However, the difference in temperature between the two ends of the bed could reach a value of 10°C (during the purge step). As can be seen in Figures 3 and 4 giving the adsorption

TABLE 3: Adsorbent and bed characteristics and operating conditions used in the Aspen Adsorption simulations.

| Parameter | Value |
|---|---|
| Number of adsorbent bed | 1 |
| VOC mixture | Dichloromethane/acetone |
| Concentration (ppm) | 3000/1500 |
| Bed height (m) | 1 |
| Bed diameter (m) | 0.2 |
| Packing density ($kg \cdot m^{-3}$) | 250 |
| Particle diameter (mm) | 1 |
| Interparticle voidage, $\varepsilon$ | 0.4 |
| Void of pellets, $\varepsilon_p$ | 0.36 |
| $\Delta H_{dichloromethane}$ (kJ/mol) | −40 |
| $\Delta H_{acetone}$ (kJ/mol) | −50 |
| $D_z$ ($m^2 \cdot s^{-1}$) | $1.83\ 10^{-3}$ |
| $h_f$ ($W \cdot m^{-2} \cdot K^{-1}$) | 56 |
| $\lambda_L$ ($W \cdot m^{-2} \cdot K^{-1}$) | 48 |
| Feed flow rate (mol/min) | 20 |
| Purge flow rate (mol/min) | 12 |
| $T_{ads}$ (K) | 298 |
| *VSA cycle* | |
| Adsorption pressure (atm) | 1 |
| Desorption pressure (atm) | 0.5 |
| $t_{ad}$ (s) | 9000 |
| $t_{evac}$ (s) | 60 |
| $t_{pg}$ (s) | 40000 |
| $t_{PR}$ (s) | 80 |
| Cycle time (s) | 49140 |
| *TSA cycle* | |
| Adsorption pressure (atm) | 1 |
| $T_{des}$ (K) | 443 |
| $t_{ad}$ (s) | 9000 |
| $t_{heat}$ (s) | 7600 |
| $t_{cool}$ (s) | 5500 |
| Cycle time (s) | 22100 |

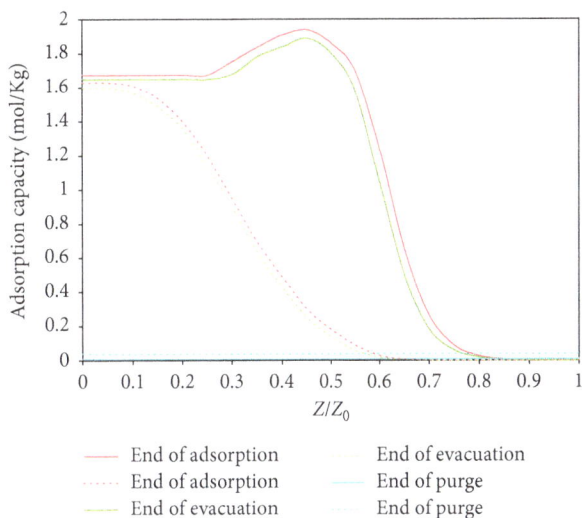

FIGURE 8: Pressure history of the adsorption column during one cycle for a VOCs mixture of dichloromethane/acetone (3000 ppm/1500 ppm).

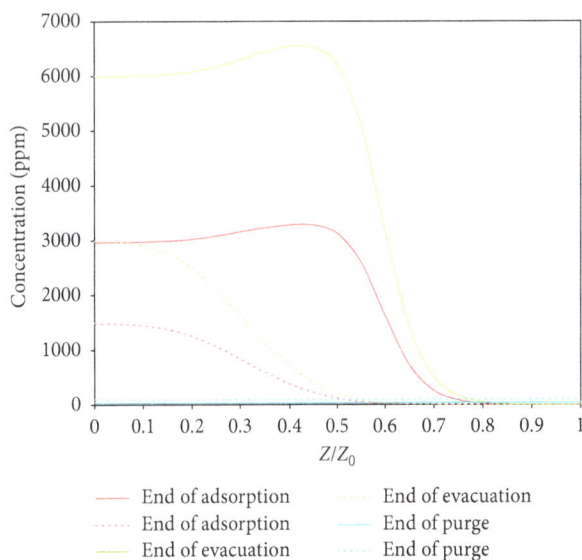

FIGURE 9: Simulated axial profiles of the adsorbed amount of dichloromethane and acetone at the end of different steps of the cycle at steady state. Solid lines: dichloromethane and dotted lines: acetone.

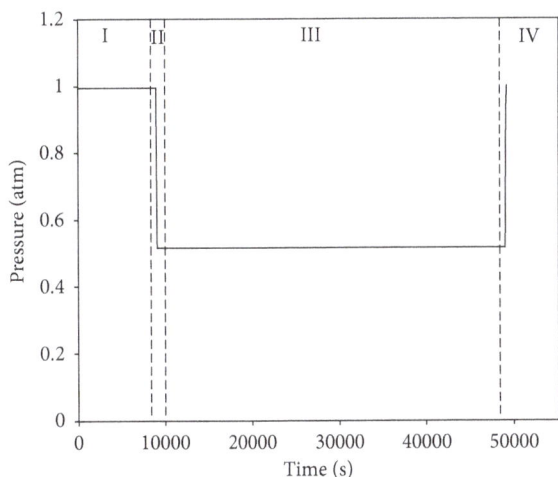

FIGURE 10: Simulated axial concentration profiles of dichloromethane and acetone at the end of different steps of the cycle at steady state. Solid lines: dichloromethane and dotted lines: acetone.

equilibrium isotherms of dichloromethane and acetone for different temperatures, a deviation of 10°C can have a significant effect on the adsorbed amount of dichloromethane and acetone. This demonstrates that incorporating energy balances in modeling is a judicious choice, and hence, simulation results of models dealing with VOCs adsorption assuming isothermal conditions should be considered with a great precaution.

Figure 12 gives histories of dichloromethane and acetone concentration at the bed exit during evacuation and purge steps at cyclic steady state. The concentrations of the two VOCs diminish gradually from the highest values attained at the end of the evacuation step until values approaching zero at the end of purge step. The desorbed VOCs in the outlet stream could be pumped to a condenser where they could be separated from $N_2$ and recovered as a liquid. This process involves cooling the outlet stream to a temperature below the dew point of each VOC.

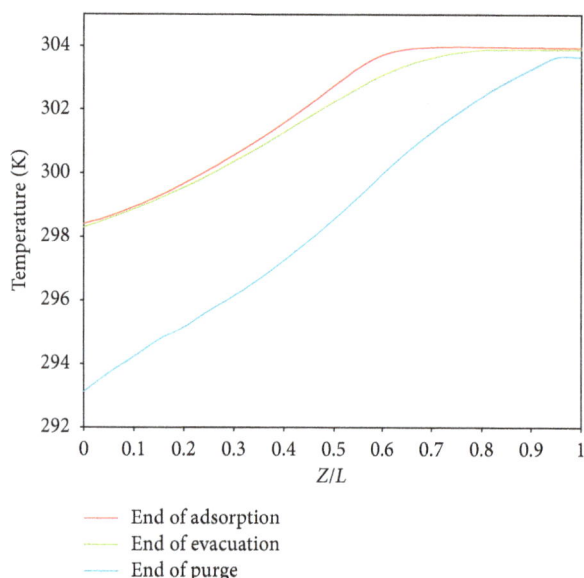

FIGURE 11: Axial Temperature profiles at the end of different steps of the cycle at steady state.

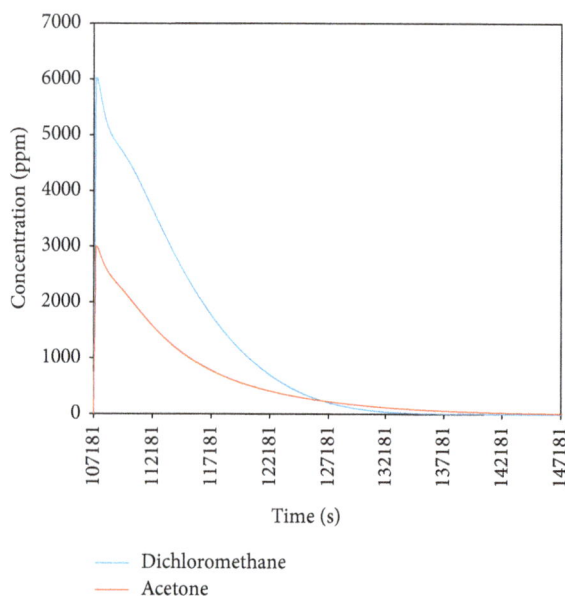

FIGURE 13: Recovery of dichloromethane and acetone with the adsorption step duration.

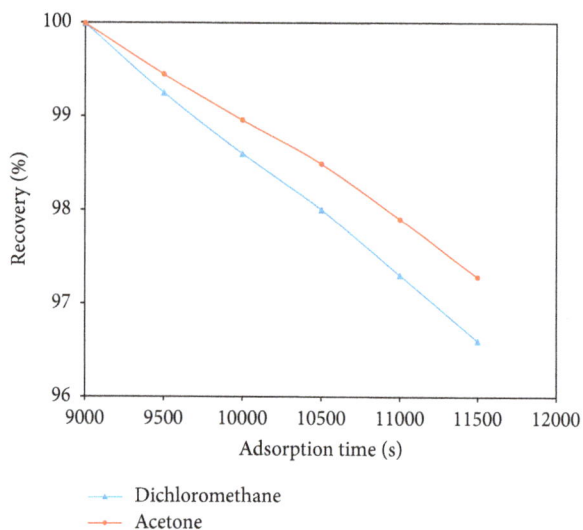

FIGURE 12: Histories of dichloromethane and acetone concentration at the bed exit during evacuation and purge steps at cyclic steady state.

The improvement of the performance of the VSA cycle is achieved through the optimization of the operating parameters, one among these parameters is the adsorption step time. Figure 13 gives the variation of recovery of dichloromethane and acetone with the adsorption step duration $t_{ad}$. All the operating parameters of the VSA cycle are kept unchanged except for $t_{ad}$. Detailed operating conditions and cycle parameters are given in Table 3. According to simulation results, it is clear that the VSA unit can recover completely the two VOCs (100% of dichloromethane and acetone) for $t_{ad}$ equal to 9000 s. For this value

of $t_{ad}$, only pure $N_2$ exits the bed, the two VOCs are totally retained. Adsorbed VOCs are then totally desorbed during the subsequent steps (evacuation and purge), thus permitting to get a recovery of 100%. The recovery of the two VOCs decreases then gradually when $t_{ad}$ increases. This can be explained by the fact that the two VOCs break through upon saturation of the bed, if $t_{ad}$ exceeds 9000 s, as shown in Figure 14 giving the change with the adsorption step time of the maximum concentration of the two VOCs attained at the outlet of the column at cyclic steady state. The amount of VOCs exiting the bed during the adsorption step constitutes a loss and causes a decrease in recovery. The adsorption step duration should be determined precisely so as not to affect the recovery and alter the quality of air being purified due to the breakthrough of VOCs.

The vacuum pressure is also a major operating parameter in the VSA process and affects significantly its performance. It has to be noted that more than 70% of the power consumption of the VSA cycle is attributed to the vacuum pump [32]. Thus, the evacuation pressure has a tremendous effect on the energy performance of the process, and its choice should be optimized so as to lower the power consumption of the system. If the vacuum pressure chosen is too low, the power consumption becomes too high. On the contrary, if the vacuum pressure selected is not low enough, the duration of the purge step will be too long. This depends of course on the shape of the adsorption isotherm. The greater the slope of the adsorption isotherm, the greater the effect of the decrease in pressure on the process performance.

Figure 15 shows the variation of purge step time with vacuum pressure. For the different simulations, the $N_2$ molar flow rate is maintained constant (12 mol/mn) and the duration of the purge step is chosen so as to obtain a recovery of 100% for dichloromethane and acetone. The other parameters are maintained unchanged (as indicated in Table 3). For $P_{des} = 1$ atm, the VSA cycle is only composed of two steps at the same pressure, adsorption, and purge, and there is no

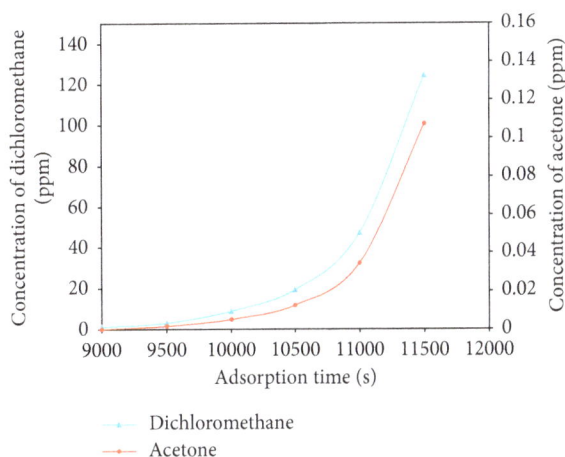

FIGURE 14: Maximum concentration of dichloromethane and acetone at the end the adsorption step for different.

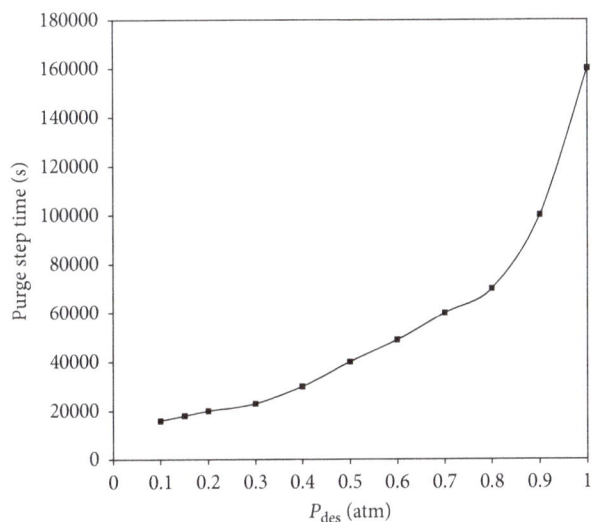

FIGURE 15: Variation of purge step time with vacuum pressure for mixtures dichloromethane/acetone vapor at 3000 ppm/1500 ppm exposure concentration.

pressure change. For the case being studied, decreasing $P_{des}$ leads to a substantial decrease in the duration of the step. In fact, the purge step time is divided by 10 when the vacuum pressure goes from 1 to 0.1 atm. For $P_{des} = 0.1$ atm, 16000 s are sufficient to regenerate the bed while 160000 s are necessary for $P_{des} = 1$ atm. As expected, the vacuum level has an appreciable effect on the VSA performance. However, the choice of the vacuum pressure has an impact on energy consumption depending on the value of the vacuum pressure and also on the duration of the purge step. The higher the vacuum pressure, the shorter the duration of the purge step but the higher the energy consumption. A compromise must be found between the energy consumption and the duration of the step. The total amount of gas used to purge the adsorber which is proportional to the duration of the purge step, given that its molar flow rate at the bed inlet is constant, increases with vacuum pressure. The amount of $N_2$

consumed for desorbing the bed increases notably with the vacuum pressure, it varies from 3 to 32 kmol when vacuum pressure increases from 0.1 to 1 atm, respectively. The vacuum pressure should be carefully chosen in order to both reduce the energy consumption and shorten the purge step duration.

*4.4. TSA Performance.* This section deals with the TSA process for the recovery of the same VOCs. The detailed cycle conditions are presented in Table 3. Simulations carried out show that cyclic steady state of the process is reached after approximately 12 cycles.

Figure 16 shows the dichloromethane and acetone concentration history at the column outlet during the heating step. The evolution with time of gas temperature at the exit of the bed for one cycle at cyclic steady state is given in Figure 17.

As can be seen from Figure 16, the increase of the bed temperature due to heating results in a more rapid regeneration of the bed in comparison with the VSA process. In the case being studied, it appears clearly that an increase in temperature during the regeneration step is more efficient than a decrease of pressure. Compared to the concentrations obtained for a VSA process, much higher concentrations of dichloromethane and acetone are obtained (approximately 32000 (6000 for VSA) and 12000 (3000 for VSA) ppm resp.). This shows that the desorbed amount of the two VOCs is more sensitive to a variation of temperature than to a variation of pressure.

During the heating step, the concentration waves of the 2 VOCs and temperature wave propagate together through the column. In terms of desorption, the higher temperature is, the weaker the van der Waals forces between VOCs and the surface of activated carbon becomes, which leads to an increasing regeneration at high desorption temperature. Since desorption temperature is the most important factor concerning the performance of the regeneration step, the effects of this operating parameter on recovery are investigated for the TSA process. Figure 18 shows the variation of the heating step duration versus desorption temperature. For the various simulations, the hot nitrogen stream used in the heating step is maintained constant (12 mol/min) and the necessary duration of this step is optimized to totally recover the 2 VOCs (100% recovery). Also, all the operating parameters of the 3-step TSA cycle (as indicated in Table 3) are kept unchanged except for $T_{des}$ and $t_{heat}$.

When desorption temperature increases from 403 to 443 K, the purge duration decreases from 8600 to 7600 s. The corresponding N2 amount consumed for the regeneration of the column decreases slightly from 1.72 to 1.52 kmol. These values are much lower than those obtained for the VSA process.

The simulation results obtained for the VSA and TSA processes show that it is possible to achieve high dichloromethane and acetone recovery (a value of 100% for the 2 VOCs could be obtained).

For the VSA process, it has been shown that the duration of the regeneration step for a 100% recovery of the two

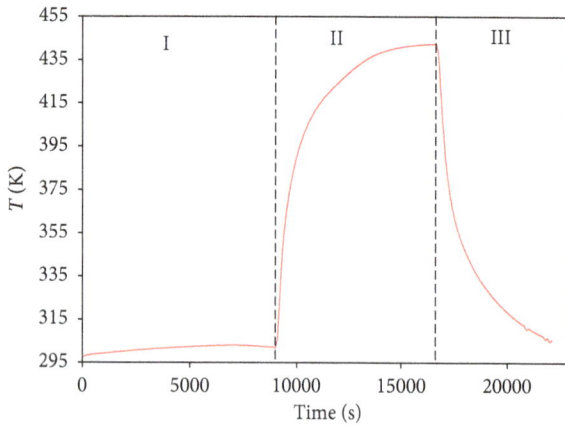

FIGURE 16: Gas temperature history along the bed during one cycle.

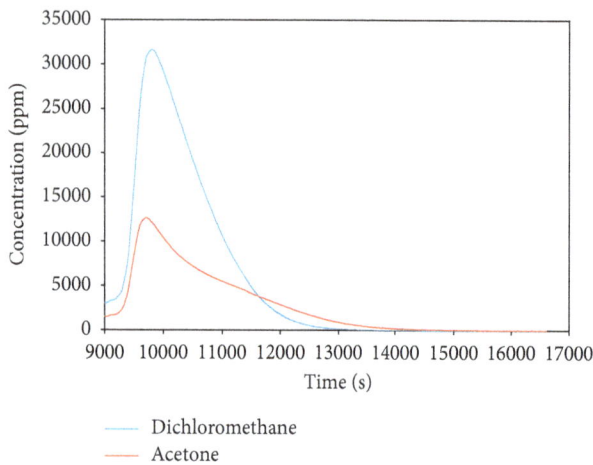

FIGURE 17: Concentration of dichloromethane and acetone profiles at the column outlet during the heating step.

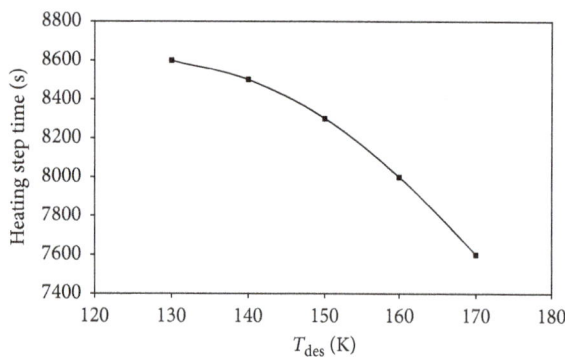

FIGURE 18: Variations of heat duration step versus desorption temperature.

VOCs varies from 20000 to 160000 s for desorption pressures varying from 0.1 to 1 atm (Figure 15). For the TSA process, the regeneration step durations vary from 8600 to 7600 s when the desorption temperature changes from 403 to 443 K at a pressure of 1 atm. Thus, there is a substantial decrease in the total duration of the TSA cycle compared to

the VSA cycle. Regeneration by hot nitrogen stream seems to be more efficient than regeneration by reducing pressure.

## 5. Conclusion

Volatile organic compounds are air pollutants that should be removed to avoid serious effects on human health and environment. A particular interest has been given to the recovery of dichloromethane and acetone used as solvents in oil sands process for extraction of bitumen. Experimental adsorption isotherms of the two VOCs are determined at three different temperatures by the dynamic column breakthrough method. The developed simulation model using Aspen Adsorption software has been used successfully to predict the experimental breakthrough curves for pure VOCs and different binary mixtures for various conditions. The mass transfer coefficients $k_{\mathrm{LDF}}$ used in the linear driving model for dichloromethane and acetone are assessed by adjusting simulation results to experimental breakthrough curves. The validated simulation model has been used for optimizing vacuum and temperature swing adsorption processes so as to achieve the highest performances. Emphasis has been given to the effect of many operating parameters, that is, the adsorption step duration, the vacuum pressure, and the desorption temperature, on the recovery of dichloromethane and acetone. It has been shown that a recovery of 100% of the two VOCs could be attained. However, the adsorption step duration should be determined precisely so as not to affect the recovery and alter the quality of air being purified due to the breakthrough of VOCs. The vacuum pressure and the desorption temperature should be carefully chosen in order to both reduce the energy consumption and shorten the purge step duration. It has to be noted that regeneration by hot nitrogen stream is more efficient than regeneration by reducing pressure.

## Nomenclature

| | |
|---|---|
| $a_s$: | Express the ratio of the particle external surface area to volume, $\mathrm{m^2 \cdot m^{-3}}$ |
| $B$: | Langmuir–Freundlich equation parameter |
| $C$: | Concentration of the VOC, $\mathrm{mol \cdot m^{-3}}$ |
| $C_{ps}$: | Heat capacity of the adsorbent, $\mathrm{J \cdot kg^{-1} \cdot K^{-1}}$ |
| $C_{O}$: | Initial inlet concentration, ppm |
| $C_{pg}$: | Gas heat capacity, $\mathrm{J \cdot kg^{-1} \cdot K^{-1}}$ |
| $D_p$: | Particle diameter, m |
| $d_{int}$: | Column internal diameter, m |
| $D_e$: | Effective diffusivity of the VOC, $\mathrm{m^2 \cdot s^{-1}}$ |
| $D_z$: | Axial dispersion coefficient, $\mathrm{m^2 \cdot s^{-1}}$ |
| $k_1, k_2, k_3$: | Langmuir–Freundlich equation parameter |
| $k_{\mathrm{LDF}}$: | Mass transfer coefficient, $\mathrm{s^{-1}}$ |
| $K_g$: | Gas thermal conductivity, $\mathrm{W \cdot m^{-1} \cdot K^{-1}}$ |
| $k_s$: | Effective axial solid phase thermal conductivity, $\mathrm{W \cdot m^{-2} \cdot K^{-1}}$ |
| $h_f$: | Film heat transfer, $\mathrm{W \cdot m^{-2} \cdot K^{-1}}$ |
| $L$: | Bed height, m |
| $m_{ad}$: | Mass of the sample of activated carbon in the bed, g |
| $Q$: | Average adsorbed concentration, $\mathrm{mol \cdot kg^{-1}}$ |

| | |
|---|---|
| $q*$: | Equilibrium adsorption concentration, mol·kg$^{-1}$ |
| $Q_m$: | Adsorption capacity of the amount, mol·kg$^{-1}$ |
| $R_p$: | Particle radius, m |
| $P$: | Pressure, Pa |
| $P_{ad}$: | Feed pressure in adsorption step, atm |
| $P_{des}$: | Vacuum pressure in evacuation and purge step, atm |
| $\dot{Q}_{feed}$: | Molar flow, mol·min$^{-1}$ |
| $Q_{pg}$: | Molar flow of purge step, mol·min$^{-1}$ |
| $t$: | Time, s |
| $t_f$: | Adsorption end time, min |
| $t_{ad}$: | Adsorption step time, s |
| $t_{cool}$: | Cool step time, s |
| $t_{evac}$: | Evacuation step time, s |
| $t_{heat}$: | Heat step time, s |
| $t_{pg}$: | Purge step time, s |
| $t_{PR}$: | Pressurization step time, s |
| $T$: | Temperature, K |
| $T_{ad}$: | Adsorption temperature, K |
| $T_{des}$: | Desorption temperature, K |
| $Z$: | Axial co-ordinate along the bed, m |
| $v_g$: | Interstitial velocity, m·s$^{-1}$ |
| $a_s$: | Express the ratio of the particle external surface area to volume, m$^2$·m$^{-3}$ |
| $\rho_g$: | Gas density, kg·m$^{-3}$ |
| $\rho_p$: | Particle density, kg·m$^{-3}$ |
| $\varepsilon$: | Void fraction |
| $\varepsilon_p$: | Void of pellets |
| $\mu$: | Dynamic viscosity, Pa·s |
| $\lambda_L$: | Axial heat dispersion, W·m$^{-2}$·K$^{-1}$ |
| $\Delta H$: | Isosteric heat of adsorption, J·mol$^{-1}$ |
| I: | Component |
| J: | Component |
| I, II, III, VI: | Step numbers. |

## Conflicts of Interest

The authors declare that they have no conflicts of interest.

## References

[1] F. I. Khan and A. K. Ghosal, "Removal of volatile organic compounds from polluted air," *Journal of Loss Prevention in the Process Industries*, vol. 13, no. 6, pp. 527–45, 2000.

[2] I. K. Shah, P. Pré, and B. J. Alappat, "Effect of thermal regeneration of spent activated carbon on volatile organic compound adsorption performances," *Journal of the Taiwan Institute of Chemical Engineers*, vol. 45, no. 4, pp. 1733–1738, 2014.

[3] J. Read, D. Whiteoak, and S. Bitumen, *The Shell Bitumen Handbook*, Thomas Telford Publishers, London, UK, 5th edition, 2003.

[4] J. Sundell, "On the history of indoor air quality and health," *Indoor Air*, vol. 14, no. s7, pp. 51–58, 2004.

[5] T. Dobre, O. C. Pârvulescu, A. Jacquemet, and V. A. Ion, "Adsorption and thermal desorption of volatile organic compounds in a fixed bed—experimental and modelling,"

*Chemical Engineering Communications*, vol. 203, no. 12, pp. 1554–1561, 2016.

[6] G. S. Cooper, C. S. Scott, and A. S. Bale, "Insights from epidemiology into dichloromethane and cancer risk," *International Journal of Environmental Research and Public Health*, vol. 8, no. 8, pp. 3380–3398, 2011.

[7] P. M. Bos, M. J. Zeilmaker, and J. C. van Eijkeren, "Application of physiologically based pharmacokinetic modeling in setting acute exposure guideline levels for methylene chloride," *Toxicological Sciences*, vol. 91, no. 2, pp. 576–585, 2006.

[8] F. Zeinali, A. A. Ghoreyshi, and G. D. Najafpour, "Adsorption of dichloromethane from aqueous phase using granular activated carbon :isotherm and breakthrough curve measurments," *Middle-East Journal of Scientific Research*, vol. 5, no. 4, pp. 191–198, 2010.

[9] L. Gales, A. Mendes, and C. Costa, "Equilibrium and heat of adsorption for organic vapors and activated carbons," *Carbon*, vol. 38, no. 7, pp. 1083–1088, 2000.

[10] C. Long, Y. Li, W. Yu, and A. Li, "Removal of benzene and methyl ethyl ketone vapor: comparison of hypercross linked polymeric adsorbent with activated carbon," *Journal of Hazardous Materials*, vol. 203-204, pp. 251–256, 2012.

[11] L. Fournel, P. Mocho, R. Brown, and P. le Cloirec, "Modeling breakthrough curves of volatile organic compounds on activated carbon fibers," *Adsorption*, vol. 16, no. 3, pp. 147–153, 2010.

[12] C. A. Grande, "Advances in pressure swing adsorption for gas separation," *ISRN Chemical Engineering*, vol. 2012, Article ID 982934, 13 pages, 2012.

[13] K. Rambabu, L. Muruganandam, and S. Velu, "CFD Simulation for separation of carbon dioxide-methane mixture by pressure swing adsorption," *International Journal of Chemical Engineering*, vol. 2014, Article ID 402756, 7 pages, 2014.

[14] J. A. Wurzbacher, C. Gebald, and A. Steinfeld, "Separation of $CO_2$ from air by temperature vacuum swing adsorption using diamine-functionalized silica gel," *Energy & Environmental Science*, vol. 4, no. 9, p. 3584, 2011.

[15] A. L. Chaffee, G. P. Knowles, Z. Liang, J. Zhang, P Xiao, and P. A. Webley, "$CO_2$ capture by adsorption: material and process development," *International Journal of Greenhouse Gas Control*, vol. 1, no. 1, pp. 11–18, 2007.

[16] F. Salvador, N. Martin-Sanchez, R. Sanchez-Hernandez, M. J. Sanchez-Montero, and C. Izquierdo, "Regeneration of carbonaceous adsorbents. Part I: thermal regeneration," *Microporous and Mesoporous Materials*, vol. 202, pp. 259–276, 2015.

[17] M. A. Sidheswarana, H. Destaillats, D. P. Sullivan, S. Cohn, and W. J. Fisk, "Energy efficient indoor VOC air cleaning with activated carbonfiber (ACF) filters," *Building and Environment*, vol. 47, pp. 357–367, 2011.

[18] M. J. Jeon and Y. W. Jeon, "Characteristic evaluation of activated carbon applied to a pilot-scale VSA system to control VOCs," *Process Safety and Environment Protection*, vol. 112, pp. 327–334, 2017.

[19] A. Kane, S. Giraudet, J. B. Vilmain, and P. Le Cloirec, "Intensification of the temperature swing adsorption process with a heat pump for the recovery of dichloromethane," *Journal of Environmental Chemical Engineering*, vol. 3, no. 2, pp. 734–743, 2015.

[20] K. S. Hwang, D. K. Choi, S. Y. Gong, and S. Y. Cho, "Adsorption and thermal regeneration of methylene chloride

vapor on activated carbon bed," *Chemical Engineering and Processing*, vol. 46, pp. 1111-1123, 1998.

[21] L. Li, Z. Liu, Y. Qin, Z. Sun, J. Song, and L. Tang, "Estimation of volatile organic compound mass transfer coefficients in the vacuum desorption of acetone from activated carbon," *Journal of Chemical & Engineering Data*, vol. 55, no. 11, pp. 4732-4740, 2010.

[22] S. Brunauer, P. H. Emmett, and E. Teller, "Adsorption of gases in multimolecular layers," *Journal of the American Chemical Society*, vol. 60, no. 2, pp. 309-319, 1938.

[23] J. A. Delgada and A. E. Rodrigues, "Analysis of the boundary conditions for the simulation of the pressure equalization step in PSA cycles," *Chemical Engineering Science*, vol. 63, no. 18, pp. 4452-4463, 2008.

[24] M. F. Edwards and J. F. Richardson, "Gas dispersion in packed beds," *Chemical Engineering Science*, vol. 23, no. 2, pp. 109-123, 1968.

[25] R. B. Bird, W. E. Stewart, and E. N. Lightfoot, "Transport phenomena," *AIChE Journal*, vol. 7, no. 2, pp. 5J-6J, 1961.

[26] M. Lei, C. Vallires, G. Grevillot, and M. A. Latifi, "Modeling and simulation of a thermal swing adsorption process for $CO_2$ capture and recovery," *Industrial and Engineering Chemistry Research*, vol. 52, no. 22, pp. 7526-7533, 2013.

[27] J. Xiao, Y. Peng, P. Benard, and R. Chahine, "Thermal effects on breakthrough curves of pressure swing adsorption for hydrogen purification," *International Journal of Hydrogen Energy*, vol. 41, no. 19, pp. 8236-8245, 2015.

[28] F. Rezaei, S. Subramanian, J. Kalyanaraman, R. P. Lively, Y. Kawajiri, and M. J. Realff, "Modeling of rapid temperature swing adsorption using hollow fiber sorbents," *Chemical Engineering Science*, vol. 113, pp. 62-76, 2014.

[29] A. P. D. Wasch and G. F. Froment, "Heat transfer in packed beds," *Chemical Engineering Science*, vol. 27, no. 3, pp. 567-576, 1972.

[30] S. Yagi, D. Kunii, and N. Wakao, "Studies on axial effective thermal conductivities in packed beds," *AIChE Journal*, vol. 6, no. 4, pp. 543-546, 1960.

[31] K. Daeho, S. Ranjani, and T. B. Lorenz, "Optimization of pressure swing adsorption and fractionated vacuum pressure swing adsorption processes for $CO_2$ capture," *Industrial & Engineering Chemistry Research*, vol. 44, pp. 8084-8094, 2005.

[32] J. Zhang, P. A. Webley, and P. Xiao, "Effect of process parameters on power requirements of vacuum swing adsorption technology for $CO_2$ capture from flue gas," *Energy Conversion and Management*, vol. 49, no. 2, pp. 346-356, 2008.

# 17

# Combined Effects of Thermal Radiation and Nanoparticles on Free Convection Flow and Heat Transfer of Casson Fluid over a Vertical Plate

M. G. Sobamowo ⓘ

*Department of Mechanical Engineering, University of Lagos, Akoka, Lagos, Nigeria*

Correspondence should be addressed to M. G. Sobamowo; mikegbeminiyi@gmail.com

Academic Editor: Iftekhar A. Karimi

The influences of thermal radiation and nanoparticles on free convection flow and heat transfer of Casson nanofluids over a vertical plate are investigated. The governing systems of nonlinear partial differential equations of the flow and heat transfer processes are converted to systems of nonlinear ordinary differential equations through similarity transformations. The resulting systems of fully coupled nonlinear ordinary differential equations are solved using the differential transformation method with Padé-approximant technique. The accuracies of the developed analytical methods are verified by comparing their results with the results of past works as presented in the literature. Thereafter, the analytical solutions are used to investigate the effects of thermal radiation, Prandtl number, nanoparticle volume fraction, shape, and type on the flow and heat transfer behaviour of various nanofluids over the flat plate. It is observed that both the velocity and temperature of the nanofluid as well as the viscous and thermal boundary layers increase with increase in the thermal radiation parameter. The velocity of the nanofluid decreases and the temperature of the nanofluid increase, respectively, as the Prandtl number and volume fraction of the nanoparticles in the base fluid increase. The decrease in velocity and increase in temperature are highest in lamina-shaped nanoparticle and followed by platelet-, cylinder-, brick-, and sphere-shaped nanoparticles, respectively. Using a common base fluid to all the nanoparticle types, it is established that the decrease in velocity and increase in temperature are highest in $TiO_2$ and followed by CuO, $Al_2O_3$, and SWCNT nanoparticles, in that order. It is hoped that the present study will enhance the understanding of free convection boundary layer problems of Casson fluid under the influences of thermal radiation and nanoparticles as applied in various engineering processes.

## 1. Introduction

The importance and the wide applications of free convection flow and heat transfer in extrusion, melt spinning, glass-fibre production processes, food processing, mechanical forming processes etc. have in recent times aroused various renewed research interests and explorations. In the study of free convection and heat transfer problems, the analysis of incompressible laminar flow of viscous fluid in a steady-state, two-dimensional free convection boundary layer has over the years been a common area of increasing research interests following the experimental investigations of Schmidt and Beckmann [1] and the pioneering theoretical work of Ostrach [2]. In their attempts to study the laminar free convection flow and heat transfer problem in 1953, Ostrach

[2] applied method of iterative integration to analyze free convection over a semi-infinite isothermal flat plate. The author obtained the numerical solutions for a wide range of Prandtl numbers from 0.01 to 1000 and validated their numerical results using experimental data of Schmidt and Beckmann [1]. Five years later, Sparrow and Gregg [3] presented a further study on numerical solutions for laminar free convection from a vertical plate with uniform surface heat flux. Considering the fact that the major part of low Prandtl number boundary layer of free convection is inviscid, Lefevre [4] examined laminar free convection of an inviscid flow from a vertical plane surface. In a further work, Sparrow and Gregg [5] developed similar solutions for free convection from a nonisothermal vertical plate. Meanwhile, a study on fluid flow over a heated vertical plate at high Prandtl

number was presented by Stewartson and Jones [6]. Due to the disadvantages in the numerical methods in the previous studies [2, 3], Kuiken [7] adopted method of matched asymptotic expansion and established asymptotic solutions for large Prandtl number free convection. In the subsequent year, the same author applied the singular perturbation method and analyzed free convection flow of fluid at low Prandtl numbers [8]. Also, in another work on the asymptotic analysis of the same problem, Eshghy [9] studied free convection boundary layers at large Prandtl number while Roy [10] investigated free convection boundary layer problem for a uniform surface heat flux at high Prandtl number. With the development of asymptotic solution, a combined study of the effects of small and high Prandtl numbers on the viscous fluid flow over a flat vertical plate was submitted by Kuiken and Rotem [11]. In the succeeding year, Na and Habib [12] utilized parameter differentiation method to solve the free convection boundary layer problem. Few years later, Merkin [13] presented the similarity solutions for free convection on a vertical plate while Merkin and Pop [14] used finite difference method to develop numerical solutions for the conjugate free convection problem of boundary-layer flow over a vertical plate. Also, Ali et al. [15] submitted a study on numerical investigation of free convective boundary layer in a viscous fluid.

The various analytical and numerical studies of the past works have shown that the boundary layer problems are very difficult to solve. This is because, besides having very thin regions where there is rapid change of the fluid properties, they are defined on unbounded domains. Although, approximate analytical methods are being used to solve boundary layer problems, they converge very slowly for some boundary layer problems, particularly for those with very large parameters. The numerical methods used in the flow process also encounter some problems in resolving the solutions of the governing equations in the very thin regions and in some cases where singularities or multiple solutions exist. Moreover, in numerical analysis, it is absolutely required that the stability and convergence analysis is carried out so as to avoid divergence or inappropriate results. Such analysis in the mathematical methods increases the computation time and cost. Therefore, in the quest for presenting symbolic solutions to the flow and heat transfer problem using one of the recently developed semianalytical methods, Motsa et al. [16] adopted homotopy analysis of free convection boundary layer flow with heat and mass transfer. In another work, the authors used spectral local linearization approach for solving the natural convection boundary layer flow [17]. Ghotbi et al. [18] investigated the application of homotopy analysis method to natural convection boundary layer flow. Although, homotopy analysis method (HAM) is a reliable and efficient semianalytical technique, it suffers from a number of limiting assumptions such as the requirements that the solution ought to conform to the so-called rule of solution expression and the rule of coefficient ergodicity. Also, the use of HAM in the analysis of linear and nonlinear equations requires the determination of auxiliary parameter which will increase the computational cost and time. Furthermore, the lack of rigorous theories or proper guidance for choosing initial approximation, auxiliary linear

operators, auxiliary functions, and auxiliary parameters limits the applications of HAM. Moreover, such method requires high skill in mathematical analysis and the solution comes with large number of terms. Nonetheless, various analyses of nonlinear models and fluid flow problems under the influences of some internal and external factors using different approximate analytical and numerical methods have been presented in the literature [19–47]. Also, the relative simplicity coupled with ease of applications of the differential transformation method (DTM) has made the method to be more effective than most of the other approximate analytical methods. The method was introduced by Zhou [48] and it has fast gained ground as it appeared in many engineering and scientific research papers. This is because, with the applications of DTM, a closed form series solution or approximate solution can be provided for nonlinear integral and differential equations without linearization, restrictive assumptions, perturbation, and discretization or round-off error. It reduces complexity of expansion of derivatives and the computational difficulties of the other traditional or recently developed methods. Therefore, Lien-Tsai and Cha'o-Kuang [49] applied the differential transformation method to provide approximate analytical solutions to the Blasius equation. Also, Kuo [50] adopted the same method to determine the velocity and temperature profiles of the Blasius equation of forced convection problem for fluid flow passing over a flat plate. An extended work on the applications of differential transformation method to free convection boundary layer problem of two-dimensional steady and incompressible laminar flow passing over a vertical plate was presented by the same author [51]. However, in the later work, the nonlinear coupled boundary value governing equations of the flow and heat transfer processes is reduced to initial value equations by a group of transformations, and the resulting coupled initial value equations are solved by means of the differential transformation method. The reduction or the transformation of the boundary value problems to the initial value problems was carried out due to the fact that the developed systems of nonlinear differential equations contain an unbounded domain of infinite boundary conditions. Moreover, in order to obtain the numerical solutions that are valid over the entire large domain of the problem, Ostrach [2] estimated the values of $f''(0)$ and $\theta'(0)$ during the analysis of the developed systems of fully coupled nonlinear ordinary differential equations. Following Ostrach's approach, most of the subsequent solutions provided in the literature [3, 9, 10, 12, 14, 15, 50, 51] were based on the estimated boundary conditions given by Ostrach [2]. Additionally, the limitations of power series solutions to small domain problems have been well established in the literature. Nevertheless, in some recent studies, the use of power series methods coupled with Padé-approximant technique has shown to be very effective way of developing accurate analytical solutions to nonlinear problems of large or unbounded domain problems of infinite boundary conditions. The application of Padé-approximant technique with power series method increases the rate and radius of convergence of power series solution. Therefore, in a recent work, Rashidi et al. [52] applied differential transformation method coupled with the Padé-approximant technique to develop a novel analytical solution for mixed convection about an inclined flat plate embedded in a porous medium.

Casson fluid is a non-Newtonian fluid that was first introduced by Casson in 1959 [53]. It is a shear thinning liquid which is assumed to have an infinite viscosity at zero rate of shear, a yield stress below which no flow occurs, and a zero viscosity at an infinite rate of shear [54]. If the yield stress is greater than the shear stress, then it acts as a solid, whereas if the yield stress is less than the applied shear stress, then the fluid would start to move. The fluid is based on the structure of liquid phase and interactive behaviour of solid of a two-phase suspension. It is able to capture complex rheological properties of a fluid, unlike other simplified models like the power law [55] and second-, third-, or fourth-grade models [56]. Some examples of Casson fluid are jelly, honey, tomato sauce, and concentrated fruit juices. Human blood is also treated as a Casson fluid in the presence of several substances such as fibrinogen, globulin in aqueous base plasma, protein, and human red blood cells. Concentrated fluids like sauces, honey, juices, blood, and printing inks can be well described using this model. It has various applications in fibrinogen, cancer homeotherapy, protein, and red blood cells, forming a chain-type structure. Due to these applications, many researchers are concentrating on characteristics of Casson fluid. Application of Casson fluid for flow between two rotating cylinders is studied in [57]. The effect of magnetohydrodynamic (MHD) Casson fluid flow in a lateral direction past linear stretching sheet was explained by Nadeem et al. [58].

The role of thermal radiation is very important in some industrial applications, such as glass production and furnace design, and also in space technology applications such as comical flight aerodynamics rocket, space vehicles, propulsion systems, plasma physics, and space craft reentry aerodynamics which operates at high temperatures, in the flow structure of atomic plants, combustion processes, internal combustion engines, ship compressors, and solar radiations. The effect of thermal radiation on magnetohydrodynamic flow was examined by Raptis and Perdikis [59] while Seddeek [60] investigated the impacts of thermal radiation and variable viscosity on magnetohydrodynamics in free convection flow over a semi-infinite flat plate. In another study, Mabood et al. [61] analyzed unsteady stretched flow of Maxwell fluid in the presence of nonlinear thermal radiation and convective condition while Hayat et al. [62] addressed the effects of nonlinear thermal radiation and magnetohydrodynamics on viscoelastic nanofluid flow. Farooq et al. [63] addressed the effects of nonlinear thermal radiation on stagnation point flow. Also, Shehzad et al. [64] presented a study on MHD three-dimensional flow of Jeffrey nanofluid with internal heat generation and thermal radiation.

The previous studies on fluid flow over stretching under investigation are based on viscous fluid flow as shown in the above-reviewed works. To the best of the author's knowledge, a study on the influence of thermal radiation and nanoparticle shape, size, and type on the free convection boundary layer flow and heat transfer of Casson nanofluids over a vertical plate at low and high Prandtl numbers using the differential transformation method coupled with Padé-approximant technique has not been investigated. Therefore, the present study focuses on the

application of differential transformation method coupled with Padé-approximant technique to develop approximate analytical solutions and carry out parametric studies of the effects of thermal radiation and nanoparticles on free convection boundary layer flow and heat transfer of nanofluids of different nanosize particles over a vertical plate at low and high Prandtl numbers. Another novelty of the present study is displayed in the development of approximate analytical solutions for the free convection boundary layer problem without the use of the estimated boundary conditions $f''(0)$ and $\theta'(0)$ during the analysis of the problem.

## 2. Problem Formulation and Mathematical Analysis

Consider a laminar free convection flow of an incompressible Casson nanofluid over a vertical plate parallel to the direction of the generating body force, as shown in Figure 1. The rheological equation for an isotropic and incompressible Casson fluid, reported by Casson [65], is

$$\tau = \tau_0 + \mu\dot{\sigma} \tag{1}$$

or

$$\tau = \left\{ 2\left(\mu_B + \frac{p_y}{\sqrt{2\pi}}\right)e_{ij}, \quad \pi > \pi_c \right\}$$
$$= \left\{ 2\left(\mu_B + \frac{p_y}{\sqrt{2\pi_c}}\right)e_{ij}, \quad \pi_c < \pi \right\}, \tag{2}$$

where $\tau$ is the shear stress; $\tau_0$ is the Casson yield stress; $\mu$ is the dynamic viscosity; $\dot{\sigma}$ is the shear rate; $\pi = e_{ij}e_{ij}$ in which $e_{ij}$ is the $(i, j)$th component of the deformation rate and $\pi$ is the product of the component of deformation rate with itself; $\pi_c$ is a critical value of this product based on the non-Newtonian model; $\mu_B$ is the plastic dynamic viscosity of the non-Newtonian fluid; and $p_y$ is the yield stress of the fluid. The velocity and the temperature are functions of $y$, $t$ only.

Assuming that the flow in the laminar boundary layer is two-dimensional and steady, the heat transfer from the plate to the fluid is proportional to the local surface temperature $T$, using the Boussinesq approximation along with the assumption that the pressure is uniform across the boundary layer, and the equations for continuity, motion, and energy are given as

$$\frac{\partial u}{\partial x} + \frac{\partial v}{\partial y} = 0, \tag{3}$$

$$\rho_{nf}\left(u\frac{\partial u}{\partial x} + v\frac{\partial u}{\partial y}\right) = \left(1 + \frac{1}{\gamma}\right)\mu_{nf}\frac{\partial^2 u}{\partial y^2} + g(\rho\beta)_{nf}(T - T_\infty), \tag{4}$$

$$(\rho c_p)_{nf}\left(u\frac{\partial T}{\partial x} + v\frac{\partial T}{\partial y}\right) = k_{nf}\frac{\partial^2 T}{\partial y^2} - \frac{\partial q_r}{\partial y}. \tag{5}$$

Assuming no slip conditions, the appropriate boundary conditions are given as

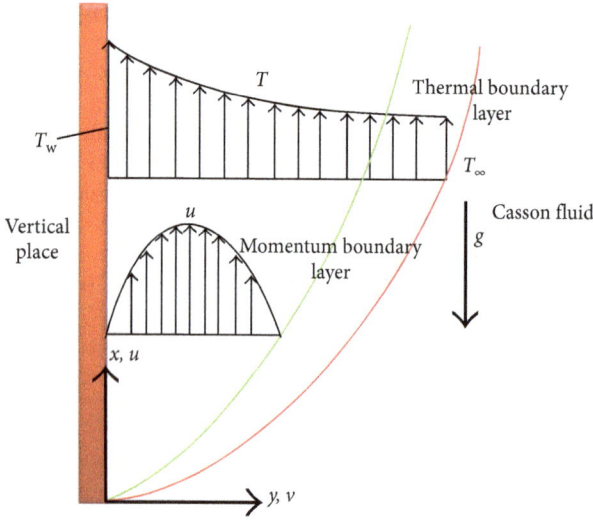

FIGURE 1: Velocity and temperature profiles in free convection flow over a vertical plate.

TABLE 1: The values of different shapes of nanoparticles [41, 66].

| S/N | Name | Shape | Shape factor ($m$) | Sphericity ($\psi$) |
|-----|------|-------|--------------------|--------------------|
| 1 | Sphere | | 3.0 | 1.000 |
| 2 | Platelet | | 5.7 | 0.526 |
| 3 | Cylinder | | 4.8 | 0.625 |
| 4 | Lamina | | 16.2 | 0.185 |
| 5 | Brick | | 3.7 | 0.811 |

TABLE 2: Physical and thermal properties of the base fluid [41, 66–70].

| Base fluid | $\rho$ (kg/m$^3$) | $c_p$ (J/kg K) | $k$ (W/mK) |
|------------|-------------------|----------------|------------|
| Pure water | 997.1 | 4179 | 0.613 |
| Ethylene glycol | 1115 | 2430 | 0.253 |
| Engine oil | 884 | 1910 | 0.144 |
| Kerosene | 783 | 2010 | 0.145 |

$$u = 0, \ v = 0, \ T = T_s \quad \text{at } y = 0, \tag{6a}$$

$$u = 0, \ T = T_w, \quad \text{at } y \to \infty, \tag{6b}$$

where the various physical and thermal properties in (3)–(5) are given as

$$\rho_{nf} = \rho_f (1 - \phi) + \rho_s \phi, \tag{7a}$$

$$(\rho c_p)_{nf} = (\rho c_p)_f (1 - \phi) + (\rho c_p)_s \phi, \tag{7b}$$

$$(\rho \beta)_{nf} = (\rho \beta)_f (1 - \phi) + (\rho \beta)_s \phi, \tag{7c}$$

$$\mu_{nf} = \frac{\mu_f}{(1 - \phi)^{2.5}}, \tag{7d}$$

$$k_{nf} = k_f \left[ \frac{k_s + (m-1)k_f - (m-1)\phi(k_f - k_s)}{k_s + (m-1)k_f + \phi(k_f - k_s)} \right], \tag{8}$$

$$\frac{\partial q_r}{\partial y} = -\frac{4\sigma}{3K} \frac{\partial T^4}{\partial y}$$
$$\cong -\frac{16\sigma T_s^3}{3K} \frac{\partial^2 T}{\partial y^2} \quad \text{(using Rosseland's approximation),} \tag{9}$$

where $m$ in the above Hamilton–Crosser model in (8) is the shape factor and its numerical values for different shapes are given in Table 1. It should be noted that the shape factor, $m = 3/\lambda$, where $\lambda$ is the sphericity (the ratio of the surface area of the sphere and the surface area of the real particles with equal volumes). Sphericity of sphere, platelet, cylinder, laminar, and brick are 1.000, 0.526, 0.625, 0.185, and 0.811, respectively. The Hamilton–Crosser model becomes a Maxwell–Garnett model, when the shape factor of the nanoparticle is 3 ($m = 3$).

Tables 2 and 3 present the physical and thermal properties of the base fluid and the nanoparticles, respectively. SWCNTs mean single-walled carbon nanotubes.

Going back to (3)–(5), if one introduces a stream function, $\psi(x, y)$, such that

$$u = \frac{\partial \psi}{\partial y},$$
$$v = -\frac{\partial \psi}{\partial x}, \tag{10}$$

and uses the following similarity and dimensionless variables:

$$\eta = \left[ \frac{\rho_f^2 (g\beta_f (T_w - T_\infty))}{4\mu_f^2 x} \right]^{1/4} y,$$

$$\psi = \frac{4\mu_f}{\rho_f} \left[ \frac{\rho_f^2 (g\beta_f (T_w - T_\infty))x^3}{4\mu_f^2} \right]^{1/4} f(\eta),$$

$$\theta = \frac{T - T_\infty}{T_w - T_\infty}, \tag{11}$$

$$\Pr = \frac{\mu_f c_p}{k_f},$$

$$R = \frac{4\sigma T_\infty^3}{3kK},$$

one arrives at fully coupled third- and second-order ordinary differential equations:

TABLE 3: Physical and thermal properties of the nanoparticles [41, 66–70].

| Nanoparticles | $\rho$ (kg/m$^3$) | $c_p$ (J/kg K) | $k$ (W/mK) |
|---|---|---|---|
| Copper (Cu) | 8933 | 385 | 401 |
| Aluminum oxide (Al$_2$O$_3$) | 3970 | 765 | 40 |
| SWCNTs | 2600 | 42.5 | 6600 |
| Silver (Ag) | 10500 | 235.0 | 429 |
| Titanium dioxide (TiO$_2$) | 4250 | 686.2 | 8.9538 |
| Copper(II) oxide (CuO) | 783 | 540 | 18 |

TABLE 4: Operational properties of differential transformation method.

| S/N | Function | Differential transform |
|---|---|---|
| 1 | $u(t) \pm v(t)$ | $U(p) \pm V(p)$ |
| 2 | $\alpha u(t)$ | $\alpha U(p)$ |
| 3 | $du(t)/dt$ | $(p+1)U(p+1)$ |
| 4 | $u(t)v(t)$ | $\sum_{r=0}^{p} V(r)U(p-r)$ |
| 5 | $u^m(t)$ | $\sum_{r=0}^{p} U^{m-1}(r)U(p-r)$ |
| 6 | $d^n u(t)/dx^n$ | $(p+1)(p+2)\cdots(p+n)U(p+n)$ |
| 7 | $\sin(\omega t + \alpha)$ | $(\omega^p/p!)\sin((\pi p/2!) + \alpha)$ |
| 8 | $\cos(\omega t + \alpha)$ | $Z(p) = (\omega^p/p!)\cos((\pi p/2!) + \alpha)$ |

$$\left(1 + \frac{1}{\gamma}\right)f''' + (1-\phi)^{2.5}\left\{\left[(1-\phi) + \phi\left(\frac{\rho_s}{\rho_f}\right)\right]\left(3ff'' - 2(f')^2\right) + \left[(1-\phi) + \phi\left[\frac{(\rho\beta)_s}{(\rho\beta)_f}\right]\right]\theta\right\} = 0, \tag{12}$$

$$\left(1 + \frac{4}{3}R\right)\theta'' + 3\left[\frac{1}{\left[(1-\phi) + \phi\left[(\rho C_p)_s/(\rho C_p)_f\right]\right]}\right]\left[\frac{k_s + (m-1)k_f - (m-1)\phi(k_f - k_s)}{k_s + (m-1)k_f + \phi(k_f - k_s)}\right]\Pr f\theta' = 0, \tag{13}$$

and the boundary conditions as

$$f = 0, \; f' = 0, \; \theta = 1, \quad \text{when } \eta = 0,$$
$$f' = 0, \; \theta = 0, \quad \quad \text{when } \eta = \infty. \tag{14}$$

It should be noted that, for a viscous fluid which does not have nanoparticles with negligible radiation, the nanoparticle volume fraction is zero, that is, $\phi = 0$, $R = 0$, and $\gamma \to \infty$, then one recovers the earlier models [2–15] from (12) and (13), which are

$$f''' + 3ff'' - 2(f')^2 + \theta = 0, \tag{15}$$

$$\theta'' + 3\Pr f\theta' = 0, \tag{16}$$

and the boundary conditions remain the same as in (14).

## 3. Method of Solution: Differential Transform Method

The relatively new semi-analytical method, differential transformation method introduced by Zhou [48], has proven very effective in providing highly accurate solutions to differential equations, difference equation, differential-difference equations, fractional differential equation, pantograph equation, and integrodifferential equation. Therefore, this method is applied in the present study. The basic definitions and the operational properties of the method are as follows.

If $u(t)$ is analytic in the domain $T$, then the function $u(t)$ will be differentiated continuously with respect to time $t$:

$$\frac{d^p u(t)}{dt^p} = \varphi(t, p) \quad \text{for all } t \in T. \tag{17}$$

If $t = t_i$, then $\varphi(t, p) = \varphi(t_i, p)$, where $p$ belongs to the set of nonnegative integers, denoted as the $p$-domain. We can therefore write (17) as

$$U(p) = \varphi(t_i, p) = \left[\frac{d^p u(t)}{dt^p}\right]_{t=t_i}, \tag{18}$$

where $U_p$ is called the spectrum of $u(t)$ at $t = t_i$.

Express $u(t)$ in Taylor's series as

$$u(t) = \sum_{p}^{\infty}\left[\frac{(t - t_i)^p}{p!}\right]U(p), \tag{19}$$

where (19) is the inverse of $U(k)$ with symbol "$D$" denoting the differential transformation process and combining (18) and (19), we have

$$u(t) = \sum_{p=0}^{\infty}\left[\frac{(t - t_i)^p}{p!}\right]U(p) = D^{-1}U(p). \tag{20}$$

Table 4 contains the differential transform of some functions. Using the operational properties of the differential transformation method, the differential transformation of the governing differential (12) is given as

$$\left(1 + \frac{1}{\gamma}\right)(p+1)(p+2)(p+3)F(p+3) + (1-\phi)^{2.5}\left\{\left[(1-\phi) + \phi\left(\frac{\rho_s}{\rho_f}\right)\right]\left[3\sum_{l=0}^{p}(p-l+1)(p-l+2)F(l)F(p-l+2)\right]\right.$$

$$\left. - 2\sum_{l=0}^{p}(l+1)(p-l+1)F(l+1)F(p-l+1)\right] + \left[(1-\phi) + \phi\left[\frac{(\rho\beta)_s}{(\rho\beta)_f}\right]\right]\Theta(p)\right\} = 0. \tag{21}$$

Equivalently, one can write the recursive relation for (21) in DTM domain as

$$F(p+3) = \frac{(1-\phi)^{2.5}}{(1+(1/\gamma))(p+1)(p+2)(p+3)} \left\{ \left[(1-\phi) + \phi\left(\frac{\rho_s}{\rho_f}\right)\right] \left[2\sum_{l=0}^{p}(l+1)(p-l+1)F(l+1)F(p-l+1)\right.\right.$$

$$\left.\left. -3\sum_{l=0}^{p}(p-l+1)(p-l+2)F(l)F(p-l+2)\right] - \left[(1-\phi) + \phi\left[\frac{(\rho\beta)_s}{(\rho\beta)_f}\right]\right]\Theta(p)\right\}. \tag{22}$$

For (13), the recursive relation in differential transform domain is given as

$$\left(1 + \frac{4}{3}R\right)(p+1)(p+2)\Theta(p+2) + \left\{ 3\,\text{Pr}\left[\frac{1}{\left[(1-\phi) + \phi\left[(\rho C_p)_s/(\rho C_p)_f\right]\right]}\right]\left[\frac{k_s + (m-1)k_f - (m-1)\phi(k_f - k_s)}{k_s + (m-1)k_f + \phi(k_f - k_s)}\right]\right.$$

$$\left. \times \sum_{l=0}^{p}(l+1)\Theta(l+1)F(p-l)\right\} = 0, \tag{23}$$

which can be written as

$$\Theta(p+2) = \frac{-3\,\text{Pr}}{(1+(4/3)R)(p+1)(p+2)} \left\{ \left[\frac{1}{\left[(1-\phi) + \phi\left[(\rho C_p)_s/(\rho C_p)_f\right]\right]}\right]\left[\frac{k_s + (m-1)k_f - (m-1)\phi(k_f - k_s)}{k_s + (m-1)k_f + \phi(k_f - k_s)}\right]\right.$$

$$\left. \times \sum_{l=0}^{p}(l+1)\Theta(l+1)F(p-l)\right\}. \tag{24}$$

Also, the recursive relation for the boundary conditions in (15) is

$$F(p) = 0 \Rightarrow F(0) = 0,$$

$$(p+1)F(p+1) = 0 \Rightarrow F(1) = 0,$$

$$\theta(p) = 1 \Rightarrow \theta(0) = 1, \tag{25}$$

$$F(2) = \frac{a}{2},$$

$$\theta(1) = b,$$

where $a$ and $b$ are unknown constants which will be found later.

From (25), the following boundary conditions in differential transform domain are established:

$$F(0) = 0,$$

$$F(1) = 0,$$

$$\theta(0) = 1, \tag{26}$$

$$F(2) = \frac{a}{2},$$

$$\theta(1) = b.$$

Using $p = 0, 1, 2, 3, 4, 5, 6, 7, \ldots$ in the above recursive relations in (21), the following equations are developed:

$$F[3] = \frac{-(1-\phi)^{2.5}}{6(1+(1/\gamma))}\left\{(1-\phi) + \phi\left[\frac{(\rho\beta)_s}{(\rho\beta)_f}\right]\right\}, \tag{27}$$

$$F[4] = \frac{-(1-\phi)^{2.5}}{24(1+(1/\gamma))}\left\{(1-\phi) + \phi\left[\frac{(\rho\beta)_s}{(\rho\beta)_f}\right]\right\}b, \tag{28}$$

$$F[5] = \frac{(1-\phi)^{2.5}}{120\,(1+(1/\gamma))}\left\{(1-\phi)+\phi\left(\frac{\rho_s}{\rho_f}\right)\right\}a^2,$$

(29)

$$F[6] = 0,$$

(30)

$$
\begin{aligned}
F[7] = \frac{(1-\phi)^{2.5}}{210\,(1+(1/\gamma))}\Bigg( & 2\left\{(1-\phi)+\phi\left(\frac{\rho_s}{\rho_f}\right)\right\}\left(-\left(\frac{1}{3}\right)(1-\phi)^{2.5}\left\{(1-\phi)+\phi\left[\frac{(\rho\beta)_s}{(\rho\beta)_f}\right]\right\}ab\right. \\
& +\left(\frac{1}{4}\right)(1-\phi)^5\left\{(1-\phi)+\phi\left[\frac{(\rho\beta)_s}{(\rho\beta)_f}\right]\right\}^2\right) -3\left\{(1-\phi)+\phi\left(\frac{\rho_s}{\rho_f}\right)\right\}\left(\frac{-7(1-\phi)^{2.5}}{24}\left\{(1-\phi)+\phi\left[\frac{(\rho\beta)_s}{(\rho\beta)_f}\right]\right\}ab\right. \\
& +\frac{(1-\phi)^5}{6}\left\{(1-\phi)+\phi\left[\frac{(\rho\beta)_s}{(\rho\beta)_f}\right]\right\}^2\right) +\frac{\left\{(1-\phi)+\phi\left[(\rho\beta)_s/(\rho\beta)_f\right]\right\}\mathrm{Pr}}{8} \\
& \times\left\{\frac{1}{\left[(1-\phi)+\phi\left[(\rho c_p)_s/(\rho c_p)_f\right]\right]}\cdot\left[\frac{k_s+(m-1)k_f-(m-1)\phi\left(k_f-k_s\right)}{k_s+(m-1)k_f+\phi\left(k_f-k_s\right)}\right]\right\}ab\Bigg),
\end{aligned}
$$

(31)

$$
\begin{aligned}
F[8] = \frac{(1-\phi)^{2.5}}{336\,(1+(1/\gamma))}\Bigg( & 2\left\{(1-\phi)+\phi\left(\frac{\rho_s}{\rho_f}\right)\right\}\left(\frac{(1-\phi)^{2.5}}{12}\left\{(1-\phi)+\phi\left(\frac{\rho_s}{\rho_f}\right)\right\}a^3 +\frac{(1-\phi)^5}{6}\left\{(1-\phi)+\phi\left[\frac{(\rho\beta)_s}{(\rho\beta)_f}\right]\right\}^2 b\right) \\
& -3\left\{(1-\phi)+\phi\left(\frac{\rho_s}{\rho_f}\right)\right\}\left(\frac{11(1-\phi)^{2.5}}{20}\left\{(1-\phi)+\phi\left(\frac{\rho_s}{\rho_f}\right)\right\}a^3 +\frac{(1-\phi)^5}{8}\left\{(1-\phi)+\phi\left[\frac{(\rho\beta)_s}{(\rho\beta)_f}\right]\right\}^2 b\right) \\
& -\left(\frac{\left\{(1-\phi)+\phi\left[(\rho\beta)_s/(\rho\beta)_f\right]\right\}^2\mathrm{Pr}\,(1-\phi)^{2.5}}{40}\right) \\
& \times\left\{\frac{1}{\left[(1-\phi)+\phi\left[(\rho c_p)_s/(\rho c_p)_f\right]\right]}\cdot\left[\frac{k_s+(m-1)k_f-(m-1)\phi\left(k_f-k_s\right)}{k_s+(m-1)k_f+\phi\left(k_f-k_s\right)}\right]\right\}b\Bigg),
\end{aligned}
$$

(32)

$$
\begin{aligned}
F[9] = \frac{(1-\phi)^{2.5}}{540\,(1+(1/\gamma))}\Bigg( & 2\left\{(1-\phi)+\phi\left(\frac{\rho_s}{\rho_f}\right)\right\}\left(-\frac{(1-\phi)^5}{24}\left\{(1-\phi)+\phi\left[\frac{(\rho\beta)_s}{(\rho\beta)_f}\right]\right\}\left\{(1-\phi)+\phi\left(\frac{\rho_s}{\rho_f}\right)\right\}a^2\right. \\
& +\frac{(1-\phi)^5}{36}\left\{(1-\phi)+\phi\left[\frac{(\rho\beta)_s}{(\rho\beta)_f}\right]\right\}^2 b^2\right) -3\left\{(1-\phi)+\phi\left(\frac{\rho_s}{\rho_f}\right)\right\}\left(-\left(\frac{13(1-\phi)^5}{360}\right)\left\{(1-\phi)+\phi\left[\frac{(\rho\beta)_s}{(\rho\beta)_f}\right]\right\}\right. \\
& \times\left\{(1-\phi)+\phi\left(\frac{\rho_s}{\rho_f}\right)\right\}a^2 +\left(\frac{(1-\phi)^5}{48}\right)\left\{(1-\phi)+\phi\left[\frac{(\rho\beta)_s}{(\rho\beta)_f}\right]\right\}^2 b^2\right) -\left(\frac{\left\{(1-\phi)+\phi\left[(\rho\beta)_s/(\rho\beta)_f\right]\right\}^2}{240}\right) \\
& \times\mathrm{Pr}\left\{\frac{1}{\left[(1-\phi)+\phi\left[(\rho c_p)_s/(\rho c_p)_f\right]\right]}\cdot\left[\frac{k_s+(m-1)k_f-(m-1)\phi\left(k_f-k_s\right)}{k_s+(m-1)k_f+\phi\left(k_f-k_s\right)}\right]\right\}(1-\phi)^{2.5}b^2\Bigg),
\end{aligned}
$$

(33)

$$F[10] = \frac{(1-\phi)^{2.5}}{720\,(1+(1/\gamma))} \left( 2\left\{(1-\phi)+\phi\left(\frac{\rho_s}{\rho_f}\right)\right\} \left( \left(\frac{a\,(1-\phi)^{2.5}}{15}\right) \left( 2\left\{(1-\phi)+\phi\left(\frac{\rho_s}{\rho_f}\right)\right\} \left(-\left(\frac{a\,(1-\phi)^{2.5}}{3}\right) \right. \right. \right.$$

$$\times \left\{(1-\phi)+\phi\left[\frac{(\rho\beta)_s}{(\rho\beta)_f}\right]\right\}b + \left(\frac{(1-\phi)^{2.5}}{4}\right)\left\{(1-\phi)+\phi\left[\frac{(\rho\beta)_s}{(\rho\beta)_f}\right]\right\}^2\right) - 3\left\{(1-\phi)+\phi\left(\frac{\rho_s}{\rho_f}\right)\right\}\left(-\left(\frac{7a\,(1-\phi)^{2.5}}{24}\right)\right.$$

$$\times \left\{(1-\phi)+\phi\left[\frac{(\rho\beta)_s}{(\rho\beta)_f}\right]\right\}b + \left(\frac{(1-\phi)^5}{6}\right)\left\{(1-\phi)+\phi\left[\frac{(\rho\beta)_s}{(\rho\beta)_f}\right]\right\}^2\right) + \left(\frac{\{(1-\phi)+\phi\,[\,(\rho\beta)_s/(\rho\beta)_f\,]\}\mathrm{Pr}}{8}\right.$$

$$\times \left\{\frac{1}{\left[(1-\phi)+\phi\left[(\rho c_p)_s/(\rho c_p)_f\right]\right]} \cdot \left[\frac{k_s+(m-1)k_f-(m-1)\phi(k_f-k_s)}{k_s+(m-1)k_f+\phi(k_f-k_s)}\right]\right\}ab\right)$$

$$-\left(\frac{(1-\phi)^5\{(1-\phi)+\phi\,[\,(\rho\beta)_s/(\rho\beta)_f\,]\}}{72}\right)\left\{(1-\phi)+\phi\left(\frac{\rho_s}{\rho_f}\right)\right\}a^2b\right) - 3\left\{(1-\phi)+\phi\left(\frac{\rho_s}{\rho_f}\right)\right\}$$

$$\times \left(\left(\frac{11a\,(1-\phi)^{2.5}}{105}\right)\left(2\left\{(1-\phi)+\phi\left(\frac{\rho_s}{\rho_f}\right)\right\}\left(-\left(\frac{1}{3}\right)(1-\phi)^{2.5}\left\{(1-\phi)+\phi\left[\frac{(\rho\beta)_s}{(\rho\beta)_f}\right]\right\}ab + \left(\frac{1}{4}\right)(1-\phi)^5\right.\right.$$

$$\times \left\{(1-\phi)+\phi\left[\frac{(\rho\beta)_s}{(\rho\beta)_f}\right]\right\}^2\right) - 3\left\{(1-\phi)+\phi\left(\frac{\rho_s}{\rho_f}\right)\right\}\left(-\left(\frac{7}{24}\right)(1-\phi)^{2.5}\right.$$

$$\times \left\{(1-\phi)+\phi\left[\frac{(\rho\beta)_s}{(\rho\beta)_f}\right]\right\}ab + \left(\frac{(1-\phi)^5}{6}\right)\left\{(1-\phi)+\phi\left[\frac{(\rho\beta)_s}{(\rho\beta)_f}\right]\right\}^2\right) + \left(\frac{\{(1-\phi)+\phi\,[\,(\rho\beta)_s/(\rho\beta)_f\,]\}\mathrm{Pr}}{8}\right.$$

$$\times \left\{\frac{1}{\left[(1-\phi)+\phi\left[(\rho c_p)_s/(\rho c_p)_f\right]\right]} \cdot \left[\frac{k_s+(m-1)k_f-(m-1)\phi(k_f-k_s)}{k_s+(m-1)k_f+\phi(k_f-k_s)}\right]\right\}ab\right)$$

$$-\left(\frac{(1-\phi)^5}{90}\right)\left\{(1-\phi)+\phi\left[\frac{(\rho\beta)_s}{(\rho\beta)_f}\right]\right\}a^2b\right) + \left(\frac{\{(1-\phi)+\phi\,[\,(\rho\beta)_s/(\rho\beta)_f\,]\}\mathrm{Pr}}{14}\right.$$

$$\times \left\{\frac{1}{\left[(1-\phi)+\phi\left[(\rho c_p)_s/(\rho c_p)_f\right]\right]} \cdot \left[\frac{k_s+(m-1)k_f-(m-1)\phi(k_f-k_s)}{k_s+(m-1)k_f+\phi(k_f-k_s)}\right]\right\}$$

$$\times \left(-\left(\frac{a^2\mathrm{Pr}}{4}\right)\left\{\frac{1}{\left[(1-\phi)+\phi\left[(\rho c_p)_s/(\rho c_p)_f\right]\right]} \cdot \left[\frac{k_s+(m-1)k_f-(m-1)\phi(k_f-k_s)}{k_s+(m-1)k_f+\phi(k_f-k_s)}\right]\right\}b\right.$$

$$+\left(\frac{(1-\phi)^{2.5}}{120}\right)\left\{(1-\phi)+\phi\left(\frac{\rho_s}{\rho_f}\right)\right\}a^2b\right).$$

$$(34)$$

In the same manner, the expressions for $F[11]$, $F[12]$, $F[13]$, $F[14]$, $F[15]$ were found, which are too large expressions to be included in this paper.

Also, using $p = 0, 1, 2, 3, \ldots$ in the above recursive relations in (24), one arrives at

$$\Theta[2] = 0, \qquad (35)$$

$$\Theta[3] = 0, \qquad (36)$$

$$\Theta[4] = -\frac{\mathrm{Pr}}{8\,(1+(4/3)R)}\left\{\frac{1}{\left[(1-\phi)+\phi\left[(\rho c_p)_s/(\rho c_p)_f\right]\right]}\right.$$

$$\times \left[\frac{k_s+(m-1)k_f-(m-1)\phi(k_f-k_s)}{k_s+(m-1)k_f+\phi(k_f-k_s)}\right]\right\}ab,$$

$$(37)$$

$$\Theta[5] = \frac{\text{Pr}}{40(1+(4/3)R)} \left\{ \left\{ \frac{1}{\left[(1-\phi)+\phi\left[(\rho c_p)_s/(\rho c_p)_f\right]\right]} \left[\frac{k_s+(m-1)k_f-(m-1)\phi(k_f-k_s)}{k_s+(m-1)k_f+\phi(k_f-k_s)}\right] \right\} \right.$$
$$\left. \cdot \frac{(1-\phi)^{2.5}}{(1+(1/\gamma))}\left[(1-\phi)+\phi\left[\frac{(\rho\beta)_s}{(\rho\beta)_f}\right]\right]b \right\},$$

(38)

$$\Theta[6] = \frac{\text{Pr}}{120(1+(4/3)R)} \left\{ \left\{ \frac{1}{\left[(1-\phi)+\phi\left[(\rho c_p)_s/(\rho c_p)_f\right]\right]} \left[\frac{k_s+(m-1)k_f-(m-1)\phi(k_f-k_s)}{k_s+(m-1)k_f+\phi(k_f-k_s)}\right] \right\} \right.$$
$$\left. \cdot \frac{(1-\phi)^{2.5}}{(1+(1/\gamma))}\left[(1-\phi)+\phi\left[\frac{(\rho\beta)_s}{(\rho\beta)_f}\right]\right]b^2 \right\},$$

(39)

$$\Theta[7] = -\frac{\text{Pr}}{14(1+(4/3)R)} \left\{ \left\{ \frac{1}{\left[(1-\phi)+\phi\left[(\rho c_p)_s/(\rho c_p)_f\right]\right]} \left[\frac{k_s+(m-1)k_f-(m-1)\phi(k_f-k_s)}{k_s+(m-1)k_f+\phi(k_f-k_s)}\right] \right\} \right.$$
$$\cdot \left( -\frac{\text{Pr}}{4}\left\{ \frac{1}{\left[(1-\phi)+\phi\left[(\rho c_p)_s/(\rho c_p)_f\right]\right]} \left[\frac{k_s+(m-1)k_f-(m-1)\phi(k_f-k_s)}{k_s+(m-1)k_f+\phi(k_f-k_s)}\right] \right\}a^2b \right.$$
$$\left. \left. +\frac{(1-\phi)^{2.5}}{120(1+(1/\gamma))}\left[(1-\phi)+\phi\left[\frac{(\rho)_s}{(\rho)_f}\right]\right]a^2b \right) \right\},$$

(40)

$$\Theta[8] = -\frac{\text{Pr}^2}{128(1+(4/3)R)} \left\{ \left\{ \frac{1}{\left[(1-\phi)+\phi\left[(\rho c_p)_s/(\rho c_p)_f\right]\right]} \left[\frac{k_s+(m-1)k_f-(m-1)\phi(k_f-k_s)}{k_s+(m-1)k_f+\phi(k_f-k_s)}\right] \right\}^2 \right.$$
$$\left. \cdot \frac{(1-\phi)^{2.5}}{(1+(1/\gamma))}\left[(1-\phi)+\phi\left[\frac{(\rho\beta)_s}{(\rho\beta)_f}\right]\right]ab \right\},$$

(41)

$$\Theta[9] = \frac{-\text{Pr}}{24(1+(4/3)R)} \left\{ \left\{ \frac{1}{\left[(1-\phi)+\phi\left[(\rho c_p)_s/(\rho c_p)_f\right]\right]} \left[\frac{k_s+(m-1)k_f-(m-1)\phi(k_f-k_s)}{k_s+(m-1)k_f+\phi(k_f-k_s)}\right] \right\} \right.$$
$$\cdot \left( \left\{ \frac{\text{Pr}}{30}\left\{ \frac{1}{\left[(1-\phi)+\phi\left[(\rho c_p)_s/(\rho c_p)_f\right]\right]}\frac{k_s+(m-1)k_f-(m-1)\phi(k_f-k_s)}{k_s+(m-1)k_f+\phi(k_f-k_s)} \right\} \cdot \frac{(1-\phi)^{2.5}}{(1+(1/\gamma))}\left[(1-\phi)+\phi\left[\frac{(\rho\beta)_s}{(\rho\beta)_f}\right]\right]ab^2 \right\} \right.$$
$$-\frac{\text{Pr}}{48}\left\{ \left\{ \frac{1}{\left[(1-\phi)+\phi\left[(\rho c_p)_s/(\rho c_p)_f\right]\right]}\frac{k_s+(m-1)k_f-(m-1)\phi(k_f-k_s)}{k_s+(m-1)k_f+\phi(k_f-k_s)} \right\} \cdot \frac{(1-\phi)^{5}}{(1+(1/\gamma))}\left[(1-\phi)+\phi\left[\frac{(\rho\beta)_s}{(\rho\beta)_f}\right]\right]^2b \right\}$$
$$+\frac{(1-\phi)^{2.5}}{210(1+(1/\gamma))}\left( 2\left[(1-\phi)+\phi\left[\frac{(\rho)_s}{(\rho)_f}\right]\right]\left( -\frac{(1-\phi)^{2.5}}{3(1+(1/\gamma))}\left[(1-\phi)+\phi\left[\frac{(\rho\beta)_s}{(\rho\beta)_f}\right]\right]ab \right.$$
$$\left. +\frac{(1-\phi)^{5}}{4(1+(1/\gamma))}\left[(1-\phi)+\phi\left[\frac{(\rho\beta)_s}{(\rho\beta)_f}\right]\right]^2 \right) -3\left[(1-\phi)+\phi\left[\frac{(\rho)_s}{(\rho)_f}\right]\right]\left( -\frac{7(1-\phi)^{2.5}}{24(1+(1/\gamma))}\left[(1-\phi)+\phi\left[\frac{(\rho\beta)_s}{(\rho\beta)_f}\right]\right]ab$$
$$\left. +\frac{(1-\phi)^{5}}{6(1+(1/\gamma))}\left[(1-\phi)+\phi\left[\frac{(\rho\beta)_s}{(\rho\beta)_f}\right]\right]^2 \right) +\frac{\text{Pr}}{8}\left\{ \left\{ \frac{1}{\left[(1-\phi)+\phi\left[(\rho c_p)_s/(\rho c_p)_f\right]\right]} \right. \right.$$
$$\left. \left. \left. \left. \left[\frac{k_s+(m-1)k_f-(m-1)\phi(k_f-k_s)}{k_s+(m-1)k_f+\phi(k_f-k_s)}\right] \right\} \cdot \left[(1-\phi)+\phi\left[\frac{(\rho\beta)_s}{(\rho\beta)_f}\right]\right]ab \right\} \right)b \right) \right\},$$

(42)

$$\Theta[10] = -\frac{\text{Pr}}{4(1+(4/3)R)} \left\{ \frac{1}{\left[(1-\phi)+\phi\left[(\rho c_p)_s/(\rho c_p)_f\right]\right]} \cdot \left[\frac{k_s+(m-1)k_f-(m-1)\phi(k_f-k_s)}{k_s+(m-1)k_f+\phi(k_f-k_s)}\right]\right\}$$

$$\times \left(-\frac{a\text{Pr}}{4}\left\{\left\{\frac{1}{\left[(1-\phi)+\phi\left[(\rho c_p)_s/(\rho c_p)_f\right]\right]} \cdot \left[\frac{k_s+(m-1)k_f-(m-1)\phi(k_f-k_s)}{k_s+(m-1)k_f+\phi(k_f-k_s)}\right]\right\}\right.$$

$$\cdot\left(-\frac{\text{Pr}}{4}\left\{\frac{1}{\left[(1-\phi)+\phi\left[(\rho c_p)_s/(\rho c_p)_f\right]\right]} \cdot \left[\frac{k_s+(m-1)k_f-(m-1)\phi(k_f-k_s)}{k_s+(m-1)k_f+\phi(k_f-k_s)}\right]\right\}a^2b\right.$$

$$+\frac{(1-\phi)^{2.5}}{120(1+(1/\gamma))}\left[(1-\phi)+\phi\left[\frac{(\rho)_s}{(\rho)_f}\right]\right]a^2b\right)\right\} - \frac{3(1-\phi)^5\text{Pr}}{320(1+(1/\gamma))}\left\{\frac{1}{\left[(1-\phi)+\phi\left[(\rho c_p)_s/(\rho c_p)_f\right]\right]}\right.$$

$$\left.\cdot\left[\frac{k_s+(m-1)k_f-(m-1)\phi(k_f-k_s)}{k_s+(m-1)k_f+\phi(k_f-k_s)}\right]\right\}\left[(1-\phi)+\phi\left[\frac{(\rho\beta)_s}{(\rho\beta)_f}\right]\right]^2b^2 - \frac{(1-\phi)^{2.5}a^3\text{Pr}}{120(1+(1/\gamma))}\left[(1-\phi)+\phi\left[\frac{(\rho\beta)_s}{(\rho\beta)_f}\right]\right]$$

$$\times\left\{\frac{1}{\left[(1-\phi)+\phi\left[(\rho c_p)_s/(\rho c_p)_f\right]\right]} \cdot \left[\frac{k_s+(m-1)k_f-(m-1)\phi(k_f-k_s)}{k_s+(m-1)k_f+\phi(k_f-k_s)}\right]\right\}b$$

$$+\frac{(1-\phi)^{2.5}}{336(1+(1/\gamma))}\left(2\left\{\left[(1-\phi)+\phi\left[\frac{(\rho\beta)_s}{(\rho\beta)_f}\right]\right]\cdot\left(\frac{a^3(1-\phi)^{2.5}}{12(1+(1/\gamma))}\left[(1-\phi)+\phi\left[\frac{(\rho\beta)_s}{(\rho\beta)_f}\right]\right]\right.\right.$$

$$\left.\left.+\frac{(1-\phi)^5}{6(1+(1/\gamma))}\left[(1-\phi)+\phi\left[\frac{(\rho\beta)_s}{(\rho\beta)_f}\right]\right]^2b\right)\right\} - 3\left\{\left[(1-\phi)+\phi\left[\frac{(\rho\beta)_s}{(\rho\beta)_f}\right]\right]\cdot\left(\frac{11(1-\phi)^{2.5}}{120(1+(1/\gamma))}a^3\left[(1-\phi)+\phi\left[\frac{(\rho\beta)_s}{(\rho\beta)_f}\right]\right]\right.\right.$$

$$\left.\left.+\frac{(1-\phi)^5}{8(1+(1/\gamma))}\left[(1-\phi)+\phi\left[\frac{(\rho\beta)_s}{(\rho\beta)_f}\right]\right]^2b\right)\right\} - \frac{\text{Pr}}{40}\left\{\left\{\frac{1}{\left[(1-\phi)+\phi\left[(\rho c_p)_s/(\rho c_p)_f\right]\right]}\right.\right.$$

$$\left.\left.\cdot\left[\frac{k_s+(m-1)k_f-(m-1)\phi(k_f-k_s)}{k_s+(m-1)k_f+\phi(k_f-k_s)}\right]\right\}\cdot\left[(1-\phi)+\phi\left[\frac{(\rho\beta)_s}{(\rho\beta)_f}\right]\right]^2(1-\phi)^{2.5}b\right\}b\right).$$

$$(43)$$

Also, the expressions for $\Theta[11]$, $\Theta[12]$, $\Theta[13]$, $\Theta[14]$, $\Theta[15]$, ... are found in the same way, but they are too large expressions to be included in this paper.

Using the definition in (20), the solutions of (13) and (14) are given as

$$f(\eta) = F[0] + \eta F[1] + \eta^2 F[2] + \eta^3 F[3] + +\eta^4 F[4]$$
$$+ \eta^5 F[5] + \eta^6 F[6] + \eta^7 F[7] + \eta^8 F[8] \qquad (44)$$
$$+ \eta^9 F[9] + \eta^{10} F[10] + \cdots,$$

$$\theta(\eta) = \Theta[0] + \eta\Theta[1] + \eta^2\Theta[2] + \eta^3\Theta[3] + +\eta^4\Theta[4]$$
$$+ \eta^5\Theta[5] + \eta^6\Theta[6] + \eta^7\Theta[7] + \eta^8\Theta[8] \qquad (45)$$
$$+ \eta^9\Theta[9] + \eta^{10}\Theta[10] + \cdots.$$

The solutions of (15) and (16) for a viscous fluid which does not have nanoparticles can be developed from (44) and (45) if nanoparticle volume fraction is set to zero, that is, $\phi = 0$.

## 4. The Basic Concept and the Procedure of Padé Approximant

The limitations of power series methods to a small domain have been overcomed by domain transformation and after-treatment techniques. These techniques increase the radius of convergence and also accelerate the rate of convergence of a given series. Among the so-called after-treatment techniques, Padé-approximant technique has been widely applied in developing accurate analytical solutions to nonlinear problems of large or unbounded domain problems of infinite boundary conditions [71]. The Padé-approximant technique manipulates a polynomial approximation into a rational function of polynomials. Such a manipulation gives more information about the mathematical behaviour of the solution. The basic procedures are as follows.

Suppose that a function $f(\eta)$ is represented by a power series:

$$f(\eta) = \sum_{i=0}^{\infty} c_i \eta^i. \tag{46}$$

This above expression is the fundamental point of any analysis using Padé approximant. The notation $c_i$, $i = 0, 1, 2, \ldots$ is reserved for the given set of coefficient, and $f(\eta)$ is the associated function. $[L/M]$ Padé approximant is a rational function defined as

$$f(\eta) = \frac{\sum_{i=0}^{L} a_i \eta^i}{\sum_{i=0}^{M} b_i \eta^i} = \frac{a_0 + a_1\eta + a_2\eta^2 + \cdots + a_L\eta^L}{b_0 + b_1\eta + b_2\eta^2 + \cdots + b_M\eta^M}, \tag{47}$$

which has a Maclaurin expansion, agreeing with (46) (1) as far as possible. It is noticed that, in (47), there are $L + 1$ numerator and $M + 1$ denominator coefficients. So, there are $L + 1$ independent number of numerator coefficients, making $L + M + 1$ unknown coefficients in all. These numbers of coefficients of numerator and denominator suggest that L/M out of fit the power series in (46) through orders 1, $\eta$, $\eta^2$, $\ldots$, $\eta^{L+M}$.

In the notation of formal power series:

$$\sum_{i=0}^{\infty} c_i \eta^i = \frac{a_0 + a_1\eta + a_2\eta^2 + \cdots + a_L\eta^L}{b_0 + b_1\eta + b_2\eta^2 + \cdots + b_M\eta^M} + \mathrm{O}\left(\eta^{L+M+1}\right), \tag{48}$$

which gives

$$\left(b_0 + b_1\eta + b_2\eta^2 + \cdots + b_M\eta^M\right)\left(c_0 + c_1\eta + c_2\eta^2 + \cdots\right)$$
$$= a_0 + a_1\eta + a_2\eta^2 + \cdots + a_L\eta^L + \mathrm{O}\left(\eta^{L+M+1}\right). \tag{49}$$

Expanding the LHS and equating the coefficients of $\eta^{L+1}$, $\eta^{L+2}$, $\ldots$, $\eta^{L+M}$, we get

$$b_M c_{L-M+1} + b_{M-1} c_{L-M+2} + b_{M-2} c_{L-M+3}$$
$$+ \cdots + b_2 c_{L-1} + b_1 c_L + b_0 c_{L+1} = 0,$$

$$b_M c_{L-M+2} + b_{M-1} c_{L-M+3} + b_{M-2} c_{L-M+4}$$
$$+ \cdots + b_2 c_L + b_1 c_{L+1} + b_0 c_{L+2} = 0,$$

$$b_M c_{L-M+3} + b_{M-1} c_{L-M+4} + b_{M-2} c_{L-M+5} \tag{50}$$
$$+ \cdots + b_2 c_{L+1} + b_1 c_{L+2} + b_0 c_{L+3} = 0,$$

$$\vdots$$

$$b_M c_L + b_{M-1} c_{L+1} + b_{M-2} c_{L+2} + \cdots + b_2 c_{L+M-2}$$
$$+ b_1 c_{L+M-1} + b_0 c_{L+M} = 0,$$

If $i < 0$, $c_i = 0$ for consistency. Since $b_0 = 1$, (50) becomes a set of $M$ linear equations for $M$ unknown denominator coefficients:

$$\begin{pmatrix} c_{L-M+1} & c_{L-M+2} & c_{L-M+3} & \cdots & c_L \\ c_{L-M+2} & c_{L-M+3} & c_{L-M+4} & \cdots & c_{L+1} \\ c_{L-M+3} & c_{L-M+4} & c_{L-M+4} & \cdots & c_{L+2} \\ \vdots & \vdots & \vdots & \vdots & \vdots \\ c_L & c_{L+1} & c_{L+2} & \cdots & c_{L+M-1} \end{pmatrix} \begin{pmatrix} b_M \\ b_{M-1} \\ b_{M-2} \\ \vdots \\ b_1 \end{pmatrix} \tag{51}$$

$$= - \begin{pmatrix} c_{L+1} \\ c_{L+2} \\ c_{L+3} \\ \vdots \\ c_{L+M} \end{pmatrix}.$$

From the above (51), $b_i$ may be found. The numerator coefficients $a_0$, $a_1$, $a_2$, $\ldots$, $a_L$ follow immediately from (49) by equating the coefficient of 1, $\eta$, $\eta^2$, $\ldots$, $\eta^{L+M}$ such that

$$a_0 = c_0,$$
$$a_1 = c_1 + b_1 c_0,$$
$$a_2 = c_2 + b_1 c_1 + b_2 c_0,$$
$$a_3 = c_3 + b_1 c_2 + b_2 c_1 + b_3 c_0,$$
$$a_4 = c_4 + b_1 c_3 + b_2 c_2 + b_3 c_1 + b_4 c_0,$$
$$a_5 = c_5 + b_1 c_4 + b_2 c_3 + b_3 c_2 + b_4 c_1 + b_5 c_0, \tag{52}$$
$$a_6 = c_6 + b_1 c_5 + b_2 c_4 + b_3 c_3 + b_4 c_2 + b_5 c_1 + b_6 c_0,$$
$$\vdots$$
$$a_L = c_L + \sum_{i=1}^{\min[L/M]} b_i c_{L-i}.$$

Equations (51) and (52) normally determine the Padé numerator and denominator and are called Padé equations. The $[L/M]$ Padé approximant is constructed which agrees with the equation in the power series through the order $\eta^{L+M}$. To obtain a diagonal Padé approximant of order $[L/M]$, the symbolic calculus software Maple is used.

It should be noted as mentioned previously that $\Delta$ and $\Omega$ in the solutions are unknown constants. In order to compute their values for extended large domains solutions, the power series as presented in (44) and (45) are converted to rational functions using [18/18] Padé approximation through the software Maple, and then, the infinite boundary conditions, that is, $\eta \to \infty$, $f' = 0$, $\theta = 0$ are applied. The resulting simultaneous equations are solved to obtain the values of $\Delta$ and $\Omega$ for the respective values of the physical and thermal properties of the nanofluids under considerations.

## 5. Flow and Heat Transfer Parameters

In addition to the determination of the velocity and temperature distributions, it is often desirable to compute other physically important quantities (such as shear stress, drag, heat transfer rate, and heat transfer coefficient) associated with the free convection flow and heat transfer problem.

Consequently, two parameters, a flow parameter and a heat transfer parameter, are computed.

### 5.1. Fluid Flow Parameter.
Skin friction coefficient is given by the following equation:

$$
c_f = \frac{\tau_w}{\rho_{nf} u^2} = \frac{\mu_{nf} (\partial u / \partial y)|_{y=0}}{\rho_{nf} u^2}
$$

$$
= \frac{\mu_{nf} ((\partial u / \partial \eta) \cdot (\partial \eta / \partial y))|_{y=0}}{\rho_{nf} u^2}.
$$

(53)

After the dimensionless exercise,

$$
c_f (\mathrm{Re}_x)^{1/2} = \frac{f''(0)}{(1-\phi)^{2.5}},
$$

(54)

$$
c_f (\mathrm{Re}_x)^{1/2} \frac{\tau_w}{(4\mathrm{Gr}_x^3)^{1/4} (\nu\mu)} = f''(0) \frac{f''(0)}{(1-\phi)^{2.5}}.
$$

### 5.2. Heat Transfer Parameter.
Heat transfer coefficient is given by the following equation:

$$
h_x = -\frac{k_{nf}}{T_w - T_\infty} \left(\frac{\partial T}{\partial y}\right)_{y=0} = -k_{nf} \theta'(0) \frac{1}{x} \left(\frac{1}{4}\mathrm{Gr}_x\right)^{1/4}.
$$

(55)

The local heat transfer coefficient at the surface of the vertical plate is obtained from (55). The local Nusselt number is

$$
Nu_x = \frac{h_x x}{k_{nf}} = -\left(\frac{x}{T_w - T_\infty}\right)\left(\frac{\partial T}{\partial y}\right)_{y=0} = -\theta'(0)\left(\frac{1}{4}\mathrm{Gr}_x\right)^{1/4},
$$

$$
Nu_x = -\frac{\theta'(0)}{\sqrt{2}} \mathrm{Gr}_x^{1/4} = f\left((\mathrm{Pr})\mathrm{Gr}_x^{1/4}\right),
$$

(56)

where $\phi(\mathrm{Pr}) = -(\theta'(0)/\sqrt{2})$ is a function of Prandtl number. The dependence of $\phi$ on the Prandtl number is evidenced by (56).

It could also be shown that

$$
\frac{Nu_x}{(\mathrm{Re}_x)^{1/2}} = -\frac{k_{nf}}{k_f}\theta'(0)
$$

$$
= -\left[\frac{k_s + (m-1)k_f - (m-1)\phi(k_f - k_s)}{k_s + (m-1)k_f + \phi(k_f - k_s)}\right]\theta'(0),
$$

(57)

where $\mathrm{Re}_x$ and $\mathrm{Gr}_x$ are the local Reynold and Grashof numbers, which are defined as

$$
\mathrm{Re}_x = \frac{ux}{\nu_{nf}},
$$

(58)

$$
\mathrm{Gr}_x = \frac{g\beta(T_w - T_\infty) x^3}{\nu^3}.
$$

## 6. Results and Discussion

The solutions of DTM with [18/18] Padé approximation is presented in this section. Tables 5–7 present various comparisons of results of the present study and the past works for viscous fluid, that is, when the volume fraction of the nanoparticles, thermal radiation, and Casson parameters are set to zero, that is, $\phi = R = \beta = 0$. It could be seen from the tables that there are excellent agreements between the past results and the present study. Moreover, the tables present the effects of Prandtl number on the flow and heat transfer processes.

Although, the nonlinear partial differential equations in Mosta et al. [17] are the same in all aspects to the present problems under investigation, there are slight differences between the transformed nonlinear ordinary differential equations in (7) of Mosta et al. [17] and (13) and (14) developed in this present study (where the volume fraction of the nanoparticles, radiation, and Casson parameters are set to zero) due to the differences in the adopted similarity variables. It is shown that, using the DTM-Padé approximant as applied in this work to the transformed nonlinear ordinary differential equations in Mosta et al. [17], excellent agreements are recorded between the results of the present study and that of Mosta et al. [17] and Kuiken [72] as shown in Tables 6 and 7.

The variations in nanoparticle volume fraction with dynamic viscosity and thermal conductivity ratios of copper(II) oxide-water nanofluid are shown in Figures 2 and 3, respectively. Also, Figure 3 show the effects of nanoparticle shape on thermal conductivity ratio. It is depicted in the figure that the thermal conductivity of the nanofluid varies linearly and increases with increase in nanoparticle volume fraction. It is also observed that the suspensions of particles with high shape factor or low sphericity have higher thermal conductivity ratio of the nanofluid. Spherical-shaped nanoparticles have the lowest thermal conductivity ratio, and lamina-shaped nanoparticles have the highest thermal conductivity ratio.

The effects of the flow and heat transfer controlling parameters on the velocity and temperature distributions are shown in Figures 4–17 for different shapes, type, and volume fraction of the nanoparticles at Prandtl number of 0.01–1000.

### 6.1. Effect of Casson Parameter on Casson Nanofluid Velocity and Temperature Distributions.
Figures 4(a) and 4(b) depict the effects of Casson parameters on the velocity and temperature profiles of the Casson nanofluid, respectively. It is obvious from the figure that the Casson parameter has influence on axial velocity. From Figure 4(a), the magnitude of velocity near the plate for Casson nanofluid parameter decreases for increasing value of the Casson parameter, while temperature increases for increase in Casson fluid parameter as shown in Figure 4(b). Physically, increasing values of the Casson parameter develop the viscous forces. These forces have tendencies to decline the thermal boundary layer.

TABLE 5: Comparison of results of $f''(0)$ and $\theta'(0)$ at different Prandtl numbers.

| | Kuo [51] | | Na and Habib [12] | | Present study | |
|---|---|---|---|---|---|---|
| Pr | $f''(0)$ | $\theta'(0)$ | $f''(0)$ | $\theta'(0)$ | $f''(0)$ | $\theta'(0)$ |
| 0.72 | 0.6760 | −0.5046 | 0.6760 | −0.5046 | 0.6760 | −0.5046 |
| 0.60 | 0.6947 | −0.4721 | 0.6946 | −0.4725 | 0.6947 | −0.4721 |
| 0.50 | 0.7132 | −0.4411 | 0.7131 | −0.4420 | 0.7132 | −0.4411 |
| 0.40 | 0.7356 | −0.4053 | 0.7354 | −0.4066 | 0.7356 | −0.4053 |
| 0.30 | 0.7636 | −0.3623 | 0.7633 | −0.3641 | 0.7636 | −0.3623 |
| 0.20 | 0.8015 | −0.3078 | 0.8009 | −0.3101 | 0.8015 | −0.3078 |
| 0.10 | 0.8600 | −0.2298 | 0.8590 | −0.2326 | 0.8600 | −0.2298 |
| 0.06 | 0.8974 | −0.1834 | 0.8961 | −0.1864 | 0.8974 | −0.1834 |
| 0.04 | 0.9233 | −0.1526 | 0.9221 | −0.1556 | 0.9233 | −0.1526 |
| 0.01 | 0.9845 | −0.0832 | 0.9887 | −0.0817 | 0.9885 | −0.0832 |
| 1.00 | 0.6421 | −0.5671 | 0.6421 | −0.5671 | 0.6421 | −0.5671 |
| 1.10 | 0.6323 | −0.5862 | 0.6323 | −0.5860 | 0.6323 | −0.5862 |
| 1.20 | 0.6233 | −0.6040 | 0.6234 | −0.6036 | 0.6233 | −0.6040 |
| 1.30 | 0.6151 | −0.6208 | 0.6152 | −0.6202 | 0.6151 | −0.6208 |
| 1.40 | 0.6075 | −0.6365 | 0.6076 | −0.6358 | 0.6075 | −0.6365 |
| 1.50 | 0.6005 | −0.6515 | 0.6006 | −0.6506 | 0.6005 | −0.6515 |
| 1.60 | 0.5939 | −0.6656 | 0.5940 | −0.6646 | 0.5939 | −0.6656 |
| 1.70 | 0.5877 | −0.6792 | 0.5879 | −0.6780 | 0.5877 | −0.6792 |
| 1.80 | 0.5819 | −0.6921 | 0.5821 | −0.6908 | 0.5819 | −0.6921 |
| 1.90 | 0.5764 | −0.7045 | 0.5767 | −0.7031 | 0.5764 | −0.7045 |
| 2.00 | 0.5712 | −0.7164 | 0.5715 | −0.7149 | 0.5712 | −0.7164 |
| 2.00 | 0.5712 | −0.7164 | 0.5713 | −0.7165 | 0.5712 | −0.7164 |
| 3.00 | 0.5308 | −0.8154 | 0.5312 | −0.8145 | 0.5308 | −0.8154 |
| 4.00 | 0.5029 | −0.8914 | 0.5036 | −0.8898 | 0.5029 | −0.8914 |
| 5.00 | 0.4817 | −0.9539 | 0.4827 | −0.9517 | 0.4817 | −0.9539 |
| 6.00 | 0.4648 | −1.0073 | 0.4660 | −1.0047 | 0.4648 | −1.0073 |
| 7.00 | 0.4507 | −1.0542 | 0.4522 | −1.0512 | 0.4507 | −1.0542 |
| 8.00 | 0.4387 | −1.0961 | 0.4405 | −1.0930 | 0.4387 | −1.0961 |
| 9.00 | 0.4283 | −1.1342 | 0.4304 | −1.1309 | 0.4283 | −1.1342 |
| 10.00 | 0.4191 | −1.1692 | 0.4215 | −1.1658 | 0.4191 | −1.1692 |

TABLE 6: Comparison of results of $f''(0)$ at different Prandtl numbers.

| | $f''(0)$ | | |
|---|---|---|---|
| Pr | Kuiken [72] | Mosta et al. [17] | Present study |
| 0.001 | 1.12313813 | 1.12313813 | 1.12313813 |
| 0.01 | 1.06338086 | 1.06338086 | 1.06338086 |
| 0.1 | 0.92408304 | 0.92408304 | 0.92408304 |
| 1 | 0.69321163 | 0.69321163 | 0.69321163 |
| 10 | 0.44711652 | 0.44711652 | 0.44711652 |
| 100 | 0.26452354 | 0.26452354 | 0.26452354 |
| 1000 | 0.15129020 | 0.15129020 | 0.15129020 |
| 10000 | 0.08554085 | 0.08554085 | 0.08554085 |

TABLE 7: Comparison of results of $-\theta'(0)$ at different Prandtl numbers.

| | $-\theta'(0)$ | | | |
|---|---|---|---|---|
| Pr | Kuiken [72] | Mosta et al. [16] | Mosta et al. [17] | Present study |
| 0.001 | 0.04680746 | 0.04680746 | 0.04680746 | 0.04680746 |
| 0.01 | 0.13576074 | 0.13576074 | 0.13576074 | 0.13576074 |
| 0.1 | 0.35005967 | 0.35005967 | 0.35005967 | 0.35005967 |
| 1 | 0.76986120 | 0.76986120 | 0.76986119 | 0.76986120 |
| 10 | 1.49709921 | 1.49709921 | 1.49709921 | 1.49709921 |
| 100 | 2.74688550 | 2.74688550 | 2.74688549 | 2.74688550 |
| 1000 | 4.93494763 | 4.93494763 | 4.93494756 | 4.93494763 |
| 10000 | 8.80444927 | 8.80444927 | 8.80444960 | 8.80444960 |

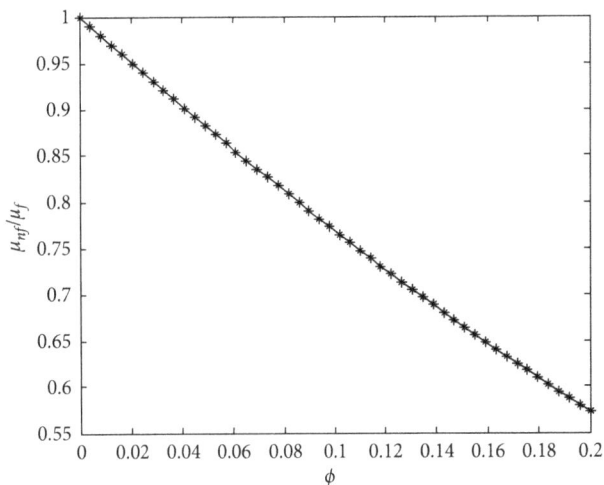

FIGURE 2: Variation in nanofluid dynamic viscosity ratio with nanoparticle volume fraction.

6.2. *Effect of Thermal Radiation Parameter on Casson Nanofluid Velocity and Temperature Distributions.* It is depicted that both viscous and thermal boundary layers increase with the increase in radiation parameter, $R$. Figure 5(a) depicts the effect of thermal radiation parameter on the velocity profiles. From the figure, it is shown that increase in radiation parameter causes the velocity of the fluid to

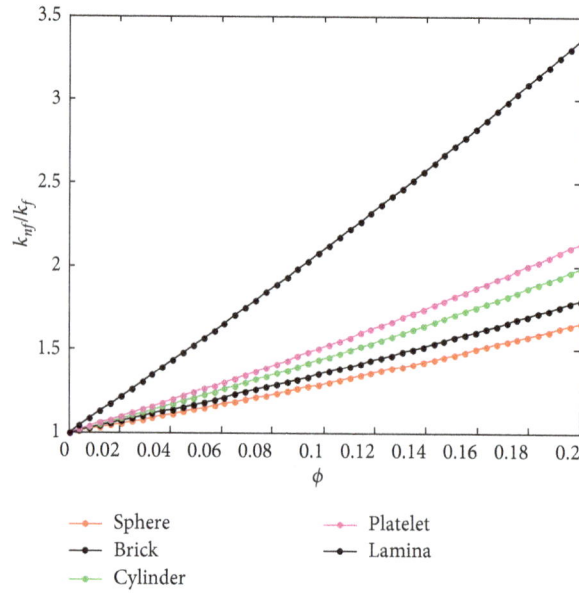

FIGURE 3: Effects of nanoparticle shape on thermal conductivity ratio of the nanofluid.

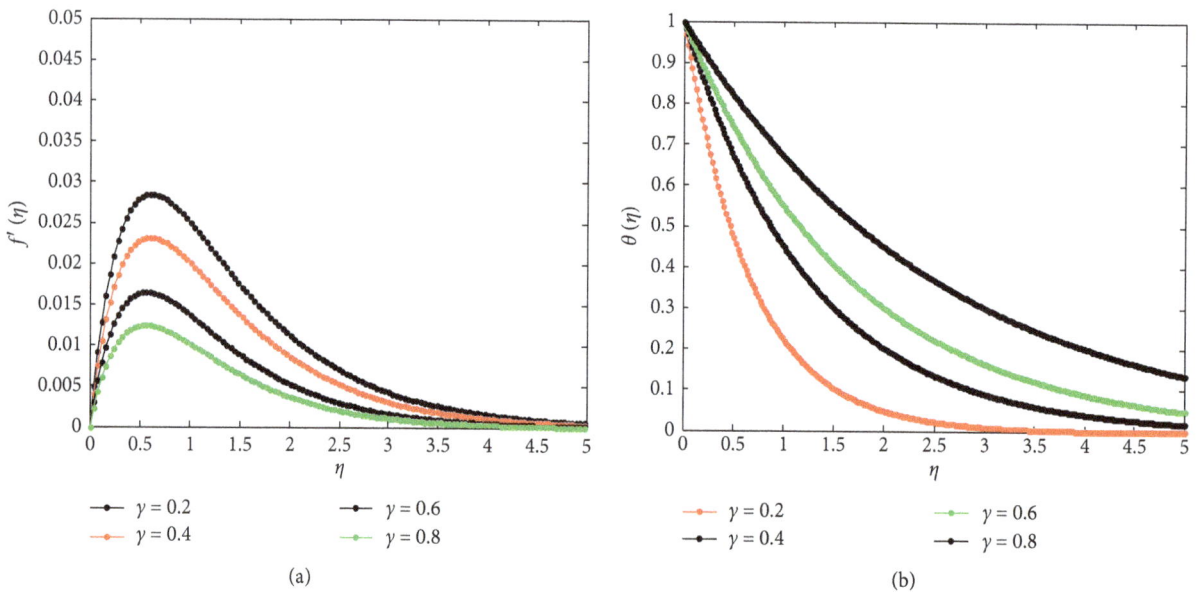

(a)

(b)

FIGURE 4: (a) Effects of Casson parameter on the velocity profile of the Casson nanofluid. (b) Effects of Casson parameter on the temperature profile of the Casson nanofluid.

increase. This is because, as the radiation parameter is augmented, the absorption of radiated heat from the heated plate releases more heat energy to the fluid and the resulting temperature increases the buoyancy forces in the boundary layer which also increases the fluid motion and the momentum boundary layer thickness. This is expected because the considered radiation effect within the boundary layer increases the motion of the fluid which increases the surface frictions. The maximum velocity for all values of $R$ is at the approximated value of $\eta = 0.5$. Therefore, it can be concluded that the inner viscous layer does not increase for variation of radiation parameter. Only the outer layer thickness has

a great influence on thermal radiation, $R$. Although the velocity gradient at the surface increases with the increase in radiation parameter, a reverse case has been established in the literature when water is used as the fluid under the study of the flow of viscous fluid over a flat surface.

Using a constant value of the Prandtl number, the influence of radiation parameter on the temperature field is displayed in Figure 5(b). Increase in the radiation parameter contributes in general to increase in the temperature. This is because, as the thermal radiation increases, the absorption of radiated heat from the heated plate releases heat energy to the fluid. Consequently, the thermal boundary layer of fluid

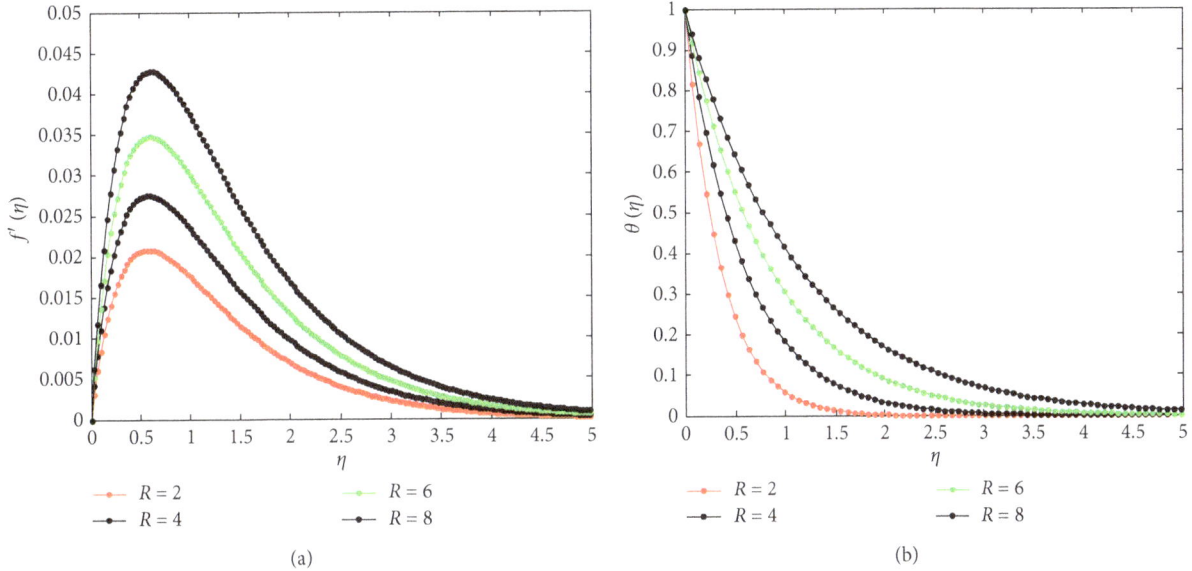

FIGURE 5: Effects of radiation parameter on the (a) velocity profile and (b) temperature profile of the Casson nanofluid.

increases as the temperature near the boundary is enhanced. This shows that influence of radiation is more effective when high temperature is required for the desired thickness of the end product. It is observed that the effect of the radiation parameter is not significant as we move away from the boundary. Also, it is observed that as the temperature of the fluid increases for increasing thermal radiation, the temperature difference between the plate and the ambient fluid reduces which turns to decrease the heat transfer rate in flow region.

6.3. Effect of Nanoparticle Volume Fraction on Casson Nanofluid Velocity and Temperature Distributions for Different Values of Prandtl Number. Figures 6–9 show the effects of nanoparticle concentration/volume fraction and Prandtl number on velocity and temperature profiles of copper(II) oxide-water Casson nanofluid. It is indicated in the figures that as the volume fraction or concentration of the nanoparticle in the nanofluid increases, the velocity decreases. However, an opposite trend or behaviour in the temperature profile is observed; that is, the nanofluid temperature increases as the volume fraction of the nanoparticles in the base fluid increases. This is because the solid volume fraction has significant impacts on the thermal conductivity. The increased volume fraction of the nanoparticles in the base fluid results in higher thermal conductivity of the base fluid which increases the heat enhancement capacity of the base fluid. Also, one of the possible reasons for the enhancement on heat transfer of nanofluids can be explained by the high concentration of nanoparticles in the thermal boundary layer at the wall side through the migration of nanoparticles. It should also be stated that the thickness of the thermal boundary layer rises with increase in the values of nanoparticle volume fraction. This consequently reduces the velocity of the nanofluid as

the shear stress and skin friction are increased. The figures also show the effects of Prandtl number (Pr) on the velocity and temperature profiles. It is indicated that the velocity of the nanofluid decreases as the Pr increases, but the temperature of the nanofluid increases as the Pr increases. This is because the nanofluid with higher Prandtl number has a relatively low thermal conductivity, which reduces conduction and thereby reduces the thermal boundary layer thickness and, as a consequence, increases the heat transfer rate at the surface. For the case of the fluid velocity that decreases with the increase in Pr, the reason is that fluid of higher Prandtl number means more viscous fluid, which increases the boundary layer thickness and thus reduces the shear stress and consequently, retards the flow of the nanofluid. Also, it can be seen that the velocity distribution for small value of Prandtl number consists of two distinct regions: a thin region near the wall of the plate where there are large velocity gradients due to viscous effects and a region where the velocity gradients are small compared with those near the wall. In the later region, the viscous effects are negligible and the flow of fluid in the region can be considered to be inviscid. Also, such region tends to create uniform accelerated flow at the surface of the plate.

The use of nanoparticles in the fluids exhibited better properties relating to the heat transfer of fluid than heat transfer enhancement through the use of suspended millimeter- or micrometer-sized particles which potentially cause some severe problems, such as abrasion, clogging, high pressure drop, and sedimentation of particles. The very low concentration applications and nanometer size properties of nanoparticles in the base fluid prevent the sedimentation in the flow that may clog the channel. It should be added that the theoretical prediction of enhanced thermal conductivity of the base fluid and prevention of clogging, abrasion, high pressure drop, and sedimentation through the addition of

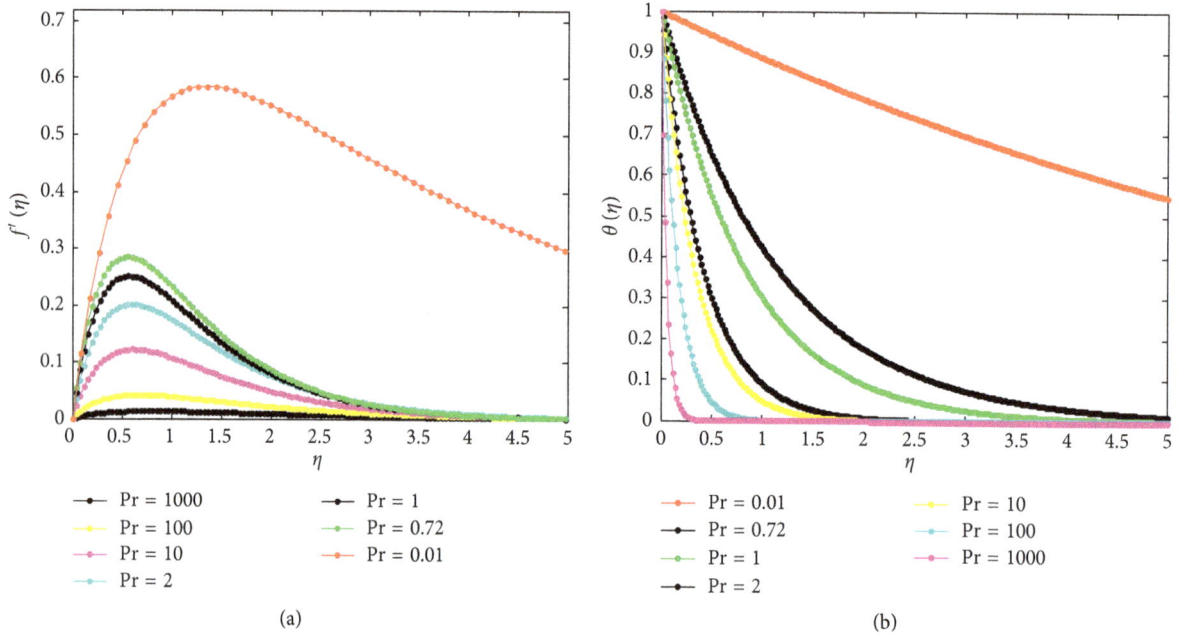

FIGURE 6: (a) Effects of Prandtl number on the velocity profile when $\phi = 0.020$. (b) Effects of Prandtl number on the temperature profile when $\phi = 0.020$.

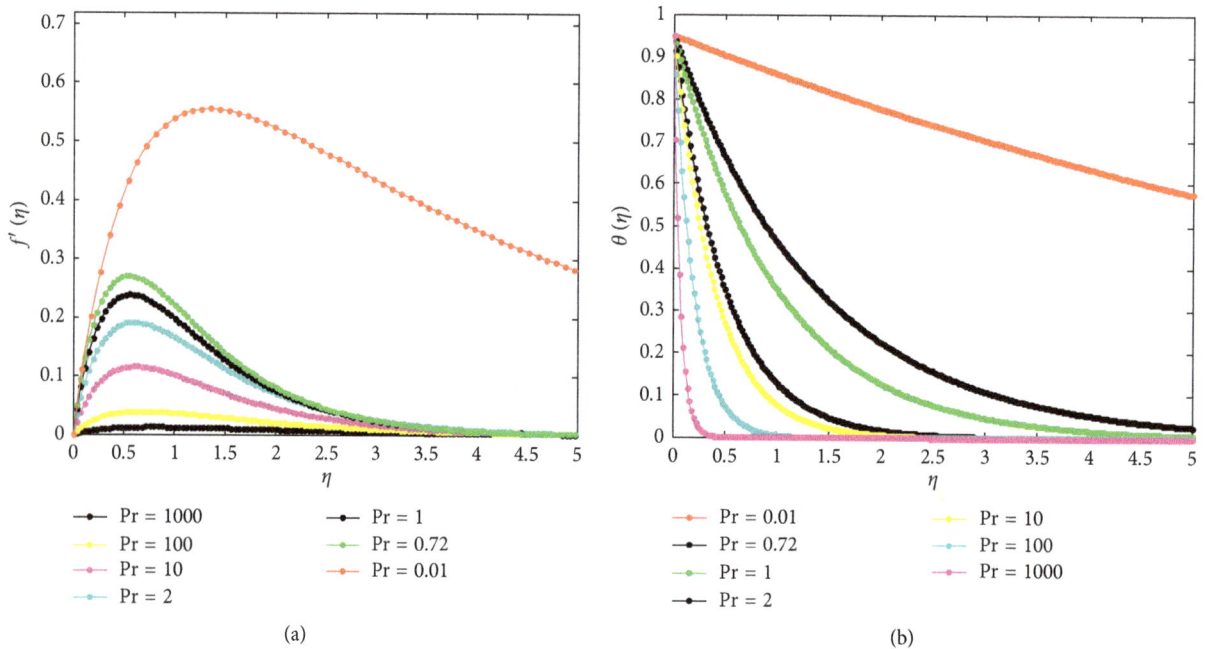

FIGURE 7: (a) Effects of Prandtl number on the velocity profile when $\phi = 0.040$. (b) Effects of Prandtl number on the temperature profile when $\phi = 0.040$.

nanoparticles in the base fluid have been supported with experimental evidences in the literature.

### 6.4. Effect of Nanoparticle Shape on Casson Nanofluid Velocity and Temperature Distributions for Different Values of Prandtl Number.
It is observed experimentally that the nanoparticle shape has significant impacts on the thermal conductivity. Therefore, the effects of nanoparticle shape at different values of

Prandtl number on the velocity and temperature profiles of copper(II) oxide-water nanofluid are shown in Figures 10–15. It is indicated that the maximum decrease in velocity and maximum increase in temperature are caused by lamina, platelets, cylinder, bricks, and sphere, respectively. It is observed that lamina-shaped nanoparticle carries maximum velocity, whereas spherical-shaped nanoparticle has better enhancement on heat transfer than other nanoparticle shapes.

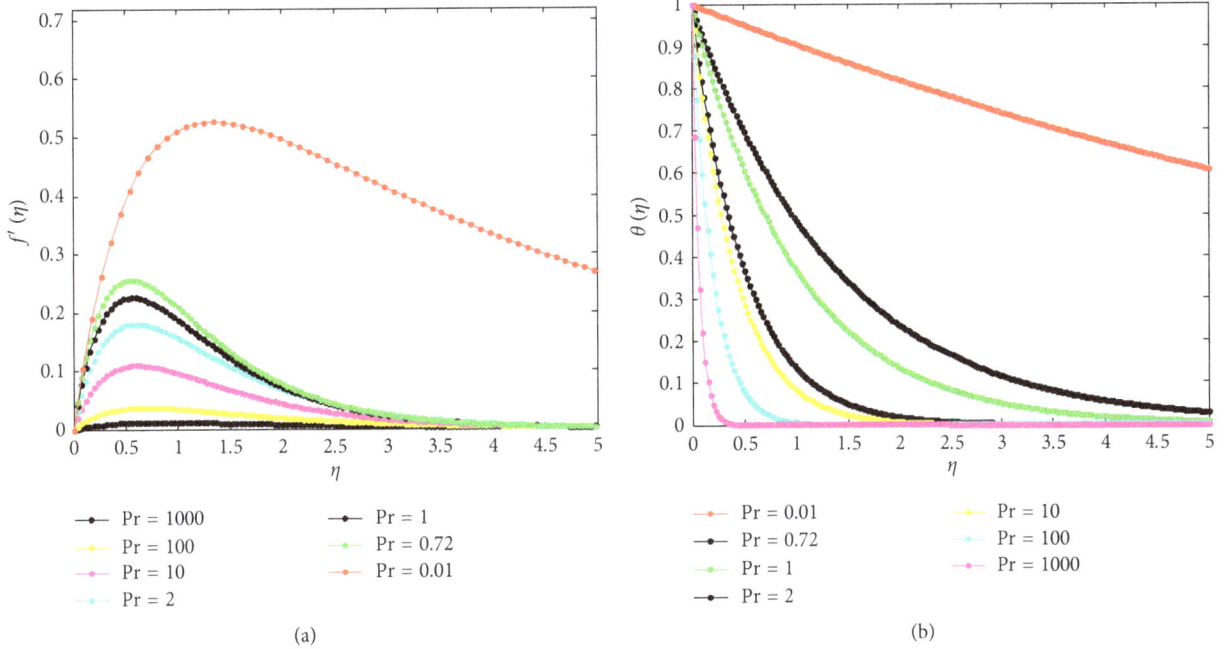

FIGURE 8: (a) Effects of Prandtl number on the velocity profile when $\phi = 0.060$. (b) Effects of Prandtl number on the temperature profile when $\phi = 0.060$.

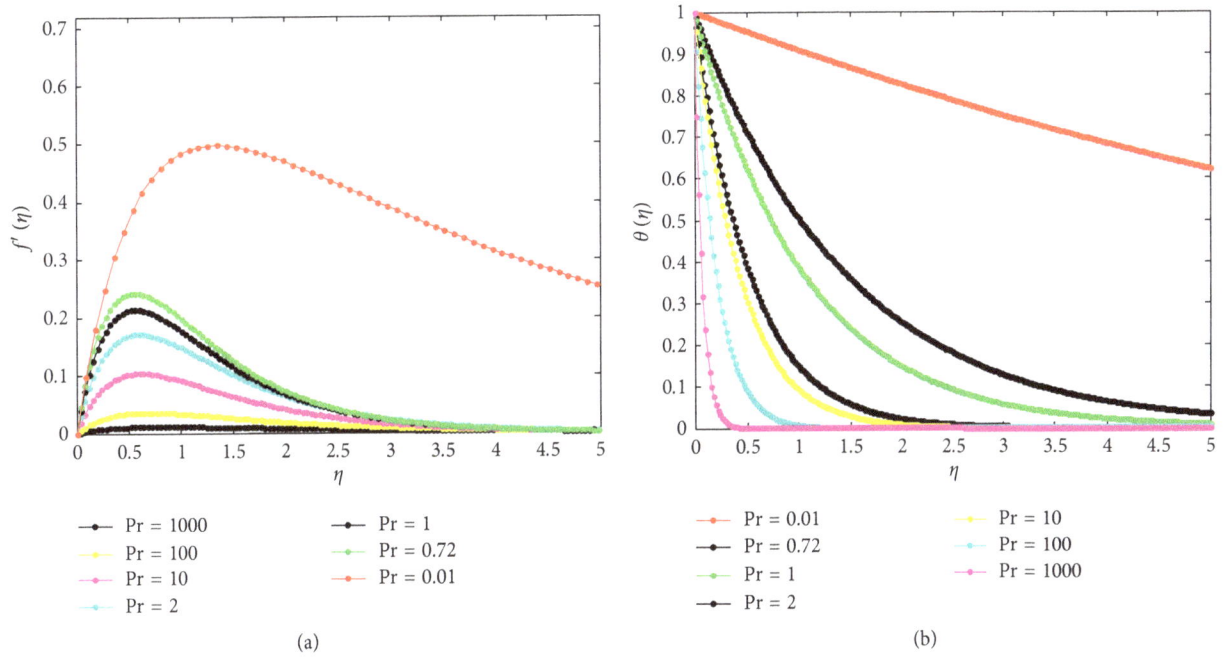

FIGURE 9: (a) Effects of Prandtl number on the velocity profile when $\phi = 0.080$. (b) Effects of Prandtl number on the temperature profile when $\phi = 0.080$.

In fact, it is in accordance with the physical expectation since it is well known that the lamina nanoparticle has greater shape factor than other nanoparticles of different shapes; therefore, the lamina nanoparticle comparatively gains maximum temperature than others. The decrease in velocity is highest in spherical nanoparticles as compared with other shapes. The enhancement observed at lower volume fractions for

nonspherical particles is attributed to the percolation chain formation, which perturbs the boundary layer and thereby increases the local Nusselt number values.

It is evident from this study that proper choice of nanoparticles will be helpful in controlling velocity and heat transfer. It is also observed that irreversibility process can be reduced by using nanoparticles, especially the spherical

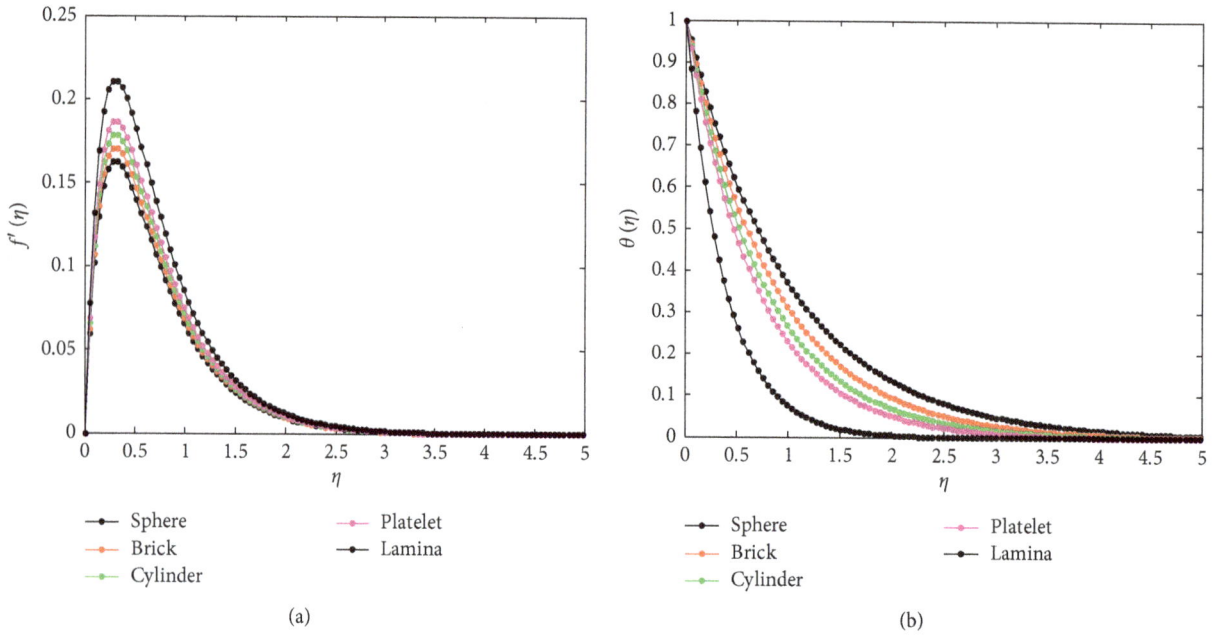

FIGURE 10: (a) Effect of nanoparticle shape on the velocity distribution of the nanofluid. (b) Effects of nanoparticle shape on the temperature distribution of the nanofluid.

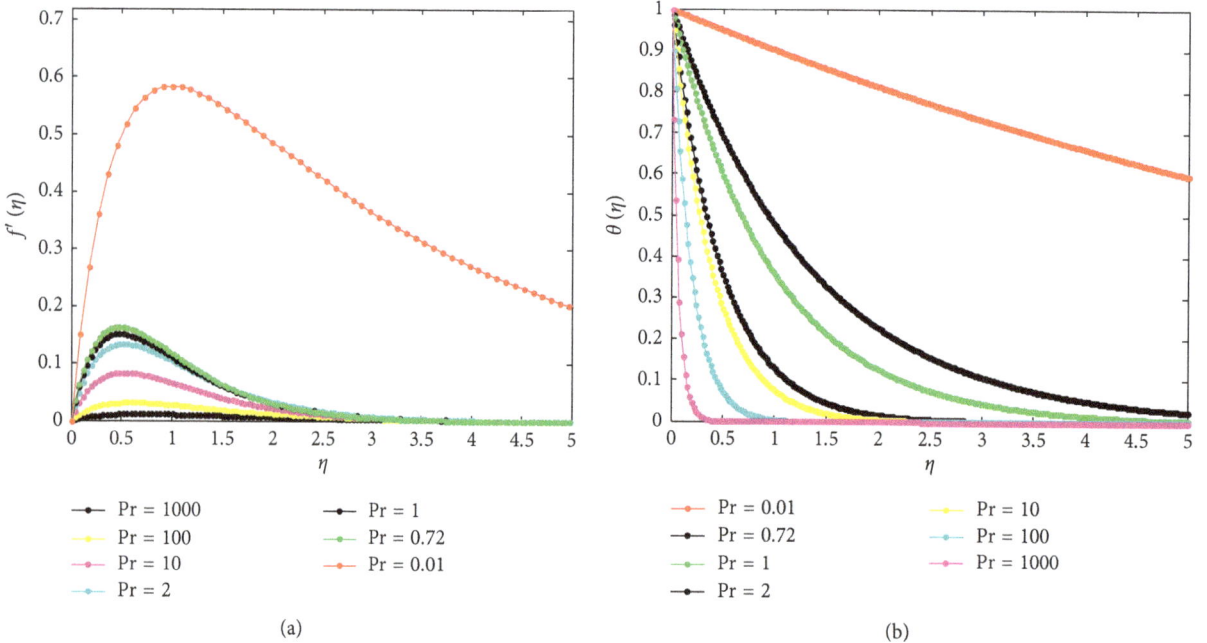

FIGURE 11: (a) Effects of Prandtl number on the velocity profile for spherical-shaped nanoparticle. (b) Effects of Prandtl number on the temperature profile for spherical-shaped nanoparticle.

particles. This can potentially result in higher enhancement in the thermal conductivity of a nanofluid containing elongated particles compared to the one containing spherical nanoparticle, as exhibited by the experimental data in the literature.

*6.5. Effect of Type of Nanoparticle on Casson Nanofluid Velocity and Temperature Distribution for Different Values of Prandtl Number.* The variations of the velocity and temperature

profiles against $\eta$ for various types of nanoparticles (TiO$_2$, CuO, Al$_2$O$_3$, and SWCNTs) are shown in Figures 16–19. Using a common base fluid for all the nanoparticle types, it is observed that the maximum decrease in velocity and maximum increase in temperature are caused by TiO$_2$, CuO, Al$_2$O$_3$, and SWCNTs, respectively. It is observed that SWCNT nanoparticle carries maximumdecrease in velocity but has better enhancement on heat transfer than other nanoparticle shapes. In accordance with the physical

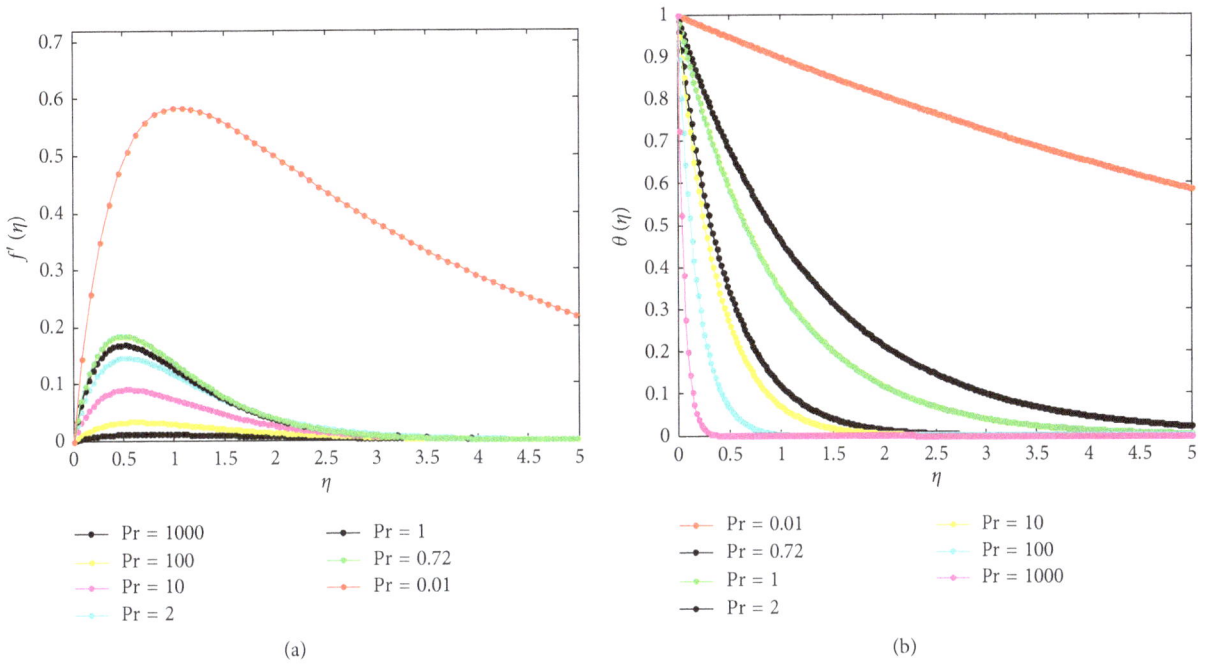

FIGURE 12: (a) Effects of Prandtl number on the velocity profile for brick-shaped nanoparticle. (b) Effects of Prandtl number on the temperature profile for brick-shaped nanoparticle.

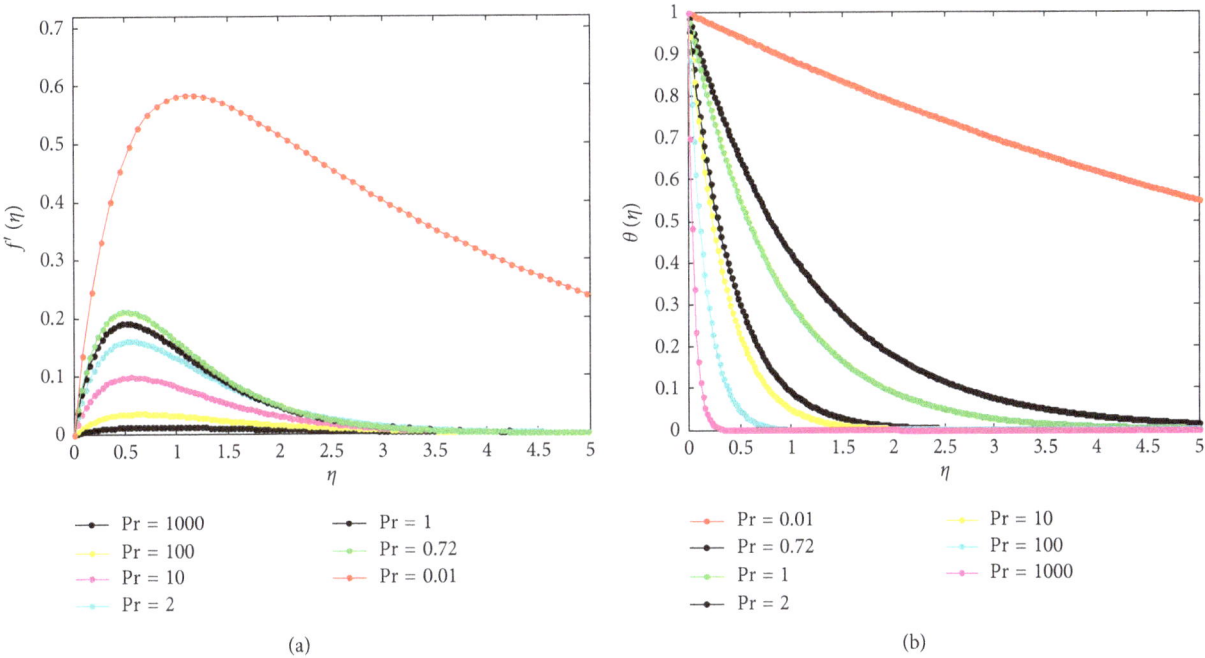

FIGURE 13: (a) Effects of Prandtl number on the velocity profile for cylindrical-shaped nanoparticle. (b) Effects of Prandtl number on the temperature profile for cylindrical-shaped nanoparticle.

expectation well, the SWCNT nanoparticle has higher thermal conductivity than other types of nanoparticles; therefore, the SWCNT nanoparticle comparatively gains maximum temperature than others. The increased thermal conductivity of the base fluid due to the use of nanoparticle of higher thermal conductivity increases the heat enhancement capacity of the base fluid.

Also, it is observed that the velocity decrease is maximum in SWCNT nanoparticles when compared with other types of nanoparticles. This is because the solid thermal conductivity has significant impacts on the momentum boundary layer of the nanofluid. The thickness of the momentum boundary layer increases with the increase in thermal conductivity. It is observed that the thickness of the

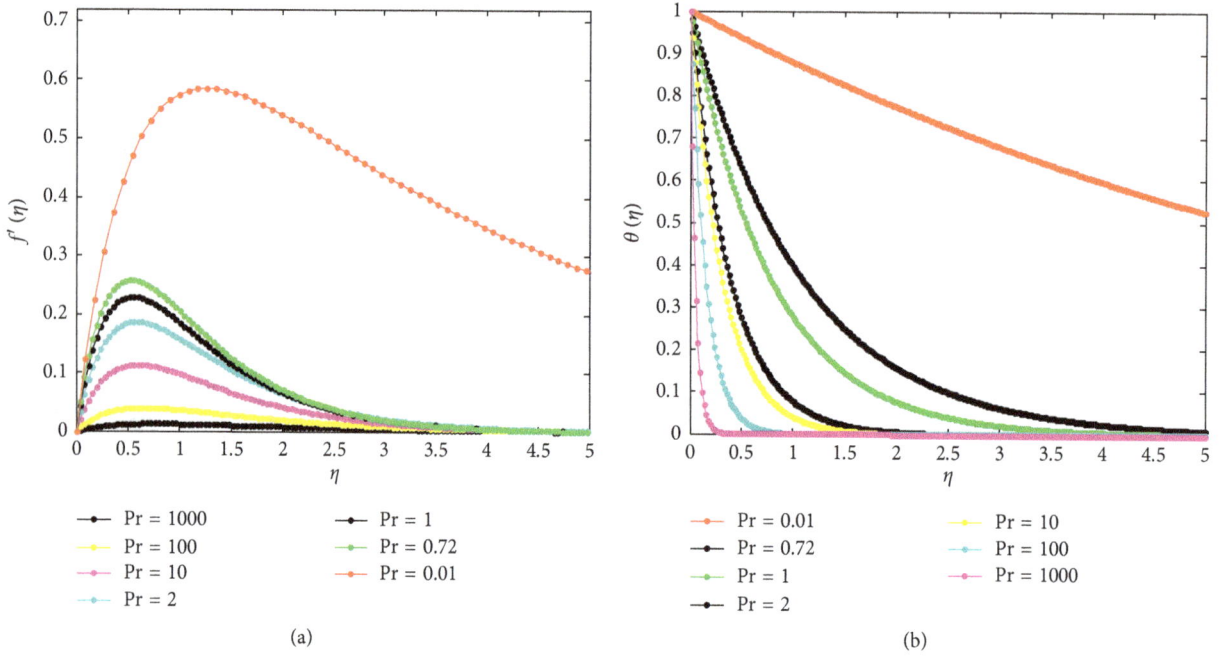

FIGURE 14: (a) Effects of Prandtl number on the velocity profile for platelet-shaped nanoparticle. (b) Effects of Prandtl number on the temperature profile for platelet-shaped nanoparticle.

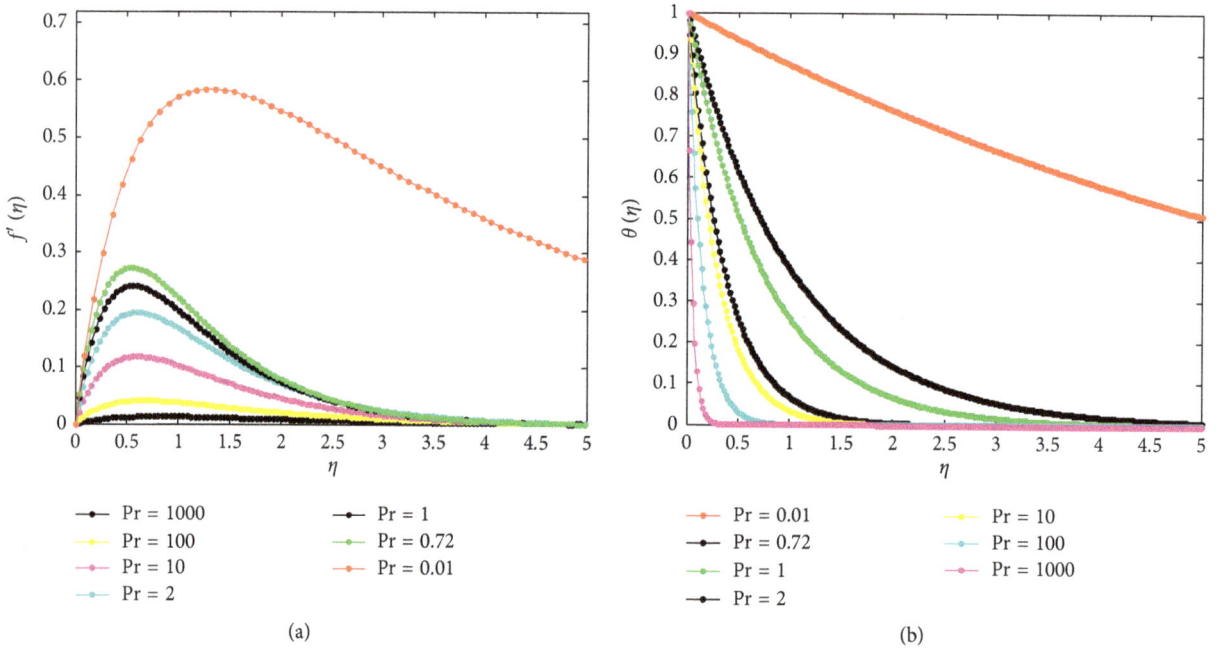

FIGURE 15: (a) Effects of Prandtl number on the velocity profile for lamina-shaped nanoparticle. (b) Effects of Prandtl number on the temperature profile for lamina-shaped nanoparticle.

thermal boundary layer enhances in the presence of higher thermal conductivity nanoparticle. Therefore, the sensitivity of the boundary layer thickness to the type of nanoparticle is correlated with the value of the thermal conductivity of the nanoparticle used, which consequently leads to enhancement of thermal conductivity of the nanofluid.

## 7. Conclusion

In this work, the influences of thermal radiation and nanoparticles on free convection flow and heat transfer of Casson nanofluids over a vertical plate have been analyzed. The governing systems of nonlinear partial differential

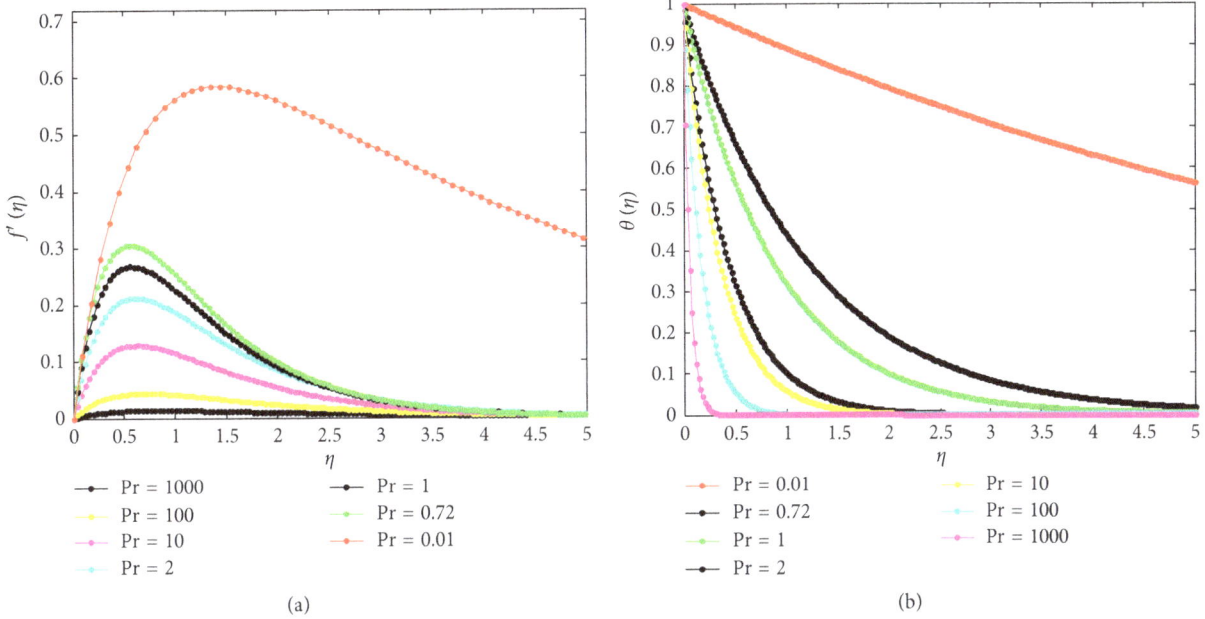

FIGURE 16: (a) Effects of Prandtl number on the velocity profile for TiO$_2$ nanoparticle. (b) Effects of Prandtl number on the temperature profile for TiO$_2$ nanoparticle.

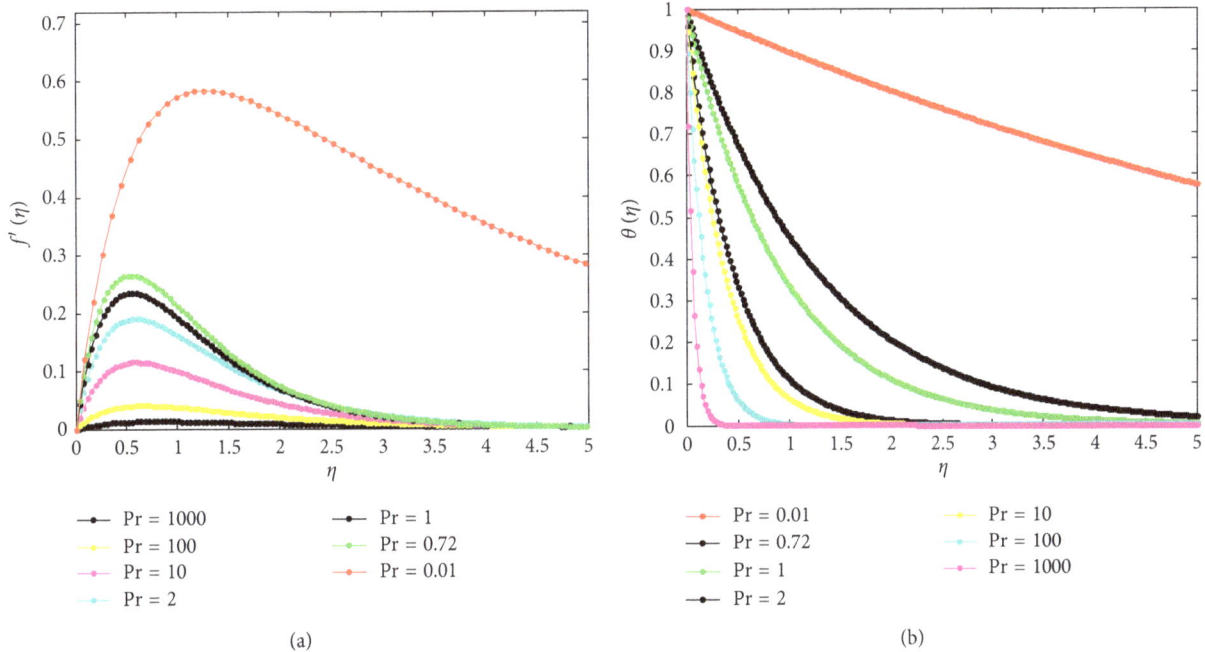

FIGURE 17: (a) Effects of Prandtl number on the velocity profile for CuO nanoparticle. (b) Effects of Prandtl number on the temperature profile for CuO nanoparticle.

equations of the flow and heat transfer processes are transformed to system of nonlinear ordinary differential equation through similarity variables. The systems of fully coupled nonlinear ordinary differential equations have been solved using differential transformation method with Padé-approximant technique. The accuracies of the developed analytical solutions were verified with the results generated by some other methods as presented in the past works. The developed analytical solutions were used to investigate the effects of Casson parameter, thermal radiation parameter, Prandtl number, nanoparticle size, and nanoparticle shapes on the flow and heat transfer behaviour of various Casson nanofluids. From the parametric studies, the following observations were established:

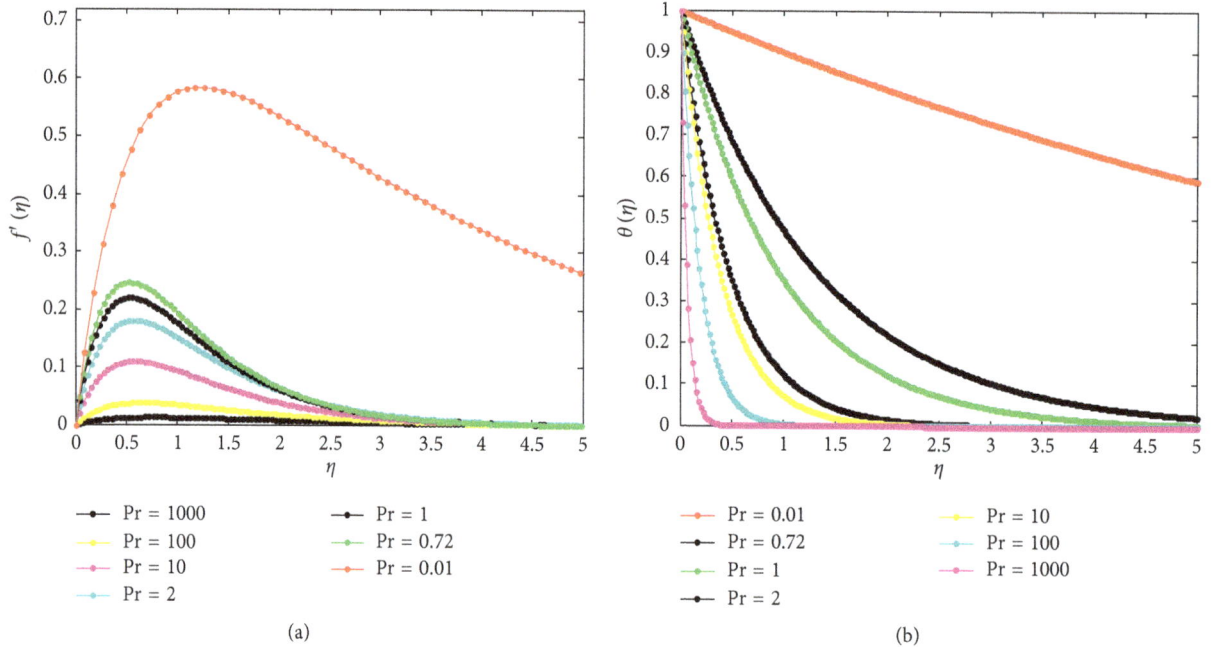

FIGURE 18: (a) Effects of Prandtl number on the velocity profile for $Al_2O_3$ nanoparticle. (b) Effects of Prandtl number on the temperature profile for $Al_2O_3$ nanoparticle.

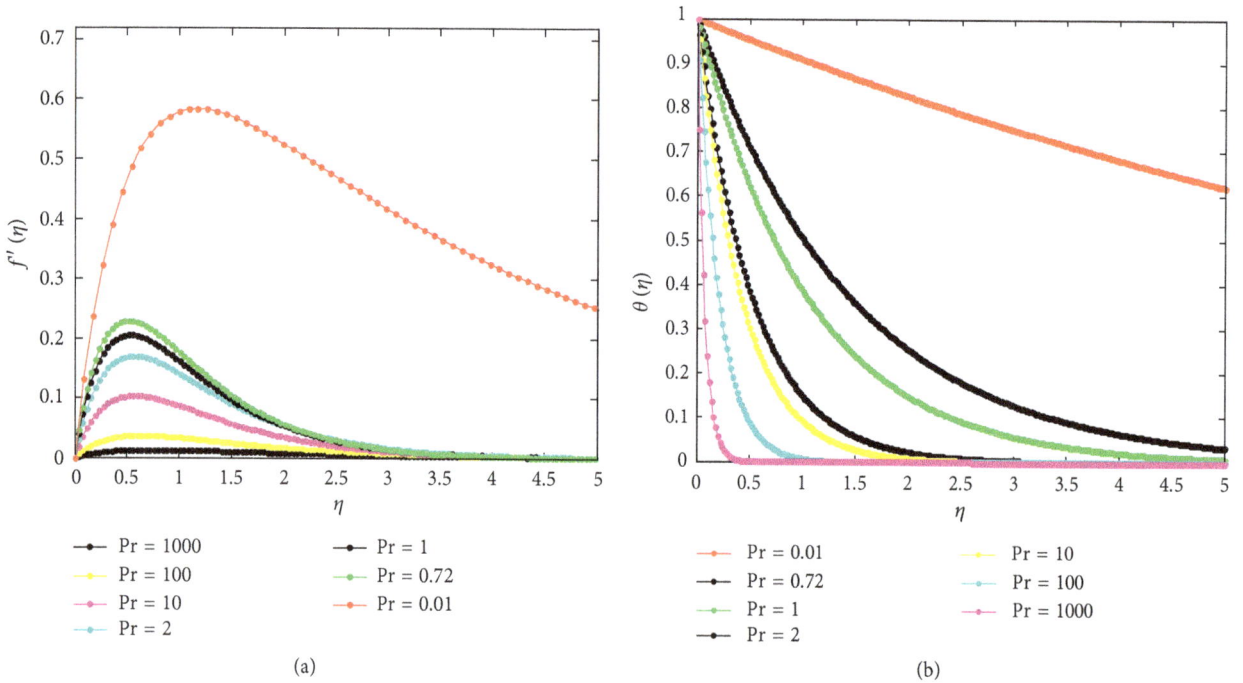

FIGURE 19: (a) Effects of Prandtl number on the velocity profile for SWCNT nanoparticle. (b) Effects of Prandtl number on the temperature profile for SWCNT nanoparticle.

(i) The magnitude of velocity near the plate for the Casson nanofluid parameter decreases for increasing value of the Casson parameter, while temperature increases for increase in Casson fluid parameter.

(ii) Both the velocity and temperature of the nanofluid as well viscous and thermal boundary layers increase with increase in the radiation parameter.

(iii) The velocity of the nanofluid decreases as the Prandtl number increases, but the temperature of the nanofluid increases as the Prandtl number increases.

(iv) The velocity of the nanofluid decreases as the volume fraction or concentration of the nanoparticle in the base fluid increases. However, an opposite trend or behaviour in the temperature profile was observed which showed that as the nanofluid temperature increases, the volume fraction of the nanoparticles in the base fluid increases.

(v) The lamina-shaped nanoparticle carries maximum velocity, whereas spherical-shaped nanoparticle has better enhancement on heat transfer than other nanoparticle shapes. The maximum decrease in velocity and maximum increase in temperature are caused by lamina-shaped nanoparticle and followed by platelet-, cylinder-, brick-, and sphere-shaped nanoparticles, respectively.

(vi) Using a common base fluid to all the nanoparticle types considered in this work, it was observed that SWCNT nanoparticle carries maximum decrease in velocity but has better enhancement on heat transfer than other nanoparticle shapes. Also, it was observed that the maximum decrease in velocity and maximum increase in temperature are caused by $TiO_2$ and followed by CuO, $Al_2O_3$, and SWCNT nanoparticles, in that order.

The present study reveals and exposes the predominant factors affecting the boundary layer of free convection flow and heat transfer of Casson nanofluids. Moreover, the high level of accuracy and versatility of differential transformation method with Padé-approximate technique have been demonstrated. It is hoped that the present study will enhance the understanding as it provides physical insights into the free convection boundary layer problems of Casson nanofluid under various parameters.

## Abbreviations

*Nomenclature*
$c_p$:    Specific heat capacity
$k$:    Thermal conductivity
$K$:    The absorption coefficient
$m$:    Shape factor
$p$:    Pressure
$p_y$:    Yield stress of the fluid
Pr:    Prandtl number
$u$:    Velocity component in the $x$-direction
$v$:    Velocity component in the $z$-direction
$y$:    Axis perpendicular to plates
$x$:    Axis along the horizontal direction
$y$:    Axis along the vertical direction

*Symbols*
$\beta$:    Volumetric extension coefficients
$\rho$:    Density of the fluid
$\mu$:    Dynamic viscosity
$\eta$:    Similarity variable
$\gamma$:    Casson parameter
$\lambda$:    Sphericity
$\phi$:    Volume fraction or concentration of the nanofluid
$\theta$:    Dimensionless temperature
$\tau$:    Shear stress
$\tau_0$:    Casson yield stress
$\mu$:    Dynamic viscosity

$\dot{\sigma}$:    Shear rate
$e_{ij}$:    The $(i, j)$th component of the deformation rate
$\pi$:    Product of the component of deformation rate with itself
$\pi_c$:    Critical value of this product based on the non-Newtonian model
$\mu_B$:    Plastic dynamic viscosity of the non-Newtonian fluid

*Subscript*
$f$:    Fluid
$s$:    Solid
$n_f$:    Nanofluid.

## Conflicts of Interest

The author declares that there are no conflicts of interest regarding the publication of this paper.

## Acknowledgments

The author expresses sincere appreciation to the University of Lagos, Nigeria, for providing material supports and good environment for this work.

## References

[1] E. Schmidt and W. Beckmann, "Das temperatur-und geschwindigkeitsfeld vor einer wärme abgebenden senkrecher platte bei natürelicher convention," *Technische Mechanik und Thermodynamik*, vol. 1, no. 10, pp. 341–349, 1930.

[2] S. Ostrach, "An analysis of laminar free-convection flow and heat transfer about a flat plate parallel to the direction of the generating body force," NACA Report, University of North Texas, Denton, TX, USA, 1953.

[3] E. M. Sparrow and J. L. Gregg, "Laminar free convection from a vertical plate with uniform surface heat flux," *Transactions ASME*, vol. 78, pp. 435–440, 1956.

[4] E. J. Lefevre, "Laminar free convection from a vertical plane surface," in *Proceedings of the 9th International Congress on Applied Mechanics*, vol. 168, Brussels, Belgium, 1956.

[5] E. M. Sparrow and J. L. Gregg, "Similar solutions for free convection from a nonisothermal vertical plate," *Transactions*, vol. 80, pp. 379–386, 1958.

[6] K. Stewartson and L. T. Jones, "The heated vertical plate at high Prandtl number," *Journal of the Aeronautical Sciences*, vol. 24, pp. 379-380, 1957.

[7] H. K. Kuiken, "An asymptotic solution for large Prandtl number free convection," *Journal of Engineering Mathematics*, vol. 2, no. 4, pp. 355–371, 1968.

[8] H. K. Kuiken, "Free convection at low Prandtl numbers," *Journal of Fluid Mechanics*, vol. 37, no. 4, pp. 785–798, 1969.

[9] S. Eshghy, "Free-convection layers at large Prandtl number," *Zeitschrift für angewandte Mathematik und Physik*, vol. 22, no. 2, pp. 275–292, 1971.

[10] S. Roy, "High Prandtl number free convection for uniform surface heat flux," *Journal of Heat Transfer*, vol. 95, pp. 124–126, 1973.

[11] H. K. Kuiken and Z. Rotem, "Asymptotic solution for the plume at very large and small Prandtl numbers," *Journal of Fluid Mechanics*, vol. 45, no. 3, pp. 585–600, 1971.

[12] T. Y. Na and I. S. Habib, "Solution of the natural convection problem by parameter differentiation," *International

*Journal of Heat and Mass Transfer*, vol. 17, no. 3, pp. 457–459, 1974.

[13] J. H. Merkin, "A note on the similarity solutions for free convection on a vertical plate," *Journal of Engineering Mathematics*, vol. 19, no. 3, pp. 189–201, 1985.

[14] J. H. Merkin and I. Pop, "Conjugate free convection on a vertical surface," *International Journal of Heat and Mass Transfer*, vol. 39, no. 7, pp. 1527–1534, 1996.

[15] F. M. Ali, R. Nazar, and N. M. Arifin, "Numerical investigation of free convective boundary layer in a viscous fluid," *American Journal of Scientific Research*, vol. 5, pp. 13–19, 2009.

[16] S. S. Motsa, S. Shateyi, and Z. Makukula, "Homotopy analysis of free convection boundary layer flow with heat and mass transfer," *Chemical Engineering Communications*, vol. 198, no. 6, pp. 783–795, 2011.

[17] S. S. Motsa, Z. G. Makukula, and S. Shateyi, "Spectral local linearisation approach for natural convection boundary layer flow," *Mathematical Problems in Engineering*, vol. 2013, Article ID 765013, 7 pages, 2013.

[18] A. R. Ghotbi, H. Bararnia, G. Domairry, and A. Barari, "Investigation of a powerful analytical method into natural convection boundary layer flow," *Communications in Nonlinear Science and Numerical Simulation*, vol. 14, no. 5, pp. 2222–2227, 2009.

[19] S. Mosayebidorcheh and T. Mosayebidorcheh, "Series solution of convective radiative conduction equation of the nonlinear fin with temperature dependent thermal conductivity," *International Journal of Heat and Mass Transfer*, vol. 55, no. 23-24, pp. 6589–6594, 2012.

[20] M. Sheikholeslami, "CVFEM for magnetic nanofluid convective heat transfer in a porous curved enclosure," *European Physical Journal Plus*, vol. 131, p. 413, 2016.

[21] M. Sheikholeslami and D. D. Ganji, *Hydrothermal Analysis in Engineering Using Control Volume Finite Element Method*, Elsevier, New York, NY, USA, 1st edition, 2015.

[22] M. Sheikholeslami, *Application of Control Volume Based Finite Element Method (CVFEM) for Nanofluid Flow and Heat Transfer*, Elsevier, New York, NY, USA, 1st edition, 2018.

[23] M. Sheikholeslami, P. Rana, and S. Soleimani, "Numerical study of MHD Natural convection liquid metal flow and heat transfer in a wavy enclosure using CVFEM," *Heat Transfer Research*, vol. 48, no. 2, pp. 121–138, 2017.

[24] S. Mosayebidorcheh, M. Rahimi-Gorji, D. D. Ganji, T. Moayebidorcheh, O. Pourmehran, and M. Biglarian, "Transient thermal behavior of radial fins of rectangular, triangular and hyperbolic profiles with temperature-dependent properties using DTM-FDM," *Journal of Central South University*, vol. 24, no. 3, pp. 675–682, 2017.

[25] S. Mosayebidorcheh, O. D. Makinde, D. D. Ganji, and M. A. Chermahini, "DTM-FDM hybrid approach to unsteady MHD Couette flow and heat transfer of dusty fluid with variable properties," *Thermal Science and Engineering Progress*, vol. 2, pp. 57–63, 2017.

[26] S. Mosayebidorcheh, M. Farzinpoor, and D. D. Ganji, "Transient thermal analysis of longitudinal fins with internal heat generation considering temperature-dependent properties and different fin profiles," *Energy Conversion and Management*, vol. 86, pp. 365–370, 2014.

[27] S. Mosayebidorcheh, T. Moayebidorcheh, and M. M. Rashidi, "Analytical solution of the steady state condensation film on the inclined rotating disk by a new hybrid method," *Scientific Research and Essays*, vol. 9, no. 12, pp. 557–565, 2014.

[28] S. Mosayebidorcheh, M. M. Vatani, D. D. Ganji, and T. Moayebidorcheh, "Investigation of the viscoelastic flow and species diffusion in a porous channel with high permeability," *Alexandria Engineering Journal*, vol. 53, no. 4, pp. 779–785, 2014.

[29] S. Mosayebidorcheh, "Analytical investigation of the micropolar flow through a porous channel with changing walls," *Journal of Molecular Liquids*, vol. 196, pp. 113–119, 2014.

[30] M. Hatami, S. Mosayebidorcheh, and D. Jing, "Thermal performance evaluation of alumina-water nanofluid in an inclined direct absorption solar collector (IDASC) using numerical method," *Journal of Molecular Liquids*, vol. 231, pp. 632–639, 2017.

[31] S. Mosayebidorcheh, "Solution of the boundary layer equation of the power-law pseudoplastic fluid using differential transform method," *Mathematical Problems in Engineering*, vol. 2013, Article ID 685454, p. 8, 2013.

[32] S. Mosayebidorcheh, M. Hatami, D. D. Ganji, T. Mosayebidorcheh, and S. M. Mirmohammadsadeghi, "Investigation of transient MHD Couette flow and heat transfer of dusty fluid with temperature-dependent properties," *Journal of Applied Fluid Mechanics*, vol. 8, no. 4, pp. 921–929, 2015.

[33] M. Sheikholeslami and S. A. Shehzad, "CVFEM for influence of external magnetic source on $Fe_3O_4$-$H_2O$ nanofluid behavior in a permeable cavity considering shape effect," *International Journal of Heat and Mass Transfer*, vol. 115, pp. 180–191, 2017.

[34] M. Sheikholeslami and S. A. Shehzad, "Magnetohydrodynamic nanofluid convective flow in a porous enclosure by means of LBM," *International Journal of Heat and Mass Transfer*, vol. 113, pp. 796–805, 2017.

[35] M. Sheikholeslami and M. Seyednezhad, "Nanofluid heat transfer in a permeable enclosure in presence of variable magnetic field by means of CVFEM," *International Journal of Heat and Mass Transfer*, vol. 114, pp. 1169–1180, 2017.

[36] M. Sheikholeslami and H. B. Rokni, "Simulation of nanofluid heat transfer in presence of magnetic field: a review," *International Journal of Heat and Mass Transfer*, vol. 115, pp. 1203–1233, 2017.

[37] M. Sheikholeslami, T. Hayat, and A. Alsaedi, "On simulation of nanofluid radiation and natural convection in an enclosure with elliptical cylinders," *International Journal of Heat and Mass Transfer*, vol. 115, pp. 981–991, 2017.

[38] M. Sheikholeslami and H. B. Rokni, "Melting heat transfer influence on nanofluid flow inside a cavity in existence of magnetic field," *International Journal of Heat and Mass Transfer*, vol. 114, pp. 517–526, 2017.

[39] M. Sheikholeslami and M. K. Sadoughi, "Simulation of CuO-water nanofluid heat transfer enhancement in presence of melting surface," *International Journal of Heat and Mass Transfer*, vol. 116, pp. 909–919, 2018.

[40] M. Sheikholeslami and M. Sadoughi, "Mesoscopic method for MHD nanofluid flow inside a porous cavity considering various shapes of nanoparticles," *International Journal of Heat and Mass Transfer*, vol. 113, pp. 106–114, 2017.

[41] M. Sheikholeslami and M. M. Bhatti, "Forced convection of nanofluid in presence of constant magnetic field considering shape effects of nanoparticles," *International Journal of Heat and Mass Transfer*, vol. 111, pp. 1039–1049, 2017.

[42] M. Sheikholeslami and M. M. Bhatti, "Active method for nanofluid heat transfer enhancement by means of EHD," *International Journal of Heat and Mass Transfer*, vol. 109, pp. 115–122, 2017.

[43] M. Sheikholeslami and H. B. Rokni, "Nanofluid two phase model analysis in existence of induced magnetic field," *In-*

*ternational Journal of Heat and Mass Transfer*, vol. 107, pp. 288–299, 2017.

[44] M. Sheikholeslami and S. A. Shehzad, "Magnetohydrodynamic nanofluid convection in a porous enclosure considering heat flux boundary condition," *International Journal of Heat and Mass Transfer*, vol. 106, pp. 1261–1269, 2017.

[45] M. Sheikholeslami, T. Hayat, and A. Alsaedi, "Numerical simulation of nanofluid forced convection heat transfer improvement in existence of magnetic field using lattice Boltzmann method," *International Journal of Heat and Mass Transfer*, vol. 108, pp. 1870–1883, 2017.

[46] M. Sheikholeslami, T. Hayat, and A. Alsaedi, "Numerical study for external magnetic source influence on water based nanofluid convective heat transfer," *International Journal of Heat and Mass Transfer*, vol. 106, pp. 745–755, 2017.

[47] M. Sheikholeslami, T. Hayat, A. Alsaedi, and S. Abelman, "Numerical analysis of EHD nanofluid force convective heat transfer considering electric field dependent viscosity," *International Journal of Heat and Mass Transfer*, vol. 108, pp. 2558–2565, 2017.

[48] J. K. Zhou, *Differential Transformation and Its Applications for Electrical Circuits*, Huazhong University Press, Wuhan, China, 1986, in Chinese.

[49] Y. Lien-Tsai and C. Cha'o-Kuang, "The solution of the Blasius equation by the differential transformation method," *Mathematical and Computer Modelling*, vol. 28, no. 1, pp. 101–111, 1998.

[50] B. L. Kuo, "Thermal boundary-layer problems in a semi-infinite flat plate by the differential transformation method," *Applied Mathematics and Computation*, vol. 150, no. 2, pp. 143–160, 2004.

[51] B. L. Kuo, "Application of the differential transformation method to the solutions of the free convection problem," *Applied Mathematics and Computation*, vol. 165, no. 1, pp. 63–79, 2005.

[52] M. M. Rashidi, N. Laraqi, and S. M. Sadri, "A novel analytical solution of mixed convection about an inclined flat plate embedded in a porous medium using the DTM-Pade," *International Journal of Thermal Sciences*, vol. 49, no. 12, pp. 2405–2412, 2010.

[53] N. Casson, *Rheology of Dispersed System*, Vol. 84, Pergamon Press, Oxford, UK, 1959.

[54] R. K. Dash, K. N. Mehta, and G. Jayaraman, "Casson fluid flow in a pipe filled with a homogeneous porous medium," *International Journal of Engineering Science*, vol. 34, no. 10, pp. 1145–1156, 1996.

[55] H. I. Andersson and B. S. Dandapat, "Flow of a power-law fluid over a stretching sheet," *Applied Analysis of Continuous Media*, vol. 1, no. 339, 1992.

[56] M. Sajid, I. Ahmad, T. Hayat, and M. Ayub, "Unsteadyflow and heat transfer of a second grade fluid over a stretching sheet," *Communications in Nonlinear Science and Numerical Simulation*, vol. 14, no. 1, pp. 96–108, 2009.

[57] N. T. M. Eldabe and M. G. E. Salwa, "Heat transfer of MHD non-Newtonian Casson fluid flow between two rotating cylinder," *Journal of the Physical Society of Japan*, vol. 64, p. 4164, 1995.

[58] S. Nadeem, R. L. Haq, N. S. Akbar, and Z. H. Khan, "MHD three-dimensional Casson fluid flow past a porous linearly stretching sheet," *Alexandria Engineering Journal*, vol. 52, no. 4, pp. 577–582, 2013.

[59] A. Raptis and C. Perdikis, "Viscoelastic flow by the presence of radiation," *Zeitschrift für Angewandte Mathematik und Mechanik*, vol. 78, no. 4, pp. 277–279, 1998.

[60] M. A. Seddeek, "Effects of radiation and variable viscosity on

a MHD free convection flow past a semi-infinite flat plate with an aligned magnetic field in the case of unsteady flow," *International Journal of Heat and Mass Transfer*, vol. 45, no. 4, pp. 931–935, 2002.

[61] F. Mabood, M. Imtiaz, A. Alsaedi, and T. Hayat, "Unsteady convective boundary layer flow of Maxwell fluid with nonlinear thermal radiation: a numerical study," *International Journal of Nonlinear Sciences and Numerical Simulation*, vol. 17, no. 5, pp. 221–229, 2016.

[62] T. Hayat, T. Muhammad, A. Alsaedi, and M. S. Alhuthali, "Magnetohydrodynamic three-dimensional flow of viscoelastic nanofluid in the presence of nonlinear thermal radiation," *Journal of Magnetism and Magnetic Materials*, vol. 385, pp. 222–229, 2015.

[63] M. Farooq, M. I. Khan, M. Waqas, T. Hayat, A. Alsaedi, and M. I. Khan, "MHD stagnation point flow of viscoelastic nanofluid with non-linear radiation effects," *Journal of Molecular Liquids*, vol. 221, pp. 1097–1103, 2016.

[64] S. A. Shehzad, Z. Abdullah, A. Alsaedi, F. M. Abbasi, and T. Hayat, "Thermally radiative three-dimensional flow of Jeffrey nanofluid with internal heat generation and magnetic field," *Journal of Magnetism and Magnetic Materials*, vol. 397, pp. 108–114, 2016.

[65] N. Casson, "A flow equation for the pigment oil suspension of the printing ink type," in *Rheology of Disperse Systems*, pp. 84–102, Pergamon, New York, NY, USA, 1959.

[66] N. S. Akbar and A. W. Butt, "Ferro-magnetic effects for peristaltic flow of Cu-water nanofluid for different shapes of nano-size particles," *Applied Nanoscience*, vol. 6, no. 3, pp. 379–385, 2016.

[67] R. Ul Haq, S. Nadeem, Z. H. Khan, and N. F. M. Noor, "Convective heat transfer in MHD slip flow over a stretching surface in the presence of carbon nanotubes," *Physica B: Condensed Matter*, vol. 457, pp. 40–47, 2015.

[68] L. D. Talley, G. L. Pickard, W. J. Emery, and J. H. Swift, *Descriptive Physical Oceanography, Physical Properties of Sea Water*, Elsevier Ltd., New York, NY, USA, 6th edition, 2011.

[69] M. Pastoriza-Gallego, L. Lugo, J. Legido, and M. Piñeiro, "Thermal conductivity and viscosity measurements of ethylene glycol-based $Al_2O_3$ nanofluids," *Nanoscale Research Letters*, vol. 6, p. 221, 2011.

[70] S. Aberoumand and A. Jafarimoghaddam, "Experimental study on synthesis, stability, thermal conductivity and viscosity of Cu–engine oil nanofluid," *Journal of the Taiwan Institute of Chemical Engineers*, vol. 71, pp. 315–322, 2017.

[71] G. A. Baker and P. Graves-Morris, *Pade Approximants*, Cambridge University Press, Cambridge, UK, 1996.

[72] H. K. Kuiken, "On boundary layers in fluid mechanics that decay algebraically along stretches of wall that are not vanishingly small," *IMA Journal of Applied Mathematics*, vol. 27, no. 4, pp. 387–405, 1981.

# 18

# Cell Disruption of *Chaetoceros calcitrans* by Microwave and Ultrasound in Lipid Extraction

**Daniela Almeida Nogueira** ⓘ**, Juliane Machado da Silveira** ⓘ**, Évelin Mendes Vidal, Natália Torres Ribeiro, and Carlos André Veiga Burkert** ⓘ

*Bioprocess Engineering Laboratory, School of Chemistry and Food, Federal University of Rio Grande, 96203-900 Rio Grande, RS, Brazil*

Correspondence should be addressed to Daniela Almeida Nogueira; nogueiradaniali@yahoo.com.br

Academic Editor: Xunli Zhang

Downstream processing, such as cell disruption and extraction, constitutes a key step in microalgal-based industrial bioprocesses, mainly due to high costs and environmental impact. In this context, extraction technologies need to be improved, including the use of nonconventional cell disruption techniques suitable for scale-up, such as microwave and ultrasound. Therefore, this study aimed at investigating the effects of different methods of cell disruption (microwave and ultrasound) on lipid extraction from biomass of the diatom *Chaetoceros calcitrans* cultured in mixotrophic conditions in a medium with natural sea water and residual glycerol, with different treatment times. Both techniques applied to the biomass were efficient; that is, the results were $24.6 \pm 1.3\%$ lipids (ultrasound for 5 min) and $24.2 \pm 0.9\%$ lipids (microwave for 40 s), with no significant differences between them ($p \geq 0.05$). Likewise, there was no significant difference regarding the chemical disruption with hydrochloric acid 2 M as control ($24.2 \pm 1.0\%$). The ultrasound method consumed less energy than the microwave method. Both cell disruption methods applied to the biomass resulted in changes in the fatty acid profiles, that is, percentages of saturated fatty acids increased from 7.7% (control) to 16.6% (microwave) and 15.5% (ultrasound), whereas polyunsaturated ones increased from 12.8% (control) to 22.8% (microwave) and 21.8% (ultrasound). Concerning monounsaturated fatty acids, percentages decreased from 79.5% (control) to 60.6% (microwave) and 62.7% (ultrasound).

## 1. Introduction

Population growth and the constant search for quality of life have led to greater demand for products obtained in sustainable ways. Microalgae stand out in the industrial sector due to their potential for yielding several metabolites, such as carbohydrates, pharmaceuticals, pigments, and enzymes [1, 2]. Furthermore, microalga biomass can easily be yielded in any season, since it neither depends on harvest periods nor competes with agriculture for arable land [3].

Several studies have focused on lipid production by microalgae, as some species have high contents of lipids, such as *Chlorella* sp. [4, 5], *Phaeodactylum tricornutum* [5–7], *Tetraselmis suecica* [7], *Nannochloropsis oculata* [5, 7], *Isochrysis galbana* [5, 7], and *Chaetoceros calcitrans* [5, 7]. Moreover, they may become a sustainable source to be used

by both food and biofuel industries, depending on their contents of saturated and unsaturated fatty acids [7, 8]. The marine diatom *Chaetoceros calcitrans*, which belongs to the class *Bacillariophyceae*, stands out among the species of microalgae as a potential producer of intracellular lipids [9]. Diatoms are peculiar microalgae not only because they do not have any flagella but also because their cell wall is composed of overlapping halves called frustules, formed by polymerized opaline silica [10].

In the extraction of lipids from microalgae, cell disruption techniques must be carefully chosen, since lipids are yielded intracellularly. The method of cell disruption should be effective in order to liberate the intracellular lipids from microalgal cells and take into account the amount of energy needed for the process [11]. However, in general, conventional techniques (i.e., mechanical, chemical, and thermal/thermochemical methods) used for cell disruption and extraction are expensive and are often hindered by low efficiency levels [12]. Regarding emerging cell disruption techniques, treatments that use ultrasound and microwave have been viewed as promising ones because they are simple, easy, and efficient methods of lipid extraction that require little time and maintain the quality of the extracts [13–16]. For example, Menéndez et al. [17] verified that lipid extraction from *Nannochloropsis gaditana* biomass with microwave and ultrasound has many advantages, requiring a lower amount of solvent, lower energy consumption, and a shorter extraction time, leading to a lower environmental impact of the process.

In cell disruption by ultrasound, the phenomenon called cavitation forms gas microbubbles that grow and collapse violently [18]. As a result, the pressure and temperature of the adjacent tissue increase and cause the disruption of the cell wall, release soluble compounds, improve mass transference, and enable the solvent to access the cell content [16, 19]. In cell disruption by microwave, non-ionizing electromagnetic waves with a frequency of 300 MHz to 300 GHz are used, and electromagnetic energy is converted into calorific energy by two mechanisms: ionic conduction and dipole rotation [18]. Efficiency of disruption by ultrasound and microwave usually depends on the cell structure, the lipid content, and the microwave and ultrasound conditions [14, 16, 19]. However, despite these advantages, there is a lack of information on these techniques applied to *Chaetoceros calcitrans* biomass.

Therefore, this study aimed to evaluate techniques of cell disruption by microwave and ultrasound in the recovery and profile of fatty acids of lipids from *Chaetoceros calcitrans* cultured in mixotrophic conditions.

## 2. Materials and Methods

The marine microalga *Chaetoceros calcitrans*, which was used in the experiments, was donated by the Laboratório de Biologia Marinha e Biomonitoramento (LABIOMAR), a laboratory that belongs to the Universidade Federal da Bahia (UFBA), located in Salvador, Bahia, Brazil. Residual glycerol (82.09% purity) was supplied by BSBIOS Indústria e Comércio de Biodiesel Sul Brasil S/A, a company located in Passo Fundo, Rio Grande do Sul, Brazil. It is the byproduct of the alkaline-catalyzed transesterification of soybean oil and methanol.

The Conway medium [20] was used for the inoculum (10% of final volume), and it was prepared with sterile sea water with the addition of saline solution ($2\,mL\cdot L^{-1}$), vitamin solution ($0.1\,mL\cdot L^{-1}$), and silicate solution ($2\,mL\cdot L^{-1}$). The saline solution contained $45\,g\cdot L^{-1}$ $C_{10}H_{14}O_8Na_2\cdot 2H_2O$ ($Na_2EDTA$), $33.6\,g\cdot L^{-1}$ $H_3BO_3$, $100\,g\cdot L^{-1}$ $NaNO_3$, $0.36\,g\cdot L^{-1}$ $MnCl_2\cdot 4H_2O$, $1.3\,g\cdot L^{-1}$ $FeCl_3\cdot 6H_2O$, $20\,g\cdot L^{-1}$ $NaH_2PO_4\cdot 2H_2O$, and $1\,mL\cdot L^{-1}$ of trace metals solution containing $21\,g\cdot L^{-1}$ $ZnCl_2$, $20\,g\cdot L^{-1}$ $CoCl_2\cdot 6H_2O$, $9\,g\cdot L^{-1}$ $(NH_4)_6Mo_7O_{24}\cdot 4H_2O$, and $20\,g\cdot L^{-1}$ $CuSO_4\cdot 5H_2O$. The vitamin solution contained $0.05\,g\cdot L^{-1}$ vitamin B12 and $1\,g\cdot L^{-1}$ vitamin B1, and the silicate solution contained $40\,g\cdot L^{-1}$ sodium silicate. The sea water was collected at the Marine Aquaculture Station (FURG), located at the Cassino Beach (Rio Grande, Brazil). It was previously filtered on a qualitative filter paper, and the salinity was adjusted to 28 PSU (practical salinity units) using a manual salinometer (Biobrix, Model 211, Brazil). The sea water, the saline solution, and the silicate solution were sterilized by autoclaving at 121°C for 15 min. The vitamin solution was sterilized separately by filtration through a $0.22\,\mu m$ filter.

In cultures aimed at biomass to be used in cell disruption assays, the microalga was cultured in 2 L Erlenmeyer flasks with 1800 mL Conway medium, in accordance with Walne [20], with modifications. The medium was prepared with the addition of $5.61\,g\cdot L^{-1}$ residual glycerol, and concentrations of silicates and sodium nitrate in the solutions were modified to $70\,g\cdot L^{-1}$ and $50\,g\cdot L^{-1}$, respectively, as recommended by previous studies (unpublished data).

The inoculum (10% of final volume) containing $0.55\,g\cdot L^{-1}$ of biomass was added to the flasks, and these were placed in an incubator with a photoperiod (Eletrolab EL-202, Brazil), on a 12 h light/dark cycle, at 30°C. Light was provided by fluorescent light bulbs simulating natural daylight, with irradiance of 3000 lx. Atmospheric air was directly and constantly injected at a flow rate of $0.2\,L\cdot min^{-1}$ by a pump and glass wool filter system to ensure air sterility. After 10 days, the biomass was recovered by centrifugation (18,800 ×g; 15 min), washed with distilled water, centrifuged again (18,800 ×g; 15 min), and lyophilized.

In cell disruption assays, portions of 0.3 g lyophilized biomass were used for testing: (1) chemical disruption (control), with the addition of 5 mL HCl 2 M, followed by immersion in a water bath for 1 h at 80°C; (2) disruption by microwave, with a frequency of 2.45 MHz and power of 1.4 kW (Brastemp, model BMS35BBHNA, Brazil), with the addition of 20 mL distilled water to the sample, in accordance with Lee et al. [14], and exposure times of 40, 105, and 300 s; (3) disruption by ultrasound, with a frequency of 20 kHz and power of 130 W (Cole Parmer, model CPX 130, USA), with the addition of 20 mL distilled water, in accordance with Lee et al. [14], and exposure times of 5, 10, and 20 min. At the end of the treatments, lipids were extracted and quantified, as proposed by Bligh and Dyer [21], with amounts of solvent recommended by Manirakiza et al. [22].

The consumption of energy (EC) (expressed as $GJ\cdot m^{-3}$) was estimated by using the following equation, based on Alagöz et al. [23]:

$$EC = P \times \frac{t}{V}, \tag{1}$$

where $P$ is the power, $t$ is the time treatment, and $V$ is the volume of the sample.

The specific energy (SE) was defined as the energy supplied per unit of mass of microalgae [23]:

$$SE = P \times \frac{t}{V \times C}, \tag{2}$$

where $C$ is the concentration of cell suspension. In both equations, the corresponding units were converted properly.

Esterification of the lipid fraction extracted from the biomass was conducted according to the method adapted from Metcalfe et al. [24]. The fatty acid profile was determined by gas chromatography. In order to separate and quantify the fatty acid mixture, a gas chromatograph (Shimadzu, model 2010 Plus, Japan), equipped with a split/splitless injector, a capillary column RTX®-1 (30 m × 0.25 mm internal diameter × 0.25 $\mu$m particle diameter), and a flame ionization detector (FID), was used. Helium, at a flow rate of 1.25 mL·min$^{-1}$, was the carrier gas. Temperatures of the injector and the detector were adjusted to 260°C; the injected volume was 1 $\mu$L. The chromatographic conditions of separation were as follows: initial temperature of the column was 50°C, which was raised to 200°C at 6°C·min$^{-1}$, and maintained for 4 min. In the second temperature ramp, the temperature was raised to 240°C at 2°C·min$^{-1}$ and maintained for 10 min. Comparison between retention times and methyl ester standards was used to identify fatty acids, which were quantified by area normalization.

The assays were performed in triplicate, and the results were submitted to analysis of variance (ANOVA) and Tukey's test at a 95% confidence level ($p \leq 0.05$) or the $t$-test at a 95% confidence level ($p \leq 0.05$). Statistica 5.0 (Stat Soft Inc., USA) software was used.

## 3. Results and Discussion

*3.1. Disruption by Microwave and Ultrasound.* The content of lipids extracted after the biomass was submitted to ultrasound waves decreased significantly ($p \leq 0.05$) when exposure time increased from 5 min to 10 min, with no significant differences ($p \geq 0.05$) between 10 min and 20 min (Figure 1). Consequently, the best condition corresponded to the 5 min treatment of biomass by ultrasound, since total lipids reached 24.6 ± 1.3% and did not differ significantly ($p \geq 0.05$) from the values of the chemical disruption (24.2 ± 1.0%). On the other hand, the longest exposure times led to significantly different ($p \leq 0.05$) percentages of total lipids, in relation to the chemical disruption. Values decreased to 20.7 ± 1.7% (10 min) and 20.4 ± 0.3% (20 min).

The pretreatment using microwave for 300 s (Figure 1) resulted in the lipid content (16.3 ± 1.0%) differing significantly ($p \leq 0.05$) from 105 s (22.3 ± 1.6%) and 40 s (24.2 ± 0.9%), which differed ($p \leq 0.05$) neither from each other nor from the chemical disruption (24.2 ± 1%).

A comparison of both methods (Figure 1) led to the conclusion that they achieved the best results at the lowest exposure times, differing neither from each other (microwave for 40 s with 24.2 ± 0.9% lipids and ultrasound for 5 min with 24.6 ± 1.3% lipids) nor from the control (24.2 ± 1.0%). Decreases in the recovery of lipids from *Chaetoceros calcitrans* due to the increase in exposure times in both methods under evaluation may be associated with lipid degradation in drastic conditions of biomass treatment. According to Naghdi et al. [25], prolonged exposure to sonication can produce free radicals that may deteriorate the quality of lipids through oxidation, while Prommuak et al. [26] found that increased exposure time led to a decrease in lipid yield from *Chlorella vulgaris* and *Haematococcus pluvialis* caused by lipid oxidation. Furthermore, Kalil et al. [18] observed that treatments such as microwave and

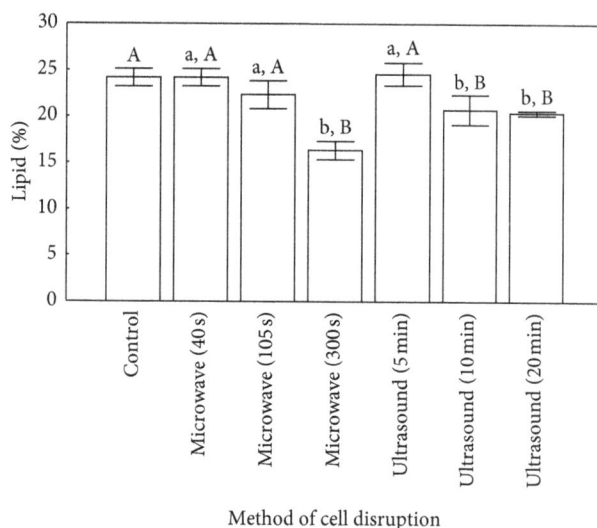

FIGURE 1: Lipid content of the microalga *Chaetoceros calcitrans* in different methods of cell disruption (mean values (bars) ± standard deviation (whiskers); $n = 3$). Equal small letters show that there was no significant difference ($p \geq 0.05$) between times of the same method of cell disruption. Equal capital letters show that there was no significant difference ($p \geq 0.05$) between methods of cell disruption in comparison with the method of chemical disruption.

ultrasound could lead to the thermal degradation of some microbial compounds by temperature increase.

According to Lee et al. [14], efficiency in lipid extraction from microalgae depends on the species and the method of extraction. Viswanathan et al. [13] studied the effects of three methods of cell disruption (autoclave, ultrasound, and high-pressure homogenization) to recover lipids from a consortium of microalgae (*Chlorella minutissima*, *Chlamydomonas globosa*, and *Scenedesmus bijuga*) and found an increase in the lipid content, from 10.78% to 12.22%, when ultrasound was applied, in comparison with the sample with no treatment.

Lee at al. [14] evaluated five methods of cell disruption (autoclave, bead milling, microwave, ultrasound, and osmotic shock with 10% NaCl) when they studied the microalgae *Botryococcus* spp., *Chlorella vulgaris*, and *Scenedesmus* spp. They found that the highest lipid contents were achieved by microwave. However, in the case of the microalga *Botryococcus* spp., ultrasound yielded low values of lipid content (8.8%), evidence that different microalgae have distinct behaviors when disruption techniques are applied because of differences in the constitution of their cell walls. Kurokawa et al. [27] established the most effective ultrasound frequency for the cell disruption of *Chaetoceros gracilis* (2.2 MHz), *Chaetoceros calcitrans* (3.3 MHz), and *Nannochloropsis* spp. (4.3 MHz) based on the rate of algal cell disruption measured by hemocytometry.

Ma et al. [16] compared cell disruption of the microalga *Chlorella* spp. by microwave and ultrasound and showed that the former was faster and more efficient in cell disruption aiming at lipid extraction than the latter. The authors attributed the results not only to the high pressure that was generated but also to fast heating and humidity inside the cell in the microwave process, whereas in the ultrasound

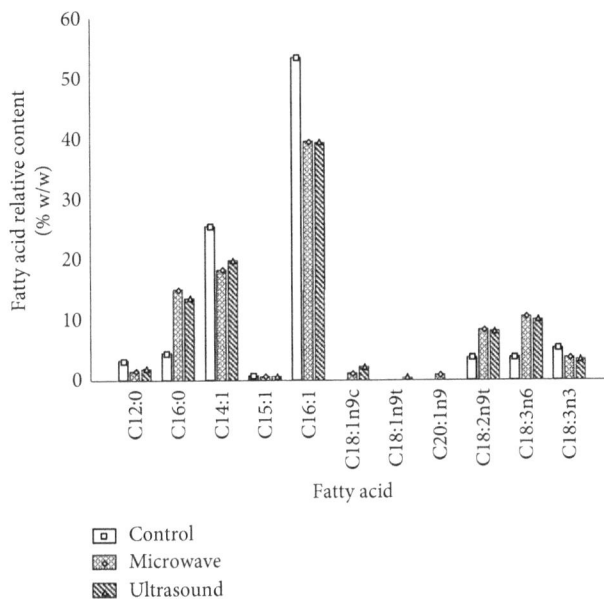

FIGURE 2: Fatty acid profiles (%) of *Chaetoceros calcitrans* in different methods of cell disruption. C12:0, lauric acid; C16:0, palmitic acid; C14:1, tetradecenoic acid; C15:1, pentadecenoic acid; C16:1, palmitoleic acid; C18:1n9c, oleic acid; C18:1n9t, elaidic acid; C20:1n9, eicosenoic acid; C18:2n9t, linoleic acid; C18:3n6, γ-linolenic acid; C18:3n3, α-linolenic acid.

TABLE 1: Energy consumption of experimental methods of cell disruption applied to the microalga *Chaetoceros calcitrans*.

| Method | Suspension volume (mL) | Concentration of cell suspension ($kg \cdot m^{-3}$) | Power of the equipment (W) | Time | Energy consumed by cell suspension ($GJ \cdot m^{-3}$) | Specific energy ($GJ \cdot kg^{-1}$) |
|---|---|---|---|---|---|---|
| Microwave | 20 | 0.15 | $1.4 \times 10^3$ | 40 (s) | 2.80 | 18.7 |
| Ultrasound | 20 | 0.15 | 130 | 5 (min) | 1.95 | 13.0 |

process, cells exploded due to the shock of bubbles formed by cavitation. In the work of Moura et al. [28], the microwave method was successfully applied to 4 different microalgae species in a shorter treatment time in comparison with the ultrasound method. This behaviour was similar to that in our results, since the same lipid extraction performance was observed for both methods. However, while the treatment time using the microwave method was 40 s, for ultrasound, it was 5 min.

Considering the above results, in relation to the best values for lipid recovery, the treatments with microwave for 40 s and ultrasound for 5 min were chosen to evaluate fatty acid profile and energy consumption.

*3.2. Fatty Acid Profiles.* Observation of fatty acids (Figure 2) shows that both treatments resulted in an increase in saturated fatty acids (SFAs) and polyunsaturated fatty acids (PUFAs). Concerning SFAs, palmitic acid had a 3.3-fold increase after disruption by microwave and a 3-fold increase after disruption by ultrasound, compared with the chemical disruption. Regarding PUFAs, linoleic acid (C18:2n9t) had approximately a 2.3-fold increase, whereas γ-linolenic acid (C18:3n6) had an increase of approximately 2.8-fold. In the case of α-linolenic acid (C18:3n3), the highest percentage was achieved by the chemical disruption. In terms of monounsaturated fatty acids (MUFAs), tetradecenoic acid (C14:1) was reduced by 28% when the biomass was

submitted to microwave and by 21.6% when it underwent ultrasound. Likewise, palmitoleic acid (C16:1) was reduced by 25.8% and 26.2%, respectively.

Li et al. [29], in their study of soybean oil extraction by ultrasound, found that a decrease in contents of unsaturated fatty acids and increase in contents of saturated fatty acids is used for evaluating the extent of oxidation because the former are more susceptible to oxidation, whereas the latter are more stable. This may explain the significant increase in contents of palmitic acid and the decrease in contents of palmitoleic acid that were found by this study when ultrasound and microwave were applied.

Ma et al. studied cell disruption in the microalga *Chlorella* spp. and found 1.3% and 6.6% increases in palmitic and linoleic acids, respectively, when treated by microwave. The authors also observed decreases of 10% and 3% in palmitoleic and α-linolenic acids, respectively, in disruption by microwave, whereas the use of ultrasound led to decreases of 7.7% and 8.4%, respectively. According to Pingret et al. [30], this behaviour may be attributed to lipid oxidation due to hydroxyl free radicals (OH) which are generated during exposure to ultrasound and microwave, since they cleave double bonds of unsaturated fatty acids as the result of their strong activity and oxidation capacity.

*3.3. Electric Energy Consumption.* Table 1 shows the estimated energy consumed by microwave (40 s) and ultrasound

(5 min) in cell disruption. There is an expressive difference in relation to energy consumption, which corresponds to an increase of approximately 50% when microwave is used instead of ultrasound (from 1.95 GJ·m³ to 2.80 GJ·m³, resp.). Halim et al. [11] studied cell disruption of the microalga *Chlorococcum* spp. by ultrasound. 200 mL cell suspension containing the biomass was sonicated at a maximum power of 130 W during 25 min, which represents an energy consumption of 0.975 GJ·m⁻³. Lee et al. [14] studied several methods of cell disruption in three species of microalgae. The microwave technique resulted in the highest lipid extraction, that is, 30% for *Botryococcus* spp., 10% for *Chlorella vulgaris*, and 10% for *Scenedesmus* spp. The energy consumed by microwave was 2.1 GJ·m⁻³ (100 mL cell suspension, 5 kg·m⁻³ cell concentration, 700 W power, and 5 min).

Both techniques are suitable for scale-up, require a single unit, and are environmentally friendly [12]. Despite the higher treatment time, the ultrasound technique was chosen because it required less energy than that of the microwave. However, further optimization of its process parameters is required to make the overall process economically sustainable, that is, concentration of cell suspension, in order to reduce the specific energy consumption.

## 4. Conclusion

Techniques of ultrasound (5 min) and microwave (40 s) applied to cell disruption of the microalga *Chaetoceros calcitrans* were efficient, since lipid content values of 24.6% and 24.2% w/w (dry basis) were obtained, respectively, with no significant differences ($p > 0.05$) in relation to the chemical disruption (24.2%). However, they used fewer chemicals and were less time-consuming. Regarding fatty acid profiles, both disruption techniques led to higher percentages of SFAs and PUFAs. Considering that the treatment of ultrasound consumed less energy than microwave, this technique is recommended for extracting lipids from *Chaetoceros calcitrans*.

## Conflicts of Interest

The authors declare that they have no conflicts of interest.

## Acknowledgments

The authors wish to thank the CAPES (Coordination of Superior Level Staff Improvement) and CNPq (National Council for Scientific and Technological Development) for the awarding of scholarships.

## References

[1] S. C. Foo, F. M. Yusoff, M. Ismail et al., "Production of fucoxanthin-rich fraction (FxRF) from a diatom, *Chaetoceros calcitrans* (Paulsen) Takano 1968," *Algal Research*, vol. 12, pp. 26–32, 2015.

[2] T.-S. Lin and J.-Y. Wu, "Effect of carbon sources on growth and lipid accumulation of newly isolated microalgae cultured under mixotrophic condition," *Bioresource Technology*, vol. 184, pp. 100–107, 2015.

[3] R. A. Lira, M. A. Martins, M. F. Machado, L. P. Corrêdo, and A. T. Matos, "As microalgas como alternativa à produção de biocombustíveis," *Revista Engenharia na Agricultura-REVENG*, vol. 20, no. 5, pp. 389–403, 2012.

[4] Y. Feng, C. Li, and D. Zhang, "Lipid production of *Chlorella vulgaris* cultured in artificial wastewater medium," *Bioresource Technology*, vol. 102, no. 1, pp. 101–105, 2011.

[5] S. Li, J. Xu, J. Chen, J. Chen, C. Zhou, and X. Yan, "The major lipid changes of some important diet microalgae during the entire growth phase," *Aquaculture*, vol. 428-429, pp. 104–110, 2014.

[6] N. Yodsuwan, S. Sawayama, and S. Sirisansaneeyakul, "Effect of nitrogen concentration on growth, lipid production and fatty acid profiles of the marine diatom *Phaeodactylum tricornutum*," *Agriculture and Natural Resources*, vol. 51, no. 3, pp. 190–197, 2017.

[7] S.-J. Lee, S. Go, G.-T. Jeong, and S.-K. Kim, "Oil production from five marine microalgae for the production of biodiesel," *Biotechnology and Bioprocess Engineering*, vol. 16, no. 3, pp. 561–566, 2011.

[8] P. Spolaore, C. Joannis-Cassan, E. Duran, and A. Isambert, "Commercial applications of microalgae," *Journal of Bioscience and Bioengineering*, vol. 101, no. 2, pp. 87–96, 2006.

[9] R. Kwangdinata, I. Raya, and M. Zakir, "Production of biodiesel from lipid of phytoplankton *Chaetoceros calcitrans* through ultrasonic method," *Scientific World Journal*, vol. 2014, Article ID 231361, 5 pages, 2014.

[10] M. R. Goulart, C. B. Silveira, M. L. Campos, J. A. Almeida, S. Manfredi-Coimbra, and A. F. Oliveira, "Methodology for the reused of the diatomite earth residue, originating from the filtration and clarification of the beer," *Química Nova*, vol. 34, no. 4, pp. 625–629, 2011.

[11] R. Halim, R. Harun, M. K. Danquah, and P. A. Webley, "Microalgal cell disruption for biofuel development," *Applied Energy*, vol. 91, no. 1, pp. 116–121, 2012.

[12] R. V. Kapoore, T. O. Butler, J. Pandhal, and S. Vaidyanathan, "Microwave-assisted extraction for microalgae: from biofuels to biorefinery," *Biology*, vol. 7, no. 1, p. 18, 2018.

[13] T. Viswanathan, S. Mani, K. C. Das et al., "Effect of cell rupturing methods on the drying characteristics and lipids compositions of microalgae," *Bioresource Techonology*, vol. 126, pp. 131–136, 2012.

[14] J.-Y. Lee, C. Yoo, S.-Y. Jun, C.-Y. Ahn, and H.-M. Oh, "Comparison of several methods for effective lipid extraction from microalgae," *Bioresource Technology*, vol. 101, no. 1, pp. S75–S77, 2010.

[15] G. S. Araújo, L. J. B. L. Matos, J. O. Fernandes et al., "Extraction of lipids from microalgae by ultrasound application: prospection of the optimal extraction method," *Ultrasonics Sonochemistry*, vol. 20, no. 1, pp. 95–98, 2013.

[16] Y.-A. Ma, Y.-M. Cheng, J.-W. Huang, J.-F. Jen, Y.-S. Huang, and C.-C. Yu, "Effects of ultrasonic and microwave pretreatments on lipid extraction of microalgae," *Bioprocess and Biosystems Engineering*, vol. 37, no. 8, pp. 1543–1549, 2014.

[17] J. M. B. Menéndez, A. Arenillas, J. A. M. Díaz et al., "Optimization of microalgae oil extraction under ultrasound and microwave irradiation," *Journal of Chemical Technology and Biotechnology*, vol. 89, no. 11, pp. 1779–1784, 2014.

[18] S. J. Kalil, C. C. Moraes, L. Sala, and C. A. V. Burkert, "Bio-product extraction from microalgal cells by conventional and nonconventional techniques," in *Handbook of Food Engineering-Volume 2-Food Bioconversion*, A. M. Grumezescu and A. M. Holban, Eds., pp. 179–206, Elsevier Inc., London, UK, 2017.

[19] G. Cravotto, L. Boffa, S. Mantegna, P. Perego, M. Avogadro, and P. Cintas, "Improved extraction of vegetable oils under high-intensity ultrasound and/or microwaves," *Ultrasonics Sonochemistry*, vol. 15, no. 5, pp. 898–902, 2008.

[20] P. R. Walne, "Experiments in the large scale culture of the larvae of *Ostrea edulis* L.," *Fishery Investigations*, vol. 25, no. 4, pp. 1–53, 1966.

[21] E. G. Bligh and W. J. Dyer, "A rapid method of total lipid extraction and purification," *Canadian Journal of Bio-chemistry and Physiology*, vol. 37, no. 8, pp. 911–917, 1959.

[22] P. Manirakiza, A. Covaci, and P. Schepens, "Comparative study on total lipid determination using Soxhlet, Roese-Gottlieb, Bligh & Dyer and modified Bligh & Dyer extraction methods," *Journal of Food Composition and Analysis*, vol. 14, no. 1, pp. 93–100, 2001.

[23] B. A. Alagöz, O. Yenigün, and A. Erdinçler, "Ultrasound assisted biogas production from co-digestion of wastewater sludges and agricultural wastes: comparison with microwave pre-treatment," *Ultrasonics Sonochemistry*, vol. 40, pp. 193–200, 2018.

[24] L. D. Metcalfe, A. A. Schmitz, and J. R. Pelka, "Rapid preparation of fatty acid esters from lipids for gas chromatography analysis," *Analytical Chemistry*, vol. 38, no. 3, pp. 514-515, 1966.

[25] F. G. Naghdi, L. M. G. González, W. Chan, and P. M. Schenk, "Progress on lipid extraction from wet algal biomass for biodiesel production," *Microbial Biotechnology*, vol. 9, no. 6, pp. 718–726, 2016.

[26] C. Prommuak, P. Pavasant, A. T. Quitain, M. Goto, and A. Shotipruk, "Microalgal lipid extraction and evaluation of single-step biodiesel production," *Engineering Journal*, vol. 16, no. 5, pp. 157–166, 2012.

[27] M. Kurokawa, P. M. King, X. Wu, E. M. Joyce, T. J. Mason, and K. Yamamoto, "Effect of sonication frequency on the disruption of algae," *Ultrasonics Sonochemistry*, vol. 31, pp. 157–162, 2016.

[28] R. R. Moura, B. J. Etges, E. O. Santos et al., "Microwave-assisted extraction of lipids from wet microalgae paste: a quick and efficient method," *European Journal of Lipid Science and Technology*, vol. 120, no. 7, article 1700419, 2018.

[29] H. Li, L. Pordesimo, and J. Weiss, "High intensity ultrasound-assisted extraction of oil from soybeans," *Food Research International*, vol. 37, no. 7, pp. 731–738, 2004.

[30] D. Pingret, G. Durand, A.-S. Fabiano-Tixier, A. Rockenbauer, C. Ginies, and F. Chemat, "Degradation of edible oil during food processing by ultrasound: electron paramagnetic resonance, physicochemical, and sensory appreciation," *Journal of Agricultural and Food Chemistry*, vol. 60, no. 31, pp. 7761–7768, 2012.

# 19

# Pre-Reducing Process Kinetics to Recover Metals from Nickel Leach Waste using Chelating Resins

**Amilton Barbosa Botelho Junior** ⓘ,[1] **David Bruce Dreisinger,**[2] **Denise Crocce Romano Espinosa,**[1] **and Jorge Alberto Soares Tenório**[1]

[1]*Department of Chemical Engineering, Polytechnic School of University of São Paulo, Sao Paulo, Brazil*
[2]*Department of Materials Engineering, The University of British Columbia, Vancouver, Canada*

Correspondence should be addressed to Amilton Barbosa Botelho Junior; amilton.barbosa20@gmail.com

Academic Editor: Eric Guibal

The main problem of the separation process from nickel mining using the ion exchange technique is the presence of iron, which precipitates in pH above 2.00 and causes coprecipitation of copper and cobalt. Chelating resins have the main advantage of being selected for a specific metal present in solution. Studies have been developed to increase the efficiency of metals recovery using chemical reduction and the ion exchange process to recover metals. The aim of this work was to use sodium sulfite as a reducing agent to convert Fe(III) to Fe(II). Chelating resins Lewatit® TP 207, selective for copper, and Lewatit® TP 220, selective for nickel and cobalt, were studied. Batch experiments were performed to study the effect of pH with and without sodium sulfite. Results indicated that the industrial process has increased efficiency when the reducing process is applied.

## 1. Introduction

Nickel laterite represents 70% of nickel reserves and 40% of nickel production, mostly processed by the hydrometallurgical process, due to the fact that the nickel laterite process is more expensive and difficult than other ores [1–6]. The main problem for metals recovery, such as nickel, copper, and cobalt, from these ores is the high concentration of iron. Limonite layer, the first layer of the nickel laterite ore and processed by high-pressure acid leaching or atmospheric acid leaching using sulfuric acid, has 40–50% of iron approximately, while nickel concentration is 0.8–1.5% and cobalt concentration is 0.1–0.2% [7].

In order to separate iron from nickel laterite leach, Jiménez Correaé et al. studied chemical precipitation of Fe (III) and Fe(II) in a solution of nickel laterite using hydroxides. Results obtained show that, at pH 2.50, 30% of iron and 20% of cobalt precipitate. At pH 3.00, 100% of iron, 60% of cobalt, and 20% of copper precipitate [8]. Another study realized by Chang et al. performed experiments to precipitate iron from nickel laterite leach by the oxidation process. Results show that there was loss of nickel to the residue with all iron. Nickel can be recovered from the residue using the weak acid solution, but more steps can turn the process impracticable [9].

Ion exchange process using chelating resins can be a solution to selectively recover metals. For this reason, Jiménez Correa et al. studied copper and nickel recovery using the chelating resin Dowex M4195 from nickel laterite leach waste. The chelating resin has the bis-picolylamine functional group, and results show that copper recovery was highly influenced by iron due to its high concentration ($150 \, mg \cdot L^{-1}$ of copper and $18000 \, mg \cdot L^{-1}$ of iron) [10]. Littlejohn and Vaughan [11] studied nickel and cobalt recovery using the chelating resin Lewatit® TP 220, with the same functional group of M4195, from nickel laterite tailings. Ferric iron was the most significant impurity adsorbed by resin, and results were obtained by Jiménez Correa et al. [10].

Zainol and Nicol studied five chelating resins with the iminodiacetate functional group to recover nickel and cobalt from nickel laterite leach tailings. The presence of iron, as well as chromium and aluminum, decreased resins efficiency, due to strong adsorption of these functional groups. In spite of all resins studied having the same functional group, results were different, and TP 207 MonoPlus had better results for nickel recovery [12]. Comparing metals recovery using Lewatit® TP 207 with the iminodiacetate functional group, copper recovery is higher than nickel in all pH values studied by Rudnicki et al. [13]. Metals recovery, both copper and nickel, increased when pH increases [13].

According to Littlejohn and Vaughan and Mendes and Martins, chelating resins with the iminodiacetate functional group are better to recover copper, and resins with the bispicolylamine functional group were more selective for nickel and cobalt than the others metals [14, 15].

In spite of chelating resins used to selectively recover metals, iron still is a problem to overcome. Botelho Junior et al. studied the difference between ferric iron and ferrous iron in nickel laterite leach waste for copper recovery using Lewatit® TP 207, in which chelating resin efficiency increases when iron is present as ferrous iron. In solution with Fe(III), copper recovery was 48.72%, while solution with Fe (II) copper recovery was 61.32% [16–18]. A way to convert ferric iron to ferrous iron is using a reducing agent. Sodium dithionite and sodium metabisulfite were studied to convert ferric iron to ferrous iron of nickel laterite leach waste reducing the potential of the solution until 590 mV (pH 0.50–2.00) and 240 mV (pH 2.50–3.50). However, the problem is these reducing agents are dangerous in acid solutions, which can release hydrogen sulfide ($H_2S$), and these reducing agents can be added if dissolved in water before. The other problem is they are expensive [19–21].

Other reducing agents that can be used such as sodium thiosulfate or microorganismos have the same problem in acid solutions [22–26]. Nevertheless, sodium sulfite is a reducing agent that can be used in acid solutions. Liu et al. studied reductive stripping of ferric iron using sulfuric acid and sodium sulfite [27]. Luo et al. studied atmospheric leaching of nickel limonite with sulfuric acid using sodium sulfite as a reducing agent to facilitate nickel extraction, comparing the leaching process using only sulfuric acid. Results indicated that nickel extraction increases in presence of sodium sulfite, as well as iron extraction. Though the nickel extraction increased, the increase of the iron extraction still keeps the problem [28].

The goal of this work was to study the batch industrial process to recover copper, nickel, and cobalt using the ion exchange technique. Chelating resins Lewatit® TP 207, selective for copper with the iminodiacetate functional group, and Lewatit® TP 220, selective for nickel and cobalt with bispicolylamine functional group, were studied. Sodium sulfite was used to reduce the potential of solution, converting ferric iron to ferrous iron, while the reducing process was studied before using sodium dithionite and sodium metabisulfite [19, 20, 29]. Three synthetic solutions were prepared to simulate real conditions of nickel laterite leach waste from the limonite ore. Solution 1 was used to study copper recovery, Solution 2 without copper was used to study nickel recovery; and Solution 3 without copper and nickel was used to study cobalt recovery. Effect of pH was studied between

pH 0.50 and 2.00, for solutions without sodium sulfite, and between pH 0.50 and 3.50, with sodium sulfite. Experiments were performed in a batch at 25°C and 150 rpm. Samples were analyzed using ICP-OES (Varian 725ES).

## 2. Materials and Methods

*2.1. Materials.* Composition of synthetics solutions is present in Table 1. Sulfate salts of each metal were dissolved in deionized water, and the pH was adjusted with sulfuric acid concentrated PA. Three different solutions were prepared: Solution 1 was prepared with all metals that compose the nickel laterite leach, Solution 2 was prepared without copper, and Solution 3 was prepared without copper and nickel. Therefore, it is a possible study effect of other metals presented in the leach solution for copper, nickel, and cobalt adsorption in three steps.

Two different chelating resins were studied: Lewatit® TP 207, for experiments performed using Solution 1, and Lewatit® TP 220, for experiments performed using Solutions 2 and 3. TP 207 is a cationic resin with the iminodiacetate functional group, crosslinked polystyrene macroporous matrix, pH range 0–14 and density $1.17 g·mL^{-1}$ [30]. The theoretical selectivity order for this resin is Fe(III) > Cu(II) > Ni(II) > Zn(II) > Fe(II) > Mn(II) > Mg(II) [13]. TP 220 is also a cationic resin, but with the bis-picolylamine functional group, crosslinked polystyrene macroporous matrix, density $1.1 g·mL^{-1}$. The theoretical selectivity order for this resin is Cu(II) > Ni(II) > Fe(III) > Co(II) > Mn(II) > K(I) > Ca(II) > Na(I) > Mg(II) > Al(III) [11, 31].

### 2.2. Method

*2.2.1. Pre-Reducing Process.* Botelho Junior et.al. studied the reducing process using sodium dithionite and sodium metabisulfite to convert Fe(III) to Fe(II) from the synthetic solution of nickel laterite leach. Effect of time, pH, and temperature was studied. Temperature decreased ferric iron chemical conversion, probably due to the reducing agent composition. The effect of time was studied until 96 hours, and after 120 min, the reaction reaches equilibrium, until 480 min, and then decreased from 95% to 80% after 1144 min. Conversion of Fe(III) to Fe(II) in 100% was performed decreasing the potential until 590 mV, at pH 0.50–2.00, and 240 mV, at pH 2.50–3.50. For synthetic solution with all metals of nickel laterite leach, ferric iron conversion is 80% [19, 20, 29].

Thus, the pre-reducing process was performed during 120 min adding sodium sulfite in order to decrease the potential until 590 mV *SHE* (standard hydrogen electrode), pH 0.50–2.00, and 240 mV *SHE*, pH 2.50–3.50, at 25°C in 150 rpm.

*2.2.2. Ion Exchange Experiments.* Ion exchange experiments were performed in flasks of 100 mL with 50 mL of solution and 0.50 mL of resin in 150 rpm at 25°C during 120 min. The effect of time was studied without the pre-reducing process between 30 and 480 min at pH 0.50 for Solution 1, with

TABLE 1: Metals concentration for batch experiments for Solution 1, Solution 2, and Solution 3 in mg·L$^{-1}$.

| — | | Al | Co | Cu | Cr | Fe | Mg | Mn | Ni | Zn |
|---|---|---|---|---|---|---|---|---|---|---|
| | | | | | | Metals | | | | |
| Conc. (mg.L$^{-1}$) | Solution 1 | 4101 | 78 | 146 | 195 | 18713 | 7774 | 387 | 2434 | 36 |
| | Solution 2 | 4101 | 78 | — | 195 | 18713 | 7774 | 387 | 2434 | 36 |
| | Solution 3 | 4101 | 78 | — | 195 | 18713 | 7774 | 387 | — | 36 |

Lewatit® TP 207, and Solution 2, with Lewatit® TP 220. The effect of pH was studied between pH 0.50 and 3.50.

Resins were washed using hydrochloric acid 6 mol·L$^{-1}$ and sodium hydroxide 1 mol·L$^{-1}$ for three times using water deionized between each step. Sulfuric acid 1 mol·L$^{-1}$ was used after wash. Sulfuric acid concentrated PA and sodium hydroxide pellet were used to correct the pH [32].

In order to quantify the cations adsorbed, Equation (1) was used, where $q_t$ is the capacity of the ion adsorbed in time $t$ in mass of the ion per mass of resin (mg·g$^{-1}$), $C_0 \cdot e \cdot C_t$ are concentrations of ions in time = 0 and time = $t$ (mg·L$^{-1}$), $v$ is the volume of solution ($L$), and $m$ is the mass of the resin (g) [13, 33]. Equation (2) was used to quantify the coefficient of distribution of ion, which one is the measure of the effectiveness of ion adsorption from solution [32]. Equation (3) was used to quantify the adsorbed ion, in percentage:

$$q_t = (C_0 - C_t) \times \frac{v}{m}, \tag{1}$$

$$Kd = \frac{q_e}{C_t} \times 1000, \tag{2}$$

$$\%S = \frac{(C_0 - C_t)}{C_0} \times 100\%. \tag{3}$$

The pH was measured using electrode Ag/AgCl (Sensoglass), and electrode ORP (oxidation reduction potential) was used to measure the potential in solution. Samples were analyzed using ICP-EOS (Varian 725ES Optical Emission Spectrometer).

# 3. Results and Discussion

Figure 1 presents Pourbaix's diagram constructed using Hydra-Medusa software in experimental conditions for the Fe-S-H$_2$O system, where the conversion of Fe(III) to Fe(II) is 100% after 590 mV between pH 0.50 and 2.00, and it is after 240 mV between pH 2.50 and 3.50. However, presence of the other metals decreases the reducing process efficiency. In acid medium with sulfuric acid, sodium sulfite reacts with H$^+$ to form H$_2$SO$_3$, as presented in Equation (4), in which $k_1 = 1.54 \times 10^{-2}$ and $k_2 = 1.02 \times 10^{-7}$. Besides, sodium dithionite and sodium sulfite are different, and both dissociate to form sodium bisulfite, which is the main responsible for the reducing process in solution, but sodium dithionite also dissociates to form sodium thiosulfate [27, 34–37].

$$SO_{2(aq)} + H_2 \rightleftharpoons H^+ + HSO_3^- \rightleftharpoons H^+ + SO_3^{-2} \tag{4}$$

Another problem about sodium dithionite is that acid medium can be dissociated to form sulphur and hydrogen sulfide, being the last extremely dangerous [38]. Besides,

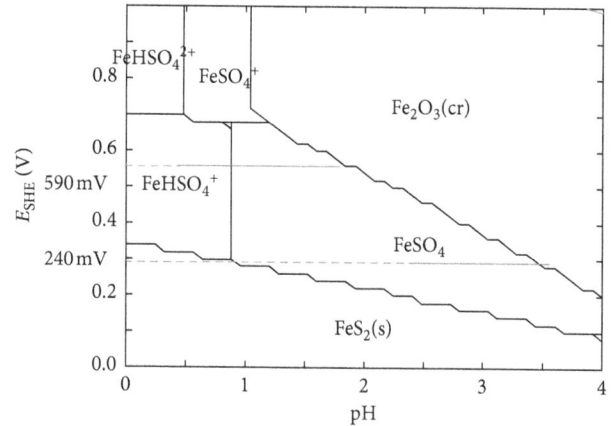

FIGURE 1: Pourbaix's diagram of Fe-S-H$_2$O constructed using Hydra-Medusa software at 25°C.

sodium dithionite, sodium thiosulfate, and sodium bisulfate are dangerous because of same problem in the acid medium, using them only in the basic medium [23, 39].

*3.1. Effect of Time.* The effect of time was studied using Solution 1 and Solution 2 without the pre-reducing process at 25°C and pH 0.50. Results for copper (Solution 1) and nickel (Solution 2) recovery are present in Figure 2 and indicate that the reaction reached equilibrium after 120 min, in which Solution 1 was in contact of Lewatit® TP 207 and Solution 2 with Lewatit® TP 220. Iron was the metal highest adsorbed in mg per g of resin (151 mg·g$^{-1}$), while copper was 1.67 mg·g$^{-1}$, due to high concentration of H$^+$ in solution at pH 0.50. For Solution 2, iron was also in this case the highest metal adsorbed (85 mg·g$^{-1}$), while nickel was 10.87 mg·g$^{-1}$. Experiments to study the effect of pH in the ion exchange process were performed during 120 min.

*3.2. Effect of pH.* The effect of pH in chelating resin with the iminodiacetate functional group can be seen in Figure 3. At pH 2, high concentration of H$^+$ in the functional group repulses cations in solution due to protonation of the functional group, where high competition between H$^+$ and cations occurs by the functional group. At pH 2–4, H$^+$ and cations present in the solution still compete for the chelating resin functional group, the latter being deprotonated. In Figure 3, at pH 7, carboxylic acid of the functional group is deprotonated, and at pH 12, the iminodiacetate functional group is totally deprotonated. However, although the last situation is the most favorable to recover cations due to no presence of H$^+$ in the functional group, metals in general

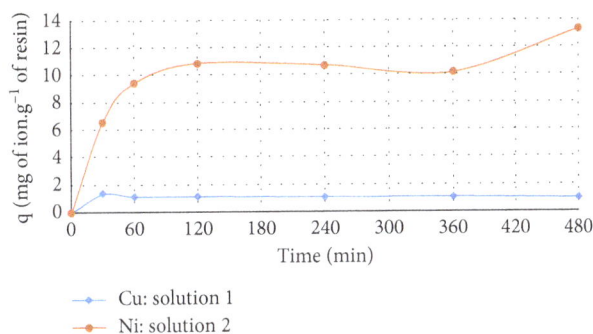

FIGURE 2: Results of effect of time of Solution 1 and Solution 2.

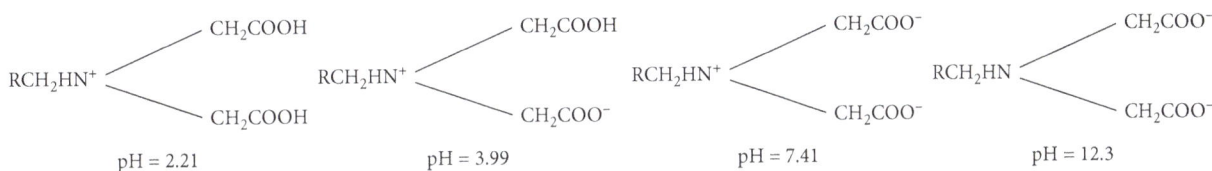

FIGURE 3: Effect of pH in the iminodiacetate functional group [40].

precipitate at pH 5, which causes working pH to be totally dependent of solution characteristics [12, 13, 40].

Results for metals recovery with and without the pre-reducing process are present below. Effect of pH without the pre-reducing process above pH 2.00 was not studied, because from this pH, iron will precipitate with copper and cobalt [8, 41]. Figure 4 presents results for copper recovery in Solution 1 by the iminodiacetate resin Lewatit® TP 207. It is observed that copper recovery increased for Solution 1 with the pre-reducing process. At pH 0.50, copper recovery was 9.66% without the pre-reducing process, while using sodium sulfite to convert Fe(III) to Fe(II), copper recovery at pH 0.50 was 16.67%. At pH 2.00, both had highest copper recovery. In solution without sodium sulfite, copper recovery was 41.43%, and with sodium sulfite as a reducing agent, copper recovery was 68.57%.

Botelho Junior et al. studied the effect of presence of Fe (III) and Fe(II) to recover copper from nickel laterite leach using resin with the iminodiacetate functional group. Results show that when iron is Fe(II), copper recovery is higher than when Fe(III) is present in solution [18]. This can occur because the iminodiacetate functional group has high affinity for Fe(III) [15, 42–44], once it was the highest metal recovery among all in mg·L$^{-1}$ in all pH studied.

Figure 5 shows results of metals recovery of Solution 2 with Lewatit® TP 220, chelating resin with the bis-picolylamine functional group. At pH 0.50–2.00, chelating resin was more selective for nickel than other metals in both solution, except for solution without the pre-reducing process at pH 0.50. At pH 2.50 and 3.00, in solution with the pre-reducing process, cobalt was the metal more selective by the resin, followed by zinc and nickel. At pH 3.50, however, chelating resin was more selective for zinc, followed by nickel and cobalt. This phenomenon was also seen in Solution 1 experiments, where chelating resins selective

order changes for different pH values and also with the pre-reducing process. The difference in the selective order for solution with and without the reducing process can be explained due to conversion of Fe(III) to Fe(II).

Chelating resin for Solution 2 with the pre-reducing process between pH 0.50 and 2.00 was more selective for nickel than cobalt and zinc. At pH 2.50 and 3.00, the resin was more selective for cobalt, and at pH 3.50, chelating resin was more selective for nickel and zinc, simultaneously.

The effect of pH for cobalt and zinc recovery from Solution 3 by the bis-picolylamine resin is shown in Figure 6. Chelating resin has high selective for cobalt between pH 0.50 and 2.00 in both situations. However, at pH 2.50 and 3.00, zinc was more selective by the resin than cobalt, and both were almost not recovered by the resin at pH 3.50. The change in the order of selectivity was also observed. Figure 7 presents coefficient distribution of copper, nickel, cobalt, and zinc in Solutions 1, 2, and 3 with the pre-reducing process.

It is possible seen that copper coefficient distribution, in Solution 1, was maximum (186 mL·g$^{-1}$) at pH 2.00, indicating that the chelating resin Lewatit® TP 207 with the iminodiacetate functional group was more selective for copper in this pH. In Solution 2, nickel coefficient distribution was maximum at pH 3.50 (121 mL·g$^{-1}$); however, at the same pH, zinc had higher coefficient distribution (160 mL·g$^{-1}$) than nickel, indicating Lewatit® TP 220 with the bis-picolylamine functional group was more selective for zinc than nickel. In the meantime, cobalt had higher coefficient distribution at pH 3.50 (61 mL·g$^{-1}$) than the other pH values, which is seen in Figure 5, where copper recovery was maximum at pH 3.5 as well as nickel and zinc. For Solution 2, the chelating resin was more selective for nickel at pH 2.00 (43 mL·g$^{-1}$ for nickel and 15 mL·g$^{-1}$ for zinc and cobalt), where nickel recovery was 32.55% and zinc and

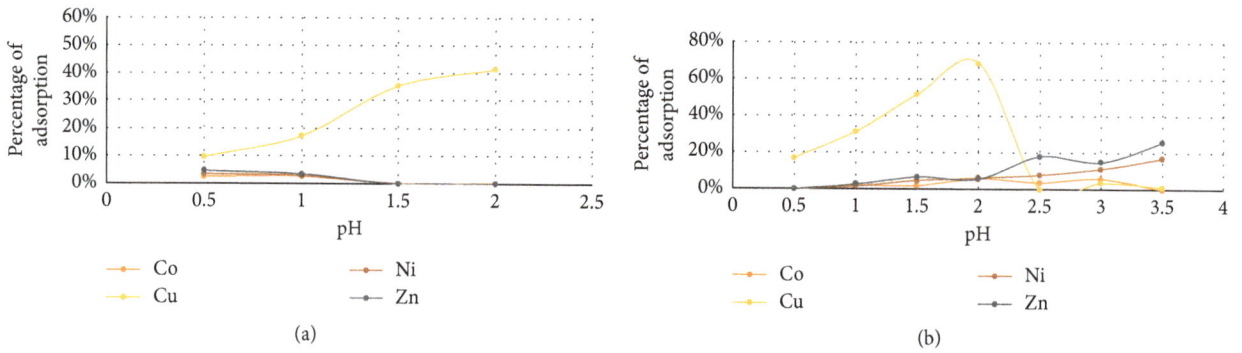

FIGURE 4: Effect of pH in Solution 1 by iminodiacetate resin (a) without and (b) with the pre-reducing process.

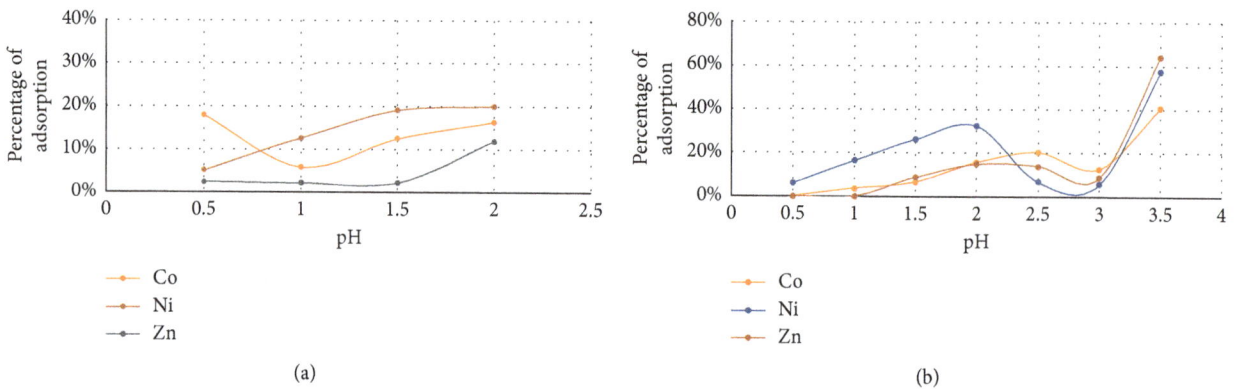

FIGURE 5: Effect of pH in Solution 2 by the bis-picolylamine resin (a) without and (b) with the pre-reducing process.

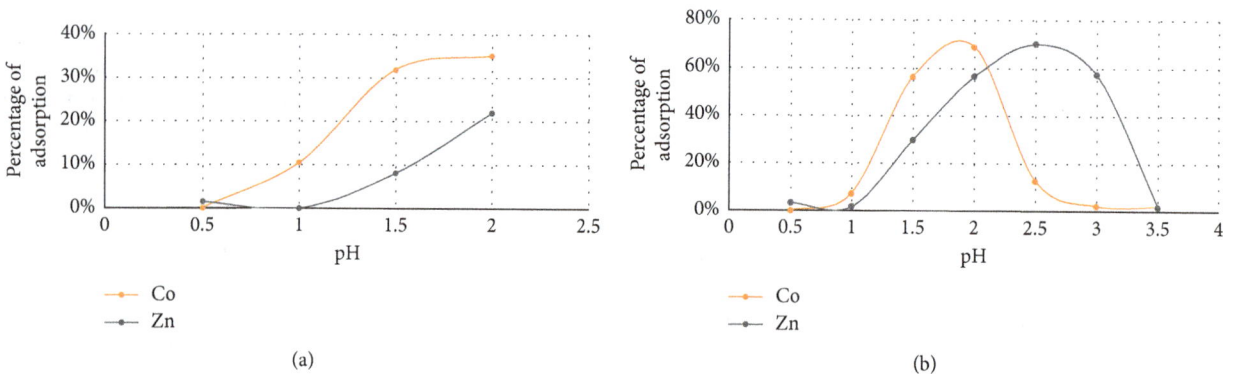

FIGURE 6: Effect of pH in Solution 3 by the bis-picolylamine resin (a) without and (b) with the pre-reducing process.

cobalt, 14.86%. For Solution 3, cobalt coefficient distribution was maximum at pH 2.00 (198 mL·g$^{-1}$), being resin more selective for this metal than others, and at pH 2.50, zinc coefficient distribution was maximum (206 mL·g$^{-1}$).

Studies to recover metals from nickel laterite leach using chelating resins indicated that resins with the iminodiacetate functional group are better to recover copper, while in order to recover nickel and cobalt resins with bis-picolylamine are better [45, 46]. In experiments performed using Solution 1, Lewatit® TP 207 was more selective for copper, while experiments performed with Solutions 2 and 3 using Lewatit®

TP 220 cobalt and nickel were selectively recovered. Zinc was also selectively recovered depending on the pH value.

## 4. Conclusion

The aim of this work was to study the batch industrial process for metals recovery using two different chelating resins from synthetic solution of nickel laterite leach. Sodium sulfite was used in order to reduce Fe(III) to Fe(II). Results indicated that Lewatit® TP 207 was more selective for copper than the other metals, due to its functional group,

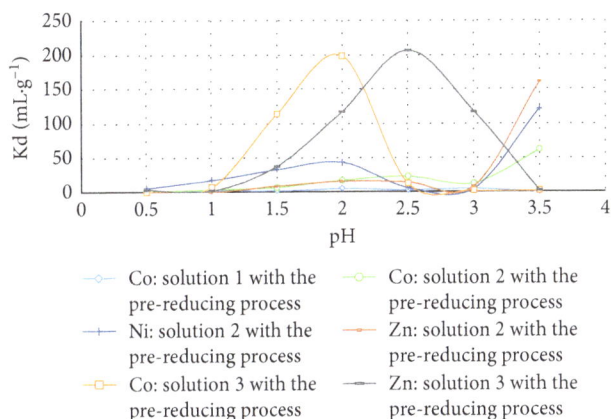

FIGURE 7: Coefficient distribution of copper, nickel, cobalt, and zinc with the pre-reducing process.

and Lewatit® TP 220 was more selective for cobalt and nickel. Sodium sulfite increased metals recovery because chelating resins were less selective for ferrous iron than ferric iron, and pH can be increased without ferric iron precipitation, due to the fact that, while increasing pH, the concentration of $H^+$ decreases, as well as competition between $H^+$ and metals in solution for the functional group of chelating resin. Another reason than can be explain the metals recovery raise is that ferrous iron occupies less active sites on the chelating resin than ferric iron. In all the solutions studied, metals recovery was higher after the pre-reducing process. A change in the selectivity order of resins was observed comparing with and without the pre-reducing process, which may be caused by conversion of Fe(III) to Fe(III) and also by changing the pH. Pre-reducing process using sodium dithionite and sodium metabisulfite was studied before, but the use of sodium sulfite as a cheap and secure option can make the process economically viable and secure. Industrial process can be a benefit for the process involving chemical reducing and ion exchange process, in which metals recovery increases comparatively without the reducing process. Column experiments are the next step to simulate fixed-bed reactors for the continuum process.

## Conflicts of Interest

The authors declare that there are no conflicts of interest regarding the publication of this paper.

## Acknowledgments

The authors acknowledge the São Paulo Research Foundation (FAPESP) (nos. 2012/51871-9, 2016/05527-5, and 2017/06563-8) and CAPES for financial support. They also acknowledge the University of British Columbia and University of São Paulo.

## References

[1] F. K. Crundwell, M. S. Moats, V. Ramachandran, T. G. Robinson, and W. G. Davenport, *Extractive Metallurgy of Nickel, Cobalt and Platinum-Group Metals*, Elsevier, Oxford, UK, 2011, http://linkinghub.elsevier.com/retrieve/pii/B9780080968094100012.

[2] G. M. Mudd, "Global trends and environmental issues in nickel mining: sulfides versus laterites," *Ore Geology Reviews*, vol. 38, no. 1-2, pp. 9–26, 2010.

[3] G. M. Mudd and S. M. Jowitt, "A detailed assessment of global nickel resource trends and endowments," *Economic Geology*, vol. 109, no. 7, pp. 1813–1841, 2014.

[4] A. D. Dalvi, W. G. Bacon, and R. C. Osborne, "The past and the future of nickel laterites," in *Proceedings of PDAC 2004 International Convention*, pp. 1–27, Toronto, Canada, January 2004.

[5] T. F. Torries, "Comparative costs of nickel sulphides and laterites," *Resources Policy*, vol. 21, no. 3, pp. 179–187, 1995.

[6] G. M. Mudd, "Nickel sulfide versus laterite: the hard sustainability challenge remains," in *Proceedings of 48th Annual Conference of Metallurgists*, pp. 1–10, Ontario, Canada, August 2009.

[7] A. Oxley and N. Barcza, "Hydro-pyro integration in the processing of nickel laterites," *Minerals Engineering*, vol. 54, pp. 2–13, 2013.

[8] M. M. Jiménez Correa, P. Aliprandini, J. A. S. Tenório, and D. C. R. Espinosa, "Precipitation of metals from liquor obtained in nickel mining," in *REWAS 2016: Towards Materials Resource Sustainability*, R. E. Kirchain, Ed., pp. 333–338, Springer, Cham, Switzerland, 2016.

[9] Y. Chang, X. Zhai, B. Li, and Y. Fu, "Removal of iron from acidic leach liquor of lateritic nickel ore by goethite precipitate," *Hydrometallurgy*, vol. 101, no. 1-2, pp. 84–87, 2010.

[10] M. M. Jiménez Correa, P. Aliprandini, F. P. C. Silvas, J. A. S. Tenório, D. Dreisinger, and D. C. R. Espinosa, "Nickel and copper adsorption from acidic sulfate medium by ion exchange," in *Proceedings of Conference of Metallurgists hosting World Gold and Nickel Cobalt*, Canadian Institute of Mining, Metallurgy and Petroleum, Vancouver, Canada, August 2017.

[11] P. Littlejohn and J. Vaughan, "Recovery of nickel and cobalt from laterite leach tailings through resin-in-pulp scavenging and selective ammoniacal elution," *Minerals Engineering*, vol. 54, pp. 14–20, 2013.

[12] Z. Zainol and M. J. Nicol, "Comparative study of chelating ion exchange resins for the recovery of nickel and cobalt from laterite leach tailings," *Hydrometallurgy*, vol. 96, no. 4, pp. 283–287, 2009.

[13] P. Rudnicki, Z. Hubicki, and D. Kołodyńska, "Evaluation of heavy metal ions removal from acidic waste water streams," *Chemical Engineering Journal*, vol. 252, pp. 362–373, 2014.

[14] P. Littlejohn and J. Vaughan, "Selectivity of commercial and novel mixed functionality cation exchange resins in mildly acidic sulfate and mixed sulfate–chloride solution," *Hydrometallurgy*, vol. 121–124, pp. 90–99, 2012.

[15] F. D. Mendes and A. H. Martins, "Selective sorption of nickel and cobalt from sulphate solutions using chelating resins," *International Journal of Mineral Processing*, vol. 74, no. 1–4, pp. 359–371, 2004.

[16] D. Kołodyńska, W. Sofińska-Chmiel, E. Mendyk, and Z. Hubicki, "DOWEX M 4195 and LEWATIT®MonoPlus TP 220 in heavy metal ions removal from acidic streams," *Separation Science and Technology*, vol. 49, no. 13, pp. 2003–2015, 2014.

[17] D. S. Stefan and I. Meghea, "Mechanism of simultaneous removal of $Ca^{2+}$, $Ni^{2+}$, $Pb^{2+}$ and $Al^{3+}$ ions from aqueous solutions using Purolite® S930 ion exchange resin," *Comptes Rendus Chimie*, vol. 17, no. 5, pp. 496–502, 2014.

[18] A. B. Botelho Junior, M. M. Jiménez Correa, D. C. R. Espinosa, and J. A. S. Tenório, "Influência do Fe(III) no lixiviado de rejeito de níquel no processo de troca-iônica," *Tecnologia em Metalurgia, Materiais e Mineração*, p. 10, 2018.

[19] A. B. Botelho Junior, M. M. Jimenez, D. C. R. Espinosa, and J. A. S. Tenório, "Redução química de Fe(III) em resíduo de mineração de níquel para recuperação de metais utilizando resinas de troca-iônica," in *Proceedings of 22° Congresso Brasileiro de Engenharia e Ciência dos Materiais*, pp. 4543–4553, Natal, Brazil, November 2016.

[20] A. B. Botelho Junior, M. M. Jiménez Correa, D. C. R. Espinosa, and J. A. S. Tenório, "Chemical reduction of Fe(III) in nickel lateritic wastewater to recover metals by ion exchange," in *The Minerals, Metals & Materials Series*, pp. 467–472, Springer, Berlin, Germany, 2017.

[21] A. B. Botelho Junior, M. M. Jiménez Correa, D. C. R. Espinosa, and J. A. S. Tenório, "Precipitação seletiva de cobre a partir de resíduo de mineração de níquel," in *Proceedings of Conference: XXVII Encontro Nacional de Tratamento de Minérios e Metalurgia Extrativa*, pp. 1–5, Belém, Brazil, November 2017.

[22] I. S. Chung and Y. Y. Lee, "Effect of oxygen and redox potential on d-xylose fermentation by non-growing cells of *Pachysolen tannophilus*," *Enzyme and Microbial Technology*, vol. 8, no. 8, pp. 503–507, 1986.

[23] K. Wejman-Gibas, T. Chmielewski, K. Borowsi, K. Gibas, M. Jeziorek, and J. Wodka, "Thiosulfate leaching of silver from a solid residue after pressure leaching of industrial copper sulfides flotation," *Physicochemical Problems of Mineral Processing*, vol. 51, no. 2, pp. 601–610, 2015.

[24] D. M. Puente-Siller, J. C. Fuentes-Aceituno, and F. Nava-Alonso, "A kinetic–thermodynamic study of silver leaching in thiosulfate–copper–ammonia–EDTA solutions," *Hydrometallurgy*, vol. 134–135, pp. 124–131, 2013.

[25] D. Zipperian, S. Raghavan, and J. P. Wilson, "Gold and silver extraction by ammoniacal thiosulfate leaching from a rhyolite ore," *Hydrometallurgy*, vol. 19, no. 3, pp. 361–375, 1988.

[26] C. Cameselle, M. José Núñez, J. M. Lema, and J. Pais, "Leaching of iron from kaolins by a spent fermentation liquor: Influence of temperature, pH, agitation and citric acid concentration," *Journal of Industrial Microbiology*, vol. 14, no. 3-4, pp. 288–292, 1995.

[27] Y. Liu, S. H. Nam, and M. Lee, "Stripping of Fe(III) from the loaded mixture of D2EHPA and TBP with sulfuric acid containing reducing agents," *Bulletin of the Korean Chemical Society*, vol. 35, no. 7, pp. 2109–2113, 2014.

[28] J. Luo, G. Li, M. Rao, Z. Peng, Y. Zhang, and T. Jiang, "Atmospheric leaching characteristics of nickel and iron in limonitic laterite with sulfuric acid in the presence of sodium sulfite," *Minerals Engineering*, vol. 78, pp. 38–44, 2015.

[29] A. B. Botelho Junior, M. M. Jiménez Correa, D. C. R. Espinosa, and J. A. S. Tenório, "Study of reducing process of iron in leachate from nickel mining waste," *Brazilian Journal of Chemical Engineering*, 2018.

[30] Lanxess, *Product Information-Lewatit® TP 207*, Lanxess, Cologne, Germany, 2011, http://www.lenntech.com/Datasheets/Lewatit-TP-207-L.pdf.

[31] Lanxess, Product Information-ewatit®, *MonoPlus TP*, Vol. 220, Lanxess, Cologne, Germany, 2011.

[32] M. L. Inamuddin, *Ion Exchange Technology I*, Springer, New York, NY, USA, 1st edition, 2012.

[33] Z. Yu, T. Qi, J. Qu, L. Wang, and J. Chu, "Removal of Ca(II) and Mg(II) from potassium chromate solution on Amberlite IRC 748 synthetic resin by ion exchange," *Journal of Hazardous Materials*, vol. 167, no. 1–3, pp. 406–412, 2009.

[34] B. Jung, R. Sivasubramanian, B. Batchelor, and A. Abdel-Wahab, "Chlorate reduction by dithionite/UV advanced reduction process," *International Journal of Environmental Science and Technology*, vol. 14, no. 5, pp. 123–134, 2017.

[35] M. Wayman and W. J. Lem, "Decomposition of aqueous dithionite. Part II. A reaction mechanism for the decomposition of aqueous sodium dithionite," *Canadian Journal of Chemistry*, vol. 48, no. 5, pp. 782–787, 1970.

[36] V. Cermak and M. Smutek, "Mechanism of decomposition of dithionite in aqueous solutions," *Collection of Czechoslovak Chemical Communications*, vol. 40, no. 11, pp. 3241–3264, 1975.

[37] C. W. Li, J. H. Yu, Y. M. Liang et al., "Ni removal from aqueous solutions by chemical reduction: impact of pH and pe in the presence of citrate," *Journal of Hazardous Materials*, vol. 320, pp. 521–528, 2016.

[38] F. A. N. G. Da Silva, F. M. D. S. Garrido, M. E. Medeiros et al., "Alvejamento químico de caulins brasileiros: efeito do potencial eletroquímico da polpa e do ajuste do pH," *Química Nova*, vol. 34, no. 2, pp. 262–267, 2011.

[39] J. L. Deutsch, *Fundamental Aspects of Thiosulfate Leaching of Silver Sulfide in the Presence of Additives*, University of British Columbia, Vancouver, Canada, 2012.

[40] Z. Zainol and M. J. Nicol, "Ion-exchange equilibria of $Ni^{2+}$, $Co^{2+}$, $Mn^{2+}$ and $Mg^{2+}$ with iminodiacetic acid chelating resin Amberlite IRC 748," *Hydrometallurgy*, vol. 99, no. 3-4, pp. 175–180, 2009.

[41] P. Aliprandini, M. Correa, A. Santanilla, J. Tenório, and D. C. R. Espinosa, "Precipitation of metals from synthetic laterite nickel liquor by NaOH," in *Proceedings of 8th International Seminar on Process Hydrometallurgy*, pp. 1–8, Santiago, Chile, June 2016.

[42] F. D. Mendes and A. H. Martins, "Recovery of nickel and cobalt from acid leach pulp by ion exchange using chelating resin," *Minerals Engineering*, vol. 18, no. 9, pp. 945–954, 2005.

[43] F. D. Mendes and A. H. Martins, "Selective nickel and cobalt uptake from pressure sulfuric acid leach solutions using column resin sorption," *International Journal of Mineral Processing*, vol. 77, no. 1, pp. 53–63, 2005.

[44] Z. Hubicki and D. Kołodyńska, "Selective removal of heavy metal ions from waters and waste waters using ion exchange methods," in *Ion Exchange Technologies*, IntechOpen Limited, London, UK, pp. 193–240, 2012.

[45] Z. Hubicki, M. Geca, and D. Kołodyńska, "The effect of the presence of metatartaric acid on removal effectiveness of heavy metal ions on chelating ion exchangers," *Environmental Technology*, vol. 32, no. 8, pp. 805–816, 2011.

[46] V. I. Kuz'Min and D. V. Kuz'Min, "Sorption of nickel and copper from leach pulps of low-grade sulfide ores using purolite S930 chelating resin," *Hydrometallurgy*, vol. 141, pp. 76–81, 2014.

# 20

# Synthesis of NiO/Y$_2$O$_3$/ZrO$_2$ Catalysts Prepared by One-Step Polymerization Method and their use in the Syngas Production from Methane

**Yvan J. O. Asencios** ⓘ[1] **and Elisabete M. Assaf**[2]

[1]*Instituto do Mar, Universidade Federal de São Paulo, 11030-100 Santos, SP, Brazil*
[2]*Instituto de Química de São Carlos, Universidade de São Paulo, 13560-970 São Carlos, SP, Brazil*

Correspondence should be addressed to Yvan J. O. Asencios; yvan.jesus@unifesp.br

Academic Editor: Bhaskar Kulkarni

In this work, the production of Syngas (H$_2$/CO) from oxidative reforming of methane (ORM) and partial oxidation of methane (POM) over NiO/Y$_2$O$_3$/ZrO$_2$ catalysts was studied. The nickel concentration was varied (ranging from 0 to 40 wt.%) aiming to optimize the performance in ORM and POM reactions; these reactions were carried out at 750°C and 1 atm for 6 hours. The catalysts were prepared by the one-step polymerization method (OSP) and characterized by different techniques. This method led to production of materials of smaller crystallite size than others of similar composition prepared under other methods; the catalysts presented good nickel dispersion, well-defined crystalline structure, and well-defined geometrical morphology. Additionally, the OSP method was advantageous because it was carried out in a single calcination step. The catalyst containing 20% wt. of nickel (20Ni20YZ sample) showed the highest methane conversion, high selectivity to H$_2$ and CO, low carbon deposition rates, and, curiously, the best geometric morphology. The results of this paper also demonstrated that the nickel concentration in the mixture strongly influenced the morphology of the catalysts; therefore, the morphology also influenced the catalytic performances during the Syngas production reactions.

## 1. Introduction

Methane is found mainly in biogas, natural gas, and shale gas. Biogas is a renewable alternative source of methane and is considered a first-generation biofuel. The production of the biogas occurs in the absence of oxygen where anaerobic bacteria break down the organic matter, producing methane and carbon dioxide as major products, and other gaseous by-products such as H$_2$S, NH$_3$, and H$_2$, in smaller amounts [1, 2].

Syngas (synthesis gas, a mixture of H$_2$/CO) is a raw material of high value which is used as the starting material in the production of synthetic fuels such as di-methyl ether (DME), methanol, and liquid hydrocarbons (by the Fischer–Tropsch process). Biogas can be transformed into Syngas through reforming reactions as reported in the literature [2–7]. Thus, the biogas can be transformed by dry

reforming of methane (DRM, reaction (1)) and partial oxidation of methane (POM, reaction (2)), resulting in the oxidative reform of methane (ORM, reaction (3)):

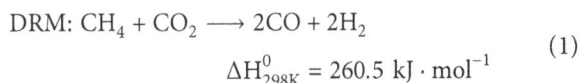

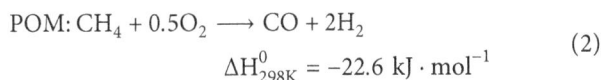

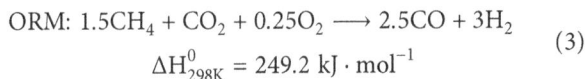

$$\text{DRM: } CH_4 + CO_2 \longrightarrow 2CO + 2H_2$$
$$\Delta H^0_{298K} = 260.5 \text{ kJ} \cdot \text{mol}^{-1} \quad (1)$$

$$\text{POM: } CH_4 + 0.5O_2 \longrightarrow CO + 2H_2$$
$$\Delta H^0_{298K} = -22.6 \text{ kJ} \cdot \text{mol}^{-1} \quad (2)$$

$$\text{ORM: } 1.5CH_4 + CO_2 + 0.25O_2 \longrightarrow 2.5CO + 3H_2$$
$$\Delta H^0_{298K} = 249.2 \text{ kJ} \cdot \text{mol}^{-1} \quad (3)$$

Carbon dioxide and methane are the two principal components of biogas, and the main greenhouse gases, so its transformation into Syngas may have very beneficial results from an environmental point of view. It is known that catalysts based on noble metals corresponding to the Group VIIIB are good catalysts for reforming methane (e.g., Pt, Rh, Pd, and Ru), demonstrating high selectivity for hydrogen, high methane conversion, and low carbon deposition rates (carbon is an undesired product). The catalysts based on nonnoble metals of Group VIIIB (Ni, Co, and Fe) have shown catalytic activity in methane reforming reactions, similarly to that of noble metals based catalysts, and they are low cost; however, nonnoble metals result in high carbon deposition rates [6, 8, 9].

The actual Syngas production by industry (for hydrogen generation) is based on the steam reforming of methane (SRM, reaction (4)), which uses a nickel catalyst supported on alumina; however, nickel catalysts present problems due to carbon deposits which deactivate the catalyst (covering the active metallic sites) and moreover, its accumulation increases the reactor pressure, leading to explosion hazards. Many efforts have been directed to decrease carbon deposits [10, 11].

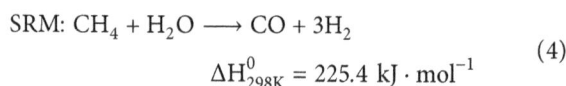

$$\text{SRM: } CH_4 + H_2O \longrightarrow CO + 3H_2$$
$$\Delta H^0_{298K} = 225.4 \text{ kJ} \cdot \text{mol}^{-1} \quad (4)$$

Several ways have been proposed for the formation of carbon deposits in methane reforming reactions; among them, the methane molecule cracking (reaction (5)) and the Boudouard reaction (reaction (6)) can lead to catalyst deactivation:

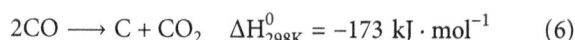

$$CH_4 \longrightarrow C + 2H_2 \quad \Delta H^0_{298K} = 75 \text{ kJ} \cdot \text{mol}^{-1} \quad (5)$$

$$2CO \longrightarrow C + CO_2 \quad \Delta H^0_{298K} = -173 \text{ kJ} \cdot \text{mol}^{-1} \quad (6)$$

The water-gas shift reaction (WGSR) is a reversible exothermic reaction, and the opposite direction is favored at high temperatures. So the reverse reaction (RWGSR, reaction (7)) is very likely to occur during methane reforming and POM reaction:

$$CO_2 + H_2 \longleftrightarrow CO + H_2O \quad \Delta H^0_{298K} = 41 \text{ kJ} \cdot \text{mol}^{-1} \quad (7)$$

The following solutions have been proposed to minimize the carbon deposits: (a) new catalyst preparation method aiming to decrease the crystallite size of the catalytic phase [7, 12, 13], (b) the use of catalytic supports based on solid solutions with the formation of oxygen vacancies [5, 6, 9, 14], (c) the use of hydrotalcites as catalyst precursor [15], and (d) the addition of noble-metal promoters to the catalysts [16, 17], among others.

Regarding to preparation methods of catalysts, Mori et al. [18] prepared NiO/YSZ catalysts (where YSZ means $Y_2O_3$-$ZrO_2$ solid solution) by impregnation method; the NiO content of these catalysts was 60% wt. They reported that the average crystallite size of NiO in their NiO/YSZ catalyst was <60 nm (calculated from XRD analysis). In the same report, the NiO size for the same mixture, but prepared by the solid-state reaction method, was 7 $\mu$m. Resini et al. [19] also prepared the NiO/YSZ catalyst by impregnation method (the nickel content was 50% wt.), and reported the formation of

heterogeneous morphology, as the former NiO particles showed large sizes (crystallites larger than 40 nm, according to TEM analysis), and YSZ reached approximately 5 nm.

Similarly, Bellido et al. [20] prepared the NiO/YSZ catalysts, and $Y_2O_3$-$ZrO_2$ support was prepared by polymerization method (Y/Z molar ratio of 12%), and the nickel was incorporated by the impregnation method, and the nickel content was 5% wt. According to them, the average crystallite size of NiO of these catalysts (calculated from the Scherrer equation) was 25.9 nm.

In our previous reports [21, 22], we reported that the one-step polymerization method improved the catalytic activity of $Co_3O_4/CeO_2$ and NiO/MgO-$ZrO_2$ catalysts in the steam reforming of bioethanol and in the oxidative reform of methane, respectively. The average crystallite sizes of NiO were smaller than those reported in [18–20], where the precipitation method was employed.

The polymerization method forms a homogeneous material with a fine dispersion of the components of the catalysts (owing to the atomistic distribution of the metal cations during polymerization) [20–22]. In view of the referenced reports, the objective of the present paper was to study the NiO/$Y_2O_3$/$ZrO_2$ mixture containing various nickel contents (ranging from 0 to 40 wt.%) in the $Y_2O_3$/$ZrO_2$ support (Y/Z ratio of constant composition) synthesized by the one-step polymerization method (OSP method). The NiO/$Y_2O_3$/$ZrO_2$ catalysts were characterized and tested in the oxidative reform of methane (ORM) and in the partial oxidation of methane (POM). In this paper, we report the optimal nickel concentration (wt.%) for a good catalytic performance in the ORM and POM reactions, and additionally, we found that (under OSP method) the nickel concentration led to catalysts with different morphologies (clear and well defined forms). Among other factors, such as crystalline structure, nickel charge, metallic dispersion, and surface area; the morphology of each catalyst influenced the catalytic performance in the reactions studied.

## 2. Methodology

*2.1. Preparation of Catalysts.* The catalyst preparation method in this study was the one-step polymerization method. In this method, the ability of the organic hydroxycarboxylic acids (citric acid in this case) to chelate with most cations was used. Ethylene-glycol (a polyhydric alcohol), when added to the chelate with adequate heating, leads to the formation of a polyester due to successive condensation reactions between the alcohol and the acid chelate. The polymerization method has been used by several researchers due to the high homogeneity of the material produced, compared to precipitation, coprecipitation, and impregnation methods [20, 21, 23]. In this sense, the present study used the one-step polymerization method, where the support precursor and the catalytic phase were mixed together (in a single step).

The catalysts were prepared by the one-step polymerization method using $Ni(NO_3)_2 \cdot 6H_2O$, $Zr(CO_3)_2 \cdot 1 \cdot 5H_2O$, $Y(NO_3)_3 \cdot 6H_2O$, citric acid, and ethylene-glycol, in accordance with the method described in the literature [5, 20, 22]. The salts of the catalyst precursor were previously dissolved (carbonate in nitric acid and nitrates in water). In this method, the precursor salts were dissolved, respectively, and mixed together forming a single solution. This solution was added to the mixture composed of ethylene-glycol and citric acid in the proportion of 1 mol of Zr per 3 mol of citric acid and a mass ratio of 60 : 40 between the citric acid and ethylene-glycol. The resulting solution produced a translucent resin. The polymerization process occurred at 120°C for 12 hours. After obtaining the polymers, they were subjected to calcination in two consecutive stages: first, under air flow at 500°C (5°C·min$^{-1}$) for 3 h and second, under the same air flow but at 750°C (5°C·min$^{-1}$) for 2 h. The resulting materials were powdered in a porcelain mortar and pestle.

The $Y_2O_3$ content was kept constant at 20 mol% relative to $ZrO_2$. The nickel content was varied as follows: 0%, 10%, 20%, and 40% relative to the total weight of the catalyst. The catalysts were named 20YZ, 10Ni20YZ, 20Ni20YZ, and 40Ni20YZ according to the percentage of Ni in the total catalyst weight. The catalyst 20YZ corresponds to the pure catalytic support (without nickel, Ni 0%).

*2.2. Characterization of the Catalysts.* The crystalline phases were identified by X-ray diffraction analysis (XRD) in a Rigaku Multiflex diffractometer (30 kV, 10 mA), in the $2\theta = 5$–$80°$ range and speed 2°min$^{-1}$ using Cu-K $\alpha$ radiation ($\lambda = 1.5406$ Å), and the diffraction patterns were identified by comparison with the database of the International Centre for Diffraction Data (JCPDS). The average crystallite sizes were determined from XRD line-broadening measurements using the Scherrer equation [24]: $d = k \cdot \lambda / (\beta_{hkl} \cdot \cos\theta)$, where $d$ is the average crystallite size, $k$ is the shape factor, taken as 0.89, $\lambda$ is the wavelength of CuK$\alpha$ radiation, $\beta_{hkl}$ is the full width at half maximum (FWHM) of the particular peak, and $\theta$ is Bragg's angle.

In the temperature programmed reduction analyzes (TPR), 100 mg of catalysts were used, a gas mixture of 1.96% $H_2$/Ar with a flow of 30 mL·min$^{-1}$ and a heating rate of 5°C min$^{-1}$ up to a temperature of 1000°C. The surface area measurements were performed on a Quantachrome Nova 1200 equipment, and the results of nitrogen adsorption were treated according to the BET method.

The analysis by SEM-EDX was particularly suitable for the study of the morphology of solids. For this analysis, a small amount of the fresh catalyst was placed in isopropyl alcohol to form a suspension. The suspension was dropped slowly onto an aluminum plate to obtain maximum dispersion of the powder in the sample holder. The apparatus used consists of LEO-440 electron microscope with Oxford detector operating at 20 kV electron beam. For each analysis, the samples were sputter-coated with gold.

The *in situ* XRD analyzes were carried out using a diffractometer (Huber) in the Brazilian Synchrotron Light Laboratory (LNLS) in Campinas, Brazil, in the D10B-XPD light-line. The wavelength used for each analysis was 1.5406 Å. Each analysis explored the region of $2\theta = 25$–$70°$. The catalyst was placed in a sample holder in a temperature programmable oven. In this system, the catalysts were placed

in contact with the respective flow gas: $N_2$ under ambient conditions and $H_2$ reduction condition, so that the first XRD pattern was obtained under ambient temperature in a $N_2$ stream (100 mL·min$^{-1}$) and the second was collected on the reduced catalyst with $H_2$ (5% $H_2$/He, 50 mL·min$^{-1}$) at a temperature of 800°C.

*2.3. Catalytic Tests.* The catalytic tests were carried out in a fixed-bed down-flow reactor (internal diameter = 10 mm) with 100 mg of catalyst. Before each reaction, the catalysts were reduced at 800°C for 1 h, under flowing $H_2$ (30 mL·min$^{-1}$). The sample was then brought to the reaction temperature (750°C for all reactions conditions) under pure $N_2$ flow.

The catalysts were tested under two conditions:

(a) Oxidative reforming of methane (ORM): the feed was a mixture of gases (60% $CH_4$ and 40% $CO_2$) and synthetic air ($O_2$: 21%, $N_2$: 79%) reaching a molar ratio of $1.5CH_4 : 1CO_2 : 0.25O_2$, giving a total flow of 107 mL·min$^{-1}$, inside the reactor. The conversion of $CH_4$ and $CO_2$ was calculated, respectively, as

$$R_{conversion} (\%) = \frac{R_{in} - R_{out}}{R_{in}} \times 100, \qquad (8)$$

where $R$ is the molar flow rate (mol·min$^{-1}$) of $CH_4$ or $CO_2$. The selectivity was calculated as

$$selectivity_{R_i} = \frac{R_{i\ produced}}{R_{CH_4\ converted} + R_{CO_2\ converted}} \times 100, \qquad (9)$$

where "$R_i$" is the molar flow rate (mol·min$^{-1}$) of the product ($H_2$ or CO).

(b) Partial oxidation of methane (POM): the feed was a mixture of gases in molar proportion of $2CH_4 : 1O_2$, stoichiometric for the POM, giving a total flow of 107.3 mL·h$^{-1}$. Oxygen was added as synthetic air (79% $N_2$, 21% $O_2$). The $CH_4$ conversion was calculated as

$$R_{conversion} (\%) = \frac{R_{in} - R_{out}}{R_{in}} \times 100, \qquad (10)$$

where $R$ is the molar flow rate (mol·min$^{-1}$) of $CH_4$. The selectivity was calculated as

$$selectivity_{R_i} = \frac{R_{i\ produced}}{R_{CH_4\ converted}} \times 100, \qquad (11)$$

where "$R_i$" is the molar flow rate (mol·min$^{-1}$) of the product ($H_2$ or CO or $CO_2$).

Carbon deposition was determined as the apparent gain in mass of the catalyst after each reaction (mmol·C·h$^{-1}$).

The gaseous reactants and products were analyzed on a gas chromatograph (Varian, Model 3800), in line with the catalytic unit test. The chromatograph includes two thermal conductivity detectors and 13X molecular sieve packed columns (this column uses $N_2$ as carrier for analysis of $H_2$

gas) and Porapak N (this column uses He as the carrier for the analysis of $CO_2$ gas, $CH_4$, and CO). The reaction temperature was controlled and measured by a thermocouple inserted into the top of the catalyst bed.

## 3. Results and Discussion

*3.1. Characterization of the Catalysts.* The XRD patterns of the catalysts were recorded *in situ* under ambient conditions and under reduction conditions in the presence of $H_2$ at 800°C, these results are shown in Figures 1 and 2, respectively. Supplementary Figure 1S shows the XRD patterns of the fresh support (20YZ) and pure $ZrO_2$ (prepared under the same conditions as the 20YZ support); in this figure, one can observe that pure $ZrO_2$ presents peaks attributed to the tetragonal (T) and monoclinic (M) phase of zirconia; the $Y_2O_3/ZrO_2$ mixture in the 20YZ support led to the stabilization of the zirconia tetragonal phase. The tetragonal phase of zirconia is stabilized due to the formation of the $Y_2O_3$-$ZrO_2$ solid solution in the mixture; in this solid solution, $Y^{3+}$ ions dissolved in the $ZrO_2$ network lead to the formation of oxygen vacancies for charge compensation effects, between $Y^{3+}$ and $Zr^{4+}$ (to maintain net neutrality in the network) [5, 9, 20].

Figure 1 presents the XRD patterns for the fresh catalysts under ambient conditions, and the average crystallite sizes (using the Scherrer equation [24]) are shown in Table 1. As expected, the higher nickel content in the catalysts favored the formation of larger NiO crystallites. Contrarily, the average crystallite sizes of zirconia decreased for catalysts with higher nickel content, suggesting that high nickel content (which formed larger NiO crystallites) hinders the $ZrO_2$ crystal growth. The YZ mixture is constant in the composition of each catalyst. In the OSP method studied in the present paper, the precursors of Ni, Y, and Zr were dissolved before the polymerization process. During the thermal treatment of the polymers performed to obtain the solid oxides with the corresponding crystalline phase, solid reactions, combustion reactions, and crystal growth took place. Therefore, we suggest that the presence of high Ni content in the catalysts hinders the crystal growth of $ZrO_2$ during the thermal treatment. This is similar to the finding by Denrya et al. [25] and Trusova et al. [26] who studied in detail the effect of thermal treatment on the crystal growth of solid oxides.

This might also explain the continuous decrease of the oxide catalyst surface area (Table 1) when the nickel content was increased.

The surface area values of each material are shown in Table 1. The surface area decreased in the following order: 10Ni20YZ $(30 \, m^2 \cdot g^{-1})$ > 20Ni20YZ $(17 \, m^2 \cdot g^{-1})$ > 40Ni20YZ $(13 \, m^2 \cdot g^{-1})$. In this context, the NiO crystallites of large size may cover the surface pores of the support 20YZ $(39 \, m^2 \cdot g^{-1})$ leading to a continued decrease of surface area in each catalyst.

In Figure 2, the reduction of NiO to $Ni°$ can be observed; the catalysts with higher nickel content led to the formation of larger $Ni°$ crystallites (Table 1). It is also observed that the $Ni°$ particles tend to sinter under reduction conditions

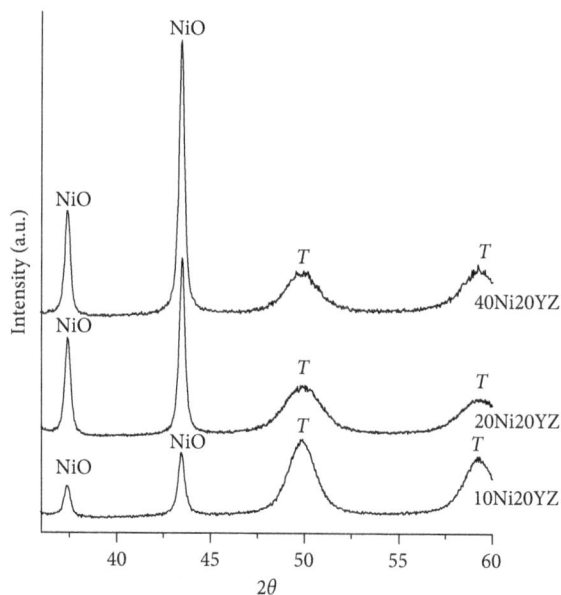

FIGURE 1: In situ XRD patterns of catalysts at room temperature (in the $2\theta = 35$–$60°$ range, speed $2° \, min^{-1}$ using Cu $K\alpha$ radiation ($\lambda = 1.5406$ Å), the diffraction patterns were identified by comparison with the database of the International Centre for Diffraction Data—JCPDS).

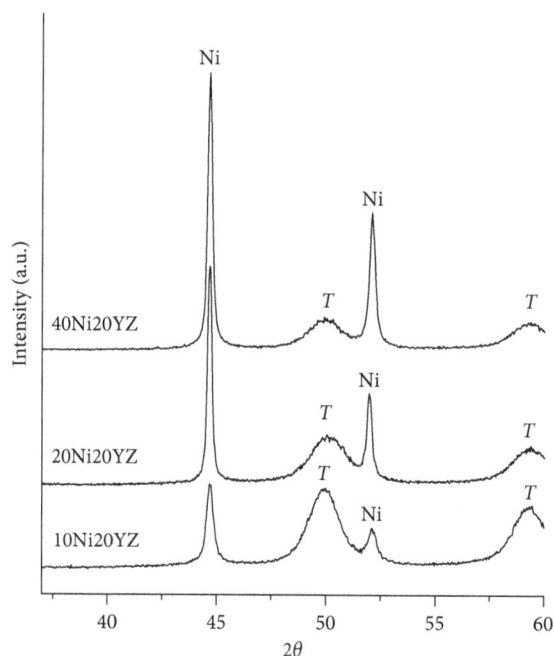

FIGURE 2: In situ XRD patterns of reduced catalysts. The XRD pattern was collected on the reduced catalyst with $H_2$ (5% $H_2$/He, $50 \, mL \cdot min^{-1}$) at a temperature of 800°C; the diffraction patterns were identified by comparison with the database of the International Centre for Diffraction Data—JCPDS.

(800°C) and form larger crystallites than their corresponding oxides (NiO) for each catalyst. The increase in crystallite size in the 10Ni20YZ catalyst shows that at this nickel concentration, the sintering is poor (when compared with the other samples with higher nickel concentration).

TABLE 1: Physical-chemical properties of the catalysts and TPR analyses results.

| Sample | Average crystallite size (nm) | | | % reduction of NiO (TPR) | Surface area (m$^2$·g$^{-1}$) | Metallic dispersion ($D_M$) |
|---|---|---|---|---|---|---|
| | ZrO$_2$ | NiO | Ni$^\circ$ | | | |
| 20YZ | 7.2 | — | — | — | 39 | — |
| 10Ni20YZ | 5.1 | 14.8 | 15.0 | 50 | 30 | 6.7 |
| 20Ni20YZ | 4.9 | 19.0 | 25.8 | 67 | 17 | 3.9 |
| 40Ni20YZ | 4.3 | 27.6 | 33.1 | 81 | 13 | 3.0 |

The smallest crystallite size of Ni$^\circ$ and NiO was obtained in sample 10Ni20YZ (15 nm).

Mori et al. [18] reported that the average crystallite size of NiO in their NiO/YSZ catalyst (where YSZ means Y$_2$O$_3$-ZrO$_2$ solid solution) was <60 nm (calculated from XRD analysis); this catalyst was obtained by the impregnation method. This NiO size for the same mixture, but prepared by the solid-state reaction method, was 7 $\mu$m. The NiO content of these catalysts was 60% wt.

Resini et al. [19] also studied the NiO/YSZ catalyst and reported the formation of heterogeneous morphology, as the former NiO particles showed large sizes (crystallites larger than 40 nm, according to TEM analysis), and YSZ reached approximately 5 nm. This catalyst was prepared by the impregnation method, and the nickel content was 50% wt.

Bellido et al. [20] reported the average crystallite size (calculated from the Scherrer equation) of NiO of 25.9 nm, for NiO/YSZ catalyst, Y$_2$O$_3$-ZrO$_2$ support was prepared by the polymerization method (Y/Z molar ratio of 12%), and the nickel was incorporated by the impregnation method, the nickel content was 5% wt.

Considering the average crystallite size of NiO reported in our study (Table 1) and compared with that obtained by Mori et al. [18], Resini et al. [19], and Bellido et al. [20]; we suggest that the OSP method produced smaller crystallite size than that reached by the referred authors. Additionally, the OSP method was very practical, since the three Ni, Zr, and Y oxides were obtained in a single step, where only one calcination procedure was needed.

Similarly, Youn et al. [27] obtained a NiO/YSZ catalyst, where the catalytic support was obtained by the templating sol-gel method, and nickel was incorporated by the incipient wetness impregnation method. The nickel concentration used was 20% wt. The Ni/ZrO$_2$ catalyst formed Ni$^\circ$ with an average crystallite size of 22 nm (calculated by the Scherrer equation); however after adding Y$_2$O$_3$ to reach the Y/Z molar ratio of 0.2 (similar molar ratio as employed in our paper for YZ support), the Ni$^\circ$ crystallite size reached 15 nm and the Y$_2$O$_3$-ZrO$_2$ support reached 6.4 nm.

Although Youn et al. [27] obtained smaller crystallite sizes than those obtained by our OSP method, the crystallite size values shown in Table 1 are relatively close to those of the referred study; and moreover, we prepared the catalysts using a single calcination step, where support and catalytic phase were obtained together (the referenced authors used a two-step calcination).

The estimated metal dispersion ($D_M$) calculated from the Ni$^\circ$ crystallites size [24, 28–30] is found in Table 1. According

to these values, as expected, the 10Ni20YZ sample has the highest dispersion Ni$^\circ$, probably owing to the low nickel concentration in the catalyst composition and due to the OSP method employed. The lowest metallic dispersion was found for the sample with the highest nickel content (40Ni20YZ).

According to the TPR profiles of the catalysts shown in Figure 3, a single reduction peak is observed below 550°C which was attributed to reduction of NiO in different interactions with the support.

In the 10Ni20YZ profile sample, the major part of the single NiO peak (located at interval 350°C–550°C) corresponds to NiO species weakly interacting with the YZ support. Furthermore, it is noted that increasing the nickel content produced the extension of this single peak to higher temperatures, suggesting the increase of NiO species that go to the bulk part and the increase of the interaction with the YZ support. This is reasonable, since Ni$^{2+}$ and Y$^{3+}$ can form the NiO-Y$_2$O$_3$ solid solution at 750°C in the presence of air, as was described in our previous report [5]. The two peaks located above 550°C correspond to the reduction of surface oxygen atoms located in the close vicinity of the oxygen vacancies of the support 20YZ [18, 22, 27].

Figures 4–6 show SEM images of the fresh catalysts; it was also possible to map the Ni and Zr elements for each SEM image by energy-dispersive X-ray spectroscopy (EDX-mapping for Ni and Zr are shown in Supplementary Figures S2–S4). Figure 4 shows that the 10Ni20YZ catalyst morphology presents irregularly shaped particles (agglomerate species) that are repeated on the surface. According to the mapping on this region of the sample (Supplementary Figure S2), the major composition of these agglomerates is Zr (the Y mapping image was very similar to Zr). Additionally, according to the Ni mapping of 10Ni20YZ catalysts (Supplementary Figure S2), the small agglomerates with diameter size ranging from 1 to 3 $\mu$m (some indicated with green circles in Figure 4) are mainly composed of Ni. These analyses suggest that the low nickel content (10%) led to a good dispersion of NiO particles on the surface of the solid solution Y$_2$O$_3$-ZrO$_2$ (YZ support).

Figure 5 shows the image of the 20Ni20YZ catalyst obtained by SEM analysis. The morphology of this catalyst shows the formation of a solid in the cubical form. According to the mapping in this region (Supplementary Figure S3), the cube-form solid is primarily formed by Zr (the Y mapping image was very similar to Zr), and the small spheres around 3 $\mu$m in diameter (well-formed) on the surface of the cube are composed principally of Ni.

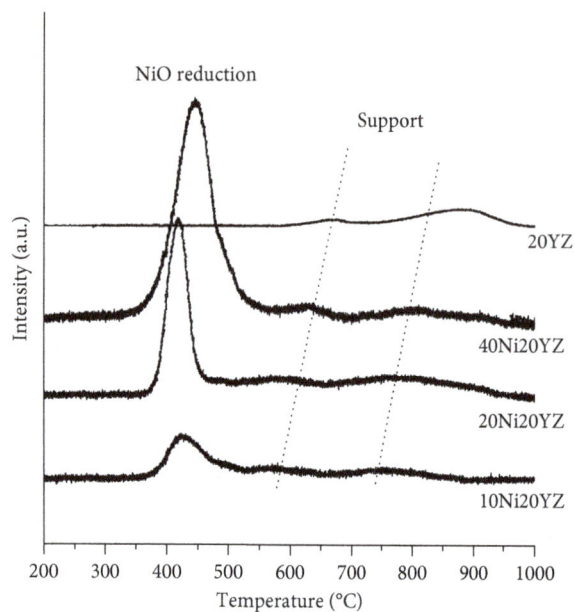

FIGURE 3: Results for temperature programmed reduction with $H_2$ ($H_2$-TPR) of the catalysts. In the $H_2$-TPR analyzes 100 mg of catalysts, a gas mixture of 1.96% $H_2$/Ar (flow at 30 mL·min$^{-1}$) and a heating rate of 5°C min$^{-1}$ (from room temperature to 1000°C) were used.

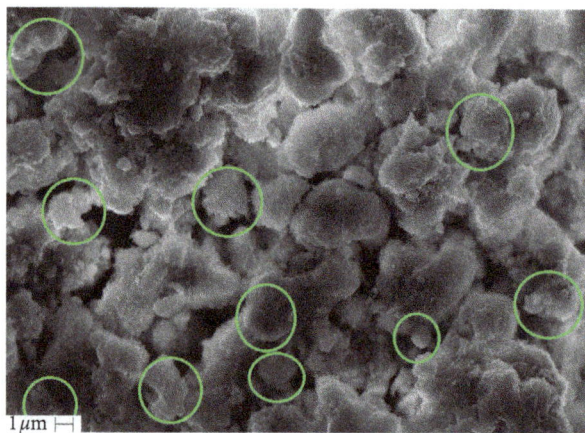

FIGURE 5: SEM image (×3000) of 20Ni20YZ catalyst before reaction.

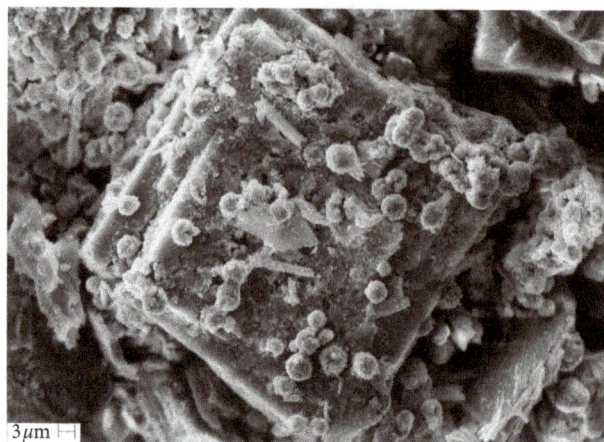

FIGURE 6: SEM image (×3000) of 40Ni20YZ catalyst before reaction.

FIGURE 4: SEM image (×10000) of 10Ni20YZ catalyst before reaction.

This analysis indicates that the 20% Ni content is a suitable concentration for the NiO/$Y_2O_3$/$ZrO_2$ mixture, as it resulted in an optimal arrangement of the catalytic phase on the support, where the NiO particles are in good dispersion on the surface of the $Y_2O_3$-$ZrO_2$ solid solution formed by 20YZ.

Figure 6 shows the SEM image of the 40Ni20YZ catalyst. In this figure, the formation of two cube-shaped solids (named YZ in the figure) with agglomerated particles in their close vicinity, in almost the entire region of the image shown, can be clearly observed. According to the mapping in this region (Supplementary Figure S4), the two cube-shaped solids (YZ) are principally composed of Zr (the image of Y mapping was very similar to that of Zr), and the agglomerate particles in their vicinity are principally composed of Ni

(named as NiO in the Figure 6). These results suggest that the large amount of nickel in the 40Ni20YZ catalyst composition led to the formation of many agglomerated NiO particles, thus resulting in poor dispersion of the active phase on the catalyst support.

The NiO particle dispersion located in different regions of each catalyst, presented in the SEM images, followed the same trend as that found for the estimated metallic dispersion (Table 1; after reduction of the catalysts with $H_2$); in this context, 10Ni20YZ and 40Ni20YZ presented the highest and the lowest nickel dispersion (resp.) among all catalysts.

### 3.2. Catalytic Tests

*3.2.1. Oxidative Reform of Methane (ORM).* The $CH_4$ and $CO_2$ conversion values and the $H_2$/CO ratio in the oxidative reform of methane over each catalyst are shown in Figure 7 and Supplementary Figure S5, respectively. The results show that the optimal nickel content was 20% wt. since the highest reactant conversions were found for the 20Ni20YZ catalyst.

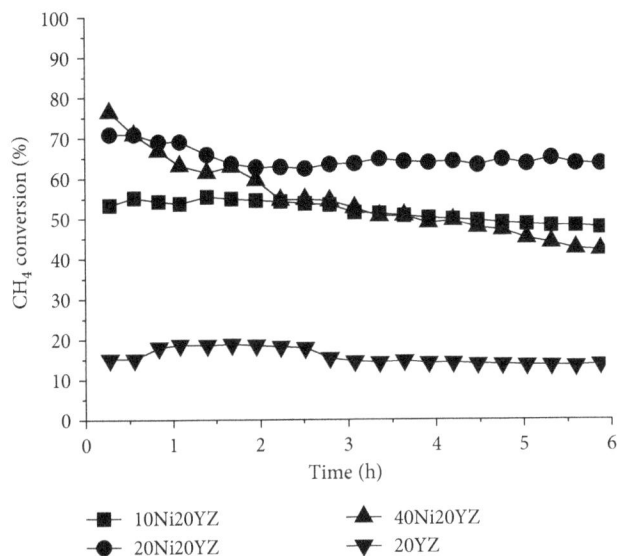

FIGURE 7: Conversion of $CH_4$ results obtained during the catalytic tests in the oxidative reform of methane at 750°C (molar ratio of reactants $1.5CH_4 : CO_2 : 0.25O_2$; molar total flow: 107 mL·min$^{-1}$).

The content of 40% wt. of nickel present in the 40Ni20YZ catalyst was too high for the catalyst, leading to catalyst low stability and deactivation during the reaction. The catalyst with 10% wt. of nickel content (10Ni20YZ) showed good stability, but low conversion values. The carbon deposition rates were 0.10 (10Ni20YZ); 0.14 (20Ni20YZ); and 0.39 (40Ni20YZ) mmolC·h$^{-1}$. These results suggest that there was a good distribution of the Ni° active centers on the $Y_2O_3$-$ZrO_2$ solid solution in the 20Ni20YZ, thus leading to a high conversion of $CH_4$, $CO_2$, and $O_2$ into $H_2$ and CO and to a low carbon deposition rate value (0.14 mmolC·h$^{-1}$). Since it is known that the $Y_2O_3$-$ZrO_2$ solid solution produce oxygen vacancies, which can help the removal of carbon deposits during the reforming reaction [5, 6, 9, 20], we suggest that oxygen vacancies of this solid solution may have favored the carbon removal in 20Ni20YZ, in a greater extent.

The conversion profile of the 40Ni20YZ catalyst (Figure 7) shows a continuous fall along the reaction. This decrease is due to the high nickel concentration in this catalyst which led to deactivation. This is expected, since the high nickel content in the 40Ni20YZ catalyst caused sintering of the Ni° particles after the reduction process, as seen in the *in situ* XRD analysis under reducing conditions with $H_2$ (see Table 1; the 40Ni20YZ catalyst had the largest crystallite size of Ni°, 33 nm). It is known that large Ni° crystal structures lead to high carbon deposition rates, producing incrustation of carbon on the active metal and causing deactivation of the catalyst-active sites. This may explain the high carbon deposition rates and poor catalytic performance recorded by the 40Ni20YZ catalyst (0.39 mmol·h$^{-1}$). Furthermore, according to the TPR analyses (Figure 3), this sample has a large proportion of NiO species interacting weakly with the catalytic support that reduced at low temperatures. These species could be facilitating their sintering. This was evidenced by the SEM analysis where the 40Ni20YZ catalyst morphology image

showed the formation of many agglomerated NiO particles situated around the cube-shaped solids of $Y_2O_3$-$ZrO_2$ solid solution (see Figure 6 and Supplementary Figure S4).

As noted in the estimation of the metallic phase dispersion, the 10Ni20YZ catalyst showed the highest metallic dispersion (6.7%) among all samples. This was also suggested by SEM analyzes of this catalyst, where the morphology showed the formation of small NiO agglomerates spaced apart (Figure 4 and Supplementary Figure S2). However, the 10% Ni content seems to be too low for higher conversion values; this high dispersion favored a good stability during the catalytic reaction.

We suggest that the $Y_2O_3$-$ZrO_2$ support and the Ni°-active sites are in good proportion in the 20Ni20YZ catalyst, thus leading to high conversion values and a relatively low carbon deposition rate. By analyzing the 20Ni20YZ images obtained by SEM analysis (Figure 5 and Supplementary Figure S3), it can be claimed that there was good conformation of the NiO particles on the catalyst support. Furthermore, in this sample, the NiO particles are present as small spheres deposited on the surface of the $Y_2O_3$-$ZrO_2$ catalytic support (which shows the symmetrical shape-cubes). This good conformation of the catalytic phase/catalytic support also explains the good catalytic performance of the 20Ni20YZ sample.

After the catalytic tests, traces of water, as a by-product, were collected and they showed the occurrence of the RWGSR reaction ($CO_2 + H_2 \longleftrightarrow CO + H_2O$). The occurrence of this reaction can explain the fact that the $CO_2$ conversion is slightly higher than the $CH_4$ conversions (Supplementary Figure S5 and Figure 7).

The $CO_2$ percentages involved in RWGSR (values calculated from the collected water) were 16%, 14%, and 19% for samples 10Ni20YZ, 20Ni20YZ, and 40Ni20YZ, respectively. These values indicate that the 20Ni20YZ sample showed the lowest water formation value and the best reactant conversion values.

Supplementary Figure S6 shows the $H_2$/CO molar ratio in the reaction products during oxidative reform of methane over the catalysts. These values are always lower than 1.2, which is the stoichiometric ratio for this reaction (reaction (3)). The low values of the $H_2$/CO molar ratio on 10Ni20YZ and 40Ni20YZ samples are consistent with the low conversion values and high contributions to RWGSR (16% and 19% of the $CO_2$ in the feed stream, resp.); this reaction uses hydrogen and carbon dioxide molecules to produce water.

### 3.2.2. Partial Oxidation of Methane (POM). 
Figure 8 and Supplementary Figures S7 and S8 show the results of the POM reaction over the catalysts. Figure 8 shows the $CH_4$ conversion profiles against reaction time; in this figure, it can be seen that the conversion of methane decreased as follows: 40Ni20YZ > 20Ni20YZ > 10Ni20YZ > 20YZ, in other words, with decreasing nickel content. Catalytic activity of sample 40Ni20YZ decreased gradually during the reaction time, demonstrating instability owing to deactivation, probably because of the high nickel content which favors sintering, as seen in characterization results. The carbon deposition rates

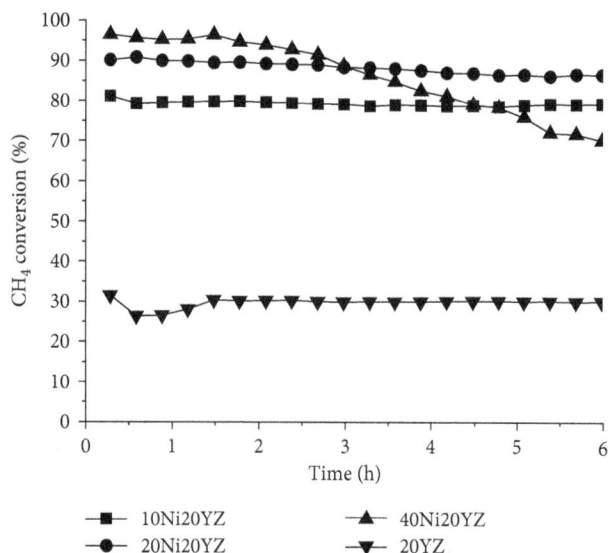

FIGURE 8: Conversion of $CH_4$ results obtained during the partial oxidation of methane over the catalysts at 750°C (molar ratio of reactants in the inlet stream: $2CH_4 : 1O_2$; molar total flow 107 mL·min$^{-1}$).

rose monotonically with nickel content on the catalysts 10Ni20YZ, 20Ni20YZ, and 40Ni20YZ; they were 0.08, 0.10, and 0.32 mmol·h$^{-1}$, respectively. These values indicate that 20Ni20YZ showed the best performance in the POM reaction, because it achieved a conversion high rate (Figure 8), relatively low carbon deposition, as well as the highest selectivity for hydrogen (Supplementary Figure S7).

The highest $H_2/CO$ ratios were produced by samples 20Ni20YZ and 10Ni20YZ and were around 1.7, which is nearly stoichiometric. The low $H_2/CO$ values for 40Ni20YZ are in agreement with the deactivation on this catalyst. The $H_2/CO$ values above the stoichiometric value (approximately 2) can be due to other reactions occurring in parallel with POM, as will be discussed in the following paragraphs.

The presence of $H_2O$ (traces collected at the end of reaction time) and $CO_2$ during each catalytic test indicates that the POM occurred through the combustion-reforming mechanism on the catalysts. This mechanism is described by the following reactions [31, 32]:

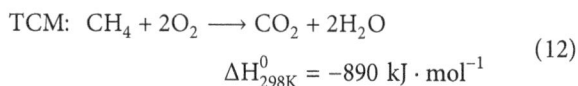

$$\text{TCM: } CH_4 + 2O_2 \longrightarrow CO_2 + 2H_2O$$
$$\Delta H^0_{298K} = -890 \text{ kJ} \cdot \text{mol}^{-1} \quad (12)$$

The global sum of reactions $(12) + (1) + 2 \times (4)$ is reaction (2), with a $H_2/CO$ ratio of 2. Additionally, under the presence of $CO_2$ and $H_2$ in the reaction products, the reverse water-gas shift reaction (RWGSR) (7) is very likely to occur; this reaction is favored at high temperature:

The occurrence of the RWGSR and the combustion-reforming mechanism explains why the $H_2/CO$ ratios for all samples were above the stoichiometric value. The selectivity results for $H_2$ and $CO_2$ are shown in Supplementary Figures S7 and S8, respectively. The selectivity for $H_2$ and CO followed a trend similar to that of $CH_4$ conversion, the selectivity for $H_2$ and CO decreased as follows:

40Ni20YZ > 20Ni20YZ > 10Ni20YZ > 20YZ, and the selectivity for $H_2$ and CO over sample 40Ni20YZ decreased continuously during the reaction, probably owing to sintering which led to the high amount of coke recorded for this sample.

Samples 10Ni20YZ and 20Ni20YZ reported low $CO_2$ selectivity; contrarily, 20YZ and 40Ni20YZ samples were the most selective for $CO_2$. The trend found in $CO_2$ selectivity indicates that the catalytic support (20YZ) strongly favors the TCM (reaction (7)); the gradual loss of Ni°-active sites of the 40Ni20YZ catalyst during the POM reaction favored the TCM reaction (expressed by the gradual increase of selectivity to $CO_2$). Probably, the oxygen vacancy formation in the 20YZ support, which activates $O_2$ molecules ($O_2$ + active site $\rightarrow O_{(s)} + O_{(s)}$), is favoring the TCM reaction and therefore the production of $CO_2$ over the 20YZ support.

From the results of the ORM and POM catalytic tests, we suggest that the catalysts 10Ni20YZ, 20Ni20YZ, and 40Ni20YZ have two types of active sites: the first is the Ni° site that dissociates the $CH_4$ molecules and the second is the $Y_2O_3$-$ZrO_2$ solid solution (which is known to form oxygen vacancies) that may promote the $O_2$ and $CO_2$ molecules [33, 34].

The SEM image of some catalyst (in a region rich in carbon, according to EDX analysis) after OPM reaction is shown in Figures 9 and 10. Figure 9 corresponds to 20Ni20YZ (our best catalyst) after POM; according to this image, very few carbon filaments were formed, the morphology of the carbon is mostly amorphous on the surface of the catalyst (the SEM image of sample 20Ni20YZ after ORM is shown in Supplementary Figure S9, and this image is very similar to the same catalyst after POM). Figure 10 corresponds to 40Ni20YZ after POM; in this case, the carbon was entirely amorphous. The results of the characterization of the spent catalysts suggest that the catalyst synthesized by the OSP method forms graphitic carbon of amorphous morphology. The variation of the nickel charge in the catalyst formulation did not markedly influence the carbon deposit morphology.

The XRD patterns of spent catalysts after 6 h of POM (see Figure S11, in the Supplementary) showed the formation of graphitic carbon, which is clearly observed (peak at 26°) over 20Ni20YZ and 40Ni20YZ catalysts. The peak related to graphitic carbon was not observed on the XRD pattern of spent 10Ni20YZ catalyst. Not significant difference was observed for coke by XRD analysis of spent catalysts after ROM.

A long-term stability test was carried out on the best catalyst in the present study (20Ni20YZ), which was tested under POM reaction conditions for 24 hours. The results are included in the Supplementary Figure S10), where it can be seen that the catalyst maintained its activity for 24 h. A slight continuous decrease in the reactant conversion rates was observed during the reaction, but this seemed to stabilize after 6 hours of reaction, as after this time, the conversion value remained constant at 85%. The coke deposition rate over 24 h was 0.12 mmol·h$^{-1}$ on 20Ni20YZ, this value being close to that found for the catalytic test over 6 h of reaction. The stability test of 20Ni20YZ catalyst in ORM was reported previously for us in [5].

FIGURE 9: SEM image of sample 20Ni20YZ (×50000) after POM reaction.

FIGURE 10: SEM image of sample 40Ni20YZ (×25000) after POM reaction.

## 4. Conclusions

The production of Syngas from methane through the POM and ORM over the catalysts rose as the Ni content increased, and the optimal nickel content in the NiO/$Y_2O_3$/$ZrO_2$ mixture prepared under the OSP method was 20% wt. (for both reactions). At this nickel concentration, high selectivity to $H_2$ and CO, and low carbon deposition rates were reported; curiously the fresh catalyst at this concentration showed the best geometric morphology. Further nickel content led to catalyst deactivation.

The variation of the charge of NiO in the NiO/$Y_2O_3$/$ZrO_2$ mixture under the OSP method led to NiO of face-centered cubic phase and to the tetragonal zirconia (demonstrated by its stabilization forming $Y_2O_3$-$ZrO_2$ solid solution in the support). According to SEM/EDX analysis, different nickel charges formed catalysts with different morphologies, NiO particles were located at different places on the surface of the $Y_2O_3$-$ZrO_2$ support, and the catalysts 20Ni20YZ and 40Ni20YZ showed well defined geometrical morphology.

From the results of the catalytic tests, we suggest that the catalysts have two types of active sites: the first is the $Ni°$ site that dissociates the $CH_4$ molecules and the second is the $Y_2O_3$-$ZrO_2$ solid-solution (present in the support) that may promote the $O_2$ and $CO_2$ molecules.

The characterization of the catalysts demonstrated that the nickel concentration in the mixture influenced the catalyst morphology; therefore, the morphology also influenced the catalytic performances during the Syngas production reactions. The variation of the nickel charge in the formulation of the catalyst influenced the carbon deposition rates, but did not markedly influence the morphology of the carbon deposit.

The OSP method led to production of materials of smaller crystallite size as others of similar composition, but prepared under other methods; additionally, this was advantageous because it was carried out in a single calcination step.

## Conflicts of Interest

The authors declare that there are no conflicts of interest regarding the publication of this paper.

## Acknowledgments

The authors thank the São Paulo Research Foundation (FAPESP) for the financial support (Grant no. 2014/24940–5), the Brazilian National Council for Scientific Development (CNPq) for the scholarship and the grant, and the Brazilian Synchrotron Light Laboratory (LNLS) for the *in-situ* XRD conducted in the D10B-XPD line.

## Supplementary Materials

The supplementary material contains Figure S1 which shows the XRD patterns of the fresh 20YZ support and pure $ZrO_2$ in the $2\theta = 20–40°$ range (speed $2°\,min^{-1}$ using Cu-K $\alpha$ radiation, $\lambda = 1.5406$ Å) to complement the XRD discussion of the manuscript. EDX mapping at the regions of SEM image of Figure 4 (10Ni20YZ), Figure 5 (20Ni20YZ), and Figure 6 (40Ni20YZ) are shown in Figure S2, Figure S3, and Figure S4, respectively. The conversion of $CO_2$ and $H_2$/CO molar ratio obtained during the catalytic tests in the oxidative reform of methane at 750°C over the catalysts is shown in Figure S5 and Figure S6, respectively. The selectivity to hydrogen and to carbon dioxide (mol produced/mol of $CH_4$ converted) obtained during the partial oxidation of methane over the catalysts at 750°C is shown in Figure S7 and Figure S8, respectively. SEM image of sample 20Ni20YZ after ORM reaction (×25 000) is shown Figure S9. Figure S10 shows the stability test of 20Ni20YZ catalysts for OPM reaction. Figure S11 shows the XRD patterns for spent catalysts after POM reaction. Finally, Figure S12 shows the graphical abstract of this manuscript. (*Supplementary Materials*)

# References

[1] S. N. Naik, V. V. Goud, P. K. Rout, and A. K. Dalai, "Production of first and second generation biofuels: a comprehensive review," *Renewable and Sustainable Energy Reviews*, vol. 14, no. 2, pp. 578–597, 2010.

[2] L. Pengmei, Z. Yuan, C. Wu, L. Ma, Y. Chen, and N. Tsubaki, "Bio-syngas production from biomass catalytic gasification," *Energy Conversion and Management*, vol. 48, no. 4, pp. 1132–1139, 2007.

[3] N. Rueangjitt, C. Akarawitoo, and S. Chavadej, "Production of hydrogen-rich syngas from biogas reforming with partial oxidation using a multi-stage AC gliding arc system," *Plasma Chemistry and Plasma Processing*, vol. 32, no. 3, pp. 583–596, 2012.

[4] M. Benito, S. García, P. Ferreira-Aparicio, L. García-Serrano, and L. Daza, "Development of biogas reforming Ni-La-Al catalysts for fuel cells," *Journal of Power Sources*, vol. 169, no. 1, pp. 177–183, 2007.

[5] Y. J. O. Asencios, C. B. Rodella, and E. M. Assaf, "Oxidative reforming of model biogas over NiO-$Y_2O_3$-$ZrO_2$ catalysts," *Applied Catalysis B: Environmental*, vol. 132-133, pp. 1–12, 2013.

[6] Y. J. O Asencios, P. A. Nascente, and E. M. Assaf, "Partial oxidation of methane on NiO-MgO-$ZrO_2$ catalysts," *Fuel*, vol. 97, pp. 630–637, 2012.

[7] A. F. Lucrêdio, J. M. Assaf, and E. M. Assaf, "Methane conversion reactions on Ni catalysts promoted with Rh: influence of support," *Applied Catalysis A: General*, vol. 400, no. 1-2, pp. 156–165, 2011.

[8] P. Kumar, Y. Sun, and R. O. Idem, "Nickel-Based ceria, zirconia, and ceria–zirconia catalytic systems for low-temperature carbon dioxide reforming of methane," *Energy and Fuels*, vol. 21, no. 6, pp. 3113–3123, 2007.

[9] J. D. A. Bellido and E. M. Assaf, "Effect of the $Y_2O_3$-$ZrO_2$ nickel catalyst support composition on evaluated in dry reforming of methane," *Applied Catalysis A: General*, vol. 352, no. 1-2, pp. 179–187, 2009.

[10] J. R. Rostrup-Nielsen and J. H. Back-Hansen, "$CO_2$ Reforming methane over transition metals," *Journal of Catalysis*, vol. 144, no. 1, pp. 38–49, 1993.

[11] V. A. Tsipouriari, A. M. Efstathiou, Z. L. Zhang, and X. E. Verykios, "Reforming of methane with carbon dioxide to synthesis gas over supported Rh catalysts," *Catalysis Today*, vol. 21, no. 2-3, pp. 579–587, 1994.

[12] Y. J. O. Asencios and E. M. Assaf, "Combination of dry reforming and partial oxidation of methane on NiO-MgO-$ZrO_2$ catalyst: effect of nickel content," *Fuel Processing Technology*, vol. 106, pp. 247–252, 2013.

[13] S. O. Soliviev, K. A. Yu, S. N. Orlyk, and E. V. Gubareni, "Carbon dioxide reforming of methane on monolithic Ni/$Al_2O_3$-based catalysts," *Journal of Natural Gas Chemistry*, vol. 20, no. 2, pp. 184–190, 2011.

[14] F. L. S. Carvalho, Y. J. O Asencios, A. M. B. Rego, and E. M. Assaf, "Hydrogen production by steam reforming of ethanol on $Co_3O_4$/$La_2O_3$/$CeO_2$ catalysts," *Applied Catalysis A: General*, vol. 483, pp. 52–62, 2014.

[15] A. R. Gonzalez, Y. J. O Asencios, E. M Assaf, and J. M Assaf, "Dry reforming of methane on Ni–Mg–Al nano-spheroid oxide catalysts prepared by the sol–gel method from hydrotalcite-like precursors," *Applied Surface Science*, vol. 280, pp. 876–887, 2013.

[16] A. F. Lucrêdio, J. M. Assaf, and E. M. Assaf, "Reforming of a model biogas on Ni and Rh Ni catalysts: effect of adding La," *Fuel Processing Technology*, vol. 102, pp. 124–131, 2012.

[17] A. F. Lucredio, G. T. Filho, and E. M. Assaf, "Co/Mg/Al hydrotalcite-type precursor, promoted with La and Ce, Studied by XPS and applied to methane steam reforming reactions," *Applied Surface Science*, vol. 255, no. 11, pp. 5851–5856, 2009.

[18] H. Mori, C. J. Wen, J. Otomo, K. Eguchi, and H. Takahashi, "Investigation of the interaction between NiO and yttria-stabilized zirconia (YSZ) in the NiO/YSZ composite by temperature-programmed reduction technique," *Applied Catalysis A: General*, vol. 245, no. 1, pp. 79–85, 2003.

[19] C. Resini, M. C. Herrera-Delgado, S. Presto et al., "Yttria-stabilized zirconia (YSZ) supported Ni–Co alloys (precursor of SOFC anodes) as catalysts for the steam reforming of etanol," *International Journal of Hydrogen Energy*, vol. 33, no. 14, pp. 3728–3735, 2008.

[20] J. D. A. Bellido, E. Y. Tanabe, and E. M. Assaf, "Carbon dioxide reforming of ethanol over Ni/$Y_2O_3$–$ZrO_2$ catalysts," *Applied Catalysis B: Environmental*, vol. 90, no. 3-4, pp. 485–488, 2009.

[21] F. L. S. Carvalho, Y. J. O. Asencios, J. D. A. Bellido, and E. M. Assaf, "Bio-ethanol steam reforming for hydrogen production over $Co_3O_4$/$CeO_2$ catalysts synthesized by one-step polymerization method," *Fuel Processing Technology*, vol. 142, pp. 182–191, 2016.

[22] Y. J. O. Asencios, J. D. A. Bellido, and E. M. Assaf, "Synthesis of NiO-MgO-$ZrO_2$ catalysts and their performance in reforming of model biogas," *Applied Catalysis A: General*, vol. 397, no. 1-2, pp. 138–144, 2011.

[23] P. J. Marcos and D. Gouvêa, "Effect of MgO segregation and solubilization on the morphology of $ZrO_2$ powders during synthesis by the Pechini method," *Cerâmica*, vol. 50, no. 313, pp. 38–42, 2004.

[24] S. Chien and W. Chiang, "Catalytic properties of NiX zeolites in the presence of cerium additives," *Applied Catalysis*, vol. 61, no. 1, pp. 45–61, 1990.

[25] I. Denry, J. A. Holloway, and P. K. Gupta, "Effect of crystallization heat treatment on the microstructure of niobium-doped fluorapatite glass-ceramics," *Journal of Biomedical Materials Research Part B: Applied Biomaterials*, vol. 100, no. 5, pp. 1198–1205, 2012.

[26] E. A. Trusova, A. Khrushcheva, and L. I. Shvorneva, "The impact of thermal treatment conditions on the formation of crystalline structure of Ce-Zr-oxide composite obtained by a modified sol-gel technique," *Journal of Physics: Conference Series*, vol. 345, article 012035, 2012.

[27] M. H. Youn, J. G. Seo, J. C. Jung, S. Park, and I. K. Song, "Hydrogen production by auto-thermal reforming of ethanol over nickel catalyst supported on mesoporous yttria-stabilized zirconia," *International Journal of Hydrogen Energy*, vol. 34, no. 13, pp. 5390–539, 2009.

[28] C. H. Bartholomew and R. B. Pannell, "The stoichiometry of hydrogen and carbon monoxide chemisorption on alumina and silica-supported nickel," *Journal of Catalysis*, vol. 65, no. 2, pp. 390–401, 1980.

[29] R. D. Jones and C. H. Barthoholomew, "Improved technique for measurement of flow hydrogen chemisorption on metal catalysts," *Applied Catalysis*, vol. 39, pp. 77–88, 1988.

[30] J. Zhang, H. Wang, and A. Dalai, "Effects of metal content on activity and stability of Ni-Co bimetallic catalysts for $CO_2$ reforming of $CH_4$," *Applied Catalysis A: General*, vol. 339, pp. 121–129, 2008.

[31] Y. Q. Song, D. H. He, and B. Q. Xu, "Effects of preparation methods of $ZrO_2$ support on catalytic performances of Ni/$ZrO_2$ catalysts in methane partial oxidation to syngas," *Applied Catalysis A: General*, vol. 337, no. 1, pp. 19–28, 2008.

[32] H. Morioka, Y. Shimizu, M. Sukenobu et al., "Partial oxidation of methane to synthesis gas over supported Ni catalysts prepared from Ni-Ca/Al-layered double hydroxide," *Applied Catalysis A: General*, vol. 215, no. 1-2, pp. 11–17, 2001.

[33] W. Dow and T. Huang, "Yttria-stabilized zirconia supported copper oxide catalyst-I. Effect of oxygen vacancy of support on copper oxide reduction," *Journal of Catalysis*, vol. 160, no. 2, pp. 155–170, 1996.

[34] L. Xiancai, L. Shuigen, Y. Yifeng, W. Min, and H. Fei, "Studies on coke formation and coke species of nickel-based catalysts in $CO_2$ reforming of $CH_4$," *Catalysis Letters*, vol. 118, no. 1-2, pp. 59–63, 2007.

# Permissions

# List of Contributors

Ruth Sánchez-Hernández, Isabel Padilla and Aurora López-Delgado
National Centre for Metallurgical Research (CSIC), Madrid 28040, Spain

Sol López-Andrés
Department of Mineralogy and Petrology, Faculty of Geology, University Complutense of Madrid, Madrid 28040, Spain

Anawe A. L. Paul and Folayan J. Adewale
Department of Petroleum Engineering, College of Chemical and Petroleum Engineering, Covenant University, Ota, Nigeria

Lennevey Kinidi, Ivy Ai Wei Tan, Noraziah Binti Abdul Wahab, Cirilo Nolasco Hipolito and Shanti Faridah Salleh
Department of Chemical Engineering and Energy Sustainability, Faculty of Engineering, Universiti Malaysia Sarawak, 94300 Kota Samarahan, Sarawak, Malaysia

Khairul Fikri Bin Tamrin
Department of Mechanical and Manufacturing, Faculty of Engineering, Universiti Malaysia Sarawak, 94300 Kota Samarahan, Sarawak, Malaysia

Karina Cardoso Valverde, Priscila Ferri Coldebella, Marcela Fernandes Silva, Letícia Nishi and Rosângela Bergamasco
Departamento de Engenharia Química, Universidade Estadual de Maringá, Av. Colombo 5790, Bloco D-90, 87020-900 Maringá, PR, Brazil

Milene Carvalho Bongiovani
Universidade Federal do Mato Grosso, Av. Alexandre Ferronato, 1200, 78557-267 Sinop, MT, Brazil

Nykolay D. Sakhnenko , Maryna V. Ved' and Ann V. Karakurkchi
National Technical University "Kharkiv Polytechnical Institute", Kyrpychova St. 2, Kharkiv 61002, Ukraine

Mariana Busto and Carlos Román Vera
Institute of Research on Catalysis and Petrochemistry, INCAPE, FIQ-UNL, CONICET, Collecting Ring, National Road 168 Km 0,El Pozo, 3000 Santa Fe, Argentina

Enrique Eduardo Tarifa
Faculty of Chemical Engineering, Universidad Nacional de Jujuy, CONICET, ítalo Palanca No. 10, San Salvador de Jujuy, Argentina

Juliana R. Machado, Emanuelle E. Severo, Janaina M. G. de Oliveira, Joana da C. Ores and Susana J. Kalil
Escola de Química e Alimentos, Universidade Federal do Rio Grande, 96203900 Rio Grande, RS, Brazil

Adriano Brandelli
Laboratório de Bioquímica e Microbiologia Aplicada, Universidade Federal do Rio Grande do Sul, 91501-970 Porto Alegre, RS, Brazil

T. Havaić, A.-M. Đumbir, M. Gretić, G. Matijašić and K. Žižek
Faculty of Chemical Engineering and Technology, University of Zagreb, Zagreb, Croatia

Luis F. Gutiérrez-Mosquera
Department of Engineering, Food and Agribusiness Research Group, Universidad de Caldas, Calle 65 No. 26–10, Manizales, Colombia

Sebastián Arias-Giraldo and Adela M. Ceballos-Peñaloza
Food and Agribusiness Research Group, Universidad de Caldas, Calle 65 No. 26–10, Manizales, Colombia

Hsinyun Hsu and Michael T. Harris
Department of Chemical Engineering, Purdue University, 480 Stadium Mall Drive, West Lafayette, IN 47907, USA

Lynne S. Taylor
Department of Industrial and Physical Pharmacy, Purdue University, 575 Stadium Mall Drive, West Lafayette, IN 47907, USA

**Caroline dos Santos Viana, Luiz Rodrigo Ito Morioka and Hélio Hiroshi Suguimoto**
Center for Research and Pos-Graduate, University of Pythagoras Unopar, Marseille Street 591, 86041-140 Londrina, PR, Brazil

**Denise Renata Pedrinho**
Anhanguera University, Uniderp, Alexandre Herculano Street 1400, 79037-280 Campo Grande, MS, Brazil

**Joshua Olusegun Okeniyi and Abimbola Patricia Idowu Popoola**
Mechanical Engineering Department, Covenant University, Ota, Nigeria
Chemical and Metallurgical Engineering Department, Tshwane University of Technology, Pretoria, South Africa

**Modupe Elizabeth Ojewumi**
Chemical Engineering Department, Covenant University, Ota, Nigeria

**Elizabeth Toyin Okeniyi**
Petroleum Engineering Department, Covenant University, Ota, Nigeria

**Jacob Olumuyiwa Ikotun**
Department of Civil Engineering and Building, Vaal University of Technology, Vanderbijlpark, South Africa

**Yasinta John and Victor Emery David Jr.**
College of Urban Construction and Environmental Engineering, Chongqing University, Chongqing 400044, China

**Daniel Mmereki**
National Centre for International Research of Low-Carbon and Green Buildings, Chongqing University, Chongqing 400045, China

**Preecha Kasikamphaiboon and Uraiwan Khunjan**
Department of Science, Faculty of Science and Technology, Prince of Songkla University, Pattani Campus, Pattani 94000, Thailand

**Rawnak Talmoudi and Amna AbdelJaoued**
Laboratoire de Recherche Génie des procédés et systèmes industriels, Ecole Nationale d'Ingénieurs de Gabès, Université de Gabès, Gabès, Tunisia

**Mohamed Hachemi Chahbani**
Laboratoire de Recherche Génie des procédés et systèmes industriels, Ecole Nationale d'Ingénieurs de Gabès, Université de Gabès, Gabès, Tunisia
Institut Supérieur des Sciences Appliquées et de Technologie de Gabès, Rue Omar Ibn Elkhattab, Zrig, Gabès 6029, Tunisia

**M. G. Sobamowo**
Department of Mechanical Engineering, University of Lagos, Akoka, Lagos, Nigeria

**Daniela Almeida Nogueira, Juliane Machado da Silveira, Evelin Mendes Vidal, Natália Torres Ribeiro and Carlos André Veiga Burkert**
Bioprocess Engineering Laboratory, School of Chemistry and Food, Federal University of Rio Grande, 96203-900 Rio Grande, RS, Brazil

**Amilton Barbosa Botelho Junior, Denise Crocce Romano Espinosa and Jorge Alberto Soares Tenório**
Department of Chemical Engineering, Polytechnic School of University of São Paulo, Sao Paulo, Brazil

**David Bruce Dreisinger**
Department of Materials Engineering, The University of British Columbia, Vancouver, Canada

**Yvan J. O. Asencios**
Instituto do Mar, Universidade Federal de São Paulo, 11030-100 Santos, SP, Brazil

**Elisabete M. Assaf**
Instituto de Química de São Carlos, Universidade de São Paulo, 13560-970 São Carlos, SP, Brazil

# Index

www.ingramcontent.com/pod-product-compliance
Lightning Source LLC
Chambersburg PA
CBHW082047190326
41458CB00010B/3481